ACS SYMPOSIUM SERIES 372

Metal Clusters in Proteins

Lawrence Que, Jr., EDITOR
University of Minnesota

Developed from a symposium sponsored by
the Division of Inorganic Chemistry
at the 194th Meeting
of the American Chemical Society,
New Orleans, Louisiana,
August 30–September 4, 1987

American Chemical Society, Washington, DC 1988

Library of Congress Cataloging-in-Publication Data

Metal clusters in proteins
Lawrence Que, Jr., editor

p. cm.—(ACS symposium series; ISSN 0097-6156; 372).

"Developed from a symposium sponsored by the Division of Inorganic Chemistry at the 194th Meeting of the American Chemical Society, New Orleans, Louisiana, August 30–September 4, 1987."

Includes bibliographies and indexes.

ISBN 0-8412-1487-5
1. Metalloproteins—Structure—Congresses. I. Que, Lawrence, Jr., 1949– . II. American Chemical Society. Division of Inorganic Chemistry. III. Series

QP552.M46M46 1988
547.7`.5—dc19 88-14504
CIP

PRINTED IN THE UNITED STATES OF AMERICA

ACS Symposium Series

M. Joan Comstock, *Series Editor*

Foreword

The ACS SYMPOSIUM SERIES was founded in 1974 to provide a medium for publishing symposia quickly in book form. The format of the Series parallels that of the continuing ADVANCES IN CHEMISTRY SERIES except that, in order to save time, the papers are not typeset but are reproduced as they are submitted by the authors in camera-ready form. Papers are reviewed under the supervision of the Editors with the assistance of the Series Advisory Board and are selected to maintain the integrity of the symposia; however, verbatim reproductions of previously published papers are not accepted. Both reviews and reports of research are acceptable, because symposia may embrace both types of presentation.

Contents

METAL–OXO CENTERS

METAL–SULFUR CLUSTERS

Preface

THE FIELD OF BIOINORGANIC CHEMISTRY HAS EXPERIENCED a great explosion of activity over the past 20 years due to the efforts of biochemists, spectroscopists, crystallographers, molecular biologists, and inorganic and organic chemists. As a result, metals have increasingly been found to play vital roles in the maintenance and regulation of many biological processes.

Plans for the symposium from which this book was developed were made because of the increasing number of metalloproteins that have active sites consisting of more than one metal center. The presence of at least one other metal center confers magnetic, spectroscopic, and chemical properties to the active site not normally observed for a single metal; thus, these proteins constitute a subset with unique characteristics. The aim of the symposium was to bring specialists in the biochemistry and spectroscopy of proteins together with chemists involved in the design and synthesis of structural and functional models for the active sites and to provide for a lively exchange of ideas among these scientists. Associated with the symposium was a tutorial in which design principles for the synthesis of analogues and the underlying basis for several spectroscopic techniques were discussed as an introduction to the field.

The chapters in this book, reflecting material derived from the symposium and the tutorial, should serve as an up-to-date summary of some important developments in this area. *Metal Clusters in Proteins* may also serve as a starting point for classroom discussions on bioinorganic chemistry. I am grateful to the authors for the time and effort they expended in assembling and presenting their views of the field. I also thank the Division of Inorganic Chemistry of the ACS for sponsoring the symposium; the Petroleum Research Fund for paying the travel expenses of the overseas speakers; and E. I. du Pont de Nemours and Company, Exxon Research and Engineering Company, and the Division of Biological Chemistry of the ACS for their financial support.

LAWRENCE QUE, JR.
University of Minnesota
Minneapolis, MN 55455

April 5, 1988

Chapter 1

Metalloprotein Crystallography

Survey of Recent Results and Relationships to Model Studies

William H. Armstrong

Department of Chemistry, University of California, Berkeley, CA 94720

The recent literature (1979–1987) pertaining to metalloprotein crystallography is reviewed. Particular systems discussed in some detail include "blue" copper proteins, iron–sulfur proteins, sulfite reductase, cytochromes c, myoglobin, hemoglobin, catalase, cytochrome c peroxidase, cytochromes P-450, hemerythrin, hemocyanin, superoxide dismutases, Zn enzymes, metallothionein, lactoferrins, ferritin, bacteriochlorophyll, and the photosynthetic reaction center. Preliminary results for nitrogenase, ribonucleotide reductase, protocatechuate 3,4-dioxygenase, hydrogenase, nitrite reductase, ascorbate oxidase, and cerulplasmin are also mentioned. The influence of metalloprotein crystallography on bioinorganic model studies and vice versa is examined.

The interrelationship between protein crystallography and small molecule model studies has been and will continue to be a very important factor contributing toward a fundamental understanding of the chemical, magnetic and spectroscopic properties of metalloproteins. Metal ions perform many different functions in biological systems including electron transfer, oxygen binding, activation of oxygen toward substrate oxidation, oxygen evolution, superoxide dismutation, peroxide disproportionation, isomerization, hydrolysis, dinitrogen reduction, sulfite and nitrite reduction, and dehydrogenation. The pursuit of low molecular weight analogs or models for the metal sites in these biological systems has received a great deal of attention in recent years. In many cases the results of these endeavors have significantly sharpened our understanding of the involvement of the metal ions in the aforementioned reactions. This branch of chemistry, commonly referred to as bioinorganic chemistry or inorganic biochemistry, was reviewed in articles appearing in 1976 and 1980 (*1,2*). In the present paper emphasis will be placed on recent (between 1979 and 1987) protein X-ray crystallographic results and the impact these results have had and will have on bioinorganic modelling. Rather than restricting the scope of this survey to metal aggregates in proteins, mononuclear centers will be examined as well, because important lessons can be learned from both structural types. This paper presents a collection of representative examples rather than an exhaustive review of the literature.

0097–6156/88/0372–0001$07.75/0

Hill defined speculative models as those compounds which are intended to mimic a protein metal site structure that has not been defined by X-ray crystallography (*1*). Complexes related to a known structure were said to be corroborative models. At first glance, it may appear that with metalloprotein X-ray crystallography becoming more routine that model studies may become obsolete. That is, if the central questions of the model study are related to the precise structure of the metal site including the arrangement of metal atoms and the identity of coordinated prosthetic groups and protein side chain residues, then a single crystal X-ray structure will answer these questions unequivocally. There are a number of problems with this oversimplified view of bioinorganic chemistry: [1] Depending on the size of the protein, X-ray crystallographic refinements may not yield a sufficiently well-defined picture of the metal site to permit a precise geometrical description. [2] Even if the structure is determined to a very high degree of accuracy, a number of important questions remain which potentially can be addressed by model studies. For example: What is the effect of the protein environment on the physical and chemical properties of the metal complex? This can only be answered if the metal center can be extruded intact away from the protein matrix (for aggregates) or if an analog complex can be prepared independently. [3] In most cases, crystal structures have been determined for only the resting state of the protein. Large structural changes may occur during the course of the catalytic cycle in an enzyme. In cases where enzyme intermediates can be prepared but not characterized by crystallography, comparison of the spectroscopic and magnetic properties of the intermediate state to those of model compounds may provide structural insight. In some cases the isolated form of the protein is actually a physiologically irrelevant form.

The points discussed above are probably best illustrated by example. Selected examples of metal site structures in metalloproteins are presented in this chapter and are summarized in Table I.

"Blue" Copper Proteins (*3–10*)

The blue or type 1 copper proteins, examples of which include azurin, plastocyanin, and stellacyanin, function as electron transfer agents (*11,12*). Among their properties that have fascinated bioinorganic chemists for years are intense charge transfer transitions which gives rise to the intense blue color, unusually positive reduction potentials, and small hyperfine coupling constants in the EPR spectrum. Both azurin and plastocyanin have been characterized by X-ray crystallography. Structure A is adapted from a recent report of the high resolution structure of oxidized azurin from *Alcaligenes denitrificans* (*3*). Structure B depicts the Cu site in oxidized poplar plastocyanin (*8*). In the former, the geometry around copper is best described as distorted trigonal planar with distant axial interactions to a methionine sulfur and a backbone carbonyl oxygen. In the latter, the structure is closer to a distorted tetrahedral configuration with coordination to two histidines, a cysteine, and a long bond to a methionine sulfur atom. Model studies prior to these crystallographic reports were predominantly directed toward obtaining Cu(II)-thiolate complexes with distorted tetrahedral geometry. While several Cu(II)-thiolate species have been isolated and characterized by X-ray crystallography (*13–15*), none of these approach tetrahedral geometry. In one case (*16*), *in situ* generation of a Cu(II) thiolate at low temperature produced a species with a visible spectrum similar to that of the protein. Discovery of the trigonal planar structure for the azurin copper site presents a difficult challenge to the bioinorganic chemist. Synthesis of a ligand with the proper donors that will maintain a three-coordinate geometry, with perhaps one or two distant axial ligands, will be a demanding task. In order to demonstrate that the unusual coordination geometry is imposed upon

Table I. Selected Metalloprotein Metal Coordination Site Structure Based on X-Ray Crystallography

Protein	Coordination Sphere	Reference
Azurin	$[Cu(N\text{-}His)_2(S\text{-}Cys)(S\text{-}Met)(O\text{-}peptide)]$	3
Plastocyanin	$[Cu(N\text{-}His)_2(S\text{-}Cys)(S\text{-}Met)]$	8
Rubredoxin	$[Fe(S\text{-}Cys)_4]$	17
Superoxide Dismutase, Mn	$[Mn(N\text{-}His)_3(O\text{-}Asp)(OH_2)]$	89
Lactoferrin	$[Fe(N\text{-}His)(O\text{-}tyr)_2(O\text{-}Asp)(OH_2)(CO_3^{2-})]$	120
Carboxypeptidase	$[Zn(N\text{-}His)_2(O\text{-}Glu)(OH_2)]$	103
Carbonic Anhydrase	$[Zn(N\text{-}His)_2(OH_2)]$	113
Insulin	$[Zn(N\text{-}His)_2(OH_2)]$	108
Thermolysin	$[Zn(N\text{-}His)_2(O\text{-}Glu]$	109-112
Liver Alcohol Dehydrogenase	$[Zn(S\text{-}Cys)_2(N\text{-}His)(OH_2)]$, [Zn(S-Cys)4]	114
Cytochromes c	$[Fe(N_4\text{-}porphyrin)(N\text{-}His)(S\text{-}Met)]$	38
Cytochrome c'	$[Fe(N_4\text{-}porphyrin)(N\text{-}His)]$	34
Myoblobin, Hemoglobin, Oxy	$[Fe(N_4\text{-}porphyrin)(N\text{-}His)(O_2)]$	39
Deoxy	$[Fe(N_4\text{-}porphyrin)(N\text{-}His)]$	39
Met	$[Fe(N_4\text{-}porphyrin)(N\text{-}His)]$	39
MetX	$[Fe(N_4\text{-}porphyrin)(N\text{-}His)(X)]$	39
CO	$[Fe(N_4\text{-}porphyrin)(N\text{-}His)(CO)]$	39
Catalase	$[Fe(N_4\text{-}porphyrin)(O\text{-}Tyr)(OH_2)]$	59
Cytochrome P_{450}	$[Fe(N_4\text{-}porphyrin)(S\text{-}Cys)(OH_2)]$	67
Cytochrome $P450_{cam}$	$[Fe(N_4\text{-}porphyrin)(S\text{-}Cys]\cdots camphor$	66
Cytochrome c Peroxidase	$[Fe(N_4\text{-}porphyrin)(N\text{-}His)(OH_2)]$	64
Hemerythrin, Met	$[(N\text{-}His)_3Fe(\mu\text{-}O)(\mu\text{-}O\text{-}Asp)(\mu\text{-}O\text{-}Glu)Fe(N\text{-}His)_2]$	76,77
Azido	$[(N\text{-}His)_3Fe(\mu\text{-}O)(\mu\text{-}O\text{-}Asp)(\mu\text{-}O\text{-}Glu)Fe(N\text{-}His)_2(N_3)]$	72,75
Oxy	$[(N\text{-}His)_3Fe(\mu\text{-}O)(\mu\text{-}O\text{-}Asp)(\mu\text{-}O\text{-}Glu)Fe(N\text{-}His)_2(O_2H)]$	74
Deoxy	$[(N\text{-}His)_3Fe(\mu\text{-}OH)(\mu\text{-}O\text{-}Asp)(\mu\text{-}O\text{-}Glu)Fe(N\text{-}His)_2]$	74
Hemocyanin	$[(N\text{-}His)_3Cu\cdots Cu(N\text{-}His)_3]$	86
Superoxide Dismutase, CuZn	$[(N\text{-}His)_3Cu(\mu\text{-}N\text{-}His)Zn(N\text{-}His)_2(O\text{-}Asp)]$	94,95
Metallothionein; Cd,Zn	$[Zn_2Cd(\mu\text{-}S\text{-}Cys)_3(S\text{-}Cys)_6]$,$[Cd_4(\mu\text{-}S\text{-}Cys)_5(S\text{-}Cys)_6]$	115
Ferredoxin, 2 Fe	$[(S\text{-}Cys)_2Fe(\mu\text{-}S)_2Fe(S\text{-}Cys)_2]$	19
Ferredoxin, 3 Fe	$[Fe_3(\mu\text{-}S)_3(S\text{-}Cys)_5(OH_2)]$	25
Ferredoxin, 4 Fe	$[Fe_4(\mu_3\text{-}S)_4(S\text{-}Cys)_4]$	25
High Potential Iron Protein	$[Fe_4(\mu_3\text{-}S)_4(S\text{-}Cys)_4]$	22
Sulfite Reductase	$[Fe(N_4\text{-}isobacteriochlorin)(\mu\text{-}S\text{-}Cys)Fe_4(\mu_3\text{-}S)(S\text{-}Cys)_3]$	30
Ferritin	Up to 4500 Fe atoms as an oxo/hydroxo oligomer	122
Photosystem I Reaction Center	4 bacteriochlorophylls, 2 bacteriopheophytins and a non-heme ferrous center: $[Fe(N\text{-}His)_4(O\text{-}Glu)]$	131

the copper atom, the structure of apoplastocyanin was solved to high resolution. Indeed, it was found that the ligating side chain residue positions for the apoprotein are almost identical to those of the holoprotein (*9*). The crystal structure of reduced (Cu(I)) poplar plastocyanin has been analyzed for six pH values in a recent report (*6*). Near physiological pH the Cu(I) coordination environment is very similar to that of the Cu(II) form. Thus the condition of minimal structural rearrangement for efficient electron transfer is satisfied.

Iron–Sulfur Proteins (*17–27*)

Iron–sulfur proteins serve predominantly as electron carriers (*28,29*). The best understood examples are those proteins with 1Fe, 2Fe, and 4Fe centers. The environment of the mononuclear iron center, rubredoxin, is shown in structure C (*17*). It consists of a distorted tetrahedral array of sulfur atoms from cysteine residues at nearly equal distances from the iron atom. Crystal structures are available for 2Fe-2S ferredoxins from *Spirulina plantensis* (*19*) and *Aphanothece sacrum* (*20*). A representation of the geometry of this site is given in structure D. The 2Fe-2S core is anchored to the polypeptide by ligation to 4 cysteine sulfur atoms, yielding distorted tetrahedral geometry for both iron atoms. Crystal structures of 4Fe-4S and 8Fe-8S ferredoxins have been carried out and yield a detailed picture of the cubane-like aggregate shown in structure E. The intensive efforts of Holm and coworkers (*29*) have produced analogs for each of these iron–sulfur protein structures. In the case of the 2Fe-2S centers, isolation of model compounds preceded the X-ray crystal structure determination. The binuclear and tetranuclear complexes afford an excellent opportunity to examine the effects of the protein milieu on the properties, especially redox behavior, of the metal centers. Important questions relating to this are how the polypeptide fine-tunes the reduction potential of an iron–sulfur site and how the high potential iron proteins (HiPIP) can access a stable oxidized ($4Fe\text{-}4S^{3+}$) cluster level whereas the ferredoxins can not. Hydrogen bonding interactions and control of ligand orientation may play key roles. From a comparison of the crystal structures of *Chromatium* HiPIP and *Pseudomonas aerogenes*, it has been postulated (*22*) that the distinct differences between the properties of these two proteins arise because the two 4Fe-4S centers are in nearly diasteriomeric environments with respect to each other.

A newly emerging class of iron sulfur proteins are those with 3Fe-XS centers. Aerobically isolated, inactive beef heart aconitase has a 3Fe-4S center which has been characterized by X-ray crystallography (*24*). The diffraction data are consistent with the presence of three Fe sites separated from each other by < 3Å. Thus, structure F represents a possible geometry for this 3Fe-4S unit. It is thought that the iron–sulfur cluster in aconitase is directly involved in substrate binding and transformation, a rare example of an iron–sulfur protein having a non-redox function (*see* Chapter 17 of this volume). Structure G shows another type of three iron-center; that characterized by X-ray crystallography for the 7Fe ferredoxin of *Azotobacter vinelandii* (*25–27*). The structure consists of a puckered six-membered ring with distorted tetrahedral geometry for each of the iron atoms. The single non-sulfur ligand is thought to be an oxygen atom from OH_2 or OH^-. Recent re-examination of the 7Fe protein from *A. vinelandii* by two groups reveals that structural type F, rather than G, is correct for the 3Fe center (*27b,c*). Three-iron aggregates of type F or G have not yet been synthesized *in vitro*, and as such remain important targets for bioinorganic chemists in this field.

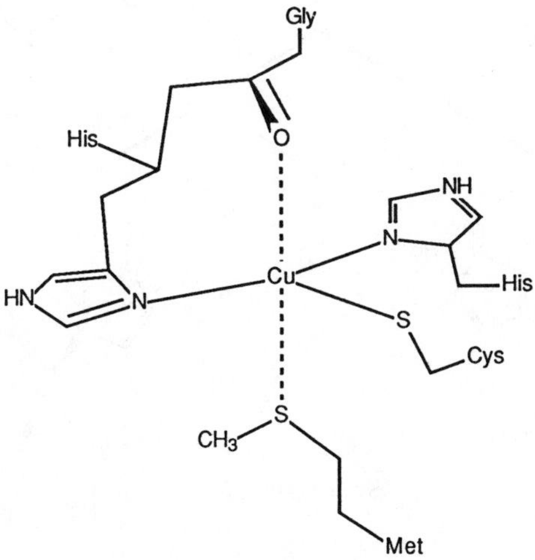

Structure A

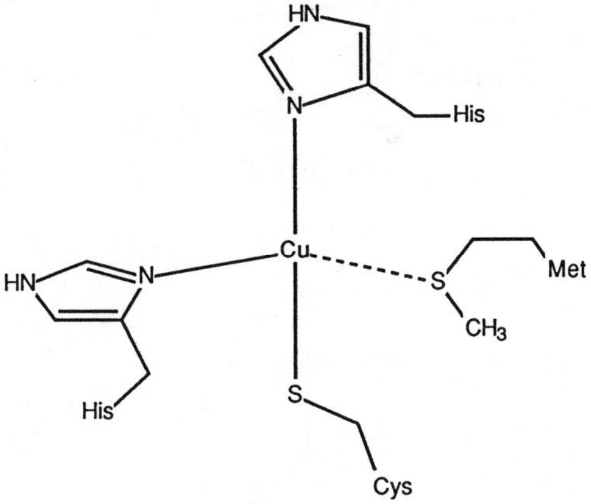

Structure B

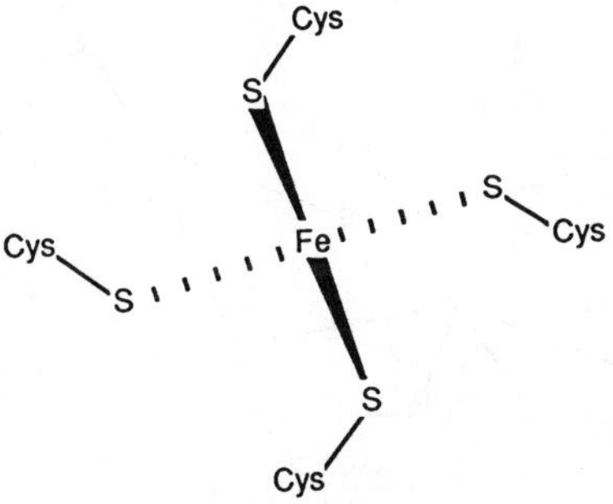

Structure C

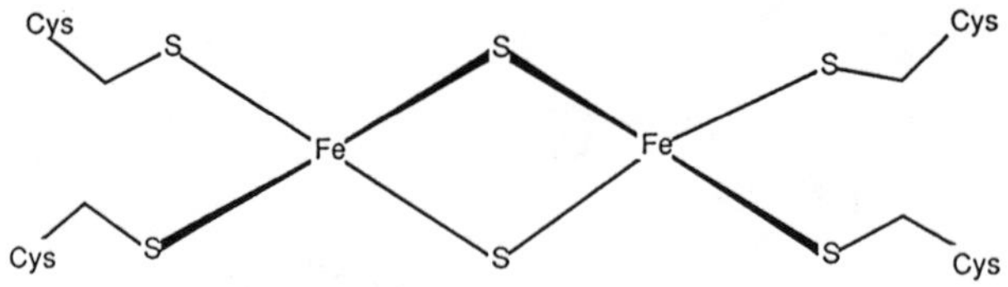

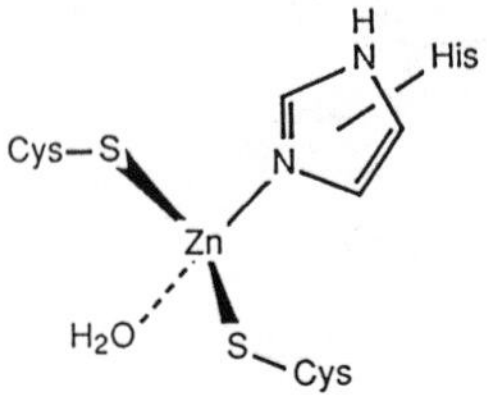

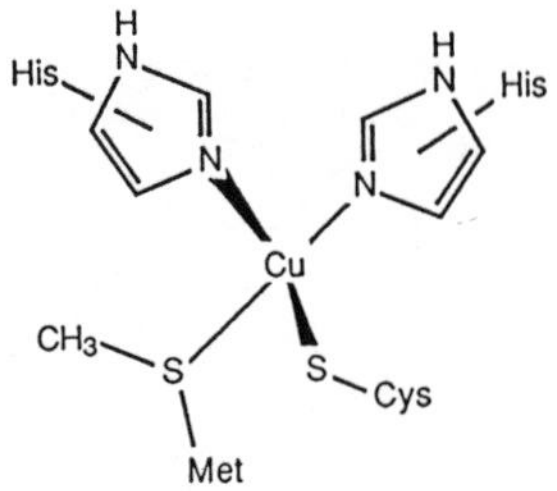

Structure D

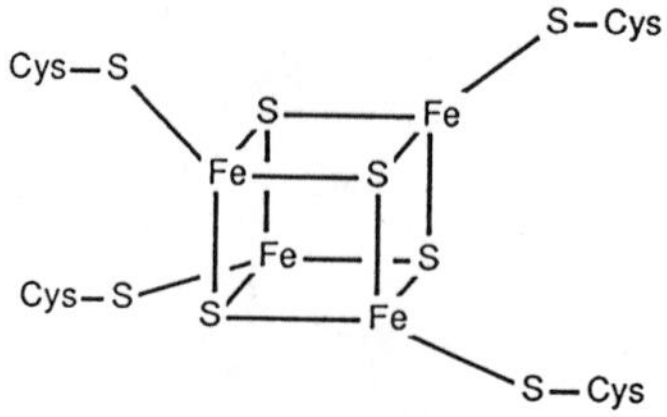

Structure E

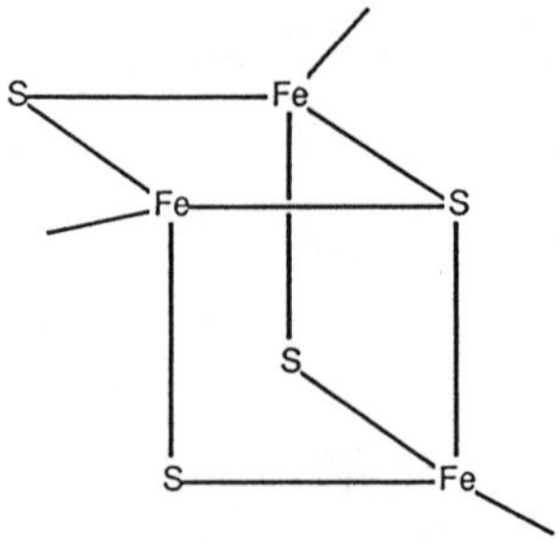

Structure F

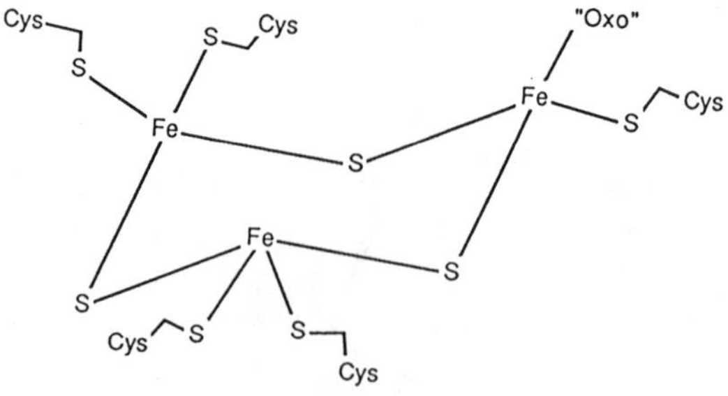

Structure G

Sulfite Reductase (SR) (*30*)

This enzyme catalyzes the six-electron reduction of sulfite to sulfide (*31a*). Both a 4Fe-4S cluster and a siroheme function as prosthetic groups in SR. Spectroscopic data indicate that these groups are closely interacting (*31b*). X-ray crystal data analysis for the *Escherichia coli* enzyme confirm this conclusion. Structure H shows a schematic diagram of the active site of SR (*30*). A cysteine sulfur atom bridges the two prosthetic groups. Presumably sulfite binds to the sixth position of the siroheme iron atom as the first step in the reduction mechanism. The difficulty in preparing models for the covalently coupled siroheme-4Fe4S active site is apparent from a recent report in which several unsuccessful attempts were made (*32*).

Cytochromes c (*33–38*)

These proteins represent a third class of electron transfer biomolecules. Protein crystal structures and models have been discussed previously in a bioinorganic context (*2*). The crystal structure of cytochrome c_5 from *Azotobacter vinelandii* was reported recently (*36*) and shown to be similar to other c-type cytochromes. In contrast, cytochrome c′ from *Rhodospirillum molischianum* has a distinctly different coordination geometry at iron than the mitochondrial c-type cytochromes (*34*). A representation of the iron site is given as structure I. The iron atom is five-coordinate with a methionine residue close to the sixth coordination site but not bound. There is also no evidence for a water molecule in the sixth position nor one hydrogen bonded to the coordinated imidazole. The absence of hydrogen bonding to the axial imidazole group is thought to make it a weaker field ligand, which in turn explains the mixed-spin state character of cytochrome c′. This is an illustration of the extent to which the peptide matrix environment can alter the iron coordination environment in a heme protein and thus alter its redox, magnetic, and spectroscopic properties. Suitable synthetic analogs for the low-spin, six-coordinate cytochromes c have been prepared using a relatively simple tetraaryl porphyrin ligand which incorporates a tethered imidazole ligand (*40*). A model for the cytochrome c′ site would require use of the same ligand and very poorly coordinating solvent and counterion (for Fe^{3+}) or a much more sterically encumbered ligand such as a "picket fence" or "strapped" porphyrin (*41–46*) that would allow imidazole coordination on only one side of the ring and no ligand on the other side.

Whether or not there are significant structural changes between the solid state and solution is an important question that is often raised when, for example, attempting to evaluate the physiological relevance of a single-crystal structure. A recent study (*33*) of the ^{1}H NMR spectrum of cytochrome c addressed this question and concluded that there were significant structural changes in the vicinity of the heme binding pocket upon dissolving the crystalline protein in solution.

Myoglobin(Mb) and Hemoglobin(Hb) (*39,48–53*)

Mb and Hb are responsible for oxygen transport in a wide variety of organisms. Several forms of both proteins have been characterized crystallographically, including deoxy, oxy, and met derivatives. The endeavors of Perutz and coworkers, recently reviewed (*39*), have made these perhaps the best characterized of metalloproteins. The oxy form of either Mb or Hb is illustrated in structure J. It consists of a distorted octahedral coordination sphere around Fe with four equatorial ligands from the protoporphyrin IX prosthetic group, and axial donors from a histidine nitrogen and the dioxygen molecule. The presence of the distal (uncoordinated) histidine ligand is important for stabilization of the coordinated dioxygen by hydrogen bonding and also sterically disfavors carbon monoxide binding. Carbon monoxide prefers a Fe-C-O angle close to 180°, in contrast to the

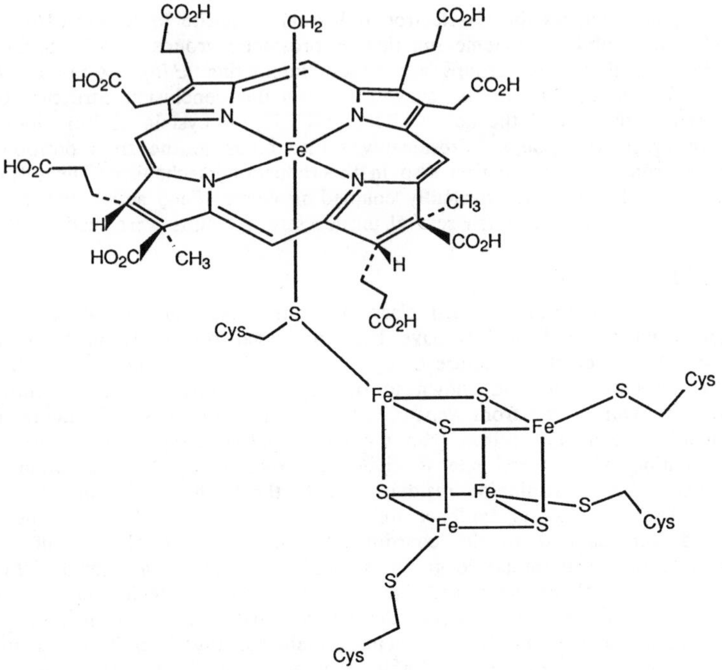

CO2H
OH2
CO2H
HO2C
CO2H
Fe
N
N
N
N
HO2C
CH3
H
CO2H
HO2C
CH3
H
CO2H
Cys
S
Fe
S
Fe
S
Cys
S
Fe
S
Cys
Fe
S
S
Cys

Structure H

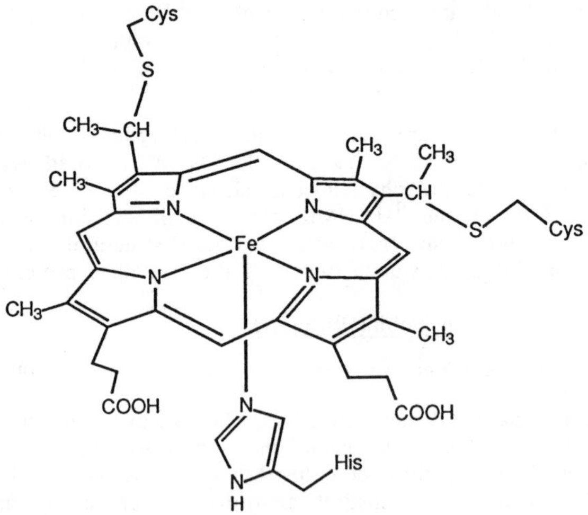

Cys
S
CH3—CH
CH3
CH3
CH3
CH
S
Cys
N
N
Fe
N
N
CH3
CH3
COOH
COOH
N
His
N
H

Structure I

bent, end-on coordination mode shown for dioxygen. Replication of the oxyMb site in synthetic systems is complicated by the tendency of simple porphyrin-Fe-O_2 species to decompose irreversibly to form oxo-bridged binuclear complexes. This decomposition pathway is not available to the protein system as the heme group is buried below the polypeptide surface. In order to mimic the peptide protection, sterically encumbered porphyrins such as the "picket fence" or "strapped" porphyrins (*41–46*) mentioned above were synthesized and oxygen binding to their Fe^{2+} complexes was examined. This strategy has provided model O_2 complexes which have been characterized by X-ray crystallography (*54*). The presence of the distal imidazole and its role of stabilizing the dioxygen complex has stimulated several workers (*55,56*) to attempt to build hydrogen bonding groups into synthetic ligand systems in order to mimic its function. Much of the interest in obtaining precise structures of Hb and Mb was generated by the desire to identify the origin of the cooperative behavior of oxygen binding (*39*). In order to understand the influence that the peptide has on the metal site, it is important to have examples of "unperturbed" model structures. A review of the structural properties of metalloporphyrin complexes has appeared recently (*57*).

Catalase (*58–60*)

This heme enzyme is thought to be involved in the protection of cells from the toxic effects of peroxides (*61*). Catalase catalyzes the decomposition of hydrogen peroxide to water and dioxygen. Crystal structures of the enzyme from beef liver (*58,59*) and *Penicillium vitale* (*60*) have been carried out. The iron coordination environment for both enzymes in the resting state is very similar and is illustrated in structure K. Instead of a proximal imidazole group in the fifth coordination site as is the case for cytochrome c′, Hb, and Mb, a tyrosine is found. Furthermore, the distal (or essential) imidazole is stacked with the heme group in contrast to the very different orientation for distal side-chain histidines in either Hb, Mb, or the cytochromes. Apparently nature differentiates the reactivities of these heme enzymes, in part, by choice of fifth ligand and by relatively subtle changes in the position of the distal imidazole group. A detailed proposal for the mechanism of action of catalase has been put forth based on the crystal structure data (*59*).

Cytochrome c Peroxidase(CCP) (*62–64*)

This enzyme catalyzes the oxidation of ferrocytochrome c by peroxides (*65*). A high resolution crystal structure for yeast CCP is available (*64*). To a first approximation, the resting state Fe^{3+} coordination environment is very similar to that of metaquoMb with four porphyrin nitrogen donors, an axial imidazole group, and a water molecule occupying the sixth position. The first step in the catalytic cycle is the reaction of CCP with a peroxide to form a semi-stable enzyme intermediate called compound I. This form of the enzyme is two electron equivalents more oxidized than the resting state. The iron atom is thought to be in the +4 oxidation state in compound I and to exist as a ferryl ion (Fe=O). The second oxidation equivalent is stored in a peptide residue to form a cation radical. The identity of the oxidized residue has been the subject of some controversy. In contrast to this situation, the active site of horseradish peroxidase compound I consists of a ferryl-porphyrin cation radical. Recently, it was found that crystals of CCP could be converted to the compound I oxidation level without cracking (*62*). Thus it was possible to collect X-ray diffraction data on this active intermediate. Based on difference maps between these data and those for the data on the resting state, it was postulated that the radical site is approximately 10Å from the heme plane, among three peptide side chains; two methionines and a tryptophan. Characterization of the CCP intermediate was made possible by recent advances in

Structure J

Structure K

crystallographic technology allowing rapid measurements of large data sets at room and low temperatures. This remarkable achievement may signal a new era in metalloenzyme crystallography in which structural information for reaction intermediates will be readily available.

Cytochromes P-450(CP450) *(66,67)*

Crystal structure refinements have been carried out for *Pseudomonas putida* CP450 both with (*66*) and without (*67*) bound substrate (camphor). CP450 are b-type heme enzymes which catalyze the hydroxylation of both aliphatic and aromatic substrates (*68*). A representation of the active site structure of the camphor-bound form is given in structure L. In the resting state, prior to binding of camphor, the Fe^{3+} atom is coordinated by the four pyrrole nitrogen atoms, a cysteine sulfur atom, and a water or hydroxide at the sixth position. Six coordination is consistent with the low spin configuration as deduced by other techniques. Examples of five-coordinate ferric porphyrin thiolate complexes had been prepared prior to the crystal structure determination (*69*). Binding of the camphor molecule causes water molecules, including the coordinated OH_2, to be excluded from the active site. The means by which the enzyme carries out stereoselective oxidation is evident by examination of the camphor orientation in the active site. The exo side of C5 of the substrate molecule is positioned adjacent to the iron atom, allowing for regio- and stereoselective hydroxylation at that site. Models for oxy-CP450 with axial thiolate ligation and a coordinated O_2 molecule have been characterized by X-ray crystallography (*70*). The chemical properties of CP450 have been mimicked with great success using iron porphyrin complexes to catalyze the oxidation of organic substrates by oxygen atom transfer agents such as iodosylbenzene (*71a*). The challenge for the bioinorganic chemist is to attach covalently a substrate recognition group into a porphyrin ligand. Use of very sterically hindered tetraaryl porphyrins such as tetra(2,4,6-triphenyl)phenyl porphyrin show regioselectivity to some extent (*71b*).

Hemerythrin(Hr) *(72–78)*

The function of dioxygen storage in several marine invertebrates is performed by the non-heme iron protein, Hr (*79,80*). Crystallographic characterization of several different forms of the protein have been carried out by two different research groups. Initial structural models for metazidomyoHr from *Themiste zostericola* and metaquoHr from *Themiste dyscritum*, proposed by independent researchers, differed substantially in terms of the detailed coordination geometry at the binuclear iron center. In the former, the two Fe atoms were bridged by only an oxo group, while in the latter they were bridged by two side-chain carboxylate groups and a water molecule. These differences stimulated futher examination of the structural models as well as a close look at the X-ray structure of metazidoHr from *T. dyscritum*. Structure M shows the configuration of the metal binding site of metazidoHr. This is in marked contrast to the earlier model for metazidomyoHr described above. Subsequently, structure M became the accepted geometry for the met azide form and the structure of metaquoHr was revised to that shown in structure N. An unexpected feature is the five coordinate iron atom with a vacant site where azide is bound in metazidoHr. As there is no exogenous ligand in the sixth coordination site, metaquoHr became known as simply metHr. Had both research groups not worked to refine their initial models to converge upon a solution consistent with both data sets, an erroneous structure may have been accepted as the correct one. At the time the metazidoHr structure was solved, there were no simple binuclear iron complexes with similar oxobis(carboxylato) bridged structures. The first question for the bioinorganic chemist was whether or not such a structure would

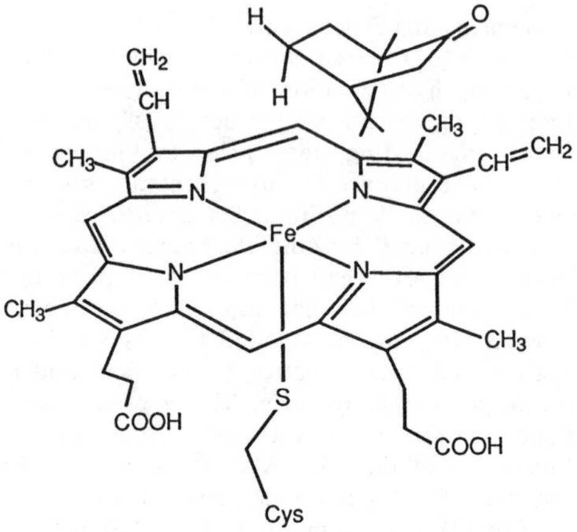

Structure L

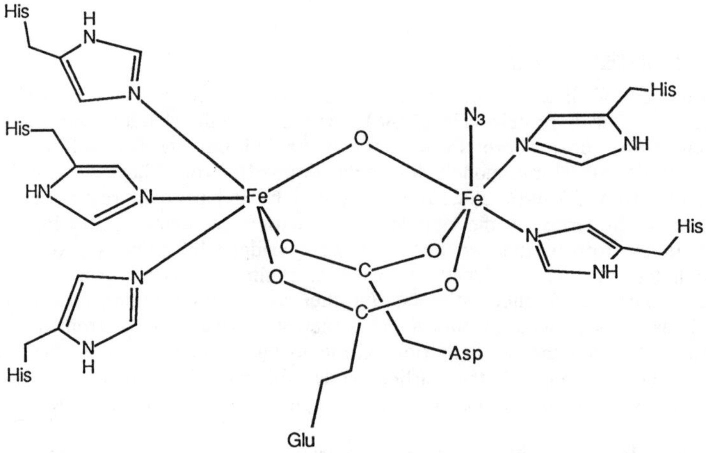

Structure M

be stable in the absence of the polypeptide environment. Synthetic efforts by two groups produced accurate replicas of the Hr active site with the requisite $Fe_2(O)(O_2CR)_2$ core and capping tripyrazolylborate ligands (*81*) on one hand, and triazacyclononane groups (*82*) on the other. The coordinates of the pyrazolylborate model complex have been used in the refinement of the binuclear iron site in azidometmyohemerythrin (*72*). Single crystals of the physiologically relevant forms of Hr, deoxyHr and oxyHr, have also been examined by X-ray diffraction techniques. The structure of deoxyHr is similar to that of metHr except that the Fe-Fe distance is somewhat longer, consistent with protonation of the bridging oxo group accompanying reduction of the diiron core by two electrons. Structure O shows the proposed structure of the binuclear iron site in oxyHr. Here, the dioxygen molecule has filled the vacant coordination site on the five coordinate iron atom. A model for deoxyHr, in which both iron atoms are six coordinate, has been prepared (*82b*). Attainment of perhaps the most important goal, a diiron model complex capable of reversible oxygen binding, has not yet been accomplished. However, Que and coworkers have been successful in preparing binuclear iron complexes which contain coordinated peroxide (*83*).

Hemocyanin(Hc) (*84–86*)

Along with hemoglobin and hemerythrin, Hc represents the third class of oxygen storage and transport proteins (*87a*). Hemocyanins are large copper-containing proteins which can be divided into two classes, molluscan and arthropodan Hc. The crystal structure of a relatively small Hc, in the deoxy state, with a molecular weight of approximately 470,000 from *Panulirus interruptus* has been determined recently. The oxygen binding site consists of a binuclear copper center with each of the copper atoms being terminally ligated by three histidine imidazole groups, shown in structure P. The Cu-Cu distance is 3.7Å and there is no evidence for a bridging side-chain residue. This latter fact was of great interest to bioinorganic chemists, because many model complexes prior to the crystallographic investigation were designed with the assumption that a tyrosine oxygen bridged the two copper atoms. A number of investigators had proposed the existence of a tyrosine bridge based on spectroscopic studies (*87a,b*). The shortest contact between a tyrosine oxygen atom and the midpoint between copper atoms indicated by the crystal structure is 17.7Å. Among the most successful of recent attempts to mimic the reversible oxygen-binding properties of Hc are those of Karlin and coworkers (*88*) (*see* Chapter 5 of this volume). A binucleating ligand that incorporates a phenolate bridge was shown to be capable of reversible O_2 binding at low temperature. However in this case Raman data indicates that the dioxygen does not bridge the two copper atoms as it is thought to in oxy Hc. The bis-Cu(I) form of another binuclear complex, in which the copper coordination spheres are bridged by a polymethylene chain, binds O_2 probably in a bridging fashion without an additional endogenous bridge (*88b*).

Manganese Superoxide Dismutase(MnSOD) (*89,90*)

Among the best characterized of manganese-containing proteins is MnSOD, which catalyzes the dismutation of superoxide (*91*). The crystal structure of the MnSOD from *Thermus thermophilus* has been determined by X-ray diffraction techniques (*89*). The active site, presented in structure Q, consists of a mononuclear manganese(III) ion surrounded by three histidines and one oxygen atom of an aspartate group. A water molecule probably occupies the fifth coordination position in a trigonal bipyramidal coordination sphere. The manganese center is rather hydrophobic. FeSOD from *Escherichia coli* has also been characterized by X-ray crystallography (*92*). There are close similarities between the structures of Fe and

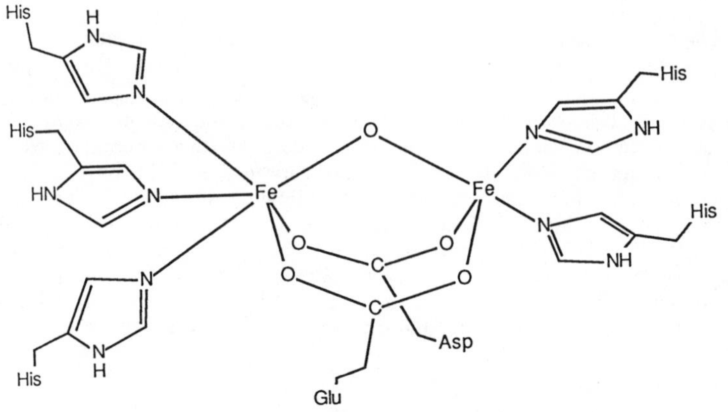

Structure N

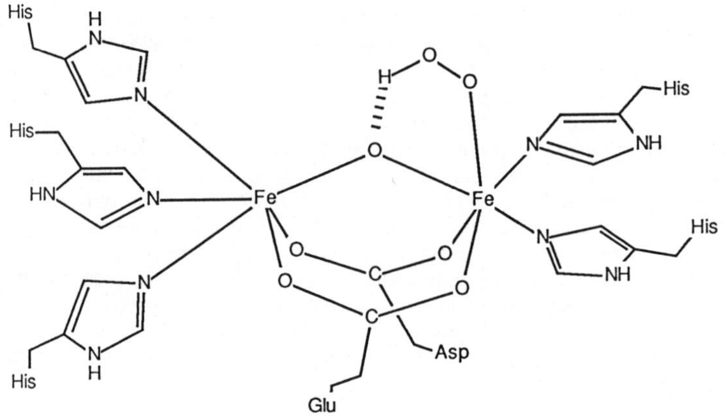

Structure O

Mn SOD enzymes (*90*). Given this fact, it is surprising that the enzymes can distinguish Mn and Fe and that the cross metallated species have greatly reduced catalytic activity (*93*). Perhaps model studies will aid in answering these important questions.

Copper, Zinc Superoxide Dismutase(CuZnSOD) (*94,95*)

In addition to the bacterial FeSOD and MnSOD, nature employs a distinctly different biomolecule for superoxide dismutation in erythrocytes, CuZnSOD. The structure of bovine erythrocyte CuZnSOD, as determined by X-ray crystallography, has been refined with high resolution data. The active site consists of copper and zinc ions bridged by an imidazolate group from a histidine side-chain residue. The copper atom is also ligated to three terminal imidazole groups, and the Zn atom to two histidines and an aspartate oxygen atom. The Cu-Zn separation is approximately 6.3Å. A representation of this binuclear active center is shown in structure R. The geometry of the four imidazole nitrogen atoms about Cu(II) is square planar with a distortion toward tetrahedral. A water molecule is coordinated in an axial position on the solvent accessible side of the copper atom (Cu-O 2.8Å). A second water molecule hydrogen bonds to a guanidinium nitrogen atom of an arginine residue (Cu-O 3.3Å). Together, these two water molecules form a "ghost" of a superoxide ion, which displaces them as it binds to the axial position on copper. Based on the crystal structure and biophysical and biochemical data for CuZnSOD, a mechanistic proposal for superoxide dismutation has been put forth (*94*). One of the roles of the Zn ion may be to increase the reduction potential of the Cu(II) site, thereby facilitating the first step of the mechanism, which is the oxidation of superoxide to dioxygen while Cu(II) is reduced to Cu(I). Model complexes incorporating bridging imidazole ligands between two copper atoms were obtained prior to the crystal structure solution (*96–98*). The stability of the imidazolate bridge was also examined as a function of pH (*99*). Details of the structure from X-ray crystallography promise to stimulate further model chemistry in this area.

Zinc-Containing Enzymes (*100–114*)

Peptidase and phosphatase activity is often associated with Zn metalloenzymes. A prototypical enzyme in this class that has been characterized extensively by X-ray crystallography is carboxypeptidase A(CPA). CPA shows preferred specificity toward peptides with bulky hydrophobic carboxy terminal residues, phenylalanine for example. The Zn coordination environment, in the absence of substrate, is shown in structure S. The Zn ion is five coordinate, with two histidine nitrogen atoms, two oxygen atoms from a glutamate, and a water molecule forming a distorted pseudo-tetrahedral first coordination sphere (if both glutamate oxygen atoms are replaced conceptually by a single atom midway between the two) (*103*). Determination of the mechanism of hydrolysis has been aided by examination of CPA-inhibitor complexes by X-ray crystallography. For example, the crystal structure of the adduct between CPA and a substrate which is slowly hydrolyzed (glycyl-L-tyrosine), demonstrates that the slow rate of hydrolysis may be related to the fact that the coordinated water molecule is displaced upon entry of the substrate (*100*). The Zn-bound water molecule is thought to be important in the hydrolysis. Details of the mechanism should provide bioinorganic chemists with clues as to how to devise low molecular weight Zn complexes that will display rates of hydrolysis approaching those of the enzyme system. Biomimetic studies of this sort have been underway for some time (*114b*). Several other examples of Zn-containing metalloenzymes are provided in Table I.

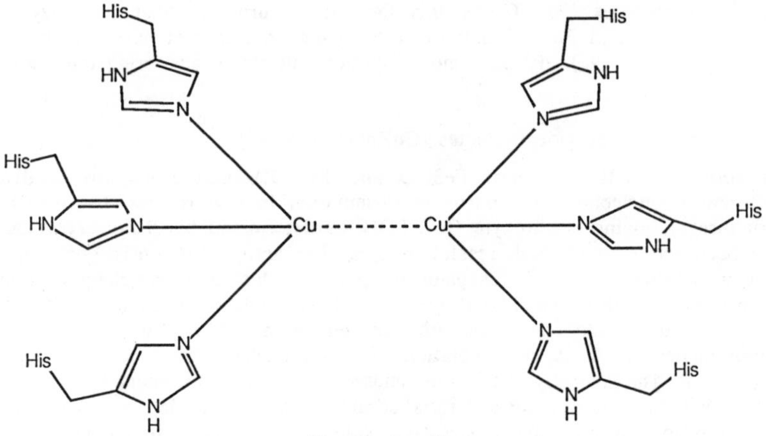

Structure P

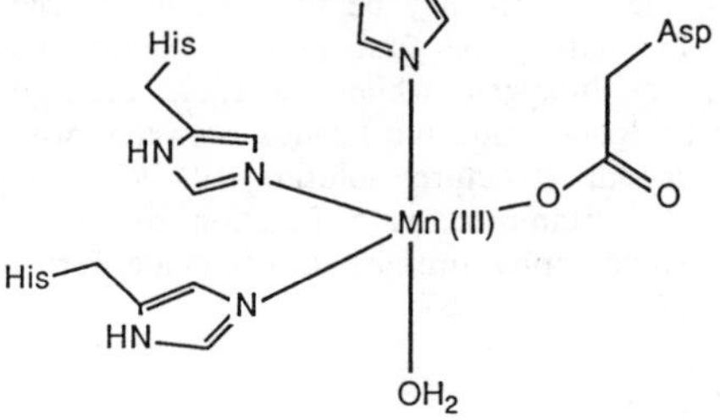

Structure Q

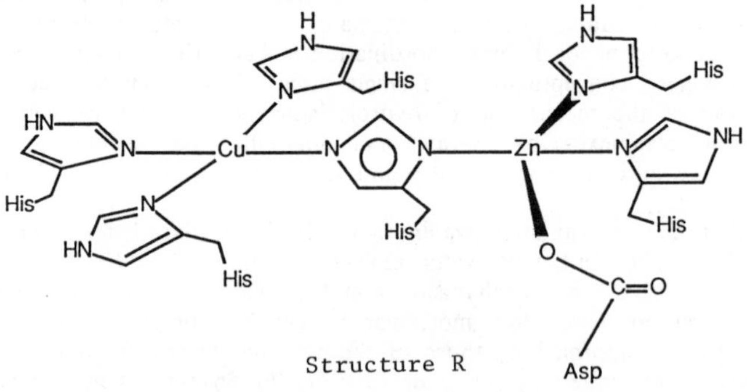

Structure R

Metallothionein(MT) (*115,116*)

These are small proteins with the ability to bind Zn, Cd, and Cu among other metals (*117*). Depending on the source, metallothioneins have 57–61 amino acids, and for the mammalian proteins there are 20 conserved cysteine residues which are all involved in coordination to the metal ions. The crystal structure of Cd, Zn MT from rat liver reveals that the metal ions are arranged in two separated domains, with four cadmium atoms in one aggregate and one cadmium and two zinc atoms in the other. Structures of the trinuclear and tetranuclear aggregates are shown in structures T and U, respectively. Thiolate-bridged ring structures similar to T have been observed in iron–sulfur chemistry (*118*). Structure U is reminiscent of, but not identical to the tetranuclear thiolates, $[M_4(SR)_{10}]^{2-}$ (*119*). More complicated thiolate ligands may be required to mimic structure U.

Lactoferrin(LT) (*120*)

Lactoferrin or lactotransferrin is one of a group of non-heme iron binding proteins which include transferrin and ovotransferrin. The crystal structure of LT indicates the bilobal structure for the polypeptide backbone. Within each lobe there is a single iron binding site. The iron atoms, which are separated by 42Å, have almost identical coordination environments, shown in structure V. Four protein side chain residues, two tyrosines, a histidine, and an aspartate, are coordinated to the iron atom leaving two *cis* positions open for binding water or the "synergistic" anion, CO_3^{2-} or HCO_3^-. From a bioinorganic viewpoint several interesting questions arise that may be approachable using model studies, including: 1) How does the protein differentiate the two binding sites? and 2) Can one devise a model system for which Fe^{3+} affinity can be drastically altered by coordination of synergistic anion as is the case for LT?

Ferritin(FT) (*122,123*)

The function of reversible iron storage in mammals is performed by ferritin (*121*). X-ray crystallographic studies on apoFT from horse spleen reveal a structure consisting of 24 cylindrical subunits symmetrically arranged around a nearly spherical 80Å diameter cavity. The protein stores up to 4500 Fe atoms as an oxo/hydroxo-bridged ferric oligomer within this polypeptide sheath. The polynuclear iron aggregate has microcrystalline properties (*124*). Inorganic chemists have been interested in FT because of its ability to solubilize Fe^{3+} at pH 7 in a form that can readily be used for the needs of the cell. Normally, under physiological conditions, iron salts hydrolyze to form insoluble polymers. Recently several groups have been successful in limiting Fe aggregation, forming discrete tetranuclear (*125*), octanuclear (*126*), and undecanuclear (*127*) complexes (*see* Chapter 10 of this volume). Through model studies of this sort, it is anticipated that a greater understanding of the growth of the ferritin iron core may be gleaned.

Bacteriochlorophyll(Bchl) (*128*)

The crystal structure of a bacteriochlorophyll *a* protein from *Prothecochloris aestuarii* has been refined with high resolution data. The Bchl *a* protein is a part of the light-harvesting apparatus of green photosynthetic bacteria (*129*). Its function is thought to be that of transmission of excitation energy to the chlorophyll special pair in the reaction center (*see* below). The Bchl *a* protein consists of three identical subunits with a total molecular weight of 150,000. Each of the subunits contains seven bacteriochlorophyll *a* molecules. All of the Mg ions are five coordinate with a water molecule, five histidines, and a carbonyl oxygen atom from the peptide backbone acting as axial ligands. Presumably, the relative orientation

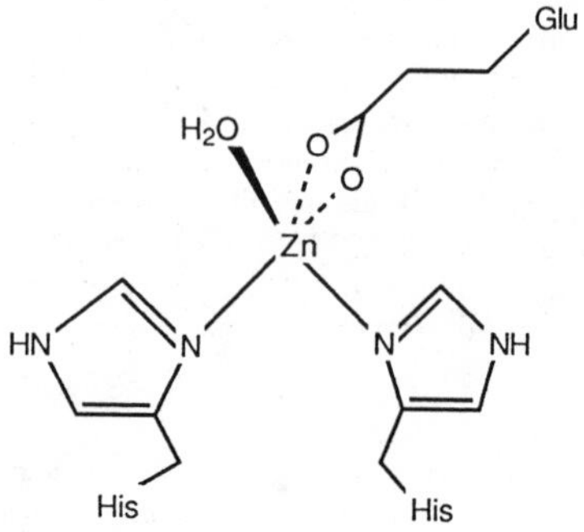

Structure S

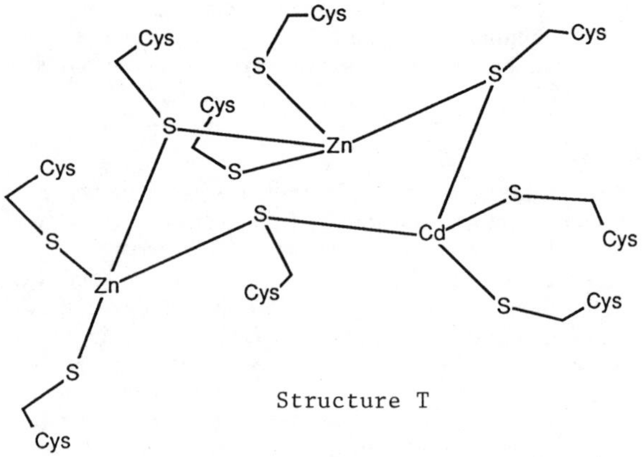

Structure T

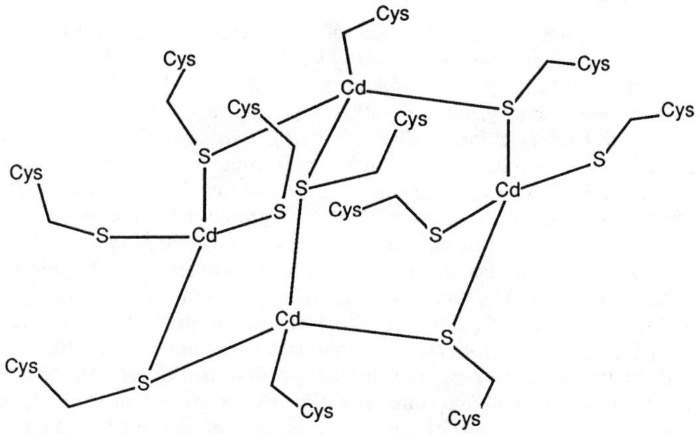

Structure U

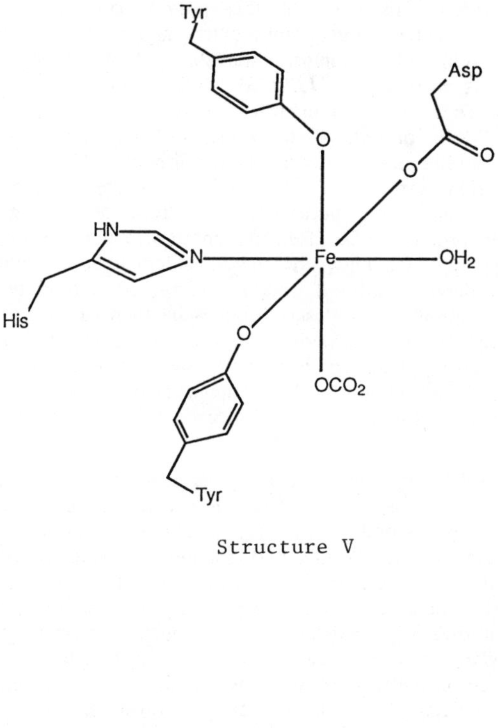

Structure V

of Bchl *a* groups is important for efficient energy transmission. Chemists have begun to approach this question by synthesis of polyporphyrin molecules and examination of ther optical properties (*130*).

Photosynthetic Reaction Center(PRC) (*131*)

Crystallization and structure determination of the membrane-bound PRC from the purple bacterium *Rhodopseudomonas viridis* was a remarkable achievement. The structure consists of four protein subunits, named L, M, H, and a c-type cytochrome with four heme groups. Prosthetic groups associated with the reaction center include four bacteriochlorophyll *b* (BChl *b*), two bacteriopheophytin *b* (BPh *b*), two quinones, carotenoids, and a non-heme ferrous iron atom. The disposition of BChl *b* and BPh *b* groups is such that a pseudo 2-fold symmetry axis is apparent. The special pair consists of two BChl *b* groups near the 2-fold axis. Two electron transfer pathways are formed on either side of the approximate 2-fold axis, and consist of a BChl *b* of the special pair next to another BChl b followed by a BPh *b* molecule. All four Mg atoms of the BChl *b* groups are ligated axially to histidine imidazole residues. One of the heme groups of the cytochrome is located near the special pair. The non-heme Fe atom is located on the 2-fold axis which relates L and M peptides, and is coordinated by four histidines and one glutamic acid residue. As mentioned above, one direction that chemists are taking in order to understand the electron transfer processes in this complicated system involves synthesis of linked tetrapyrrole molecules and investigation of their properties in comparison to unassociated tetrapyrroles.

Preliminary Data for Selected Metalloproteins

In addition to the proteins discussed to this point, many others have been characterized by X-ray cyrstallography to some extent and reported in the literature in preliminary fashion. One of the most interesting of these enzymes from a bioinorganic viewpoint is **nitrogenase** (*132,133*), which catalyzes the reduction of dinitrogen to ammonia. The enzyme is made up of two separable proteins, the Fe protein (*133*) and the MoFe protein. Both of the proteins have been crystallized and preliminary X-ray studies have appeared in the literautre. The site of nitrogen coordination and reduction is thought to be the MoFe protein, specifically at the MoFe cofactor. Two groups have undertaken the task of solving the crystal structure of the MoFe protein (*132*). Recent reports from the group in Russia indicate that the closest approach between the two independent cofactor aggregates is approximately 7Å at the α–α subunit interface. If the Mo atoms, one from each cofactor, are at these points of closest approach then a dinitrogen-bridged dimolybdenum structure for an intermediate in the course of the catalytic cycle becomes feasible. The Mo-N_2 chemistry in model systems that has been carried out to date has been at mononuclear Mo centers (*134*). Perhaps the recent crystallographic results will stimulate further investigation into binuclear systems.

Ribonucleotide reductase(RR) protein B2 from *Escherichia coli* has been studied by X-ray crystallography (*135*). This enzyme catalyzes the reduction of ribonucleotides to their corresponding deoxyribonucleotides (*136*). The B2 subunit consists of two polypeptides of molecular weight 43,000, and contains a binuclear iron center as well as a tyrosine radical. Spectroscopic properties of RR and certain synthetic binuclear iron complexes bear a strong resemblance to those of methemeryhtrin (*81*). Recent EXAFS studies are consistent with a binuclear structure containing an oxo and two carboxylate bridges in addition to nitrogen and oxygen terminal ligands (*137*). It will be interesting to see how accurately the spectroscopic and model studies have predicted the structure of the diiron site once the protein crystal structure has been solved and refined. Another important question that may be answered is how the tyrosine radical is stabilized and whether or not the iron center plays a role in this stabilization. Another non-heme iron enzyme whose structure is expected to appear in the literature in the near future is **protocatechuate 3,4-dioxygenase**, which catalyzes ring cleavage of the substrate to form β-carboxy-*cis,cis*-muconic acid (*138a,b*). Model studies of catechol dioxygenases have provided a great deal of information on the mechanism of action of certain members of this class of enzymes (*138c*). The **hydrogenase** from *Desulfovibrio vulgaris* recently has been crystallized and X-ray data have been collected (*139*). The enzyme is bidirectional, either reducing protons to dihydrogen or oxidizing dihydrogen to protons. Two [4Fe-4S] centers are the only prosthetic groups for this enzyme. Preliminary characterization of crystals of **nitrite reductase**, a copper-containing protein from *Alcaligenes faecalis*, has has been carried out (*140*). Two members of the class of copper proteins referred to as the "blue oxidases", **ascorbate oxidase** (*141*) and **ceruloplasmin** (*142*), have been studied by X-ray diffraction techniques. Both of these enzymes are multi-copper proteins with type 1, type 2, and type 3 sites. The latter two types have not previously been determined by X-ray crystallography.

Conclusions

Many of the metalloprotein crystal structures discussed in the 1980 review of Ibers and Holm (*2*) have been refined to higher precision in recent years, yielding a more accurate description of the metal binding sites. In the case of hemerythrin, further refinement of the structure of met forms of the protein by two groups have corrected the description given in the Ibers and Holm review. Similarly, the original description for the 3Fe center in the *Azotobacter vinelandii* has been corrected recently. Since 1980, new structural information has become available for many metalloproteins. In some cases, comparison of spectral properties of analogs to those of the natural systems have yielded accurate predictions for the metal site structures, and in this sense the crystallography has corroborated the structural models developed by bioinorganic chemists. Another way of stating this situation would be to say that synthetic models that were considered to be speculative prior to the solution of the X-ray crystal structure become corroborative after the fact. Solution of the protein crystal structure, despite the fact that spectroscopic evidence may have predicted certain structural features, is expected to prompt bioinorganic chemists into action, their goal being to find out whether such a metal complex can be prepared in the absence of the protein matrix and if it can be made, to find out how chemical and spectroscopic properties are perturbed by the polypeptide. Until recently, X-ray crystallography was only useful for states of the metalloprotein that were stable for very long periods of time. Cases for which enzyme intermediates could be characterized by X-ray crystallography were few and far between. For carboxypeptidase, slowly-reacting substrate adducts of the enzyme allow examination of the detailed interactions between the Zn ion and substrate and thereby provide further evidence for the mechanism of action. In the case of cytochrome c peroxidase, a reactive intermediate (compound I) was generated in a single crystal and characterized by difference X-ray techniques. Recent developments (*143–146*) in data collection techniques, including use of area detectors and high intensity sources such as synchrotron radiation, along with more powerful computers, promise to increase dramatically the rate at which new structural data becomes available and, perhaps more importantly allow useful X-ray studies to be carried out on protein species that have limited stability. Data collection rates up to 100,000 reflections per second have been reported using synchrotron radiation and Laue film techniques (*144*). Another extremely important development is the crystallization and X-ray characterization of a membrane-bound protein system, photosystem I reaction center. Cytochrome c oxidase is another membrane-bound system for which crystallographic data would be very valuable. Characterization of 2-dimensional crystals of cytochrome oxidase has been reported (*147*). Single crystal X-ray structures of these and other membrane-bound enzyme systems will greatly enhance our understanding of essential processes such as photosynthesis and respiration.

Acknowledgments

Financial support from the Searle Scholars Program/The Chicago Community Trust and Grant No. GM382751-01 from the National Institutes of General Medical Science is gratefully acknowledged. I also thank Sandhya Subramanian for work on the structures.

References

1. Hill, H. A. O. *Chem. Br.* **1976,** *12,* 119–123.
2. Ibers, J. A.; Holm, R. H. *Science (Washington, DC)* **1980,** *209,* 223–235.
3. Norris, G. E.; Anderson, B. F.; Baker, E. N. *J. Am. Chem. Soc.* **1986,** *108,* 2784–2785.
4. Adman, E. T.; Jensen, L. H. *Isr. J. Chem.* **1981,** *21,* 8–12.
5. Adman, E. T.; Stenkamp, R. E.; Sieker, L. C.; Jensen, L. H. *J. Mol. Biol.* **1978,** *123,* 35–47.
6. Guss, J. M.; Harrowell, P. R.; Murata, M.; Norris, V. A.; Freeman, H. C. *J. Mol. Biol.* **1986,** *192,* 361–387.
7. Church, W. B.; Guss, J. M.; Potter, J. J.; Freeman, H. C. *J. Biol. Chem.* **1986,** *261,* 234–237.
8. Guss, J. M.; Freeman, H. C. *J. Mol. Biol.* **1983,** *169,* 521–563.
9. Garrett, T. P. J.; Clingeleffer, D. J.; Guss, J. M.; Rogers, S. J.; Freeman, H. C. *J. Biol. Chem.* **1984,** *259,* 2822–2825.
10. Colman, P. M.; Freeman, H. C.; Guss, J. M.; Murata, M.; Norris, V. A.; Ramshaw, J. A. M.; VenKatappa, M. P. *Nature (London)* **1978,** *272,* 319–324.
11. Gray, H. B.; Solomon, E. I. In *Copper Proteins;* Spiro, T. G., Ed.; Wiley: New York, 1981; Volume 3, pp 1–39.
12. (a) Fee, J. A. *Struct. Bonding (Berlin)* **1975,** *23,* 1–60. (b) McMillin, D. R. *J. Chem. Educ.* **1985,** *62,* 997–1001.
13. Addison, A.; Sinn, E. *Inorg. Chem.* **1983,** *22,* 1225–1228.
14. John, E.; Bharadwaj, P. K.; Potenza, J. A.; Schugar, H. J. *Inorg. Chem.* **1986,** *25,* 3065–3069.
15. Andersen, O. P.; Perkins, C. U.; Brito, K. K. *Inorg. Chem.* **1983,** *22,* 1267–1273.
16. Thompson, J. A.; Marks, T. J.; Ibers, J. A. *J. Am. Chem. Soc.* **1979,** *101,* 4180–4192.
17. Watenpaugh, K. D.; Sieker, L. C.; Jensen, L. H. *J. Mol. Biol.* **1980,** *138,* 615–633.
18. Sieker, L. C.; Jensen, L. H.; Prickril, B. C.; LeGall, J. *J. Mol. Biol.* **1983,** *171,* 101–103.
19. Fukuyama, K.; Hase, T.; Matsumoto, S.; Tsukihara, T.; Katsube, Y.; Tanaka, N.; Kakudo, M.; Wada, K.; Matsubara, H. *Nature (London)* **1980,** *286,* 522–524.
20. Tsukihara, T.; Fukuyama, K.; Nakamura, M.; Katsube, Y.; Tanaka, N.; Kakudo, M.; Wada, K.; Hase, T.; Matsubara, H. *J. Biochem.* **1981,** *90,* 1763–1773.

21. Tsutsui, T.; Tsukihara, T.; Fukuyama, K.; Katsube, Y.; Hase, T.; Matsubara, H.; Nishsikawa, Y.; Tanaka, N. *J. Biochem.* **1983,** *94,* 299–302.
22. Carter, C. W., Jr. *J. Biol. Chem.* **1977,** *252,* 7802–7811.
23. Beinert, H.; Emptage, M. H.; Dreyer, J.-L.; Scott, R. A.; Hahn, J. E.; Hodgson, K. O.; Thomson, A. J. *Proc. Natl. Acad. Sci. USA* **1983,** *80,* 393–396.
24. Robbins, A. H.; Stout, C. D. *J. Biol. Chem.* **1985,** *260,* 2328–2333.
25. Ghosh, D.; O'Donnell, S.; Furey, W., Jr.; Robbins, A. H.; Stout, C. D. *J. Mol. Biol.* **1982,** *158,* 73–109.
26. Howard, J. B.; Lorsbach, T. W.; Ghosh, D.; Melis, K.; Stout, C. D. *J. Biol. Chem.* **1983,** *258,* 508–522.
27. (a) Ghosh, D.; Furey, W., Jr.; O'Donnell, S.; Stout, C. D. *J. Biol. Chem.* **1981,** *256,* 4185–4192. (b) Stout, C. D., personal communication. (c) Jensen, L. H., personal communication.
28. *Iron–Sulfur Proteins;* Lovenberg, W., Ed.; Academic: New York, 1973/1977.
29. Berg, J. R.; Holm, R. H. *Met. Ions Biol.* **1982,** *4,* 1–66.
30. McRee, D. E.; Richardson, D. C.; Richardson, J. S.; Siegel, L. M. *J. Biol. Chem.* **1986,** *261,* 10277–10281.
31. (a) Kemp, J. D.; Atkinson, D. E.; Ehret, A.; Lazzarini, R. A. *J. Biol. Chem.* **1961,** *238,* 3466–3475. (b) Cline, J. F.; Janick, P. A.; Siegel, L. M.; Hoffman, B. M. *Biochemistry* **1985,** *24,* 7942–7947, and references therein.
32. Stolzenberg, A. M.; Stershic, M. T. *Inorg. Chem.* **1985,** *24,* 3095–3098.
33. Williams, G.; Clayden, N. J.; Moore, G. R.; Williams, R. J. P. *J. Mol. Biol.* **1985,** *183,* 447–460.
34. Finzel, B. C.; Weber, P. C.; Hardman, K. D.; Salemme, F. R. *J. Mol. Biol.* **1985,** *186,* 627–643.
35. Finzel, B. C.; Salemme, F. R. *Nature (London)* **1985,** *315,* 686–688.
36. Carter, D. C.; Melis, K. A.; O'Donnell, S. E.; Burgess, B. K.; Furey, W. F., Jr.; Wang, B.-C.; Stout, C. D. *J. Mol. Biol.* **1985,** *184,* 279–295.
37. Salemme, F. R. *Annu. Rev. Biochem.* **1977,** *46,* 299.
38. Dickerson, R. E.; Timkovich, R. In *The Enzymes,* 3rd ed.; Boyer, P. D., Ed.; Academic: New York, 1975; Volume 11, p 397.
39. Perutz, M. F.; Fermi, G.; Luisi, B.; Shaanan, B.; Liddington, R. C. *Acc. Chem. Res.* **1987,** *20,* 309–321.
40. Mashiko, T.; Marchon, J.-C.; Musser, D. T.; Reed, C. A.; Kastner, M. E.; Scheidt, W. R. *J. Am. Chem. Soc.* **1979,** *101,* 3653–3655.
41. Groves, J. T. *J. Chem. Educ.* **1985,** *62,* 928–931.
42. Suslick, K. S.; Reinert, T. J. *J. Chem. Educ.* **1985,** *62,* 974–983.
43. Traylor, T. G.; Traylor, P. S. *Annu. Rev. Biophys. Bioeng.* **1982,** *11,* 105–127.
44. Smith, T. D.; Pilbrow, J. R. *Coord. Chem. Rev.* **1981,** *39,* 295–383.
45. Jones, R. D.; Summerville, D. A.; Basolo, F. *Chem. Rev.* **1979,** *79,* 139–179.
46. Tsuchida, E.; Nishide, H. *Top. Curr. Chem.* **1987,** *132,* 63–69.
47. Morgan, B.; Dolphin, D. *Struct. Bonding (Berlin)* **1987,** *64,* 115–203.
48. TenEyck, L. F.; Arnone, A. *J. Mol. Biol.* **1976,** *100,* 3–11.
49. Takano, T. *J. Mol. Biol.* **1977,** *110,* 569–584.
50. Phillips, S. E. V. *J. Mol. Biol.* **1980,** *142,* 531–554.
51. Weber, E.; Steigemann, W.; Jones, T. A.; Huber, R. *J. Mol. Biol.* **1978,** *120,* 327–336.
52. Tilton, R. F., Jr.; Kuntz, I. D., Jr.; Petsko, G. A. *Biochemistry* **1984,** *23,* 2849–2857.
53. Bolognesi, M.; Coda, A.; Giuseppina, G.; Ascenzi, P.; Brunori, M. *J. Mol. Biol.* **1985,** *183,* 113–115.
54. Jameson, G. B.; Molinaro, F. S.; Ibers, J. A.; Collman, J. P.; Brauman, J. I.; Rose, E.; Suslick, K. S. *J. Am. Chem. Soc.* **1980,** *102,* 3224–3237.
55. Walker, R. A.; Bowen, J. *J. Am. Chem. Soc.* **1985,** *107,* 7632–7635.

56. Chang, C. K.; Kondylis, M. P. *J. Chem. Soc. Chem. Comm.* **1986,** 316–318.
57. Scheidt, W. R.; Lee, Y. J. *Struct. Bonding (Berlin)* **1987,** *64,* 1–20.
58. Fita, I.; Silva, A. M.; Murthy, M. R. N.; Rossmann, M. G. *Acta Cryst.* **1986,** *B42,* 497–515.
59. Fita, I.; Rossman, M. G. *J. Mol. Biol.* **1985,** *185,* 21–37.
60. Vainshtein, B. K.; Melik-Adamyan, W. R.; Barynin, V. V.; Vagin, A. A.; Grebenko, A. I.; Borisov, V. V.; Bartels, K. S.; Fita, I.; Rossmann, M. G. *J. Mol. Biol.* **1986,** *188,* 49–61.
61. Schonbaum, G. R.; Chance, B. In *The Enzymes, 3rd ed.;* Boyer, P. D., Ed.; Academic: New York, 1976; pp 363–408.
62. Edwards, S. L.; Xuong, N. H.; Hamlin, R. C.; Kraut, J. *Biochemistry* **1987,** *26,* 1503–1511.
63. Edwards, S. L.; Poulos, T. L.; Kraut, J. *J. Biol. Chem.* **1984,** *259,* 12984–12988.
64. Finzel, B. C.; Poulos, T. L.; Kraut, J. *J. Biol. Chem.* **1984,** *259,* 13027–13036.
65. Yonetani, T.; Ray, G. S. *J. Biol. Chem.* **1966,** *241,* 700–705.
66. Poulos, T. L.; Finzel, B. C.; Gusalus, I. C.; Wagner, G. C.; Kraut, J. *J. Biol. Chem.* **1985,** *260,* 16122–16130.
67. Poulos, T. L.; Finzel, B. C.; Howard, A. *J. Biochem.* **1986,** *25,* 5314–5322.
68. *Cytochrome P-450;* Ortiz de Montellano, P. R., Ed.; Plennum: New York, 1986.
69. Tang, S. C.; Koch, S.; Papaefthymiou, G. C.; Foner, S.; Frankel, R. B.; Ibers, J. A.; Holm, R. H. *J. Am. Chem. Soc.* **1976,** *98,* 2414–2434.
70. Richard, L.; Schappacher, M.; Weiss, R.; Montiel-Montoya, R.; Bill, E.; Gonser, U.; Trautwein, A. *Nouv. J. Chim.* **1983,** *7,* 405–408.
71. (a) See for example: reference 40. (b) Cook, B. R.; Reinert, T. J.; Suslick, K. S. *J. Am. Chem. Soc.* **1986,** *108,* 7281–7286.
72. Sheriff, S.; Hendrickson, W. A.; Smith, J. L. *J. Mol. Biol.* **1987,** *197,* 273–296.
73. Sheriff, S.; Hendrickson, W. A.; Stenkamp, R. E.; Sieker, L. C.; Jensen, L. H. *Proc. Natl. Acad. Sci. USA* **1985,** *82,* 1104–1107.
74. Stenkamp, R. E.; Sieker, L. C.; Jensen, L. H.; McCallum, J. D.; Sanders-Loehr, J. S. *Proc Natl. Acad. Sci. USA* **1985,** *82,* 713–716.
75. Stenkamp, R. E.; Sieker, L. C.; Jensen, L. H. *J. Am. Chem. Soc.* **1984,** *106,* 618–622.
76. Stenkamp, R. E.; Sieker, L. C.; Jensen, L. H. *J. Inorg. Biochem.* **1983,** *19,* 247–253.
77. Stenkamp, R. E.; Sieker, L. C.; Jensen, L. H. *Acta Crystallogr.* **1982,** *B38,* 784–792.
78. Stenkamp, R. E.; Sieker, L. C.; Jensen, L. H.; Sanders-Loehr, J. *Nature (London)* **1981,** *291,* 263–264.
79. Wilkins, R. G.; Harrington, P. C. *Adv. Inorg. Biochem.* **1983,** *3,* 51–85.
80. Wilkins, P. C.; Wilkins, R. G. *Coord. Chem. Rev.* **1987,** *79,* 195–214.
81. Armstrong, W. H.; Spool, A.; Papaefthymiou, G. C.; Frankel, R. B.; Lippard, S. J. *J. Am. Chem. Soc.* **1984,** *106,* 3653–3667.
82. (a) Wieghardt, K.; Pohl, K.; Gebert, W. *Angew. Chem. Int. Ed. Engl.* **1983,** *22,* 727; (b) Hartman, J. R.; Rardin, R. L.; Chaduri, P.; Pohl, K.; Wieghardt, K.; Nuber, B.; Weiss, J.; Papaefthymiou, G. C.; Frankel, R. B.; Lippard, S. J. *J. Am. Chem. Soc.* **1987,** *109,* 7387–7396.
83. Murch, B. P.; Bradley, F. C.; Que, L., Jr. *J. Am. Chem. Soc.* **1986,** *108,* 5027–5028.
84. Gaykema, W. P. J.; Hol, W. G. J.; Vereijken, J. M.; Soeter, N. M.; Bak, H. J.; Beintema, J. J. *Nature (London)* **1984,** *309,* 23–29.
85. Linzen, B.; Soeter, N. M.; Riggs, A. F.; Schneider, H.-J.; Schartan, W.; Moore, M. D.; Yokota, E.; Behrens, P. Q.; Nakashima, H.; Takagi, T.; Nemoto, T.; Vereijken, J. M.; Bak, H. J.; Beintema, J. J.; Volbeda, A.; Gaykema, W. P. J.; Hol, W. G. J. *Science (Washington, D.C.)* **1985,** *229,* 519–524.

86. Gaykema, W. P. J.; Volbeda, A.; Hol, W. G. J. *J. Mol. Biol.* **1985,** *187,* 255–275.
87. (a) Ellerton, H. D.; Ellerton, N. F.; Robinson, H. A. *Prog. Biophys. Mol. Biol.* **1983,** *41,* 143–248. (b) Van Holde, K. E.; Miller, K. I. *Quart. Rev. Biophys.* **1982,** *15,* 1–129.
88. (a) Karlin, K. D.; Cruse, R. W.; Gultneh, Y.; Farooq, A.; Hayes, J. C.; Zubieta, J. *J. Am. Chem. Soc.* **1987,** *109,* 2668–2679. (b) Chapter 5 of this volume.
89. Stallings, W. C.; Pattridge, K. A.; Strong, R. K. ; Ludwig, M. L. *J. Biol. Chem.* **1985,** *260,* 16424–16432.
90. Stallings, W. C.; Pattridge, K. A.; Strong, R. K.; Ludwig, M. L. *J. Biol. Chem.* **1984,** *259,* 10695–10699.
91. *Superoxide and Superoxide Dismutases;* Michelson, A. M.; McCord, J. M.; Fridovich, I., Eds.; Academic: New York, 1977.
92. (a) Ringe, D.; Petsko, G. A.; Yamakura, F.; Suzuki, K.; Ohmori, D. *Proc. Natl. Acad. Sci. USA* **1983,** *80,* 3879–3883. (b) Stallings, W. C.; Powers, T. B.; Pattridge, K. A.; Fee, J. A.; Ludwig, M. L. *Proc. Natl. Acad. Sci.* **1983,** *80,* 3884–3888.
93. (a) Kirby, T.; Blum, J.; Kahane, I.; Fridovich, I. *Arch. Biochem. Biophys* **1980,** *201,* 551–555. (b) Brock, C. J.; Harris, J. I. *Biochem. Soc. Trans.* **1977,** *5,* 1537–1538.
94. Tainer, J. A.; Getzoff, E. D.; Richardson, J. S.; Richardson, D. C. *Nature (London)* **1983,** *306,* 284–287.
95. Tainer, J. A.; Getzoff, E. D.; Beem, K. M.; Richardson, J. S.; Richardson, D. C. *J. Mol. Biol.* **1982,** *160,* 181–217.
96. Kolks, G.; Lippard, S. J. *J. Am. Chem. Soc.* **1977,** *99,* 5804–5806.
97. Coughlin, P. K.; Dewan, J. C.; Lippard, S. J. *J. Am. Chem. Soc.* **1979,** *101,* 265–266.
98. Haddad, M. S.; Hendrickson, D. N. *Inorg. Chem.* **1978,** *17,* 2622.
99. O'Young, C. L.; Dewan, J. C.; Lilienthal, H. R.; Lippard, S. J. *J. Am. Chem. Soc.* **1978,** *100,* 7291–7300.
100. Christianson, D. W.; Lipscomb, W. N. *Proc. Natl. Acad. Sci. USA* **1986,** *83,* 7568–7572.
101. Christianson, D. W.; Lipscomb, W. N. *J. Am. Chem. Soc.* **1986,** *108,* 545–546.
102. Christianson, D. W.; Lipscomb, W. N. *Proc. Natl. Acad. Sci.* **1985,** *82,* 6840–6844.
103. Rees, D. C.; Lewis, M.; Lipscomb, W. N. *J. Mol. Biol.* **1983,** *168,* 367–387.
104. Rees, D. C.; Lipscomb, W. N. *J. Mol. Biol.* **1982,** *160,* 475–498.
105. Rees, D. C.; Lewis, M.; Honzatko, R. B.; Lipscomb, W. N.; Hardman, K. D. *Proc. Natl. Acad. Sci.* **1981,** *78,* 3408–3412.
106. Rees, D. C.; Lipscomb, W. N. *Proc. Natl. Acad. Aci.* **1980,** *77,* 4633–4637.
107. Smith, G. D.; Duax, W. L.; Dodson, E. J.; Dodson, G. G; de Graaf, R. A. G.; Reynolds, C. D. *Acta Cryst.* **1982,** *B38,* 3028–3032.
108. Smith, G. D.; Swenson, D. C.; Dodson, E. J.; Dodson, G. G.; Reynolds, C. D. *Proc. Natl. Acad. Sci.* **1984,** *81,* 7093–7097.
109. Tronrud, D. E.; Holden, H. M.; Matthews, B. W. *Science (Washington, D.C.)* **1987,** *235,* 571–574.
110. Tronrud, D. E.; Monzingo, A. F.; Matthews, B. W. Eur. *J. Biochem.* **1986,** *157,* 261–268.
111. Monzingo, A. F.; Matthews, B. W. *Biochemistry* **1984,** *23,* 5724–5729.
112. Holmes, M. A.; Tronrud, D. E.; Matthews, B. W. *Biochemistry* **1983,** *22,* 236–240.
113. Kannan, K. K.; Notstrand, B.; Fridborg, K.; Lovgren, S.; Ohlsson, A.; Petef, M. *Proc. Natl. Acad. Sci. USA* **1975,** *72,* 51–55.

114. (a) Eklund, H.; Nordstrom, B.; Zeppezauer, E.; Soderlund, G.; Ohlsson, I.; Boiwe, T.; Soderberg, B.-O.; Tapia, O.; Branden, C.-I. *J. Mol. Biol.* **1976,** *102,* 27–59. (b) See for example: Breslow, R. *Adv. Enzymol. Relat. Areas Mol. Biol.* **1986,** *58,* 1–60.
115. Furey, W. F.; Robbins, A. H.; Clancy, L. L.; Winge, D. R.; Wang, B. C.; Stout, C. D. *Science (Washington, D.C.)* **1986,** *231,* 704–710.
116. Melis, K. A.; Carter, D. C.; Stout, C. D.; Winge, D. R. *J. Biol. Chem.* **1983,** *258,* 6255–6257.
117. *Metallothionein;* Kägi, J. H. R.; Nordberg, M., Eds.; Birkhäuser: Basel, 1979.
118. Whitener, M. A.; Bashkin, J. K.; Hagen, K. S.; Girerd, J.-J.; Gamp, E.; Edelstein, N.; Holm, R. H. *J. Am. Chem. Soc.* **1986,** *108,* 5607–5620.
119. Hagen, K. S.; Holm, R. H. *Inorg. Chem.* **1984,** *23,* 418–427.
120. Anderson, B. F.; Baker, H. M.; Dodson, E. J.; Norris, G. E.; Rumball, S.V.; Waters, J. M.; Baker, E. N. *Proc. Natl. Acad. Sci. USA* **1987,** *84,* 1769–1773.
121. (a) Brock, J. H. *Top. Mol. Struct. Biol.* **1985,** *7,* 183–262. (b) Aisen, P.; Listowsky, I. *Annu. Rev. Biochem.* **1980,** *49,* 357–393.
122. Harrison, P. M. *Struct. Stud. Mol. Biol. Interest* **1981,** 291–309.
123. Rice, D. W.; Ford, G. C.; White, J. L.; Smith, J. M. A.; Harrison, P. M. *Adv. Inorg. Biochem.* **1983,** *5,* 39–50.
124. Harrison, P. M.; Fishbach, F. A.; Hoy, T. G.; Hoggis, G. H. *Nature (London)* **1967, 216, 1188–1190.**
125. Armstrong, W. H.; Roth, M. E.; Lippard, S. J. *J. Am. Chem. Soc.* **1987,** *109,* 6318–6326 and references therein.
126. Wieghardt, K.; Pohl, K.; Ventur, D. *Angew. Chem. Int. Ed. Engl.* **1985,** *24,* 392–393.
127. Gorun, S. M.; Papaefthymiou, G. C.; Frankel, R. B.; Lippard, S. J. *J. Am. Chem. Soc.* **1987,** *109,* 3337–3348.
128. Tronrud, D. E.; Schmid, M. F.; Matthews, B. W. *J. Mol. Biol.* **1986,** *188,* 443–454.
129. Stryer, L. S. *Biochemistry; W. H. Freeman: New York, 1981; Chapter 19.*
130. (a) Sessler, J. L.; Hugdahl, J.; Johnson, M. R. J. Org. Chem. **1986,** *51,* 2838–2840. (b) Dubowchik, G.; Heiler, D.; Hamilton, A. D. *Rev. Port. Chim.* **1985,** *27,* 358–359.
131. (a) Deisenhofer, J.; Epp, O.; Miki, K.; Huber, R.; Michel, H. *Nature (London)* **1985,** *318,* 618–624. (b) Allen, J. P.; Feher, G.; Yeates, T. O.; Rees, D. C.; Deisenhofer, J.; Michel, H.; Huber, R. *Proc. Natl. Acad. Sci. USA* **1986,** *83,* 8589–8593.
132. (a) Shilov, A. E. *J. Mol. Catal.* **1987,** *41,* 221–234. (b) Sosfenov, N. I.; Andrianov, V. I.; Vagin, A. A.; Strokopytov, B. V.; Vainstein, B. K.; Shilov, A. E.; Gvozdev, R. I.; Likhtenshtein, G. I.; Mitsova, I. Z.; Blazhchuk, I. S. *Dokl. Akad. Nauk. SSSR* **1986,** *291,* 1123–1127. (c) Flank, A. M.; Weininger, M.; Mortenson, L. E.; Cramer, S. P. *J. Am. Chem. Soc.* **1986,** *108,* 1049–1055. (d) Weininger, M. S.; Mortenson, L. E. *Proc. Natl. Acad. Sci.* **1982,** *79,* 378–380. (e) Yamane, T.; Weininger, M. S.; Mortenson, L. E.; Rossmann, M. G. *J. Biol. Chem.* **1982,** *257,* 1221–1223.
133. Rees, D. C.; Howard, J. B. *J. Biol. Chem.* **1983,** *258,* 12733–12734.
134. *Current Perspectives in Nitrogen Fixation;* Gibson, A. H.; Newton, W. E., Eds.; Australian Academy of Sciences: Canberra, 1981.
135. Joelson, T.; Uhlin, U.; Eklund, H.; Sjöberg, B.-M.; Hahne, S.; Karlsson, M. *J. Biol. Chem.* **1984,** *259,* 9076–9077.
136. Sjöberg, B.-M.; Gräslund, A. *Adv. Inorg. Biochem.* **1983,** *54,* 27–91.
137. Scarrow, R. C.; Maroney, M. J.; Palmer, S. M.; Que, L., Jr. Salowe, S. P.; Stubbe, J. *J. Am. Chem. Soc.* **1986,** *108,* 6832–6834.

138. (a) Ludwig, M. L.; Weber, L. D.; Ballou, D. P. *J. Biol. Chem.* **1984,** *259,* 14840–14842. (b) Ohlendorf, D. H.; Weber, P. C.; Lipscomb, J. D. *J. Mol. Biol.* **1987**, *195,* 225–227. (c) See for example Que, L., Jr. *J. Chem. Ed.* **1985,** *62,* 938–943.
139. Higuchi, Y.; Yasuoka, N.; Kakudo, M.; Katsube, Y.; Yagi, T.; Inokuchi, H. *J. Biol. Chem.* **1987,** *262,* 2823–2826.
140. Petratos, K.; Beppu, T.; Banner, D. W.; Tsernoglou, D. *J. Mol. Biol.* **1986,** *190,* 135.
141. Bolognesi, M.; Gatti, G.; Coda, A.; Avigliano, L.; Marcozzi, G.; Finazzi-Agro, A. *J. Mol. Biol.* **1983,** *169,* 351–352.
142. Zaitsev, V. N.; Vagin, A. A. ; Nekrasov, Y. V.; Moshkov, K. A. *Sov. Phys. Crystallogr.* **1986,** *31,* 557–559.
143. Ringe, D. *Nature (London)* **1987,** *329,* 102.
144. Hajdu, J.; Machin, P. A.; Campbell, J. W.; Greenhough, T. J.; Clifton, I. J.; Zurek, S.; Gover, S.; Johnson, L. N.; Elder, M. *Nature (London)* **1987,** *329,* 178–181.
145. Gruner, S. M. *Science (Washington, D.C.)* **1987,** *238,* 305–312.
146. Prewitt, C. T.; Coppens, P.; Phillips, J. C.; Finger, L. W. *Science (Washington, D.C.)* **1987,** *238,* 312–319.
147. See for example Capaldi, R. A.; Zhang, Y. Z. *Methods Enzymol.* **1986,** *126,* 22–31.

RECEIVED April 18, 1988

Chapter 2

X-ray Absorption Spectroscopy for Characterizing Metal Clusters in Proteins

Possibilities and Limitations

James E. Penner-Hahn

Department of Chemistry, University of Michigan, Ann Arbor, MI 48109–1055

With the development of intense synchrotron X-ray sources, it has become possible to use X-ray absorption spectroscopy to structurally characterize the metal clusters in metalloproteins. A wide variety of systems have been studied in this manner in the last ten years. The present article reviews the nature of X-ray absorption spectroscopy and the information which can be obtained from its study. The strengths and limitations of the technique are discussed with reference to the recent literature.

Since their discovery in 1895, X-rays have been one of the primary probes with which chemists and physicists have investigated the structure of matter. Although most well known for their use in crystallography, the earliest studies with X-rays emphasized spectroscopic measurements. In the early years of this century X-ray spectroscopy, in particular the work of Moseley, played a key role in the discovery and characterization of new elements. Following these early successes however, the potential chemical applications of X-ray spectroscopy were largely unappreciated until the mid-seventies. The reasons for this can be traced both to the weak interactions of X-rays with matter and to the low intensities available from conventional X-ray sources. These factors combine to limit conventional X-ray spectroscopy to relatively concentrated samples.

This situation changed dramatically in the early 1970's with the development of synchrotron X-ray sources. When a relativistic electron beam is accelerated in a circular path, for example in the storage rings used for high-energy physics research, it emits an intense, highly collimated beam of light, with energies extending into the hard X-ray region (energies up to 40 KeV, wavelengths down to .25 Å). Current synchrotron sources increase the available X-ray brightness by as much as ten orders-of-magnitude in comparison with conventional X-ray tubes. These sources have made possible, and in some cases nearly routine, experiments which were not even imagined ten years ago.

This article provides an overview of the information which can be obtained by X-ray absorption spectroscopy, with emphasis on the utilization of this technique to characterize the metal clusters in proteins. No attempt is made to provide a detailed description of either the theory or the practice of X-ray spectroscopy; these topics have been discussed in a number of excellent review articles (1). Rather, a small sample of the recent literature is used to illustrate both the potentials and the limitations of the

0097–6156/88/0372–0028$06.25/0

technique (i.e., what *can* and what *cannot* be learned). The hope is that this will permit the non-expert to read critically the X-ray spectroscopy literature and that this may, perhaps, encourage wider utilization of X-ray spectroscopy for the investigation of biological metal clusters.

Fundamentals of X-ray Absorption.

Every element shows several discontinuous jumps, or "edges" in its X-ray absorption coefficient as illustrated schematically in Figure 1. Each jump corresponds to the excitation of a core electron. To a good approximation, the energy of a particular absorption edge is determined only by the atomic number of the absorber and the identity of the excited core electron. For a multi-element cluster it is therefore possible to study, independently, the X-ray absorption spectrum of each of the constituent elements. The K-edge (1s initial state) occurs at experimentally convenient energies for elements between sulfur and cadmium. Elements with higher atomic numbers are accessible via their L absorption edges (2s and 2p initial states). Elements with atomic numbers below sulfur generally require the use of ultra-high vacuum systems.

The term X-ray absorption spectroscopy (XAS) refers to the structured absorption which is superimposed on the edge jump. Conventionally XAS spectra are divided into two regions, as shown in Figure 2. The X-ray absorption near edge structure (XANES) region refers to absorption within ~50 eV of the absorption edge while the extended X-ray absorption fine structure (EXAFS) region is taken as ~50-1000 eV above the edge. The distinction between these regions is somewhat artificial since the same physical principles govern absorption in both regions. However, as discussed below, qualitatively different information is generally obtained from analysis of the XANES and EXAFS spectra.

EXAFS Structure. In the EXAFS region the incident photon energy is much larger than the core electron binding energy, resulting in substantial kinetic energy for the excited photoelectron. The emitted photoelectron can be viewed as an outwardly propagating wave with wavelength $\lambda = h/p$. As illustrated in Figure 2, this wavefront will, upon reaching a neighboring ligand, be scattered back in the direction of the absorber. Interference between the outgoing and the backscattered waves perturbs the probability of X-ray absorption and gives rise to the EXAFS structure. This can be appreciated by considering that the absorption cross section depends on both the initial state and the final state of the absorber. The final state includes both the core hole and the photoelectron wave (2), and will therefore depend on the interference of the outgoing and backscattered waves. Constructive interference will give rise to a local maximum in absorption, while destructive interference results in a local minimum. As shown in Figure 2, the interference alternates between constructive and destructive as the X-ray energy is increased (thereby decreasing the photoelectron wavelength).

It is important to remember that, despite common usage, EXAFS is at heart not a spectroscopic measurement but rather a scattering measurement. It is impossible to correlate specific EXAFS features with particular absorber-scatterer interactions, rather it is necessary to analyze the entire EXAFS region in order to determine the absorber environment.

The EXAFS (χ) is defined as the modulation in the absorption cross-section (μ) on the high-energy side of an absorption edge: $\chi=(\mu-\mu_o)/\mu_o$, where μ_o is the absorption in the absence of any EXAFS effects. The difficulty with this definition is that it is not possible to determine μ_o experimentally. As an approximation of the true EXAFS, one therefore uses a practical definition of the general form $\chi=(\mu-\mu_s)/\mu_f$, where μ_s is a calculated spline background and μ_f is the expected falloff in X-ray absorption as a function of energy (2). For quantitative analysis, it is most convenient if the data are converted from X-ray photon energy (E) as the independent variable to photoelectron wave vector $k = 2\pi / \lambda = [2m_e(E-E_o) / h^2]^{1/2}$, where E_o is the threshold energy for excitation of the core electron and m_e is the mass of the electron.

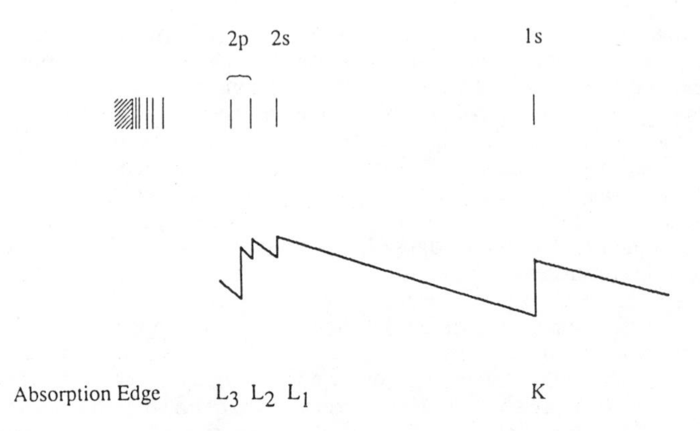

Figure 1. Schematic illustration of the energy levels associated with X-ray absorption spectroscopy. Abscissa is energy, increasing to right. Ordinate for lower trace is absorption cross-section.

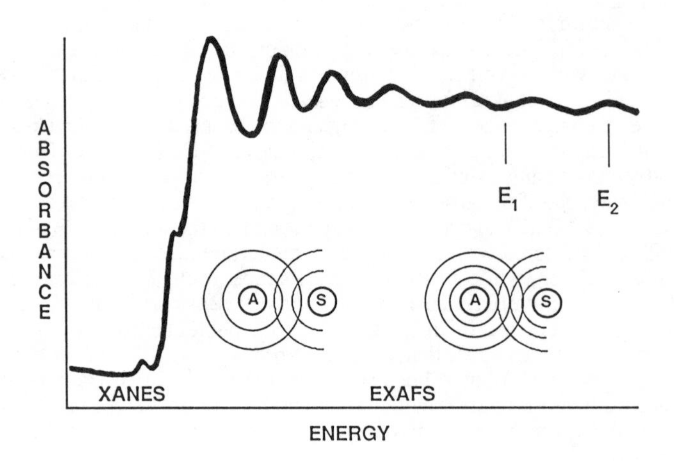

Figure 2. Typical X-ray absorption spectrum. Inset is schematic illustration of out-going and backscattered photoelectron wave for energies E_1 (left) and E_2 (right).

In a single scattering formalism (3), where the photoelectron interacts with a single scatterer before returning to the absorber, the theoretical expression for the EXAFS is (4)

$$\chi(k) = \frac{N_s A_s(k)}{R_{as}^2} \exp(-2k^2\sigma_{as}^2) \exp(-2 R_{as} / \lambda) \bullet \sin(2k R_{as} + \Phi_{as}(k)) \quad (1)$$

now be given for the EXAFS. In this expression, N_s is the number of scatterers at a distance R_{as} from the absorbing atom, and the two exponential damping terms are the so-called Debye-Waller factor containing σ_{as}^2, the mean-square deviation in R_{as}, and a factor containing λ, the mean-free-path of the photoelectron. The factors $A_s(k)$ and $\Phi_{as}(k)$ are, respectively, the backscattering strength of the scattering atom and the phase shift which the photoelectron wave undergoes when passing through the potentials of the absorber and the scatterer.

As seen from Eq. 1, the EXAFS for a single absorber-scatterer pair will consist of a damped sine wave. The chemically useful information is contained in the amplitude, frequency and phase of this wave. The amplitude factor $A_s(k)$ has a different functional dependence on k for different scatterers, thus the shape of the EXAFS amplitude envelope provides a method for identifying the scattering atom. Unfortunately, the dependence of $A_s(k)$ on atomic number is relatively weak and thus in practice it is difficult to distinguish scatterers from the same row of the periodic table. For biologically relevant ligands, this means that C,N, and O cannot be distinguished although N and S are readily distinguishable. If $A_s(k)$ is known, then the amplitude of the EXAFS can be used to determine the number of scattering atoms around the absorber. Uncertainties in $A_s(k)$ and in the precise calculation of EXAFS amplitudes generally limit determinations of N_s to $\pm$ 25%, giving, for typical coordination numbers, an uncertainty of $\pm$ one atom.

This simplified analysis ignores the exponential damping factors. The mean-free-path term is generally assumed to be constant for all of the samples being studied. This is a reasonable assumption (5) providing that all of the samples are of a similar chemical form. The Debye-Waller term is more difficult, since this varies depending on the strength of an absorber-scatterer bond. Under ideal conditions it is possible to determine absolute Debye-Waller factors using EXAFS spectra, however these conditions rarely obtain for biological metal clusters.

EXAFS amplitudes can be difficult to determine and analyze, due to uncertainties in background removal and the large number of parameters which affect the amplitude. In contrast, the frequency of an EXAFS wave is well defined experimentally and depends, to a first approximation, only on the absorber-scatterer distance. A typical EXAFS spectrum has three or more full periods, thus providing an excellent determination of the frequency. Bond length determination is complicated by uncertainty in the value of E_o, however by comparing the unknown with structurally characterized model compounds one can generally determine bond lengths with a precision of 1% ($\pm$ 0.02 Å). Phase shifts are less useful than frequencies due to their strong correlation with E_o, nevertheless they can be helpful in identifying the scattering ligand.

EXAFS Data Analysis. A key aspect of the analysis outlined above is knowledge of the correct $A_s(k)$ and $\Phi_{as}(k)$ for a particular absorber-scatterer interaction. These parameters can either be calculated *ab initio* (6) or can be determined by measuring the EXAFS of structurally characterized model compounds (7). The *ab initio* method has the advantage that one need not prepare appropriate models for all possible unknowns. Unfortunately however, the *ab initio* parameters must be adjusted by a scaling factor and an assignment of E_o (8). For this reason, one typically calibrates the calculated parameters by comparison with model compounds, thereby minimizing the distinction between *ab initio* and empirical methods.

A number of different data analysis procedures have been reported, using both empirical and *ab initio* parameters. Most of these involve essentially the same steps and, where comparisons are possible, produce essentially the same results. The disagreements which do exist in the literature can generally be attributed to differences in the measured data rather than differences in the data analysis methods. The general analysis procedure is to fit the experimental data to a model based on Eq. 1 using a non-linear least-squares algorithm. The variable parameters include some subset of the following: R_{as}, N_s, E_o, scale factor (for *ab initio* parameters), and σ_{as}.

EXAFS provides information similar to that obtained from crystallography, however EXAFS has the decided advantage of being applicable to non-crystalline systems. Indeed, the ability to study essentially any form of matter (solution, lyophillized powder, membrane, etc.) is one of the key attractions of XAS for studies of biological systems. In comparison with protein crystallography, EXAFS provides significantly better precision in bond length determination and can be measured and analyzed in a matter of weeks as compared to years for typical protein crystal structures. The mean-free-path damping factor limits EXAFS information to the immediate ($R < \sim 4$ Å) environment of the absorbing atom. This precludes the use of EXAFS to study, for example, protein tertiary structure, however it does have the advantage of simplifying EXAFS data analysis; One need not solve the entire protein structure in order to obtain useful information. A more severe limitation of EXAFS *vis a vis* crystallography is that EXAFS does not provide directly any information regarding the angular arrangement of ligands around the absorber. Some ways around this problem are discussed below (in particular XANES spectroscopy offers the promise of eventually providing such information), however at present one generally cannot determine stereochemical information from EXAFS measurements.

All of the discussion thus far has assumed that the absorber is surrounded by only a single type of scattering atom at a single distance. If there is more than one scatterer, the observed EXAFS will simply be the sum of the EXAFS contribution from each atom: $\chi(k)=\Sigma\,\chi_i(k)$, where the sum is taken over all scattering atoms. In practice, this summation is simplified by dividing the scatterers into "shells", where a shell consists of a single type of scatterer at an average distance from the absorber.

The analysis of multiple shell EXAFS data can be simplified by using standard frequency analysis methods. Thus, the Fourier transform of Eq. 1 transforms the data from *k* space ($Å^{-1}$) to R space (Å) giving a peak centered at a distance $R+\alpha$, where α depends on $\Phi_{as}(k)$. As illustrated in Figure 3, the Fourier transform of an EXAFS spectrum gives a phase-shifted radial distribution function. In some cases, simple examination of the Fourier transform provides important information regarding the structure of a site, however caution must be used in quantitatively interpreting Fourier transforms. More generally, the Fourier transform is used to simplify data analysis. In order to restrict attention to a single shell of scatterers, one selects a limited range of the R-space data for back-transformation to *k*-space, as illustrated in Figure 3B,C. In the ideal case, this procedure allows one to analyze each shell separately, although in practice many shells cannot be adequately separated by Fourier filtering (9).

XANES Structure. If the energy of the incident X-rays is close to the threshold for core-electron excitation, new spectroscopic features are observed. For energies below the threshold, one observes bound→bound transitions such as 1s→3d and 1s→4p (10). For energies slightly above the threshold, the absorption cross-section is affected by photoelectron diffraction in the same manner as EXAFS, however the low energy, and consequently long wavelength of the photoelectron in the XANES region results in qualitatively different behavior from that observed in the EXAFS region. Specifically, the photoelectron can no longer be approximated by a plane-wave and thus the simple expression given in Eq. 1 is no longer valid. Perhaps more important is the fact that in this low energy region the photoelectron has a much longer mean-free-path and a much larger scattering cross-section. Consequently multiple-scattering pathways, in which the photoelectron is scattered from two or more ligands before returning to the absorber, become increasingly important. Since EXAFS depends only on pair-wise scattering, it contains only angularly averaged radial structure information. Multiple-scattering, on the other hand, is inherently dependent on the three-dimensional arrangements of ligands around the absorber. To the extent that multiple-scattering is important, it may therefore be possible to determine geometry from XANES. Unfortunately, multiple-scattering greatly complicates the analysis of XANES data and has limited the application of this technique.

Experimentally, XANES spectra are found to be extremely sensitive to the oxidation state, geometry, and electronic structure of the absorbing site. XANES features, especially the bound→bound transitions can be sufficiently well resolved that it is possible to correlate a specific transition with a specific structural feature. An example, involving the distinction between Cu(I) and Cu(II) is discussed below.

Numerous theories have been put forward to explain XANES structure (1d). Although there is some consensus regarding the interpretation of bound-state transitions, the higher energy "continuum" region remains a subject of controversy. The number of adjustable parameters in most of the current theoretical models greatly exceeds the amount of observable data in the XANES region. These models have been useful in rationalizing the observed XANES spectra for structurally characterized compounds, however it is quite difficult to proceed in the reverse direction - that is, to use a XANES spectrum to elucidate an unknown structure. It is the author's opinion that XANES spectroscopy is one of the most promising areas of XAS. Despite recent progress, however, this is a promise that is largely unrealized today. The greatest utility of XANES spectra at present is as a fingerprint for comparing unknown compounds with structurally characterized model compounds. Several examples of this sort of empirical correlation are discussed below.

Polarization Properties. One of the key attractions of XAS for biological systems is the ability to study unoriented samples such as proteins in solution. If samples can be oriented, however, it is possible to greatly enhance the information content of XAS spectra. In particular, it is possible to obtain direct information about the relative orientation of specific structural features. These experiments are possible because the synchrotron X-ray beam is highly plane polarized. Several examples of polarized XAS of biological samples are discussed below.

Data Collection. The practical problems involved in collecting accurate XAS data have been reviewed (1) and will not be discussed here. It is worth noting however, that concentration limits range from ca. 1 mM for lighter elements (e.g., Cu) to ca. 0.2 mM for heavier elements (e.g., molybdenum). These limits are much higher if the sample is contained in a highly absorbing (e.g., high atomic number) matrix. Virtually all elements of bioinorganic interest are accessible, however certain combinations of elements can interfere. Thus, it is impossible to measure sulfur EXAFS spectra for Mo/S compounds since the Mo L(III) edge lies just to the high energy side of the S K edge.

Synchrotron beam time remains a scarce resource world-wide, however with over 30 synchrotron radiation centers either operational or under construction, most prospective users can obtain access to the facilities (11). In most cases, beam time is available free of charge for non-proprietary research. Access to a facility is often by means of a peer reviewed proposal. Adequate equipment for sample handling and data collection is available at most synchrotron radiation laboratories. Many laboratories also provide software for data analysis.

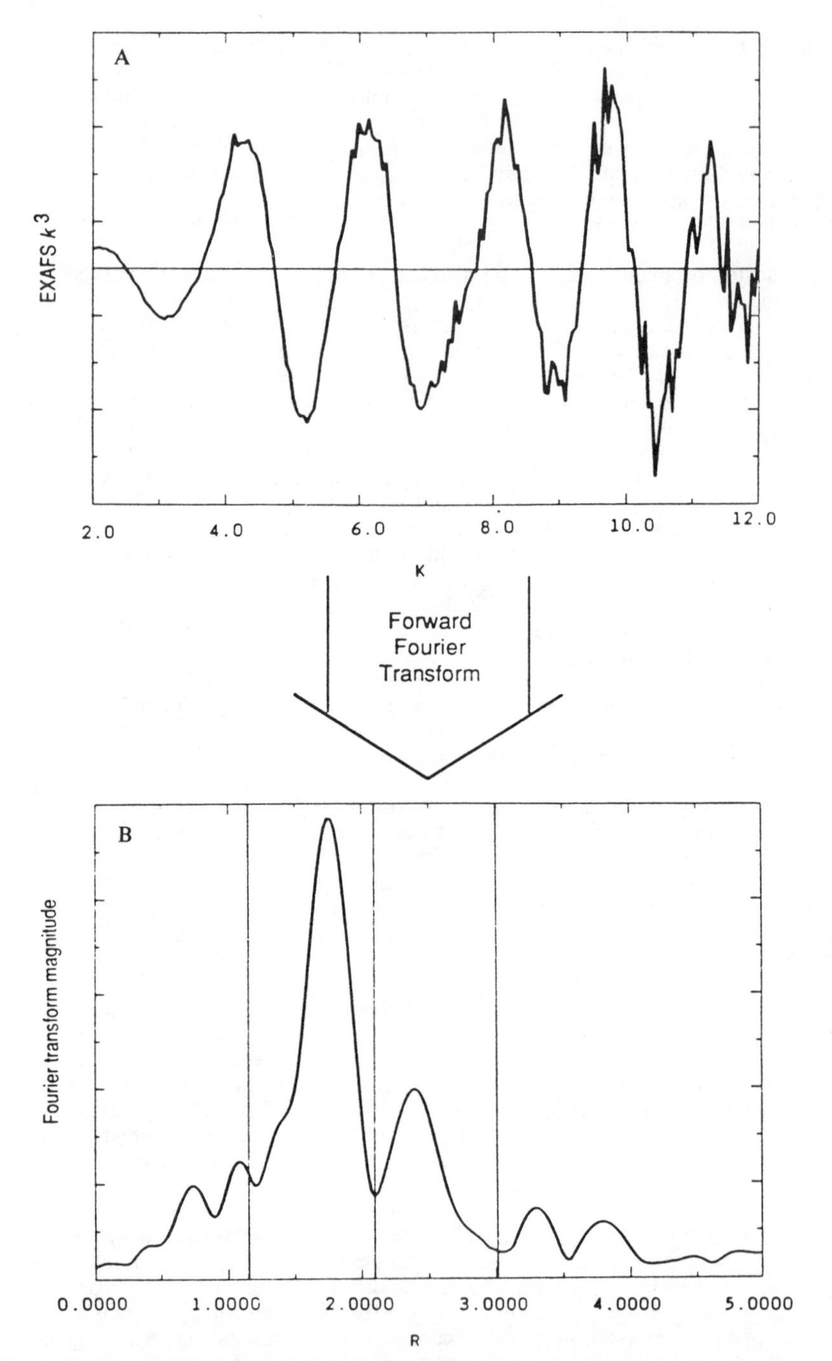

Figure 3. X-ray absorption data analysis of Fe EXAFS data for Rieske-like Fe_2S_2 cluster. A) EXAFS data; B) Fourier transform of EXAFS data showing peaks for Fe–S and Fe–Fe scattering. Thin vertical lines indicate filter windows for first and second shell.

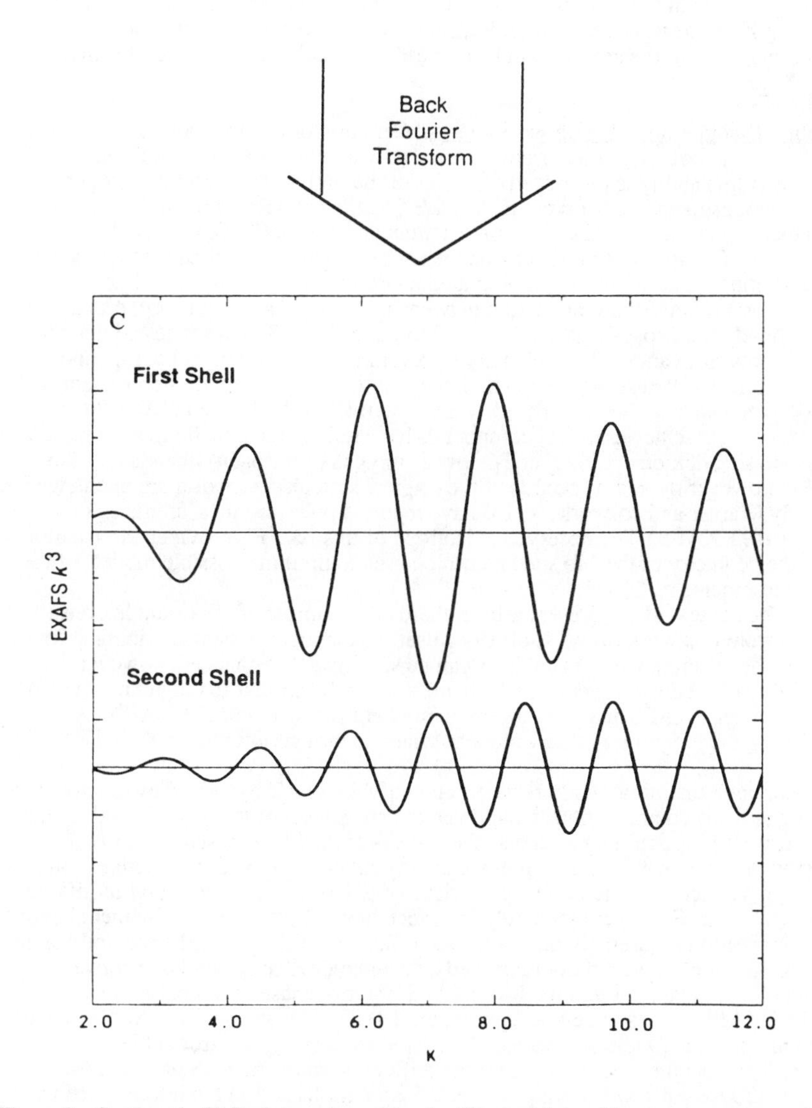

Figure 3. *Continued.* C) Back-transform (Fourier filter) of data. Upper trace corresponds to Fe–S (maximum amplitude at ca. k = 7Å $^{-1}$). Lower trace corresponds to Fe–Fe (maximum amplitude at ca. k = 10Å $^{-1}$).

Examples.

The following examples represent only a minor fraction of the published literature involving XAS measurements of biological samples. They have been chosen in order to draw attention to the specific kinds of questions which can, or cannot, be answered using XAS.

Structure Elucidation in Unknown Systems. Certainly one of the most exciting uses of XAS is as a tool for characterizing structurally unknown systems. A dramatic example of this ability is provided by studies of the molybdenum site in nitrogenase. In 1978 Hodgson and co-workers reported EXAFS results for MoFe protein from the nitrogenase enzyme system in *C. pasteurianum* and *A. vinelandii* (12). This work provided the first direct structural evidence concerning the Mo site in these proteins, demonstrating that the Mo was coordinated predominantly by sulfur. In addition, evidence was found for an interaction between the Mo and another metal, probably Fe and definitely not Mo, at ~2.7 Å from the Mo. The EXAFS data suggested that the Mo was present in an Mo-Fe-S cluster, however at the time no model compounds with the appropriate structure were known. Since that time, several models consistent with the EXAFS data have been prepared. Subsequent EXAFS (13) and XANES (14) studies using these newer model compounds have refined the structural interpretation of the Mo site, demonstrating that 2-3 low Z (oxygen or nitrogen) ligands are also present, however the main features of the original structural proposal are unchanged. Recently Cramer and co-workers (15) have reported polarized measurements of nitrogenase EXAFS. The observed anisotropy of this EXAFS provides information about the geometry of the Mo site and can be used to limit the possible models of the Mo environment.

In the context of the present article, there are a number of important lessons to be gleaned from this work. XAS is clearly a useful technique for characterizing unknown systems. The accuracy of this analysis can be significantly enhanced, however, by the availability of model compounds which are structurally similar to the unknown (13). In this manner EXAFS analyses are frequenly iterative, with initial EXAFS results stimulating the prepartion of new models, which in turn permit more detailed EXAFS analyses.

Although the initial workers did not identify the low Z ligands, they did note that such ligands would not be detectable given the noise level of the data. In addition, the initial reports proposed that a second shell of sulfur might be present at 2.5 Å, however this shell has been found in some (15) but not all (13) of the more recent work. This pattern is generally characteristic of EXAFS - basic structural motifs (in this case the Mo-Fe-S core) are rarely incorrect, however structural refinement beyond this point becomes extremely difficult. Determination of the detailed structural features generally requires substantially improved data and model compounds. Another point worth noting is that although the Mo EXAFS of nitrogenase was reported in 1978, the first Fe EXAFS results were not reported until 1982 (16), and sulfur XANES spectra have only now (1987) been reported (17). This sequence arises from at least two effects: 1) it becomes progressively more difficult to measure XAS data at lower energies [20 KeV for Mo, 7 KeV for Fe, 2.5 KeV for S] and 2) the interference from multiple sites is negligible for Mo (2 Mo per cluster), severe for Fe (~10 Fe per cluster) and possibly prohibitive for S due to the presence of adventitious sulfur.

Structure Refinement in Previously Characterized Systems. Even when a protein crystal structure is known, EXAFS measurements can provide useful information. An example is the Fe-S protein rubredoxin, for which initial crystallographic results suggested that the Fe was coordinated by three sulfurs at normal distances (2.24-2.34 Å) and one sulfur at a very short distance (2.05 Å) (18). Detailed EXAFS characterization of this Fe site was possible because of the simplicity of the site (only sulfur ligation) and the availablility of high quality, chemically relevant Fe/S model compounds. Under these conditions, it proved possible to determine the mean deviation in Fe-S bond length and to thereby test the crystallographic results. The EXAFS results (19) demonstrated that the mean deviation in Fe-S bond length was 0.03 Å, with a maximum spread of 0.04 Å in the oxidized form and 0.08 Å in the reduced form. At approximately the same time, a re-analysis of the crystal data produced new structural parameters in substantial agreement with the EXAFS results. This story emphasizes the relative merits of crystallography and EXAFS - crystallography is essential for the determination of overall three-dimensional structure while EXAFS is unsurpassed in its ability to determine bond-lengths in the first coordination sphere. Combination of the two techniques can provide a particularly detailed structural picutre.

Heme proteins have been extensively studied using XAS (20). Since the iron is known to be coordinated to four pyrrole nitrogens, the heme-iron structure is two-thirds determined before the experiment is begun. Nevertheless, there are several interesting questions which can be addressed via XAS. In the case of cytochrome P-450, EXAFS was used to demonstrate conclusively that sulfur is coordinated as an axial ligand to the Fe, and that this sulfur is retained throughout the catalytic cycle (21). Similarly, EXAFS has been used to demonstrate that a short (1.6 Å) Fe=O bond is present in the oxidized forms of horseradish peroxidase (22). In both cases, biologically important questions concerning axial ligation were answered using EXAFS spectroscopy.

Two features of these systems made them amenable to EXAFS investigation. First, the axial and equatorial ligands were clearly distinguished, either in terms of atomic number (S vs. N for P-450) or in terms of bond length (1.6 vs. 2.0 Å for HRP-I). Second, the question of interest concerned the presence or absence of a specific ligand, rather than quantitation of the number of ligands. It is relatively straightforward to determine whether or not a specific ligand contributes to an EXAFS spectrum, however it is quite difficult to determine exactly how many ligands are present in a particular shell. Examples of iron-porphyrin structural questions which would not be answered readily with EXAFS are: 1) Is an imidazole-iron-porphyrin complex 5 or 6 coordinate (23). 2) What is the distribution of bond lengths in a bis-imidazole iron porphyrin, i.e., are the axial bond lengths identical to, slightly longer than, or slightly shorter than the equatorial bond lengths.

In comparison with the success in determining the structures of P-450 and HRP, EXAFS studies of hemoglobin have been generally less informative. The question in this case is the out-of-plane distance (R_{Fe-Ct}) of the Fe atom. This distance plays an important role in some models of hemoglobin cooperativity, however the available crystallographic results are unable to determine R_{Fe-Ct} with sufficient precision to distinguish between the different models. Different EXAFS studies (24) have come to the same conclusion regarding the average R_{Fe-N}, however these results have subsequently been used to support very different pictures of R_{Fe-Ct}. The difficulty in this case lies not with the EXAFS results, but with the interpretation of these results. Specifically, R_{Fe-Ct} can only be determined by triangulation, using both R_{Fe-N} and R_{Ct-N}, however R_{Ct-N} cannot be measured experimentally and must be approximated from model compound studies. A strong case can be made that a short R_{Fe-Ct} is the most reasonable conclusion (25), however this cannot be proven experimentally using EXAFS measurements. A number of workers have recently reported evidence suggesting that weak XANES features on the high energy side of the absorption maximum can be correlated with the geometry of an Fe porphyrin. These are discussed below.

Limitations of Structural Studies. EXAFS is not well suited for identifying weak scatterers, (e.g. oxygen in the presence of sulfur) as illustrated by recent data for the Rieske-like site in *P. cepacia* phthalate oxygenase (26). This site is believed to contain an Fe_2S_2 core, and EXAFS spectroscopy has proven useful in elucidating the dimensions of this core. Although it is known from ENDOR measurements that at least one nitrogen is ligated to the Fe_2S_2 unit, the EXAFS spectra show no evidence for this ligand. The nitrogen can, of course, be included in the fit, however the improvement in EXAFS fit quality for a 3S+1N fit compared to a 4S fit is too small to demonstrate conclusively the presence of Fe-N ligation. The reasons for this are illustrated in Figure 4, where it is seen that the EXAFS from three sulfurs overwhelms that from a single nitrogen.

In comparison with nitrogenase, where EXAFS was able to detect oxygen in the presence of sulfur ligation, phthalate dioxygenase suffers from the limited *k* range and data quality of Fe vs. Mo XAS and from the disparity in coordination numbers (one nitrogen vs. three sulfurs). In the specific case of Fe/S clusters, EXAFS is an excellent technique for determining the Fe-S and Fe-Fe distances. If nitrogen coordination is known to be present, EXAFS can be used to determine the Fe-N bond lengths (although the uncertainties are substantially larger here). On the other hand, EXAFS is *not* a useful technique for determining how many, if any, nitrogens are present. Similar limitations apply to EXAFS characterization of all metal clusters.

EXAFS is a bulk technique and thus can only measure the average structure of all of the absorbing atoms. Samples must therefore have high purity, at least as far as the metal of interest is concerned, and in addition one must guard against impurities in sample cells, beam line components, etc. (this is not necessarily simple for a metal as ubiquitous as iron). Even in the absence of contaminants, EXAFS interpretation can be difficult for multimetal proteins. For example, if a protein contains two copper sites and has EXAFS corresponding to N_2S_2 ligation, it is impossible to determine whether this represents one CuN_4 site and one CuS_4 site, two identical CuN_2S_2 sites, or any intermediate combination.

Chemical perturbations can be a useful approach to this problem. In laccase, selective replacement of the type 1 copper with mercury has allowed the type 1 site in laccase to be characterized structurally, even in the presence of three other copper atoms (27). Likewise, Simolo et al. have studied independently the α and β subunits in hemoglobin by preparation of the $(\alpha\text{-M})_2(\beta\text{-M}')_2$ derivaties, where M and M' are different metals (28). An alternative to replacement is metal removal, for example Cu_a in cytochrome-c oxidase (29) or the T3 Cu in laccase (30). The concern with all of these approaches is to establish that the modified protein has the same metal-site structures as the native protein.

A final limitation of EXAFS is that under some conditions, atoms which are present in the coordination sphere will *not* make a detectable contribution to the measured spectrum. An example is plastocyanin, which has been shown crystallographically to contain a copper atom coordinated to two imidazole nitrogens at 2.0 Å, one cysteine sulfur at 2.15 Å, and one methionine sulfur at 2.9 Å (31). The original EXAFS measurements (which were carried out prior to determination of the crystal structure), correctly determined the short (N_2S) ligation environment, but found no evidence for the distant methionine (32). Possible explanations for this observation included disorder in the S(Met) position and destructive interference between the EXAFS from the imidazole carbons (at ca. 3.0 Å) and EXAFS from the methionine sulfur. The crystal structure gave no indication of unusual disorder in either the Cu or the S(Met) positions, thus ruling out the former. The second explanation was tested by means of a polarized single crystal EXAFS experiment (33).

Plastocyanin crystallizes with all of the Cu-S(Met) vectors parallel to the crystallographic **c** axis. EXAFS spectra measured with the polarization (**e**) parallel to **c** will therefore maximize the contribution from the S(Met) sulfur, while spectra measured with $\mathbf{e} \perp \mathbf{c}$ will, by symmetry, have no contribution from Cu-S(Met) EXAFS. As indicated in Figure 5, this experiment demonstrated conclusively that there was no detectable contribution of the S(Met) to the overall EXAFS at room

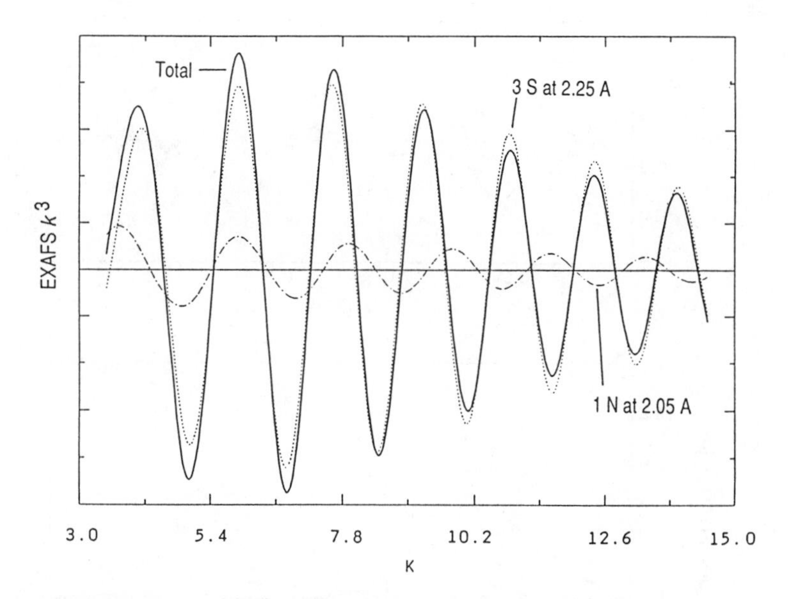

Figure 4. Simulated Fe EXAFS for Rieske-like Fe_2S_2 cluster. Dotted line: 3 sulfurs at 2.25 Å. Dot-dash line: 1 N 2.05 Å. Solid line: sum of 3S + 1N. The 3S wave differs only slightly from 3S + 1N.

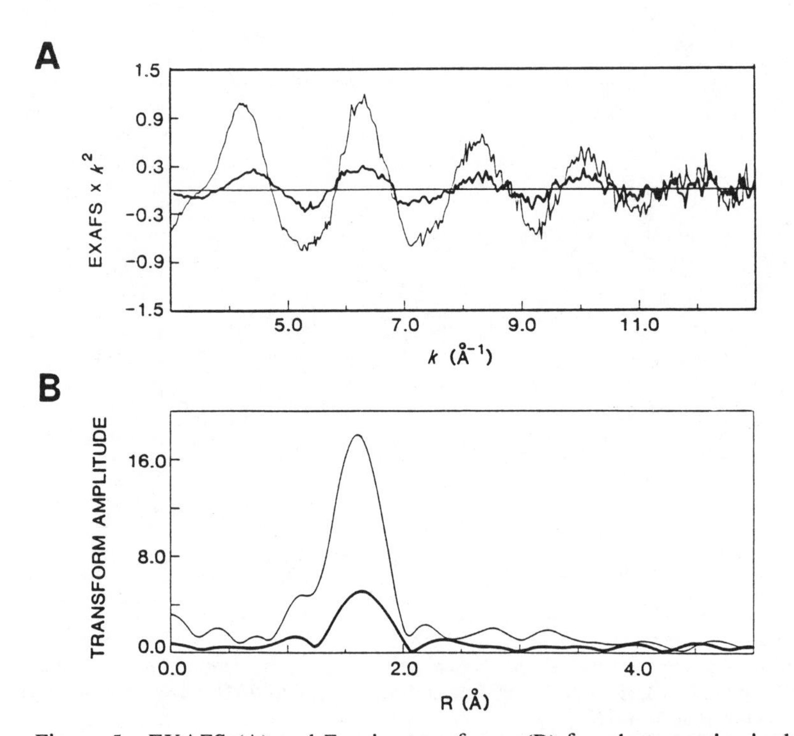

Figure 5. EXAFS (A) and Fourier transforms (B) for plastocyanin single crystals. In both traces, dark line corresponds to polarization parallel to Cu-S(Met) and light line corresponds to polarization perpendicular to Cu-S(Met). No difference is observed for R=2.5 Å, demonstrating that Cu-S(Met) makes no detectable contribution to EXAFS.

temperature. Recent results have confirmed this finding at temperatures as low as 4K (34). This unusual finding, that an atom known crystallographically to be present in the first coordination sphere does not contribute to the EXAFS, reflects the difference between EXAFS and crystallography. Crystallographic temperature factors contain information about the motion of each atom about its mean position, while the EXAFS Debye-Waller factor depends on the pair-wise motion of the absorber and scatterer. The present results thus demonstrate that while the individual mean motions of both the Cu and the S(Met) atoms are normal, their relative motion must be largely uncorrelated in order to account for the non-observation of Cu-S(Met) EXAFS. The combination of crystallographic and EXAFS information is thus able to shed light on the dynamic motion of the protein.

One of the most important points of the plastocyanin experiments is their reminder that the non-observance of a ligand in an EXAFS spectrum does not necessarily demonstrate that a ligand is absent from the coordination sphere, although it may have bearing on the strength of a metal-ligand interaction. This concern is especially important in the case of spectra having a lower signal/noise ratio or in the case of multi-metal proteins. In the latter case, a single ligand bonded to one of several metal sites may make an experimentally undetectable contribution to the EXAFS.

EXAFS non-detectability becomes more severe as R_{as} increases. Not only is σ_{as} likely to be larger, but in addition analysis is more likely to be complicated by interference from other scatterers with simlar R_{as}. In the present context, the question might be whether or not a protein contains a dinuclear metal cluster, and the difficulty is to distinguish metal-metal EXAFS from metal-carbon EXAFS (e.g. from second shell imidazole carbons). Recent EXAFS studies of a series of crystallographically characterized dinuclear Fe complexes illustrates this difficulty (35). For μ-oxo bridged complexes, the Fe-Fe EXAFS is readily distinguishable. However in μ-hydroxo bridged complexes, ones interpretation of the Fe-Fe scattering is strongly influenced by the assumptions that one makes regarding the Fe-C scattering. As with plastocyanin, the lesson is that although a dinuclear unit is clearly present, this fact cannot be unambiguously established from EXAFS data alone (35).

Oxidation State Determination. It has long been known that the energy of an absorption edge increases as the oxidation state of the absorbing metal site increases, as illustrated in Figure 6. This can be rationalized in terms of the increasing energy required to remove an electron from an oxidized atom. This correlation, known as Kunzl's law (36), has been widely used to infer oxidation state in unknown samples. In the case of metalloproteins, edge energy has been used to assign the average oxidation state of the Mn atoms in the photosynthetic oxygen evolving complex from photosystem II (37) and to investigate the oxidation state of Mo in nitrogenase (38).

The biggest concern with analyses of this sort is that they rely solely on an empirical correlation, with no good theoretical prediction of the exact dependence expected between edge energy and oxidation state. To the extent that the model compounds accurately reflect the structural environment of the unknown, this type of analysis is reliable. For less well defined systems, there is always the concern that the shape of an absorption edge, and hence the apparent energy, can depend strongly on the detailed three-dimensional structure of an absorbing site. Since it is not clear which feature of the edge should be used to define the edge energy (first inflection point, half-height, first maximum, etc.), changes in the shape of an absorption edge can introduce error into the analysis of oxidation state. For example, it has recently been reported that the edge energy for some Mn(III) compounds shifts by several eV in response to changes in the Mn nuclearity (39).

A somewhat different approach has been utilized in XANES studies of the oxidation state of Cu complexes. This work relied on an absorption band at 8984 eV which was shown to be unique to Cu(I) complexes (40). This band, which can be assigned to the 1s→4p transition, shifts to ca. 8987 eV in Cu(II) complexes. The difference of properly normalized XANES spectra for Cu(II) and Cu(I) compounds thus gives a characteristic derivative-like pattern, the amplitude of which is proportional to the percent of Cu(I) in a sample. Recent work has demonstrated that, in fact, this correlation is slightly more complicated, with different amplitudes for 2-, 3-, or 4-coordinate Cu(I) complexes. The details of this work, and its applications to the protein laccase are discussed elsewhere in this volume (41).

Geometry. It has been noted several times that the detailed shape of a XANES spectrum is sensitive to the structural environment and geometry of the absorbing atom. The most well understood of these changes involves the weak, pre-edge peak arising from the 1s→3d transition. This transition is observed for all elements of the first transition series, provided that they are in oxidation states having a vacancy in the 3d shell. This transition is forbidden according to dipole selection rules ($\Delta l=2$), nevertheless it is observed, even in centrosymmetric environments, due to direct coupling to the quadrupole component of the radiation field. In non-centrosymmetric environments, 3d+4p orbital mixing becomes possible with a consequent increase in the intensity of the 1s→3d transition. Roe et al. have quantitated this increase for a series of Fe complexes having different geometries. This work, illustrated in Figure 7, demonstrates that the 1s→3d intensity correlates nicely with the calculated percent of 4p mixed into the 3d ground state (42). This correlation was used to determine the geometry of the Fe sites in iron-tyrosinate proteins. Recently, polarized XANES measurements have been used to determine both the intensity *and* the orientation dependence of the 1s→3d transition in synthetic copper complexes and in the Cu protein plastocyanin (43). This work demonstrates that it is possible to determine both the extent of 3d+4p mixing and the identity of the p orbitals (p_x, p_y, or p_z) which are mixed into the 3d ground state.

The features on the higher energy side of the absorption edge are, unfortunately, less well understood than the 1s→3d transition. The lack of a good theoretical model for these transitions has limited their usefulness for characterizing unknown compounds. Empirical correlations can, however, be quite useful, as illustrated by XANES studies of Fe porphyrins. As noted previously, an important question in metalloporphyrin structural chemistry is the out-of-plane distance of the metal ($R_{M\text{-}Ct}$), however this parameter is not readily determined from analysis of EXAFS data.

Chance and co-workers have identified a region ca. 50 eV above the absorption edge as a "Ligand Field Indicator Region" (LFIR) (44). For Fe porphyrins, this region consists of two local maxima in absorption, (see Figure 8A) the relative height of which seems to correlate with the spin state of the iron. To date, the analysis of this region has consisted of calculating the ratio of the heights of these two peaks. Even neglecting the difficulty in determining what background to use in calculating peak heights, this analysis suffers from the loss of information inherent in reducing a complex spectrum to a single number. Nevertheless, it is clear that Chance et al. have identified spectral features which are sensitive to the geometry of the Fe site. Additional study of these features should prove quite productive. Two groups, Oyanagi et al. (45) and Bianconi et al. (46), have identified features closer to the absorption edge which can also be correlated with the spin state of the iron. In comparison with Chance's LFIR, these features are relatively subtle, as indicated in Figure 8B. Once again, an empirical correlation is observed between the XANES features and the spin state of the Fe. In both of these cases, it is unlikely that the spin state of the Fe is directly responsible for changes in the XANES structure. More likely, spin state and XANES structure are both correlated to the geometry of the Fe site. Polarized measurements of the XANES spectra for crystallographically characterized Fe porphyrins would be useful for investigating more fully the nature of these intriguing spectral features.

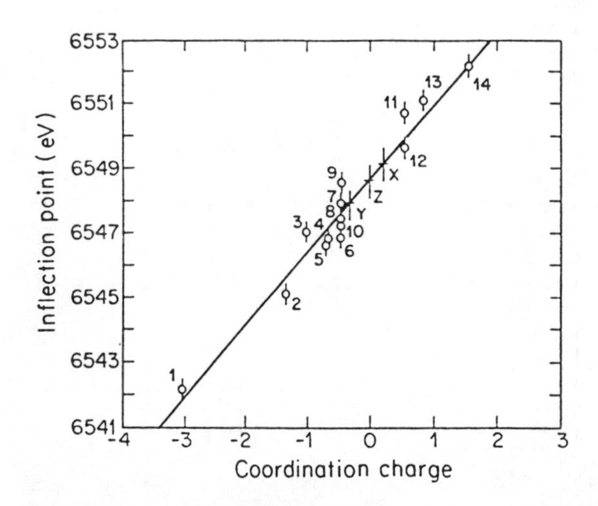

Figure 6. XANES energy (inflection point) as a function of charge for 15 Mn complexes. (Reproduced from reference 37. Copyright 1981 American Chemical Society.)

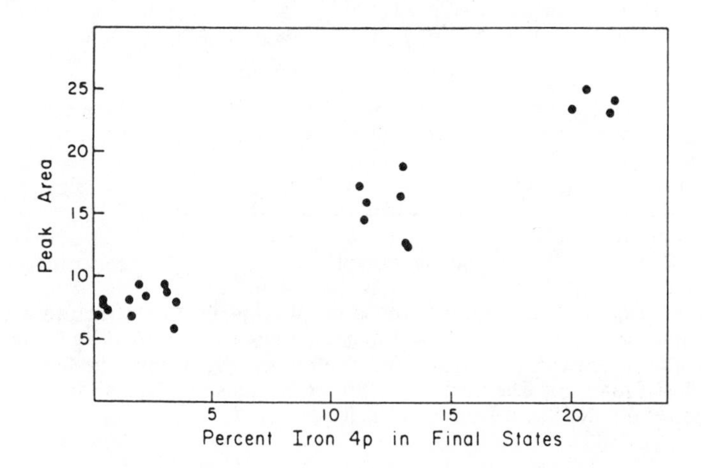

Figure 7. Intensity of the 1s→3d transition in Fe complexes as a function of the percent of 4p mixed into the 3d ground state. The clusters of points correspond to 6, 5, and 4-coordinate, from left to right. (Reproduced from reference 42. Copyright 1984 American Chemical Society.)

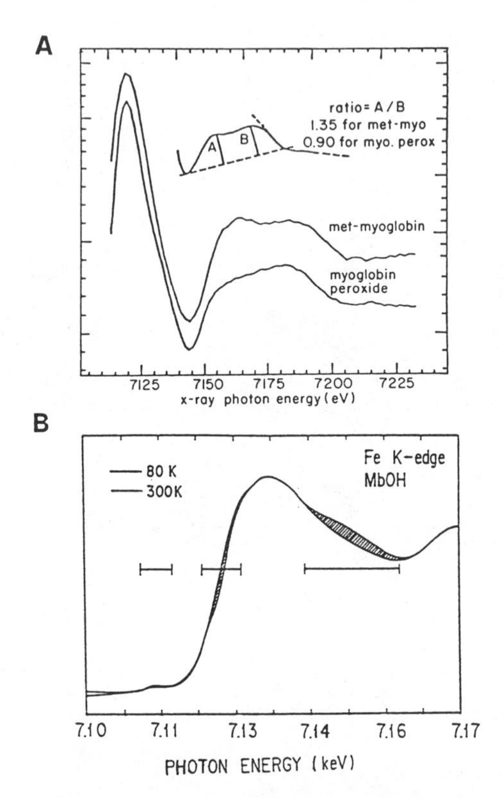

Figure 8. Empirical correlations of Fe porphyrins XANES structure with spin state. A) The "Ligand Field Indicator Region" identified by Chance et al. (Reproduced with permission from reference 44. Copyright 1986, Journal of Biological Chemistry) B) Spin state sensitive bands identified by Oyanagi et al. Redrawn from data in Reference 45.

Radiation Damage. When samples are irradiated with intense beams of highly energetic radiation sample damage may occur, for example as photoreduction mediated by X-ray generated solvated electrons. As a precaution against this possibility, most experimenters routinely measure biological XAS data at as low a temperature as possible in order to limit the diffusion of X-ray generated electrons. In addition, it is essential that one always test for the possibility of radiation damage, either by assaying the protein before and after the XAS experiment or preferably, by on-line monitoring of the sample integrity. An elaborate *in situ* optical reflectance apparatus has been used for this task, however a simpler and possibly more sensitive approach is to examine the XANES spectra as a function of time. For a dilute metalloprotein, one typically measures anywhere from 6 to 40 scans, each of approximately one-half hour duration. Given the sensitivity of XANES to oxidation state and geometry, a simple comparison of the first and last scans can provide a sensitive probe of sample integrity. This method was used, for example, to demonstrate that when Cu(II) plastocyanin is irradiated with one of the most intense X-ray sources available, the Cu undergoes slow photoreduction, even at 4K (34).

Summary and Future Outlook.

In comparison with other physical methods, EXAFS has both advantages and limitations. It is always detectable, does not require the use of special isotopes, and provides unparalleled information regarding the structure of the absorbing site. EXAFS provides only limited information on coordination number, however this information is comparable to that available from most other spectroscopic methods. Regarding ligand identity, EXAFS is useful for distinguishing nitrogen or oxygen ligation from sulfur ligation and for identifying metal-metal interactions. It is useless however, for the biologically important question of distinguishing *between* nitrogen and oxygen ligation (except indirectly, since M-O and M-N bond lengths are generally different). Finally, EXAFS provides no information regarding the electronic structure of an absorbing site. Considering these strengths and weaknesses, EXAFS should be viewed as an extremely useful complement, rather than a replacement, for the more traditional biophysical tools.

The use of XAS for structural characterization of the metal sites in metalloproteins has increased dramatically in the last ten years. Throughout most of this period, advances in biological XAS have been driven largely by the increasing availability of synchrotron radiation and by the increasing quality of synchrotron sources. With each new increase in the available synchrotron intensity, new classes of experiments have become feasible. Some of the inherent limitations of the XAS method have already been discussed. It is appropriate now to consider more practical concerns regarding the future of biological XAS, and the way in which this will be affected by the development of new synchrotron sources.

The state of the art has now advanced to the point that synchrotron sources are frequently not the limiting factor in biological XAS. In particular, there is concern that sample measurement techniques have not kept pace with the development of new sources. Several new detection schemes are under development, however it may be several years before detectors will be available which are capable of fully utilizing the intensity available from new sources. In addition, there is concern that very intense sources may compromise sample integrity. Taken together, these concerns suggest that it is not reasonable to extrapolate past experience when attempting to predict the effect of new sources on future concentrations limits. Metal ion concentrations below ca. 100 μM are likely to remain inaccessible for many years.

On the other hand, there are plans to construct new synchrotron sources optimized to provide high brightness (photons/unit area) X-ray sources. These will permit so-called micro-probe XAS measurements, involving very small sample volumes. The corresponding reduction in the amount of protein required will permit studies which are presently impractical.

In summary, XAS is and will continue to be one of the premier methods for elucidating the immediate structural environment of metal atoms in non-crystalline systems. Although some of the early claims regarding the utility of XAS have not survived the test of time, it is nevertheless true that XAS, when used in conjunction with other biophysical probes, can provide a detailed picture of the mteal clusters in a metalloprotein. EXAFS is becoming a relatively mature scientific discipline and it is likely to be utilized with increasing regularity as one of the many methods for characterizing new proteins. In comparison with EXAFS, XANES has been the subject of many more claims and far fewer experimental results. While some of these claims will no doubt prove to be ill-founded, XANES is likely to play an increasingly important role in the characterization of unknown systems. Careful, detailed study of XANES spectra are only now beginning to be undertaken - it will be several years before one can assess with confidence the utility of these spectra.

Acknowledgments.

Preparation of this manuscript was supported in part by the Camille and Henry Dreyfus Foundation and the National Institutes of Health (R29-GM38047).

References and Notes.

1. See a) Lee, P.A.; Citrin, P.H.; Eisenberger, P.; Kincaid, B.M., *Rev. Mod. Phys.*, **1981**, *53*, 769 for a review of the physical principles of EXAFS; b) Scott, R.A., *Methods Enzymol.*, **1985**, *117*, 414-59 for a discussion of methods for measuring and analyzing EXAFS data, c) Teo, B.K., "EXAFS: Basic Principles and Data Analysis", **1986** for a general discussion, and d) Bart, J.C.J., *Adv. Catal.*, **1986**, *34*, 203-296 for a review of XANES. Several dozen other reviews discuss the application of XAS to specific problems.
2. See McMaster, W.H.; Kerr Del Grande, N.; Mallett, J.H.; Hubbell, J.H., "Compilation of X-ray Cross Sections", **1969**, NTIS Report UCRL-50174 or "International Tables for X-ray Crystallography", Vol. III, **1968**, D. Reidel, 170-174.
3. Multiple scattering, where the photoelectron wave samples several scatterers before returning to the absorber, is only important in EXAFS for cases where two scatterers and the absorber are nearly collinear. In such cases, the EXAFS amplitude of the outer scatterer will be significantly enhanced. See Teo, B.K., *J. Am. Chem. Soc.*, **1981**, *103*, 3990-4001.
4. (a) Sayers, D.E.; Stern, E.A.; Lytle, F.W. *Phys. Rev. Lett.* **1971**, *27*, 1204. (b) Ashley, C.A.; Doniach, S. *Phys. Rev.* **1975**, *B11*, 1279. (c) Lee, P.A.; Pendry, J.B. *ibid.*, 2795. (d) Lee, P.A. *ibid.* **1976**, *B13*, 5261.
5. (a) Eisenberger, P.; Lengeler, B. *Phys. Rev.* **1980**, *B22*, 3551. (b) Stern, E.A.; Bunker, B.A.; Heald, S.M. *ibid.*, *B21*, 5521.
6. (a) Teo, B.K.; Lee, P.A. *J. Am. Chem. Soc.* **1979**, *101*, 2815-2832. (b) Teo, B.K. Atonio, M.R.; Averill, B.A. *ibid.* **1983**, *105*, 3751-3762.
7. Cramer, S.P.; Hodgson, K.O. *Prog. Inorg. Chem.* **1979**, *25*, 1-39.
8. An alternate theoretical model, including explicitly the multiple scattering terms has been proposed. See for example Blackburn, N.J.; Strange, R.W.; McFadden, L.M.; Hasnain, S.S. *J. Am. Chem. Soc.* **1987**, *109*, 7162-7170 and references therein.
9. In order to be clearly separable in the Fourier transform, two shells typically must be separated by at least 0.5 Å.
10. In addition, edge structure may be influenced by multi-electron transitions. See for example Gewirth, A.A.; Cohen, S.L.; Schugar, H.J.; Solomon, E.I. *Inorg. Chem.* **1987**, *26*, 1133-1146 and references therein.

11. For a world-wide census of synchrotron radiation laboratories and a list of contact persons at each laboratory, consult Mulhaupt, G. *Nucl. Instrum. Methods Phys. Res.*, **1986**, *A246*, 845-876.
12. Cramer, S.P.; Hodgson, K.O.; Gillum, W.O.; Mortenson, L.E. *J. Am. Chem. Soc.*, **1978**, *100*, 3398-3407. Cramer, S.P.; Gillum; W.O.; Hodgson, K.O.; Mortenson, L.E.; Stiefel, E.I.; Chisnell, J.R.; Brill, W.J.; Shaw, V.K., *ibid.*, 3814-3819.
13. Conradson, S.D., Ph.D. Thesis, Stanford University (1983).
14. Conradson, S.D.; Burgess, B.K.; Newton, W.E.; Hodgson, K.O.; McDonald, J.W.; Rubinson, J.F.; Gheller, S.F.; Mortenson, L.E.; Adams, M.W.W.; Mascharak, P.K.; Armstrong, W.A.; Holm, R.H., *J. Am. Chem. Soc.,* **1985**, *107*, 7935-40.
15. Flank, A.M.; Weininger, M.; Mortenson, L.E.; Cramer, S.P. *J. Am. Chem. Soc.* , **1986**, *108,* 1049-55.
16. Antonio, M.R.; Teo, B.K.; Orme-Johnson, W.H.; Nelson, M.J.; Gorh, S.E.; Lindahl, P.A.; Kauzlarich, S.M.; Averill, B.A., *J. Am. Chem. Soc.* **1982**, *104*, 4703-5.
17. Hedman, B.; Hodgson, K.O.; Roe, A.L.; Frank, P.; Tyson, T.A.; Gheller, S.F.; Newton, W.E., *Recl. de Trav. Chim.*, **1987**, *106*, 304.
18. Watenpaugh, K.D.; Sieker, L.C.; Herriott, J.R.; Jensen, L.H., *Acta Cryst.*, **1973**, *B29*, 943.
19. Shulman, R.G., Eisneberger, P.; Blumnber, W.E.; Stombaugh, N.A. *Proc. Natl. Acad. Sci. USA*, **1975**, *72*, 4002-7. Sayers, D.E.; Stern, E.A.; Herriott, J.R. *J. Chem. Phys.*, **1976**, *64*, 427-8. Bunker, B.; Stern, E.A. *Biophys. J.*, **1977**, *19*, 253-264. Shulman, R.G. et al., *J. Mol. Biol.,* **1978**, *124*, 305-21.
20. Penner-Hahn, J.E.; Hodgson, K.O. in "Physical Bioinorganic Chemistry", A.B.P. Lever and H.B. Gray (Eds.) VCH, **1987**, in press.
21. Hahn, J.E.; Hodgson, K.O.; Andersson, L.A.; Dawson, J.H., *J. Biol. Chem.,* **1982**, *257*, 10934-41. Cramer, S.P.; Dawson, J.H.; Hodgson, K.O.; Hager, L.P. *J. Am. Chem. Soc.*, **1978**, *100*, 7282-90.
22. Penner-Hahn, J.E.; Eble, K.S.; McMurry, T.J.; Renner, M.; Balch, A.L.; Groves, J.T.; Dawson, J.H.; Hodgson, K.O., *J. Am. Chem. Soc.*, **1986**, *108*, 7819-25. Chance, B.; Powers, L.; Ching, Y.; Poulos, T.; Schonbaum, G.R.; Yamazaki, I.; Paul, K.G., *Arch. Biochem. Biophys.*, **1984**, *235*, 596-611. Penner-Hahn, J.E.; McMurry, T.J.; Renner, M.; Latos-Grazynski, L.; Eble, K.S.; Davis, I.M.; Balch, A.L. ;Groves, J.T.; Dawson, J.H.; Hodgson, K.O., *J. Biol. Chem.*, **1983**, *258*, 12761-4.
23. Note however that the Fe-N bond lengths vary for 5 and 6 coordinate porphyrins, hence an EXAFS determination of bond length could in fact be used to infer coordination number, even though coordination number cannot be determined directly.
24. a) Eisenberger, P.M.; Shulman, R.G.; Kincaid, B.M.; Brown, G.S., Ogawas, S. *Nature (London)* **1978**, *274*, 30-34. b) Perutz, M.F. Hasnain, S.S. ;Duke, P.J.; Sessler, J.L.; Hahn, J.E., *ibid.* **1982**, *295*, 535-8.
25. Shulman, R.G. *Proc. Natl. Acad. Sci. USA* **1987**, *84*, 973-4.
26. Tsang, H.T.; Batie, C.J.; Ballou, D.P.; Penner-Hahn, J.E., manuscript in preparation.
27. Tsang, H.T.; Schmidt, A.; McMillin, D.R.; Penner-Hahn, J.E., manuscript in preparation.
28. Simolo, K.; Korszun, G.; Moffat, K.; McLendon, G.; Bunker, G. *Biochemistry*, **1986**, *25*, 3773.
29. Li, P.M.; Gelles, J.; Chan, S.I.; Sullivan, R.J.; Scott, R.A. *Biochemistry*, **1987**, *26*, 2091-2095.
30. Woolery, G.L.; Powers, L.; Peisach, J.; Spiro, T.G. *Biochemistry*, **1984**, *23*, 3428-3434.

31. Colman, P.M. Freeman, H.C.; Guss, J.M.; Murata, M.; Norris, V.A.; Ramshaw, J.A.M.; VenKatappa, M.P., *Nature (London)* **1978**, *272*, 319-24.
32. Tullius, T.D., Ph.D. Thesis, Stanford University, (1978).
33. Scott, R.A.; Hahn, J.E.; Doniach, S.; Freeman, H.C.; Hodgson, K.O., *J. Am. Chem. Soc.* **1982**, *104*, 5365-9.
34. Penner-Hahn, J.E.; Murata, M.; Hodgson, K.O.; Freeman, H.C., submitted to *J. Am. Chem. Soc.* **1987**.
35. Hedman, B.; Co, M.S.; Armstrong, W.H.; Hodgson, K.O.; Lippart, S.J. *Inorg. Chem.*, **1986**, *25*, 3708-3711.
36. Kunzl, C. *Collect Czech. Commun.*. **1932**, *4*, 213.
37. Kirby, J.A.; Goodin, D.; Wydrynski, T.; Robertson, A.S.; Klein, M.P. *J. Am. Chem. Soc.* **1981**, *103*, 5537.
38. Cramer, S.P.; Eccles, T.K.; Kutzler, F.W.; Hodgson, K.O.; Mortenson, L.E. *J. Am. Chem. Soc.* **1976**, *98*, 1287-8.
39. Cartier, C.; Verdagrier, M.; Menage, S.; Girerd, J.J.; Tuchagues, J.P.; Mabad, B., *J. de Phys (Paris)* **1986**, *47*, C8-623 - C8-625.
40. Hahn, J.E.; Co, M.S.; Spira, D.J.; Hodgson, K.O.; Solomon, E.I., *Biochem. Biophys. Res. Commun.* **1983**, *112*, 737-45. Penner-Hahn, J.E.; Hedman, B.; Hodgson, K.O.; Spira, D.J.; Solomon, E.I., *ibid.* **1984**, *119*, 567-74.
41. Solomon, E.I., this volume.
42. Roe, A.L.; Schnider, D.J.; Mayer, R.J.; Pyrz, J.W.; Widom, J.; Que, L. Jr.*J. Am. Chem. Soc.* **1984**, *106*, 1676.
43. Penner-Hahn, J.E.; Hodgson, K.O.; Solomon, E.I., unpublished results.
44. Chance, M.; Powers, L.; Kumar, C.; Chance, B. *Biochem.*, **1986**, *25*, 1259-65. Chance, M.R.; Parkhurst, L.J.; Powers, L.S.; Chance, B. *J. Biol. Chem.* **1986**, *261*, 5689-92.
45. a) Oyanagi, H. Iizuka, T.; Matsushita, T.; Saigo, S.; Makino, R.; Ishimura,*J. de Phys (Paris)* **1986**, *47*, C8-1147 - C8-1150. b) Oyanagi, H. Iizuka, T.; Matsushita, T.; Saigo, S.; Makino, R.; Ishimura, submitted to *J. Phys. Soc. Jpn.*, **1987**.
46. Bianconi, A., Congiu-Castellano, A.; Dell'Ariccia, M.; Giovannelli, A.; Morante, S. *FEBS Lett.* **1985**, *191*, 241-4.

RECEIVED January 20, 1988

Chapter 3

Resonance Raman Spectroscopy of Iron–Oxo and Iron–Sulfur Clusters in Proteins

Joann Sanders-Loehr

Department of Chemical and Biological Sciences, Oregon Graduate Center, Beaverton, OR 97006–1999

Resonance Raman studies of Fe- and Cu-containing proteins have led to the identification of tyrosine, histidine, cysteine, and hydroxide ligands as well as Fe-O and Fe-S clusters. For the Fe-O clusters, the frequency and oxygen isotope dependence of the Fe-O-Fe symmetric stretch relates to Fe-O-Fe bond angle, while the peak intensity relates to the disposition of the other ligands in the cluster. For the Fe-S proteins, the frequencies and sulfur isotope dependence of the Fe-S vibrational modes can be used to distinguish mononuclear, binuclear, and tetranuclear clusters. Hydrogen bonding of both Fe-O and Fe-S clusters can be detected by frequency shifts in deuterium-substituted proteins.

In resonance Raman spectroscopy, excitation within a metal-ligand charge transfer band leads to enhanced intensities of M-L and internal ligand vibrational modes, thereby allowing these modes to be selectively observed in metalloproteins. Because vibrational frequencies are sensitive to bond strength and coordination environment, this technique provides information which is complementary to that obtained by x-ray crystallography or x-ray absorption spectroscopy. This chapter will discuss the general principles of resonance Raman spectroscopy as they apply to metal centers in proteins and then focus specifically on binuclear and tetranuclear iron clusters with oxo or sulfido bridging groups.

Raman Spectroscopy

Raman spectroscopy is a form of vibrational spectroscopy which, like infrared spectroscopy, is sensitive to transitions between different vibrational energy levels in a molecule (1). It differs from infrared spectroscopy in that information is derived from a light scattering rather than a direct absorption process. Furthermore, different selection rules govern the intensity of the respective vibrational modes. Infrared absorptions are observed for vibrational modes which change the permanent dipole moment of the

0097–6156/88/0372–0049$06.00/0

molecule while Raman scattering is associated with normal modes which produce a change in the polarizability (or induced dipole moment) of the molecule. Thus, symmetric stretching modes tend to be the most intense features in Raman spectra, whereas asymmetric stretches and deformation modes tend to be more intense in infrared spectra. This differential character makes Raman spectroscopy more favorable for the study of biological materials since there is considerably less spectral interference from the intra- and intermolecular deformation modes of water molecules; these are dominant features in the infrared spectra of aqueous samples.

The predominant type of light scattering which occurs when radiation impinges upon atoms in a sample is known as Rayleigh scattering. The scattered light appears equally in all directions and is of the same frequency as the incident light. The Raman effect relates to the fact that a small fraction of the scattered radiation may have undergone a change in frequency (2). This can be explained as the result of an inelastic collision with an incoming photon ($h\nu_o$) such that some of its energy ($h\nu_V$) is transferred to a vibrational mode of the molecule. The energy of the Stokes Raman scattered photon ($h\nu_R$) is, thus, equal to $h\nu_o - h\nu_V$ (Figure 1a). Since Raman scattering is 1000-fold weaker than Rayleigh scattering, its detection requires the use of an intense monochromatic light source, a high-resolution monochromator, and a highly efficient photodetector (4). In practice, visible or uv laser lines provide the best input radiation and the instrument is calibrated for the particular ν_o being used. The scattered photons of varying frequencies (ν_R) are resolved by a scanning monochromator (Figure 2) or a diode array instrument, and their intensities are plotted as a function of frequency, $\nu_V = \nu_o - \nu_R$. Use of computer-related capabilities such as multiple scanning, signal averaging, background subtraction, and data smoothing leads to significant enhancements in spectral quality (5).

Raman spectra obtained on concentrated protein solutions (ca. 20 mM) are dominated by contributions from the repeating amide group of the polypeptide backbone and from the sidechains of aromatic amino acids (6). Phenylalanine and tryptophan exhibit particularly intense ring breathing modes at 1006 and 1014 cm^{-1}, respectively. An additional strong peak due to tryptophan is typically observed at 760 cm^{-1}.

Resonance Raman Spectroscopy

Resonance-enhanced Raman scattering occurs when the energy of the incident radiation, $h\nu_o$, is close to or within an electronic absorption band of the sample (7,8). In this case, vibronic coupling with the electronically excited state increases the probability of observing Raman scattering ($h\nu_{RR}$) from vibrational transitions in the electronic ground state (Figure 1b). The intensity of such resonance-enhanced vibrational transitions can be described in simplified terms as:

$$I_R \propto I_o \sum_e \left| \frac{(M_j)_{me} (M_i)_{en}}{\nu_e - \nu_o + i\Gamma_e} \right|^2 \quad (1)$$

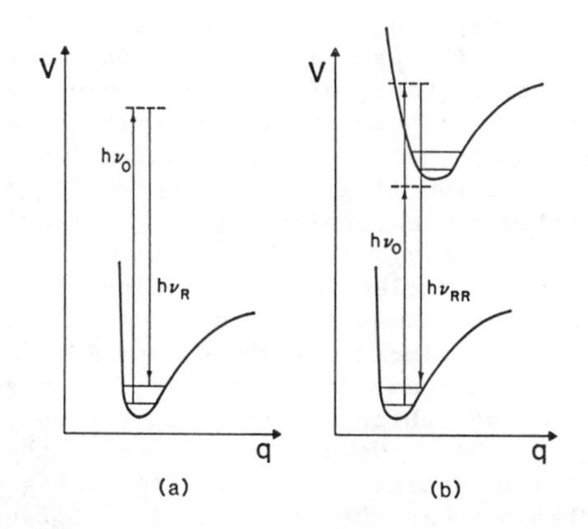

Figure 1. Diagrams of potential energy, V, versus internuclear separation, q, for a molecule undergoing vibrational excitation by (a) the Raman effect or (b) a resonance Raman effect ($h\nu_o \simeq h\nu_e$) or a pre-resonance effect ($h\nu_o < h\nu_e$).

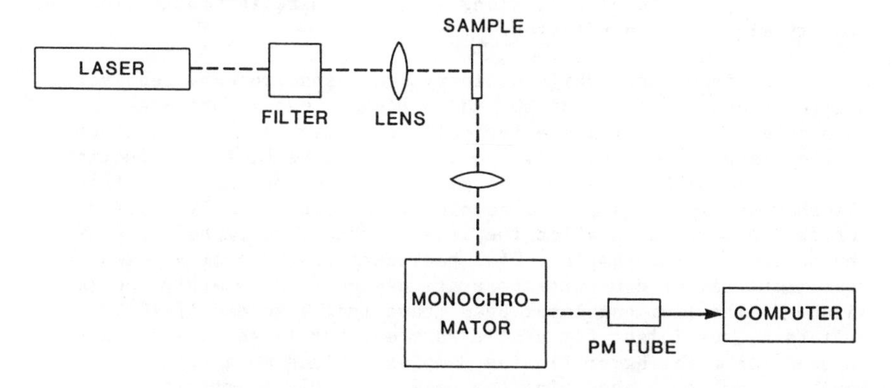

Figure 2. Experimental set-up for Raman spectroscopy. The desired laser line is isolated from other plasma lines by a narrow bandpass filter or broadband prism monochromator, then focused onto a sample in a capillary tube. A collecting lens placed at a 90° angle to the incident beam focuses the scattered light onto the entrance slit of a monochromator with output to a photomultiplier tube (in the case of a scanning instrument) or a diode array detector.

where I_0 is the intensity of the incident photon and the summation is a measure of molecular polarizability. The quantities M_j and M_i are the electric dipole transition moments connecting the initial vibrational state, m, the excited state, e, and the final vibrational state, n. This relationship predicts an exponential increase in intensity as the frequency of the incident radiation, ν_0, approaches ν_e, the frequency of the m to e electronic transition; the damping term, Γ_e, prevents the denominator from becoming zero. Thus, while vibrational frequencies in resonance Raman spectroscopy are still a function of the electronic ground state, vibrational intensities are determined by the properties of the electronic excited state (8,9).

The exact features of molecular and electronic structure which give rise to the resonance Raman effect are not well understood. It is believed that those vibrations whose normal modes most closely approximate the nuclear displacements in the electronic excited state are the ones which are the most susceptible to resonance enhancement (10). Hence, the nature of the electronic transition becomes an important factor in determining the extent of Franck-Condon overlap between the vibrational wave functions in the ground and excited electronic states. In considering metal-based chromophores, the two most common electronic transitions are metal-centered ligand field (LF) transitions and metal-ligand charge transfer (CT) transitions. Since the vibrations associated with the metal center generally involve either M-L or internal ligand modes, they are more likely to exhibit resonance-enhanced intensities in conjunction with CT transitions which redistribute electron density along M-L coordinates.

Sample Requirements. While ordinary Raman spectroscopy requires a scatterer concentration of 100 mM or higher, metalloprotein resonance Raman spectra are typically obtained in the 1 mM concentration range. For metal-ligand CT transitions having extinction coefficients of 2,000-4,000 M^{-1} cm^{-1}, this corresponds to sample absorbances of 2 to 4 (i.e., samples which are strongly colored). For frozen samples in which the laser light is scattered only off the surface of the sample (~150° backscattering geometry), two to four times higher concentrations are needed. The sharply focused nature of the incident laser beam means that a volume of 10-20 μliters is sufficient for static samples, but larger volumes must be used for rotating or flowing samples. Since metalloprotein samples are highly absorbing (as well as highly scattering), spectral contributions from amide and sidechain vibrations are suppressed. However, the strongly scattering modes of tryptophan and phenylalanine near 750 and 1000 cm^{-1}, respectively, can often still be detected in addition to the vibrations of the metal chromophore.

The use of an intense laser light source with biological materials is accompanied by the concomitant problems of localized sample heating and the possibility of protein denaturation. A further complication introduced by resonance Raman spectroscopy is the increased potential for photochemical destruction of chromophoric metal centers as a result of the absorption of large amounts of incident radiation. Both of these situations may be ameliorated by freezing samples to liquid nitrogen temperature (~90 K), while the even lower temperatures made possible with a closed-cycle helium refrigerator (10-20 K) yield even greater sample stability. Lowered sample temperatures have the additional advantage of offering significant improvements in spectral resolution and signal to noise ratios (11). Various designs for low temperature work include cold-finger Dewars capable of cooling samples in capillaries or devices in which the sample is frozen directly onto a cold finger (3,4). Since in the latter case the sample is generally farther removed from any glass windows, such devices can result in less interference from the broad glass band in the 430-500 cm^{-1} region. However, frozen samples do exhibit a strong peak at ~230 cm^{-1} and a weaker feature at ~310 cm^{-1} which are due to ice translational vibrations and must be carefully distinguished from sample vibrations (12). Concentrated buffer solutions can also contribute Raman lines, and it is wise to run a dialysate or ultrafiltrate from the protein sample as a control. Preparation of the sample as a frozen glass through the addition of glycerol or ethylene glycol does not improve spectral quality and introduces excessive detail from the solvent.

<u>Identification of Vibrational Modes</u>. Ligands containing C=C and C=N bonds exhibit Raman peaks in the 1000-1600 cm^{-1} region. These are often present in a characteristic pattern of frequency and intensity which can be identified by reference to the Raman spectra of model compounds. Metal-ligand stretching vibrations occur in a narrower range of frequencies, 200-600 cm^{-1}, and are considerably harder to assign because different ligands can give rise to M-L modes of similar energy and because internal ligand vibrations can occur in this energy range, as well. Where possible, it is wise to try to verify an assignment by atom or isotope substitution. Vibrational frequencies, ν, are dependent on the masses of the vibrating atoms and the strength of the bond force constant, k, according the harmonic oscillator relationship:

$$\nu = 1/2\pi \sqrt{k/\mu} \tag{2}$$

where μ is the reduced mass: $m_1m_2/(m_1 + m_2)$. The inverse dependence on mass predicts a decrease in frequency for an increase in mass. Atom substitutions (e.g., Se for S; Cu for Fe) alter force constants and coordination geometry as well as mass, while isotopic substitutions (e.g., O-18 for O-16; Cu-65 for Cu-63) mainly affect the reduced mass without significantly altering the force constant or the coordination environment. Thus, although isotope-dependent shifts are more difficult to detect, they are considerably easier to interpret particularly when used in conjunction with normal coordinate analysis calculations.

The most common ligands found at metal centers in metalloproteins are the imidazole group of histidine, the thiolate group of cysteine, the phenolate group of tyrosine, the carboxylate groups of aspartic and glutamic acids, solvent-derived or exchangeable groups (e.g., aquo, hydroxo, oxo, sulfido), and the tetrapyrrole macrocycle of porphyrins. The porphyrin chromophores, by virtue of their extended conjugated π systems, have very distinctive optical and resonance Raman spectra (13). The only amino acid moiety which has a sufficiently characteristic resonance Raman spectrum to be immediately identified as a metal ligand is tyrosine. Excitation within the phenolate → Fe CT band results in the selective enhancement of phenolate ring vibrations (Figure 3, Table I). Although imidazole also has characteristic ring modes in the 1000-1400 cm^{-1} region, they are only weakly resonance-enhanced and have not yet been detected in metalloprotein spectra. The location of the imidazole → Cu CT bands in the uv region, which is more difficult to probe experimentally, may be partly responsible for this difficulty. However, resonance-enhanced metal-histidine stretching modes have been observed in the 220-290 cm^{-1} region. These undergo 1 to 2 cm^{-1} shifts to lower energy when the proteins are incubated in D2O. Although a pure M-N vibration should occur closer to 400 cm^{-1} (1), the imidazole ring behaves as a rigid moiety with an effective mass of 67 (including hydrogens) and, thus, the M-N(imidazole) vibrations are found at lower energy. Substitution of the N1 and, less frequently, the C2 hydrogens of imidazole by deuterium lowers the effective mass to 66 or 65, thereby accounting for the observed shifts (15). It should be noted, however, that hydrogen bonding of ligating oxygen or sulfur atoms to the polypeptide backbone can cause similar shifts of 1 to 2 cm^{-1} in D2O (21). Thus, deuterium isotope effects do not unequivocally signify histidine ligation.

Hydroxo ligands produce resonance-enhanced Fe-OH stretching vibrations in the 550-600 cm^{-1} region (Table I). These have been verified by observing a ~25 cm^{-1} shift to lower energy in 18-O-labeled water and a somewhat lesser shift in D2O (22). The aquo group, another potential source of an M-O stretch between 300 and 500 cm^{-1} (1), has not yet been detected in resonance Raman spectra, even in such examples as Cu-substituted alcohol dehydrogenase where a water molecule is believed to be metal-coordinated (12). The M-O stretch of metal carboxylates which occurs in the 300-400 cm^{-1} region (Table I) has also not yet been observed in metalloprotein spectra. However, a peak at 530 cm^{-1} in methemerythrin (Table I) has been tentatively assigned to a carboxylate internal deformation which undergoes "intensity borrowing" by virtue of its proximity to a strongly enhanced Fe-L mode at 510 cm^{-1}.

The cysteine-ligated metal centers yield more complicated resonance Raman spectra, in part due to the coupling of different vibrations. The Fe-4Cys protein, rubredoxin, has a series of four Fe-S stretching modes between 314 and 376 cm^{-1} (Table I) arising from symmetric and asymmetric vibrations of the four sulfur ligands, with contributions from S-C-C (cysteine) deformations (19). Similarly, the 2Fe-2S-4Cys and 4Fe-4S-4Cys ferredoxins exhibit multiple Fe-S stretching modes of predominately cysteinate character from 310-370 cm^{-1} and from 360-390 cm^{-1}, respectively (23, 24). In rubredoxin the C-S stretch of the cysteinate ligand is assigned at 653 cm^{-1}, while Fe-S-Fe and S-Fe-S deformation modes account for the weaker peaks in the region between 130 and 200 cm^{-1} (19). The blue copper proteins such as azurin (Table I) with their unusually short Cu-S(Cys) bonds of ~2.1 Å offer a striking contrast. The Cu-cysteinate chromophore gives rise to a number of intense features in the 370-430 cm^{-1} range whose higher frequency is commensurate with a stronger Cu-S bond, but whose multiple bands are surprising in view of the presence of only a single cysteinate ligand. It is likely that these additional peaks represent admixtures of internal ligand motions (such as the S-C-C bend) with the

C-S stretch, a coupling phenomenon facilitated by the distorted geometry of the blue copper site (20, 25). A vibration near 750 cm^{-1} is assigned to the Cu-S stretch of cysteine. Both iron-sulfur and the blue copper proteins exhibit an additional set of bands above 600 cm^{-1} which are due to overtones and combinations of the fundamentals in the 300-450 cm^{-1} region.

Raman Spectra of Metal Clusters

The principal types of metal clusters which have been characterized in metalloproteins are binuclear copper, iron-oxo, and iron-sulfur clusters. In each case the multiple metal ions appear to be connected by one or more bridging ligands which mediate strong antiferromagnetic interactions between the metal centers. The clusters containing oxo (O^{2-}) or sulfido (S^{2-}) bridges also have intense absorption bands arising from O $\rightarrow$ Fe or S $\rightarrow$ Fe charge transfer which facilitate the observation of resonance-enhanced M-O and M-S vibrations. Conversion of an oxo bridge to a hydroxo bridge by protonation leads to a weaker chromophore which is less likely to produce a resonance Raman spectrum. Although the bridging ligand in a binuclear copper site has yet to be identified, x-ray crystallographic studies of deoxyhemocyanin indicate that it is unlikely to be an endogenous amino acid group. An attractive candidate would be a bridging hydroxide which would account for both the strong antiferromagnetic coupling in oxyhemocyanin ($-J > 550\ cm^{-1}$) and the failure to observe any resonance Raman modes attributable to a Cu-L-Cu moiety (26).

The structures of typical binuclear iron-oxo and iron-sulfur clusters are shown in Figure 4. The iron-oxo cluster, as exemplified by the respiratory protein hemerythrin (26, 29), contains octahedral iron atoms bridged by two protein carboxylates and a single, solvent-derived oxo group. The site denoted L accepts exogenous ligands and is the place where dioxygen binds in the biologically active form of the protein. In contrast, iron-sulfur clusters have tetrahedral iron atoms with multiple sulfido bridging groups and tend to be coordinatively saturated, in keeping with their role as redox-active, electron transfer proteins (30). Common features of the iron-oxo and iron-sulfur clusters are that they exhibit vibrational modes encompassing the entire Fe-O-Fe or Fe-S-Fe moiety and that these assignments can be readily verified by substitution of isotopically labeled oxo or sulfido groups.

Iron-Oxo Clusters. The Fe-O-Fe moiety has three normal modes of vibration: a symmetric stretch (ν_s), Fe$\leftarrow$O$\rightarrow$Fe, an asymmetric stretch (ν_{as}), Fe$\rightarrow$O$\rightarrow$Fe, and a bending mode (δ), Fe-$\underset{\downarrow}{O}$-Fe. All three have been observed in resonance Raman spectra, but the intense symmetric mode is by far the easiest to detect (26, 33). A typical spectrum is shown in Figure 5 for oxyhemerythrin where the only three clearly visible peaks are ν_s at 486 cm^{-1}, ν_{as} at 753 cm^{-1}, and ν_1 of sulfate at 981 cm^{-1} (an internal frequency and intensity standard). The bridging oxo group in oxyhemerythrin does not exchange with solvent when the protein is equilibrated in water containing 18-O, but will exchange when the protein is reduced to its deoxygenated state. The latter, which is believed to contain an Fe(II)-OH-Fe(II) cluster and is colorless, cannot be probed by resonance Raman spectroscopy. However, after reconversion to

Table I. Resonance-Enhanced Metal-Ligand and Ligand Vibrational Modes in Metalloproteins[a]

Type of Binding	ν(M-L)	ν(Ligand)	Example
$CH_3-C_6H_4-$O···Fe	(575)	803, 869, 1164, 1281, 1497, 1597	Purple acid phosphatase[c]
H-N(imidazole)N···Cu	245	1140, 1168, 1256, 1330, 1426, 1490	$Cu(ImH)_4Cl_2$[d]
	225, 265	n.o.	Oxyhemocyanin[d]
HO···Fe	565	n.o.	Methemerythrin(OH)[e]
-C(O···Fe)(=O···Fe)	338	662	$Fe_2O(Ac)_2(HBpz_3)_2$[f]
	n.o.	(530)	Methemerythrin[g]
$-CH_2-$S···Fe	314, 348, 363, 376	653	Rubredoxin[h]
$-CH_2-$S···Cu	(373, 401, 409, 428)[b]	753	Azurin[i]

[a]Vibrational frequencies in cm^{-1}. ν(M-L) = metal-ligand stretching mode, ν(ligand) = internal ligand mode, n.o. = not observed. Values in parentheses are tentative assignments. [b]Coupled ν(Cu-S) + δ(Cys) modes. [c](14). [d](15). [e](16). [f](17). [g](18). [h](19). [i](20).

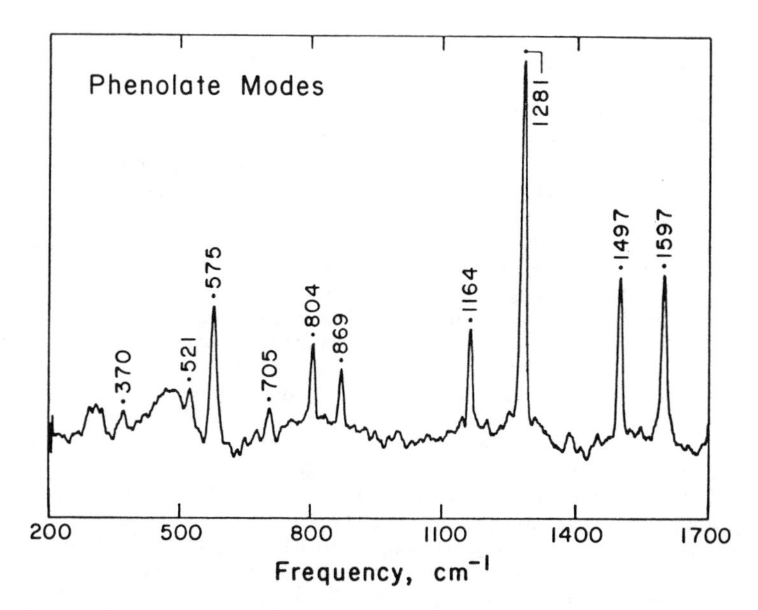

Figure 3. Resonance Raman spectrum of purple acid phosphatase. Protein (5 mM) maintained at 5°C in a glass Dewar and probed with 514.5 nm excitation (within the 560 nm phenolate → Fe(III) CT band, ε = 4,000 M^{-1} cm^{-1}). The broad, underlying feature from 400-550 cm^{-1} is due to Raman scattering from glass. (Reproduced from Ref. 14. Copyright 1987 American Chemical Society.)

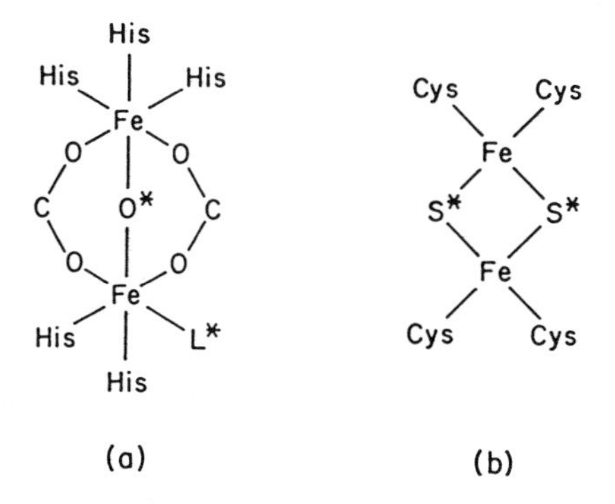

Figure 4. Structures of binuclear metal clusters. (a) Iron-oxo cluster in oxyhemerythrin from Themiste dyscrita; L denotes a coordinated hydroperoxide (27). (b) Iron-sulfur cluster in oxidized ferredoxin from Spirulina platensis (28). Starred ligands can be exchanged with appropriate species in solution.

oxyhemerythrin bridge exchange is apparent from the significant shifts in ν_s and ν_{as} (Figure 5, Table II). The isotope shift of the ν_{as} peak reveals a residual component at 757 cm^{-1} due to tryptophan.

A similar set of ν_s and ν_{as} frequencies and O-18 shifts have been observed (Table II) for the binuclear iron protein, ribonucleotide reductase, and for the hemerythrin model compound, $Fe_2O(Ac)_2(HBpz_3)_2$, which has one oxo and two acetate groups bridging the iron atoms. Treatment of the Fe-O-Fe unit as a harmonic oscillator allows it to be considered as a mechanical system of three masses connected by springs with characteristic force constants. Using normal coordinate analysis, a set of secular equations can be derived which relate vibrational frequencies to force constants and reduced masses in terms of the geometry of the participating atoms (35). This analysis predicts a marked dependence of vibrational frequency on Fe-O-Fe bond angle, with ν_s increasing and ν_{as} decreasing as the bond angle decreases from 180° to 90°. Such behavior is apparent from Table II where the decrease in bond angle from 155° in the monobridged phenanthroline complex to 135° in azidomethemerythrin (36) and to 124° in the tribridged $HBpz_3$ complex is accompanied by an increase in ν_s from 395 to 507 to 528 cm^{-1} and a decrease in ν_{as} from 827 to 768 to 751 cm^{-1}. Incorporation of the O-16 and O-18 values of ν_s into the secular equations leads to a predicted Fe-O-Fe angle of 138° for ribonucleotide reductase. The observation of such a narrow angle implies that the binuclear iron center in ribonucleotide reductase is also triply bridged.

Table II. Fe-O-Fe Modes in Proteins and Model Compounds[a]

Sample	ν_s(Fe-O-Fe)			ν_{as}(Fe-O-Fe)	
	O-16	[ΔO-18]	{ΔD}	O-16	[ΔO-18]
Oxyhemerythrin[b]	486	[-14]	{+4}	753	[-37]
Azidomethemerythrin[b]	507	[-14]	{0}	768	[-35]
Ribonucleotide reductase[c]	492	[-13]	{+3}	756	[-25]
$Fe_2O(Ac)_2(HBpz_3)_2$[d]	528	[-17]		751	[-30]
$[Fe_2O(phen)_4(H_2O)_2]^{4+}$[e]	395	[-5]		827	[-39]

[a]Peak frequencies and isotope shifts in cm^{-1}. [b]Isotope shifts with 18-O in bridge larger than in Figure 4 due to more complete isotope replacement (16). [c]Protein B2, radical-free form (22). [d]Tribridged cluster with tris(1-pyrazolyl)borate capping ligands (31). [e]Monobridged cluster with 1,10-phenanthroline and aquo ligands (32).

An unusual property of the Fe-O-Fe symmetric stretch in oxyhemerythrin and ribonucleotide reductase is its 3-4 cm^{-1} shift to higher energy when the protein is equilibrated with deuterated solvent (Table II). Since this type of deuterium-induced shift is not observed in methemerythrins with a variety of anionic ligands, it is not likely to be the result of a protein conformational change in D_2O (16). A more plausible explanation is that the bridging oxo group is behaving as a hydrogen bond acceptor and that the strength of this hydrogen bond is somewhat weaker with deuterium than with hydrogen. In the case of oxyhemerythrin, the most probable hydrogen bond donor is the hydroperoxide ligand (derived by reduction and protonation of an iron-coordinated dioxygen) (Figure 6). Further evidence for this structure is the 20-30 cm^{-1} decrease of both ν_s and ν_{as} in oxyhemerythrin relative to the non-hydrogen-bonded methemerythrins (Table II). The decrease in force constants for the Fe-O-Fe vibrations in oxyhemerythrin is ascribed to a lowering of electron density in the Fe-O-Fe moiety as a result of the hydrogen bonding interaction (16). In ribonucleotide reductase it is more likely that the hydrogen bond donor to the Fe-O-Fe moiety is an endogenous protein component (22).

The purple acid phosphatases may also contain oxo-bridged binuclear Fe(III) clusters on the basis of their strong antiferromagnetic coupling (14). However, their electronic and resonance Raman spectra (14, 37) are dominated by contributions from phenolate ligands (Figure 3), and no evidence for a resonance-enhanced Fe-O-Fe vibration has yet been obtained. A comparison of a series of oxo-bridged model compounds reveals that the intensity of the Fe-O-Fe symmetric stretch can vary over several orders of magnitude. For example, the molar scattering of the Fe-O-Fe vibration with 406.7 nm excitation is 100 times greater for the tribridged $HBpz_3$ complex than for the tribridged tacn complex (Figure 7). The strong enhancement of the former is similar to that observed with hemerythrin (Figure 8) and ribonucleotide reductase, while the weak enhancement of the latter is perhaps closer to that of purple acid phosphatase. A third $Fe_2O(Ac)_2$ model compound (39) with two unsaturated, heterocyclic nitrogens *cis* to the oxo group (as in the $HBpz_3$ complex) and one saturated nitrogen *trans* to the oxo group (as in the tacn complex) has a similarly weak enhancement. These results suggest that the absence of a strong Fe-O-Fe mode in the resonance Raman spectrum of purple acid phosphatase is due either to a smaller number of histidine ligands than in hemerythrin or to their being located *cis* to the oxo group (14). Conversely, the intensity of the Fe-O-Fe vibration in ribonucleotide reductase, which appears from EXAFS studies to also have a smaller number of histidine ligands (37), is most likely due to the presence of one histidine per Fe *trans* to the oxo bridge (22).

The resonance Raman enhancement profiles in Figures 7 and 8 show that the maximum intensity of the Fe-O-Fe symmetric stretch fails to correspond to a distinct absorption maximum in the electronic spectrum. This implies that the oxo $\rightarrow$ Fe CT transitions responsible for resonance enhancement are obscured underneath other, more intense bands. Although strong absorption bands in the 300-400 nm region ($\varepsilon > 6,000$ M^{-1} cm^{-1}) are a ubiquitous feature of Fe-O-Fe clusters, the Raman results make it unlikely that they are due to oxo $\rightarrow$ Fe CT. An alternative possibility is that they represent simultaneous pair excitations of LF transitions in both of the

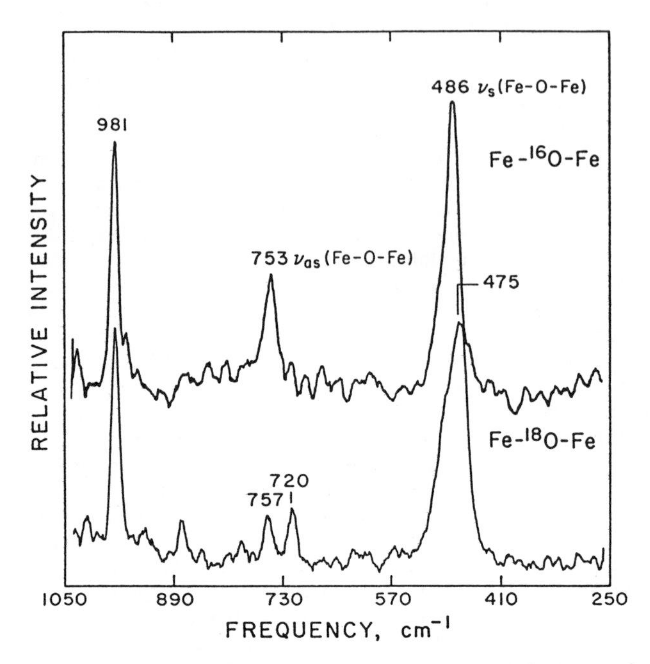

Figure 5. Resonance Raman spectra of oxyhemerythrin with O-16 (upper) and O-18 (lower) in the oxo bridge. Protein (1.2 mM) and Na_2SO_4 (0.3 M) maintained at 5°C in a flow cell and probed with 363.8 nm excitation. (Reproduced from Ref. 34. Copyright 1984 American Chemical Society.)

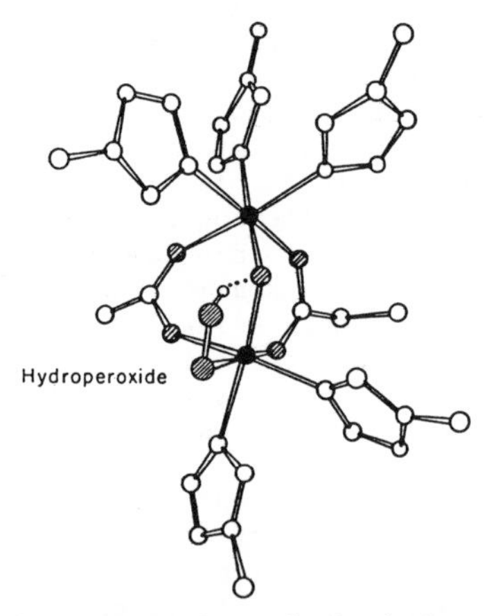

Figure 6. Proposed coordination of the hydroperoxide ligand in oxyhemerythrin (16,27). Coordination geometry taken from crystal structure of azidomethemerythrin at 2.0 Å resolution (36). Iron (●), carbon and nitrogen (○), oxygen (◍).

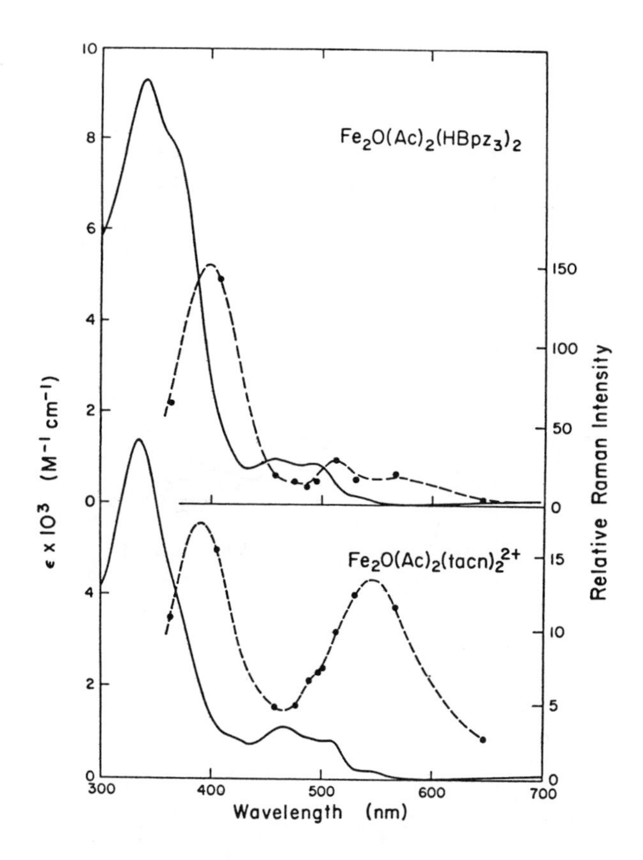

Figure 7. Electronic absorption spectra (——) and resonance Raman enhancement profiles for ν_s(Fe-O-Fe) (---) for two tribriged model compounds. (Upper) Height of 528 cm^{-1} sample peak measured relative to height of 981 cm^{-1} peak of sulfate at a series of excitation wavelengths (●) and corrected for the relative molar concentrations of sample and sulfate. (Lower) Similar treatment for 540 cm^{-1} sample peak. Tribridged cluster contains triazacyclononane capping groups (38). (From Wheeler, W. D.; Shiemke, A. K., Averill, B. A., Loehr, T. M.; Sanders-Loehr, J., unpublished results.)

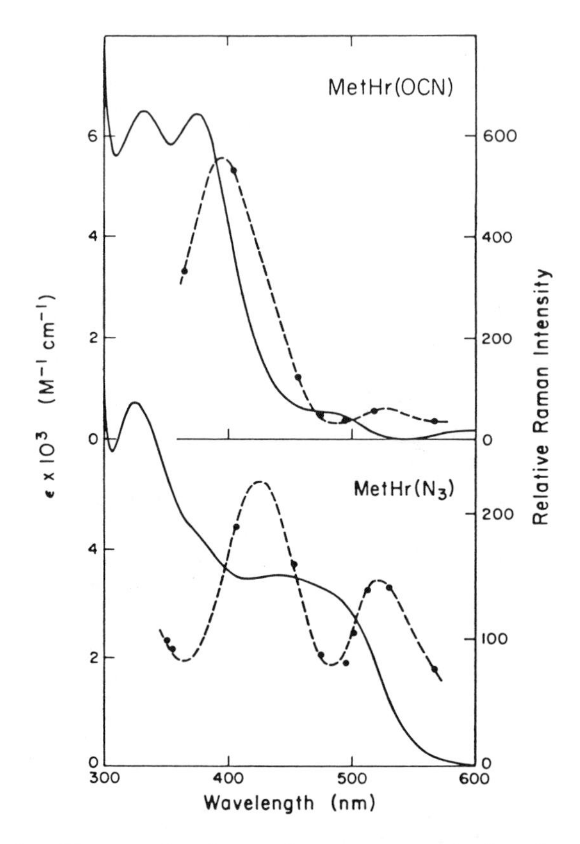

Figure 8. Electronic absorption spectra (——) and resonance Raman enhancement profiles for ν_s(Fe-O-Fe) (---) for methemerythrins with cyanate and azide as exogenous ligands. Relative Raman intensities determined as in Figure 7. (Upper) 509 cm^{-1} sample peak. (Lower) 507 cm^{-1} sample peak.

Fe(III) ions, as proposed previously for $Fe_2O(hedta)_2^{2-}$ (40). The metal-centered character of such transitions could explain their failure to yield resonance-enhanced Fe-L vibrational modes.

Iron-Sulfur Clusters. Both the binuclear and tetranuclear iron-sulfur proteins can be partially unfolded under conditions which allow cluster extrusion and replacement (41), thereby making it possible to incorporate S-34 into the bridging sulfide positions. From these experiments it appears that there are two categories of resonance Raman modes; those with substantial isotope shifts which are ascribed to vibrations of the 2Fe-2S or 4Fe-4S cores and those with lesser isotope shifts which are ascribed to vibrations of the Fe-S(Cys) terminal ligands (23,24,42,43). The peaks due to bridging and terminal sulfur ligands are identified in Figures 9 and 10 by "-S-" and "Cys", respectively. The two types of clusters have distinctive differences in peak intensities which allow them to be distinguished from one another in proteins of unknown cluster geometry as, for example, in the hydrogenases from *Clostridium* (44). In the 2Fe-2S proteins ferredoxin and adrenodoxin (23), the Raman spectrum is dominated by the peaks at ~280 and ~390 cm^{-1}, whereas in the 4Fe-4S proteins there is only weak scattering in the 280 cm^{-1} region and the most intense feature occurs at ~340 cm^{-1}. Crystal structures for both types of clusters show that the bridging and terminal sulfur ligands are extensively hydrogen bonded to amide NH groups of the polypeptide backbone (28,45). The generality of this occurrence has been corroborated for the 2Fe-2S proteins where the vibrations of the bridging and terminal sulfur groups undergo shifts of 0.5 to 2.0 cm^{-1} when the proteins are equilibrated in D_2O (21).

Normal coordinate analysis of a 2Fe-2S-4Cys cluster with D_{2h} symmetry predicts a total of four Raman-active and four infrared-active modes, and these are observed in the respective Raman and infrared spectra of model compounds such as $Fe_2S_2Cl_4^{2-}$ (23). In the resonance Raman spectrum of 2Fe-2S-4Cys ferredoxin (Figure 9), the expected features are observed at 329 and 393 cm^{-1} as well as peaks at 284, 338, 367, and 425 cm^{-1} which correspond to the infrared-active modes in model compounds. The strong Raman intensities of the latter group imply that the Fe-S clusters in ferredoxin and adrenodoxin are significantly distorted from D_{2h} symmetry (23,42). A related phenomenon is observed in the 4Fe-4S proteins. The resonance Raman spectra of oxidized ferredoxin and reduced high-potential iron protein are similar to one another. Both show a greater number of vibrational modes than predicted for pure T_d symmetry, but show good correspondence with the expected distribution for a distorted D_{2d} cube (24,43). This finding is in agreement with the x-ray crystallographic results which indicate that the 4Fe-4S clusters in these two proteins, as well as in a number of model compounds, have a compressed tetragonal structure with 4 short and 8 long Fe-S bonds (41,45).

Conclusions

The application of resonance Raman spectroscopy to the study of metalloproteins has led to the identification of metal-ligated groups based on the appearance of characteristic metal-ligand and intraligand vibrational modes. Electronic spectral transitions due

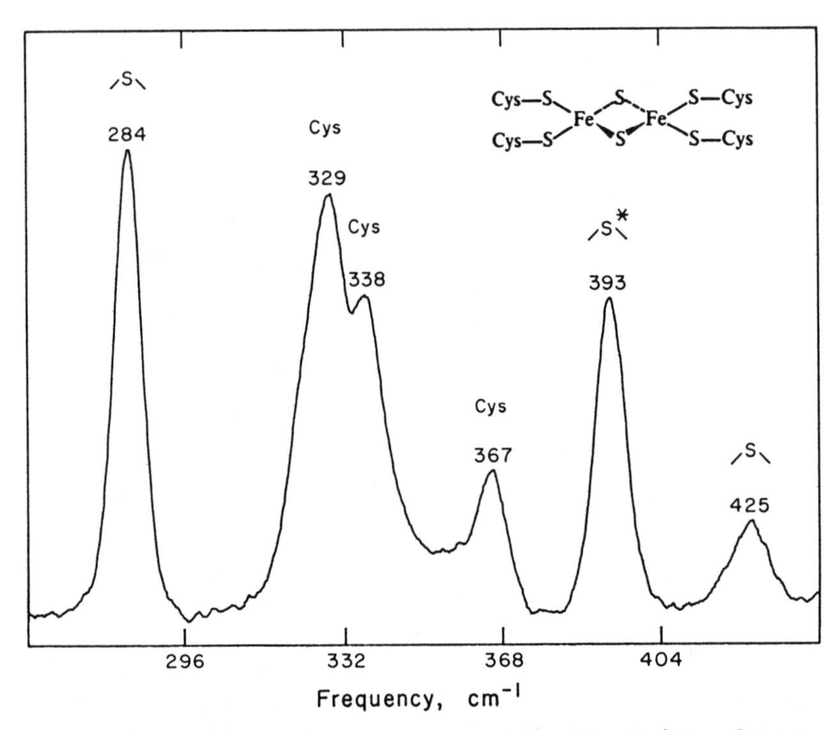

Figure 9. Resonance Raman spectrum of the 2Fe-2S-4Cys cluster in oxidized spinach ferredoxin. Protein (~2 mM) maintained at 15 K in a helium Displex and probed with 488.0 nm excitation. Spectral contribution of ice in 220-320 cm^{-1} region has been subtracted. (Data from Ref. 21). Starred features indicate peaks identified as totally symmetric A_{1g} vibrations (23,24).

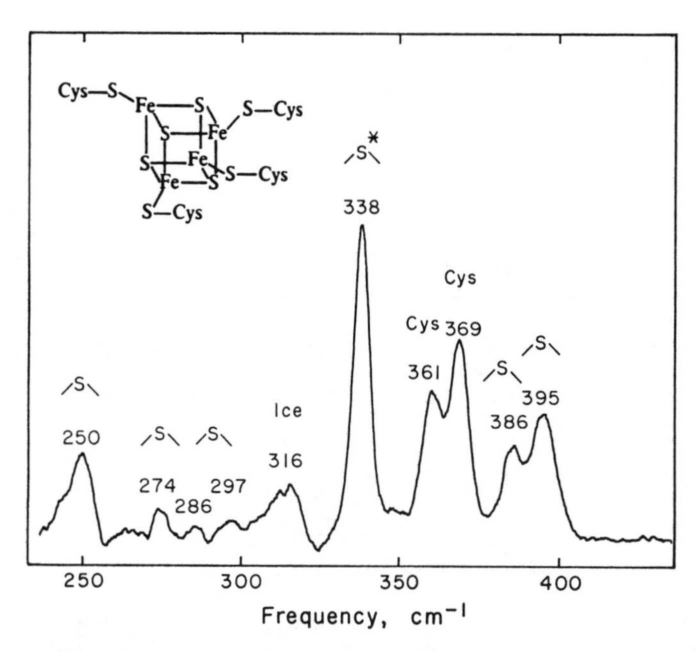

Figure 10. Resonance Raman spectrum of the 4Fe-4S-4Cys cluster in reduced high-potential iron protein from Rhodopseudomonas globiformis. Spectral conditions as in Figure 9, but spectrum of frozen solvent not subtracted out. (From Mino, Y.; Loehr, T. M.; Sanders-Loehr, J.; Cusanovich, M., unpublished results.) Assignments of bridging (-S-) and terminal (Cys) Fe-S vibrations based on S-34 isotope shifts for oxidized ferredoxin from Clostridium pasteurianum (24,43).

to ligand → metal charge transfer can be identified from resonance Raman enhancement profiles of M-L and L vibrations. For proteins containing Fe-O and Fe-S clusters, the multiple atom character of the vibrational modes provides additional information on bond angles and cluster geometry.

Acknowledgments

The author acknowledges the valuable contributions of students and coworkers whose work is cited herein and the support of the National Institutes of Health (GM 18865).

Literature Cited

1. Nakamoto, K. Infrared and Raman Spectra of Inorganic and Coordination Compounds; Wiley-Interscience: New York, 1986.
2. Long, D. A. Raman Spectroscopy; McGraw-Hill: New York, 1977.
3. Loehr, T. M.; Sanders-Loehr, J. In Copper Proteins and Copper Enzymes; Lontie, R., Ed.; CRC Press: Boca Raton, 1984; pp. 115-155.
4. Strommen, D. P.; Nakamoto, K. Laboratory Raman Spectroscopy; Wiley: New York, 1984.
5. Loehr, T. M.; Keyes, W. E.; Pincus, P. A. Anal. Biochem. **1979**, 96, 456.
6. Frushour, B. G.; Koenig, K. L. In Advances in Infrared and Raman Spectroscopy; Clark, R. J. H.; Hester, R. E., Eds.; Heyden: London, 1975; Vol. 1, pp. 35-97.
7. Spiro, T. G.; Stein, P. Annu. Rev. Phys. Chem. **1977**, 28, 501.
8. Carey, P. R. Biochemical Applications of Raman and Resonance Raman Spectroscopies; Academic Press: New York, 1982; pp. 42-47.
9. Myers, A. B. ; Mathies, R. A. In Biological Applications of Raman Spectroscopy; Spiro, T. G., Ed.; Wiley: New York, 1987; Volume 2, pp. 1-58.
10. Hirakawa, A. Y.; Tsuboi, M. Science **1975**, 188, 359.
11. Woodruff, W. H.; Norton, K. A.; Swanson, B. I.; Fry, H. A. Proc. Natl. Acad. Sci USA **1984**, 81, 1263.
12. Maret, W.; Shiemke, A. K.; Wheeler, W. D.; Loehr, T. M.; Sanders-Loehr, J. J. Am. Chem. Soc. **1986**, 108, 6351.
13. Spiro, T. G. In Iron Porphrins; Lever, A. B. P.; Gray, H. B., Eds.; Addison-Wesley: Reading, MA, 1983; Part II, Chapter 3.
14. Averill, B. A.; Davis, J. C.; Burman, S.; Zirino, T.; Sanders-Loehr, J.; Loehr, T. M.; Sage, J. T.; Debrunner, P. G. J. Am. Chem. Soc. **1987**, 109, 3760.
15. Larrabee, J. A.; Spiro, T. G. J. Am. Chem. Soc. **1980**, 102, 4217.
16. Shiemke, A. K.; Loehr, T. M.; Sanders-Loehr, J. J. Am. Chem. Soc. **1986**, 108, 2437.
17. Czernuszewicz, R. S.; Sheats, J. W.; Spiro, T. G. Inorg. Chem. **1987**, 26, 2063.
18. Shiemke, A. K. Ph.D. Thesis, Oregon Graduate Center, 1986.
19. Czernuszewicz, R. S.; LeGall, J.; Moura, I.; Spiro, T. G. Inorg. Chem. **1986**, 25, 696.
20. Woodruff, W. H.; Dyer, R. B.; Schoonover, J. R. In Biological Applications of Raman Spectroscopy; Spiro, T. G., Ed.; Wiley: New York, in press, Volume 3.

21. Mino, Y.; Loehr, T. M.; Wada, K.; Matsubara, H.; Sanders-Loehr, J. Biochemistry, in press.
22. Sjöberg, B.-M.; Sanders-Loehr, J.; Loehr, T. M. Biochemistry **1987**, 26, 4242.
23. Yachandra, V. K.; Hare, J.; Gewirth, A.; Czernuszewicz, R. S.; Kimura, T.; Holm, R. H.; Spiro, T. G. J. Am. Chem. Soc. **1983**, 105, 6462.
24. Czernuszewicz, R. S.; Macor, K. A.; Johnson, M. K.; Gewirth, A; Spiro, T. G. J. Am. Chem. Soc., in press.
25. Nestor, L.; Larrabee, J. A.; Woolery, G., Reinhammar, B.; Spiro, T.G. Biochemistry **1984**, 23, 1084.
26. Loehr, T. M.; Shiemke, A. K. In Biological Applications of Raman Spectroscopy; Spiro, T. G., Ed.; Wiley: New York, in press, Volume 3.
27. Stenkamp, R. E.; Sieker, L. C.; Jensen, L. H.; McCallum, J. D.; Sanders-Loehr, J. Proc. Natl. Acad. Sci. USA **1985**, 82, 713.
28. Tsukihara, T; Fukuyama, K.; Nakamura, M.; Katsube, Y.; Tanaka, N.; Kakudo, M.; Wada, K.; Hase, T.; Matsubara, H. J. Biochem. **1981**, 90, 1763.
29. Wilkins, R. G.; Harrington, P. C. Adv. Inorg. Biochem. **1983**, 5, 52.
30. Thompson, A. J. In Metalloproteins; Harrison, P., Ed.; Verlag Chemie: Weinheim, 1985; Part 1, pp. 79-120.
31. Armstrong, W. H.; Spool, A.; Papaefthymiou, G. C.; Frankel, R. B.; Lippard, S. J. J. Am. Chem. Soc. **1984**, 106, 3653.
32. Plowman, J. E.; Loehr, T. M.; Schauer, C. K.; Anderson, O. P. Inorg. Chem. **1984**, 23, 3553.
33. Kurtz, D. M., Jr.; Shriver, D. F.; Klotz, I. M. Coord. Chem. Rev. **1977**, 24, 145.
34. Shiemke, A. K.; Loehr, T. M.; Sanders-Loehr, J. J. Am. Chem. Soc. **1984**, 106, 4951.
35. Wing, R. M.; Callahan, K. P. Inorg. Chem. **1969**, 8, 871.
36. Stenkamp, R. E.; Sieker. L. C.; Jensen, L. H. J. Am. Chem. Soc. **1984**, 106, 618.
37. Antanaitis, B. C.; Strekas, T.; Aisen, P. J. Biol. Chem. **1982**, 257, 3766.
38. Wieghardt, K.; Pohl, K.; Gebert, W. Angew. Chem. Int. Ed. Engl. **1983**, 22, 727.
39. Toftlund, H.; Murray, K. S.; Zwack, P. R.; Taylor, L. F.; Anderson, O. P. J. Chem. Soc. Chem. Commun. **1986**, 191.
40. Schugar, H. J.; Rossman, G. R.; Barraclough, C. G.; Gray, H. B. J. Am. Chem. Soc. **1972**, 94, 2683.
41. Berg. J. M.; Holm, R. H. In Iron-Sulfur Proteins; Spiro, T. G., Ed.; Wiley: New York, 1982; pp. 1-66.
42. Meyer, J., Moulis, J.-M.; Lutz, M. Biochim. Biophys. Acta **1986**, 873, 108.
43. Moulis, J.-M., Meyer, J.; Lutz, M. Biochemistry **1984**, 23, 6605.
44. Macor, K. A.; Czernuszewicz, R. S.; Adams, M. W. W.; Spiro, T. G. J. Biol. Chem. **1987**, 262, 9945.
45. Carter, C. W., Jr. In Iron-Sulfur Proteins; Lovenberg, W., Ed.; Academic Press: New York, 1977; Chapter 6.

RECEIVED December 18, 1987

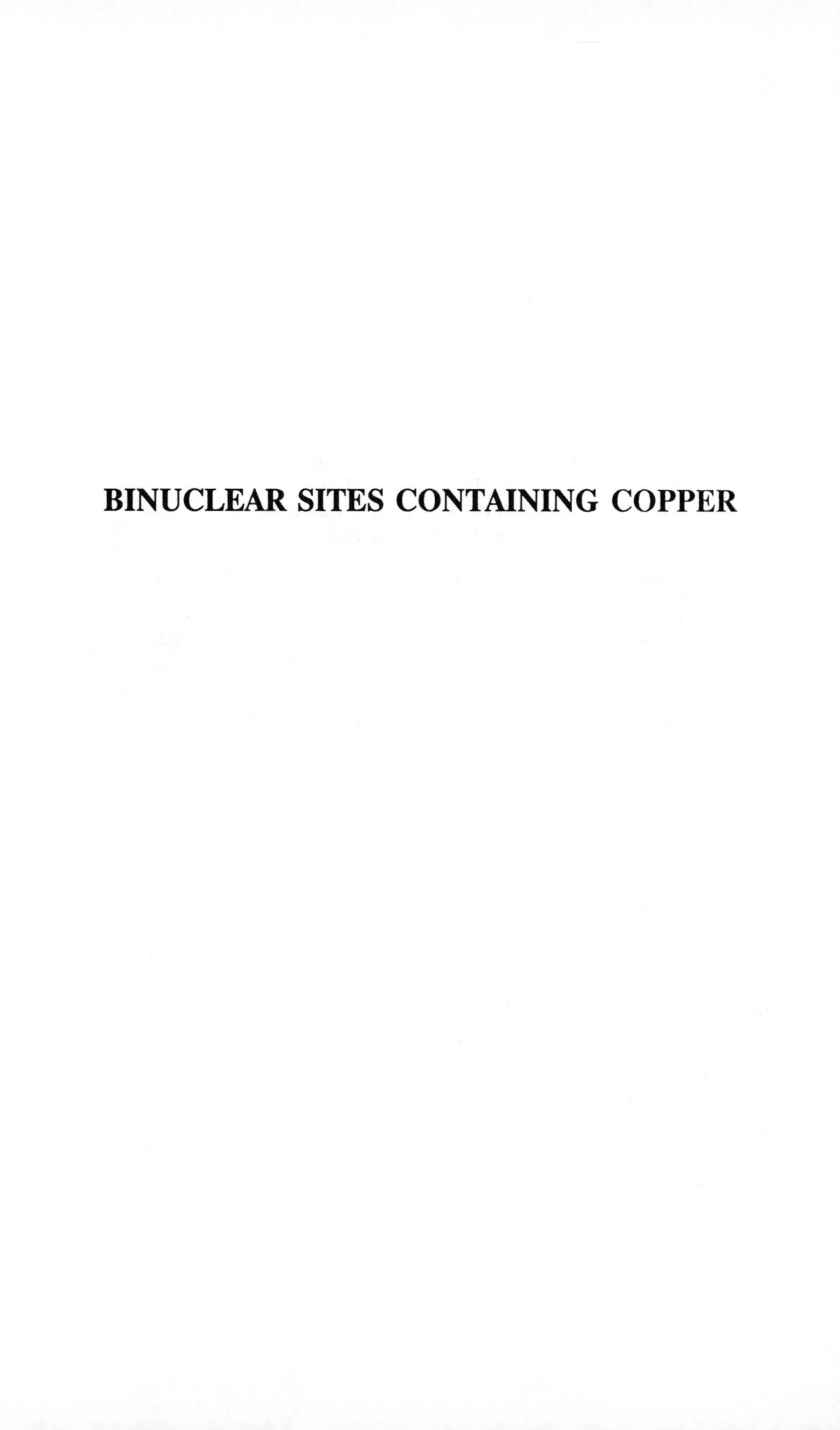

BINUCLEAR SITES CONTAINING COPPER

Chapter 4

NMR of Paramagnetic Systems

Magnetically Coupled Dimetallic Systems (Cu_2Co_2Superoxide Dismutase as an Example)

Ivano Bertini[1], Lucia Banci[1], and Claudio Luchinat[2]

[1]Department of Chemistry, University of Florence, 50121 Florence, Italy
[2]Institute of Agricultural Chemistry, University of Bologna, 40126 Bologna, Italy

Paramagnetic systems are widespread in macromolecules of biological relevance. A number of these contain dimetallic clusters with at least one paramagnetic center. The recent advances in obtaining ^1H NMR spectra of paramagnetic macromolecules and in understanding the theory for electron-nuclear coupling will be presented, with emphasis on magnetically coupled dimetallic systems. The analysis of the ^1H NMR spectra of Cu_2Co_2-superoxide dismutase will be presented in some detail to show how important structural and chemical information can be obtained using this approach. Other pertinent systems will be briefly discussed and the prospective uses of the technique will be critically evaluated.

Dimetallic systems are common among metalloproteins and their characterization is important for understanding the nature of the interaction between the two metal centers.

A dimeric unit may incorporate a metal ion that is a catalytic center and another one with some function, ranging from a structural role to regulation of enzymatic activity. Examples of such sites include erythrocyte superoxide dismutase (SOD), which contains copper(II) and zinc(II) ions, the latter being essentially structural and the former catalytic (1,2), and alkaline phosphatase, which has two zinc ions, one with a catalytic role and another one which is regulatory (3). The zinc ion can be replaced by several other metal ions, sometimes without a significant loss of enzymatic activity (4). Other examples of bimetallic systems include those formed by iron(II) and iron(III), e.g. the reduced Fe_2S_2 cluster in ferredoxins (5) and the mixed-valent dimetallic units contained in reduced uteroferrin (6,7), reduced acid phosphatases (6) and semimet–hemerythrin (7–9). This latter group of proteins have active sites which consist of a paramagnetic metal ion with fast electronic relaxation times (iron(II)) interacts with another paramagnetic metal ion with slow electronic relaxation times (iron(III)) (10). Such an interaction results in binuclear complexes with favorable NMR properties.

0097–6156/88/0372–0070$06.00/0

The proteins containing a zinc atom in the dimetallic unit can be manipulated in such a way as to achieve similar situations, like CuCo or Cu-Ni derivatives of CuZn native systems (1,11–14). The unpaired electron in an isolated copper(II) ion relaxes slowly, whereas the electrons of cobalt(II) and nickel(II) relax rapidly. The occurrence of magnetic coupling has a major effect on the electronic relaxation time of the slowly relaxing metal ion; its electronic relaxation times are largely reduced with the result that ^{1}H NMR signals of ligand protons can be readily observed, as in the case of metalloproteins containing only the fast–relaxing metal (10). The NMR parameters of the coupled system change with respect to the single ion systems, and such artificial metal pairs may provide a tool for the NMR investigation of slowly relaxing systems, like copper proteins, which would give rise to signals too broad for detection. In homodimeric systems like the oxidized form of Fe_2S_2 ferredoxins, the oxidized and reduced forms of hemerythrin, Cu_2Cu_2SOD etc., the magnetic coupling does not produce any major change in the electronic properties of the ions (15).

We will discuss here the NMR parameters in paramagnetic systems and the effect of the magnetic coupling. One example will be discussed in detail, the Cu_2Co_2SOD system, and other relevant systems will also be mentioned.

NMR Parameters in Paramagnetic Systems

As far as paramagnetic systems are concerned, the relevant NMR parameters are the isotropic shift, i.e. the chemical shift minus the shift of an analogous diamagnetic system, and the T_1^{-1} and T_2^{-1} enhancements (T_{1M}^{-1} and T_{2M}^{-1}) (10). Such enhancements are due to the coupling between the unpaired electron(s) responsible for paramagnetism and the resonating nuclei. As will be shown in more detail, electronic relaxation times are very short compared to the nuclear relaxation times and the coupling provides further pathways for nuclear relaxation whose rate depends on the value of the electronic relaxation times, among other factors.

The isotropic shift. The isotropic shift is the sum of two contributions: the contact and the dipolar contributions. The former is due to the presence of unpaired electron density on the resonating nucleus. The latter arises from the anisotropy of the magnetic susceptibility tensor, modulated by the distance between the unpaired electron and the resonating nucleus, and is also dependent on the orientation of the metal nucleus vector with respect to the principal axes of the magnetic susceptibility tensor. Some problems arise when the spin delocalization is taken into account in calculating the dipolar coupling, but we will not address this problem except when strictly necessary.

The effect of magnetic coupling on the above parameters is negligible as long as the energy levels arising from magnetic coupling are equally populated at the temperature of the experiment. If we consider, for example, two $S = 1/2$ electron spins, they give rise to two magnetically coupled $S' = 0$ and $S' = 1$ spin states. As long as the separation of these states, which is equal to J if the spin Hamiltonian for the interaction is written as

$$H = J\, S_1 \bullet S_2 \qquad (1)$$

is small compared to kT, they are equally populated regardless of the ferromagnetic or antiferromagnetic nature of the coupling. Under these circumstances the magnetic susceptibility is not altered by the magnetic coupling. Since the contact shift depends on the magnetic susceptibility, it should also remain unaltered. In other words, weak magnetic coupling does not change the contact contribution of a metal to the resonating nucleus and, if the nucleus interacts with two metal ions, the contact shift is strictly additive (16–18). This conclusion also holds in practice for the dipolar shift, although even small perturbations to the energy levels may change the orientation of the magnetic susceptibility tensor within the molecule, in principle.

When J is of the same order of magnitude as kT, the magnetic susceptibility is reduced (if the coupling is antiferromagnetic) or enhanced (if the coupling is ferromagnetic) according to the Van Vleck equation. An investigation of the temperature dependence of the shift may provide information on the value of J, if the hyperfine coupling is independently determined or assumed to be known (19) (although in principle the latter can also be parametrized).

Longitudinal relaxation rates. This section considers separately the cases of homodinuclear and heterodinuclear systems. In the former case, an unpaired electron feeling another electron with the same spin lifetime should not affect the longitudinal relaxation rate, unless the new energy levels arising from the coupling provide additional electron relaxation mechanisms. The only example available in the literature that addresses this interaction is provided by Cu_2Cu_2SOD. Water ^1H NMR relaxation experiments show that the electronic relaxation time of copper at the native copper site is not affected by the presence of the copper ion at the zinc site, despite an antiferromagnetic coupling constant of 52 cm^{-1} (20).

The problem becomes more complicated when the magnetic coupling occurs between two different metal ions, one of which has a long electronic lifetime and the other has a short electronic lifetime. It is clear that, when the slow-relaxing electron starts feeling the fast–relaxing electron, the former may start exchanging energy with the lattice through the latter and thus increase its relaxation rate. The quantitation of this effect is a difficult matter. One approach applies perturbation theory to the slow–relaxing ion (21). Using Fermi's golden rule, the effect of the magnetic coupling on the electronic relaxation rate of the slow–relaxing metal ion (where (1) represents the slow–relaxing metal ion) can be calculated (21) to be

$$T^{-1}_{1e(1)}\ (J) = T^{-1}_{1e(1)}\ (0) + (J^2/\hbar^2)\ T_{1e(1)}\ (0) \qquad (2)$$

$T_{1e(2)}$ is not relevant as it is much smaller than $T_{1e(1)}$; and $T_{1e(1)}$ (0) refers to the electronic spin–lattice relaxation time for the slow–relaxing metal ion in the absence of coupling. The approach seems quite reasonable for small J values. A J value of a fraction of 1 cm^{-1} may account for large variations in T_{1e}^{-1}. For example, in the case of low spin iron(III) interacting with copper(II), a J of 0.1 cm^{-1} could account for the experimental increase in T_{1e}^{-1} of copper from 10^9 to 10^{11} s^{-1}. For larger J values, equation 2 breaks down. However, by increasing J, the T_{1e}^{-1} of the slow–relaxing electron is expected to approach that of the fast–relaxing electron, until eventually the two systems experience the same spin lifetime. In order to set some limits, we can tentatively propose that when J is smaller than $(\hbar T_{1e(1)})^{-1}(0)$ the magnetic coupling may be quite ineffective. When J is intermediate between $(\hbar T_{1e(1)})^{-1}(0)$ and $\hbar T_{1e(2)}{}^{-1}(0)$ then the effect can be sizeable and in some cases an attempt of evaluation can be made according to equation 2. When J >> $\hbar T_{1e(2)}^{-1}(0)$, we can expect that the electronic relaxation times of the two systems are equal. More quantitative treatments are difficult to envisage owing to the presence of other effects, such as zero field splitting, which introduce additional parameters. However, the above semiquantitative considerations can guide us in developing new strategies to investigate bimetallic systems through NMR.

In principle, there are several contributions to nuclear T_{1M}^{-1}; however, the dipolar coupling term often dominates (10,22). The dipolar contribution depends on the reciprocal of the sixth power of the distance between the resonating nucleus and the relaxing electron, on the square of the magnetic moments associated with the unpaired electrons ($g_e^2\mu_B^2S(S+1)$) and with the nucleus ($\gamma_N^2\hbar^2$), on the magnetic field as expressed by the Larmor frequencies, ω_I and ω_S, and finally on the correlation time for the electron-nuclear coupling (τ_c) (Solomon equation, equation 3 (23)):

$$T_{1M}^{-1} = \frac{2}{15}\left[\frac{\mu_0}{4\pi}\right]^2 \frac{\gamma_N^2\, g_e^2\, \mu_B^2\, S(S+1)}{r^6}\left[\frac{7\tau_c}{1+\omega_S^2\tau_c^2} + \frac{3\tau_c}{1+\omega_I^2\tau_c^2}\right] \tag{3}$$

The correlation time is determined by the fastest motion among those capable of inducing a change in the interaction energy between the nuclear and electronic magnetic moments. The most common of these motions are the electronic relaxation itself (τ_s), the molecular tumbling (τ_r), and the chemical exchange between a group containing the resonating nucleus and another containing the paramagnetic metal ion (τ_m). The Solomon equation (equation 3) holds in the so–called Redfield limit, i.e. when the electron is always in equilibrium with the lattice and the motions inducing electronic relaxation are faster than the electronic relaxation itself. It has been shown that, as far as nuclear relaxation is concerned the electron–lattice system behaves as if this were always the case, even if the mechanism for electronic relaxation is not known (24). Under these circumstances the following relation holds:

$$\tau_c^{-1} = \tau_s^{-1} + \tau_r^{-1} + \tau_m^{-1} \tag{4}$$

If we restrict this discussion to nuclei belonging to coordinated ligands, the exchange rate seldom makes a significant contribution to the correlation rate. The molecular tumbling rate may contribute in small complexes ($\tau_r^{-1} \simeq 3 \times 10^{11}$ s^{-1} for the hexaaqua complexes at room temperature), but never in macromolecules ($\tau_r^{-1} \simeq 10^8$ s^{-1} for spherical molecules of MW 30,000). The electronic relaxation rate varies from 10^8 to 10^{13} s^{-1} depending on the distribution of the energy levels. It may also vary with the magnetic field, as it occurs with manganese(II), gadolinium(III) and nickel(II) (the latter for magnetic fields larger than 0.1 T). The concept of τ_s is not straightforwardly related to electronic T_1 or T_2. At low magnetic fields T_{1e} and T_{2e} are presumably equal, whereas at high magnetic fields nuclear relaxation is determined by T_{1e}. T_{1e} and T_{2e} are defined with respect to the direction of the external magnetic field. However, when there is a strong zero field splitting the molecular axes are independent of the external magnetic field and is possible that τ_s is a combination of T_{1e} and T_{2e}.

The rationale for understanding nuclear T_1^{-1} in weakly magnetically coupled dimers is that metal ions with large τ_s^{-1} which interact with metal ions with small τ_S^{-1} induce an increase of the latter values and thus a decrease in nuclear relaxation rates.

The equations relating nuclear relaxation and τ_S should take into consideration that each new orbital arising from magnetic coupling has different S' quantum number and different hyperfine coupling with the resonating nucleus. Recalculation of these effects results in a correction factor to the Solomon equation (23). These values, now available in the literature, are 1/2 for each homonuclear couple and, for example, 3/8 for Cu and 7/8 for Co in the Cu–Co couple. When J is of the order of kT, the coefficients should be calculated by using the same procedure (17) and taking into account the Boltzmann distribution of the occupied levels.

Transverse relaxation. The contributions to transverse relaxation are contact, dipolar and Curie type (10,25,26) in origin. The contact contributions are meaningful only for heteronuclei directly bound to the paramagnetic metal ion, and for protons only when τ_S is relatively very long. In the example we are going to discuss next the contact contribution is negligible. Dipolar coupling is sizeable and of the same order of magnitude as dipolar nuclear T_1. An equation similar to equation 3 describes the interaction, and the changes in cases of magnetic coupling are the same as for T_1. Curie relaxation deserves some comments. In the classical dipolar coupling mechanism, the difference in population of the various electronic spin states is neglected. Although small, this difference provides a permanent magnetic moment associated with the molecule. The nucleus sees this magnetic moment in different positions according to the rotational correlation time (25,26). In macromolecules τ_r^{-1} is small and its effect on T_2^{-1} is large; it has been shown that its effect on T_1^{-1} is negligible. Of course Curie relaxation increases with the magnetic field, which increases the difference in population among the electronic spin levels. Indeed this contribution is proportional to the square of the magnetic field, just like the magnetic susceptibility. Since the coupling is still dipolar in nature, it can provide structural information. In practice the linewidth, which is equal to $(\pi T_2)^{-1}$, is measured at various magnetic fields, i.e. 90–500 MHz, and is plotted against the square of the magnetic field. At high magnetic fields the data are on a straight line because Curie relaxation is the dominant term. The intercept at zero magnetic field provides the other contributions to the linewidth at high magnetic fields. An approximate equation to describe Curie relaxation is (25,26)

$$T_{2M}^{-1} = \frac{1}{5}\left[\frac{\mu_0}{4\pi}\right]^2 \frac{\omega_I^2\, g_e^4\, \mu_B^4\, S^2(S+1)^2}{(3kT)^2 r^6}\left[4\tau_r + \frac{3\tau_r}{1+\omega_I^2\tau_r^2}\right] \qquad (5)$$

where the symbols have the same meaning as in equation 3.

In the case of magnetically coupled systems, Curie relaxation is simply additive as long as the magnetic susceptibility is the sum of the two components. When the magnetic susceptibility is not a simple sum of the components, the magnetic susceptibility contribution of each metal ion should be evaluated.

Metal ions with small τ_S^{-1} produce a large broadening of the lines through dipolar coupling. However, metal ions with large τ_S^{-1} give rise to small NMR line broadening and therefore the ^{1}H NMR spectra can easily be observed, and the advantage of observing large isotropic shifts can be fully exploited. At high magnetic fields some advantages may be lost due to Curie line broadening, depending on the size of the molecule.

Cu_2Co_2SOD: ^1H NMR Spectra

Cu_2Zn_2SOD is a dimeric metalloprotein, each subunit containing a zinc and a copper ion (27). Zinc is coordinated to two histidine nitrogens and to an aspartate oxygen; copper is coordinated to three histidine nitrogens. In addition, a histidinato ion bridges the two metal ions. The copper ion is solvent exposed, and a water molecule completes the coordination sphere (Scheme 1). When cobalt(II) is substituted for zinc, the system consists of two paramagnetic metal ions, one with $S = 3/2$ and the other with $S = 1/2$ (1,11,12). The electronic relaxation rate of copper in the native protein has been independently estimated to be 5×10^8 s^{-1} (28). That of cobalt alone in the E_2Co_2SOD derivative (29) (E = empty) is estimated from ^1H NMR to be around 3×10^{12} s^{-1}. In the Cu_2Co_2SOD derivative each metal feels the other, although at room temperature the magnetic susceptibility is the sum of the magnetic susceptibilities of the two metal ions. Through magnetic coupling, $S' = 1$ and $S' = 2$ manifolds are formed. From magnetic susceptibility measurements at low temperatures, the separation between the two levels is estimated to be 33 cm^{-1}, with the $S' = 1$ being the ground state (30). As a result of this magnetic coupling, the electronic relaxation rates of copper increase towards those of cobalt and consequently the NMR lines of protons of histidine ligands of copper become narrow.

The ^1H NMR spectra of several Cu_2Co_2SOD isoenzymes exhibit 14 to 17 lines downfield, which are about as many as the ring protons, i.e. there are a total of 17 ring protons on the histidines bound to both cobalt and copper ions (Figure 1) (29,31,32). The assignment of the signals starts with the comparison with the spectrum obtained in D_2O after exhaustive deuteration of the apoenzyme. Under these conditions all the histidine NH signals disappear; they are indicated as shaded peaks in Figure 1. The following step is trying to distinguish between the copper and the cobalt signals. Curie relaxation is quite helpful in this respect because the linewidth should contain contributions proportional to the square of S(S+1) (Eq. 5), which is 9/16 for the protons feeling copper and 225/16 for those feeling cobalt. Curie relaxation is also sensitive to the sixth power of the metal proton distance. Another source of distance information is the analysis of the nuclear T_1 values which, according to eq. 3, are also dependent on the sixth power of the electron proton distance. The problem could be solved easily in principle. In a qualitative way, the assignments can be made for 60–70% of the signals. In order to have a complete assignment, a more quantitative treatment is needed. The main problem with this approach is that the unpaired electron is delocalized onto the ligand, and even a small amount of spin density on a carbon atom can give a sizeable dipolar contribution to the relaxation of the corresponding C–H signal due to the small distance involved. These are called ligand–centered effects; they are particularly significant for protons relatively far from the metal but connected to it through chemical bonds i.e., in the present case, for the so–called meta–like histidine protons. In fact, as the distance from the metal increases the metal–centered dipolar contribution to T_{1M}^{-1} decreases and ligand–centered effects may dominate. Therefore they make the T_1 ratio between meta and ortho–like histidine protons smaller than expected. The ortho–like protons are very close to the paramagnetic center and metal–centered dipolar coupling is still strong. Furthermore, several studies (33,34) have shown that for nuclei about 3 Å

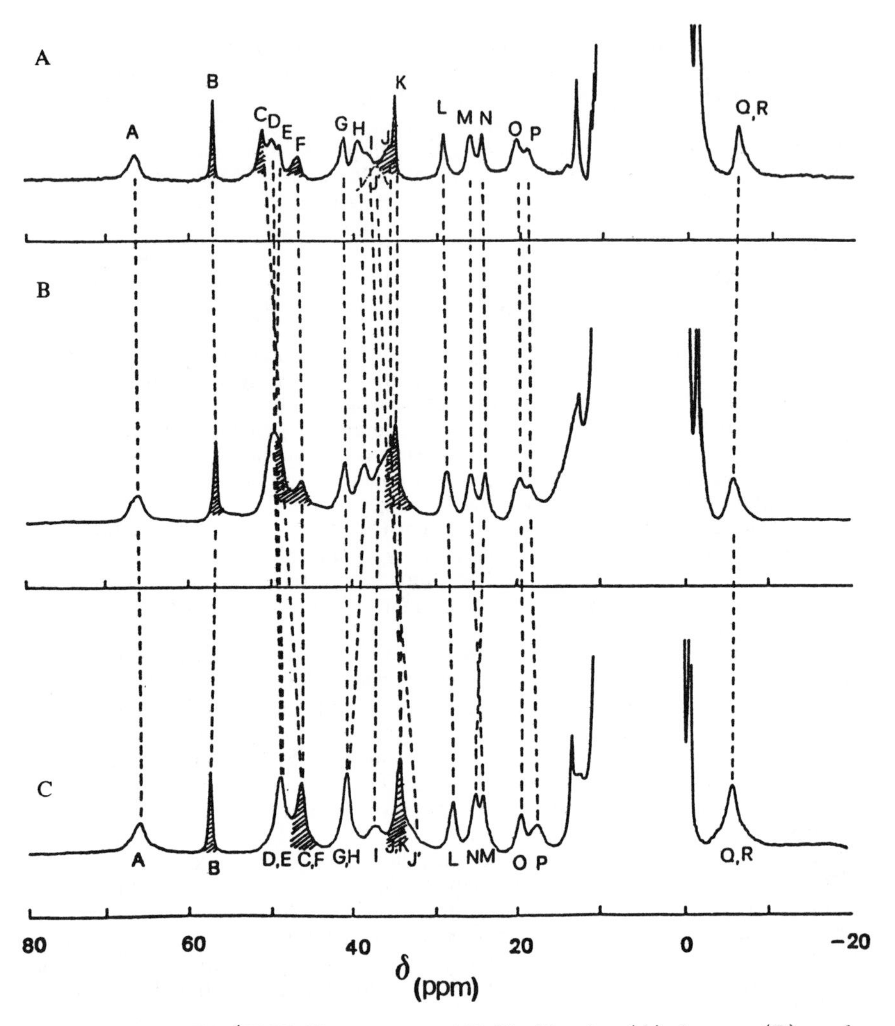

Figure 1 300 MHz ^1H NMR spectra at 303 K of bovine (A), human (B), and yeast (C) Cu_2Co_2SOD derivatives in 10 mM acetate buffer, pH 5.6. The correlation of individual signals of human and yeast derivatives with those of the bovine isoenzyme is shown by dashed lines (32). Shaded signals disappear in D_2O.

from the metal and for π–type spin delocalization the ligand–centered dipolar contribution essentially compensates for the decrease in the metal–centered contribution, so that the total T_1^{-1} value is close to the value obtained by a metal–centered point–dipole approximation (equation 3). For the meta–like protons we have then tentatively attributed a constant value to the ligand–centered dipolar contribution, covering 85–90% of the total relaxation rate. With this procedure, a satisfactory agreement between calculated and experimental distances is found. In summary, by using equation 3 corrected with the appropriate coefficients, equation 5, and the distances taken from the structure of the native protein, all the experimental data have been reproduced (i.e. T_1 and T_2, the latter at magnetic fields between 60 and 400 MHz) by using as parameters τ_s(Cu), τ_s(Co), τ_r of the protein, and a single value for the ligand–centered contribution at each meta–like proton (31). A τ_s for copper of 1.7×10^{-11} s and for cobalt of 1.2×10^{-12} s are obtained. The copper(II) ion in the native SOD has a τ_s of 2×10^{-9} s (28). Its decrease is due to magnetic coupling with the fast–relaxing cobalt(II) ion.

The proposed assignments were verified by analyzing the pattern of the chemical shifts when anions were allowed to interact with the protein; the three ring signals of every histidine residue should experience a similar behavior. When for example a solution of the Cu_2Co_2–protein is titrated with azide, the signals move just as expected for the case of rapid exchange between free and bound anion (Figure 2). One result of anion binding is that His–44 (numbering of the bovine isoenzyme) is removed from coordination since its proton signals move towards the diamagnetic region (29,31). The bridging histidine experiences larger isotropic shifts, possibly as a result of a less strained coordination. The other two histidines bound to copper undergo a small rearrangement, whereas the histidines bound to cobalt are almost insensitive to the binding of azide. NCS^- and NCO^- behave in a quite similar way (29). CN^- binds in a slow exchange regime above 200 MHz, so that the original spectrum decreases in intensity and a new spectrum appears (35). The latter is very similar to the limit spectrum obtained with azide. The assignment has been verified through saturation transfer techniques. Phosphate does not alter the coordination sphere of the two metal ions (36). In the case of detachment of His–44, the knowledge of the correlation time allowed an estimate to be made, through equation 3, of the new position of the histidine ring after anion binding.

These studies, extended to bovine, human, yeast, and to mutants human SOD Lys–143 and Ile–143 (which substitute the native Arg) have shown strict conservation of the active cavity as far as the metal ion is concerned (37).

Other Systems

The approach described above is in principle quite general. For example a Cu_2Ni_2SOD derivative in which nickel(II) replaces zinc has been recently reported (13). Nickel(II) has an $S = 1$ ground state and short electronic relaxation times when tetrahedral, even shorter than those of tetrahedral cobalt(II). It follows that the ^{1}H NMR spectra are sharper and the T_1 values are longer in this system than in the CuCo case, whereas the signals of protons feeling copper are essentially in the same position (Figure 3) (38). The analysis of copper signals provides essentially the same picture as that obtained for the CuCo derivative, but with shorter τ_s. In the case of the signals of histidines bound to nickel(II) it appears that the ligand–centered

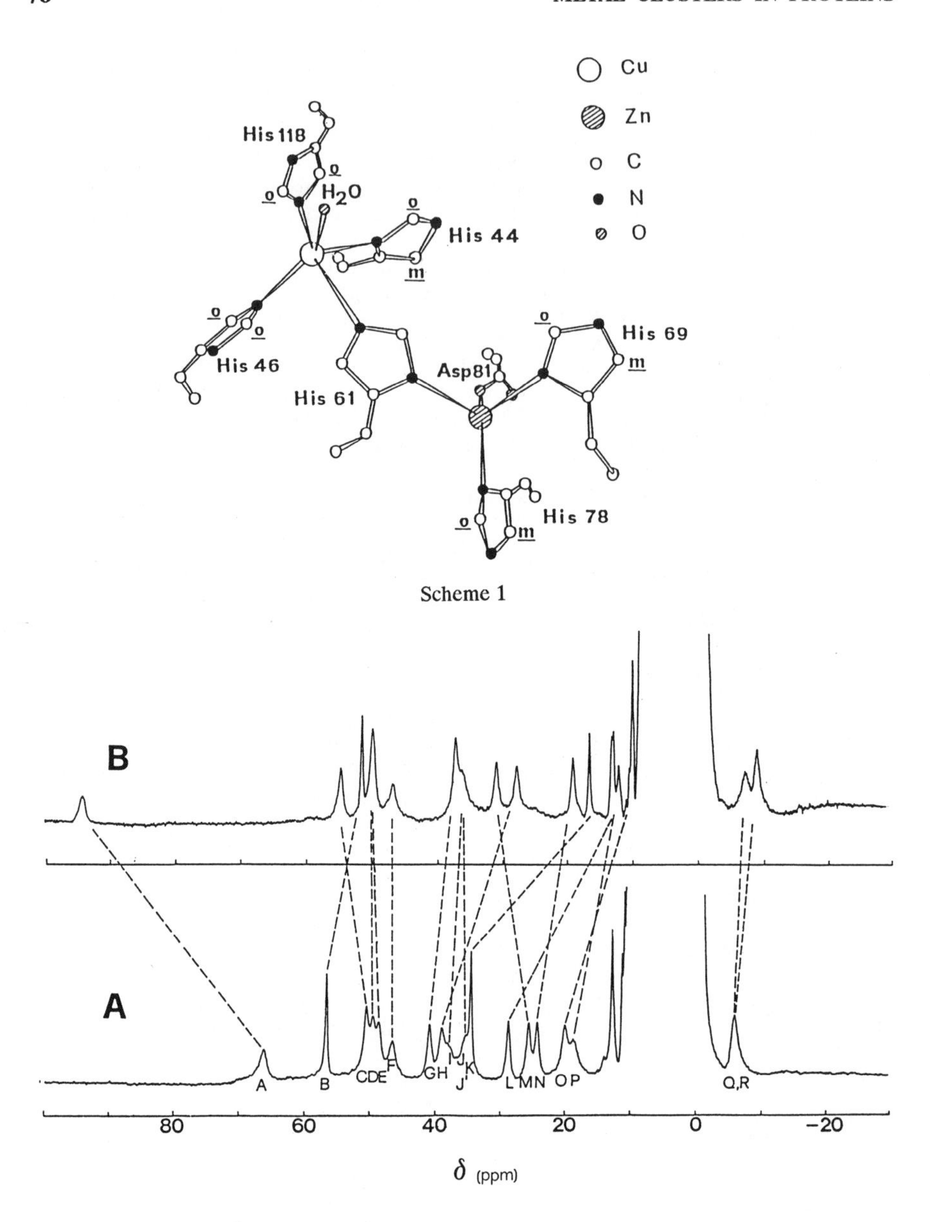

Scheme 1

Figure 2 300 MHz ^{1}H NMR spectra at 303 K in 10 mM acetate buffer, pH 5.6, of bovine Cu_2Co_2SOD derivative (A) and of bovine Cu_2Co_2SOD + N_3^- 2 x 10^{-1} M (B). The correlation of signals in the unligated and fully bound derivative is shown by dashed lines. (Reproduced from reference 31. Copyright 1987 American Chemical Society.)

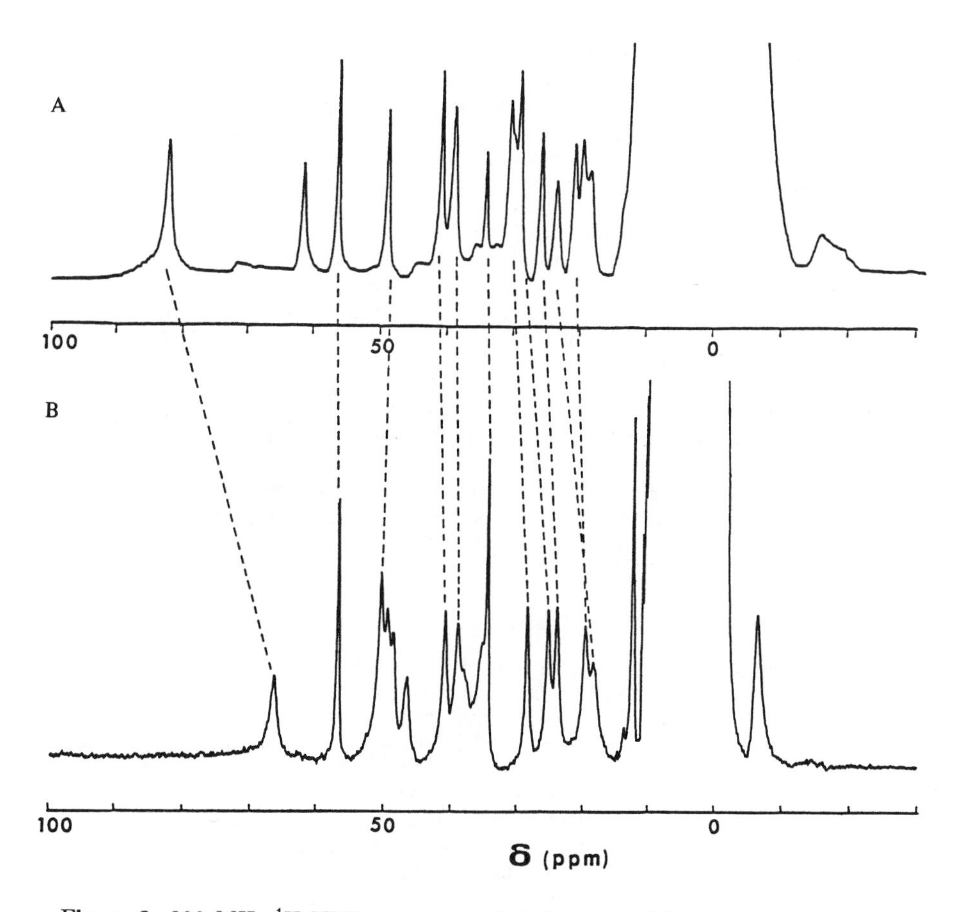

Figure 3 300 MHz ^1H NMR spectra at 303 K in 10 mM acetate buffer, pH 5.6 of bovine Cu_2Co_2SOD (A) and Cu_2Ni_2SOD (B) derivatives. The correlation of signals due to protons of residues bound to copper in the two derivatives is shown by dashed lines (38).

effects are minor compared to the case of cobalt. An interesting feature of this system is a decrease in T_1 and increase in linewidth with increasing magnetic field, as a result of a field dependence of τ_S of nickel. This property of nickel(II) had already been reported (10).

Another example is provided by the $Cu_2Co_2Mg_2$ alkaline phosphatase (AP) derivative. The copper ion is in the A–site, cobalt in the B–site and magnesium in the C–site, whose known ligands are shown in Figure 4. Since there is no bridging ligand between copper and cobalt, the magnetic coupling is expected to be small (14). Despite this, the electronic relaxation time of the copper(II) electron is strongly reduced, and the signals of histidines coordinated to copper(II) are observed, particularly the histidine NH's, which are shaded in Figure 4.

For the case of reduced Fe_2S_2 ferredoxins, the iron(II) ion shortens the electronic relaxation times of iron(III) through magnetic coupling ($J \simeq 100$ cm^{-1}) so that sharp signals for protons of both iron(II) and iron(III) ligands are observed (Figure 5) (39). Owing to the large magnetic coupling, the magnetic susceptibility of the dimer at room temperature, as well as the hyperfine coupling, is a fraction of the sum of those of the single ions. The change in population of the magnetically coupled levels, each with a hyperfine coupling value that can be easily calculated (40), accounts for the Curie temperature dependence of the shifts of the signals of protons feeling iron(III) and for the anti–Curie temperature dependence of the shifts of the signals of protons feeling iron(II) around room temperature (40,41).

These considerations are the basis for the understanding of systems like porcine uteroferrin (19), acid phosphatases and methemerythrin (42), some of which will be discussed in this book.

Concluding Remarks and Perspectives

When a metal ion with short electronic relaxation times is placed near to a metal ion with long electronic relaxation times, the T_{1e} of the latter time will become shorter. This effect is significant even for small magnetic coupling. As a result the NMR signals of nuclei feeling the latter ion will become sharper and may become detectable (even in the absence of chemical exchange). Several systems with these properties exist in nature; others can be artificially prepared in order to understand this type of interaction from the point of view of time–dependent phenomena. Erythrocyte SOD provides a nice framework to incorporate magnetically coupled pairs of metal ions. Alkaline phosphatase is another interesting system in which it has been possible to place copper close to cobalt (14). Lectins are other systems with two metal ions bridged by carboxylate ligands (43).

The aims of this study were to observe isotropically shifted signals for metal ions like copper(II), which usually give signals broadened beyond detection, and to relate the observed shifts and relaxation times to those of the uncoupled ions in order to understand the phenomena in theoretical terms. This approach allows the power of the NMR technique to fully exploit paramagnetic species and obtain information on spin delocalization, chemical bonding and so on. It is likely that the theory also applies to coupled metal ion–radical systems like those proposed for derivatives of peroxidases (compound I), which contain iron(IV) and a heme radical (44).

A major aspect of the problem is the relationship between J and the effect on the electronic relaxation rates of the slow–relaxing metal ion. More experimental data are probably necessary to make general statements. Finally, we have always assumed the zero field splitting of uncoupled ions to

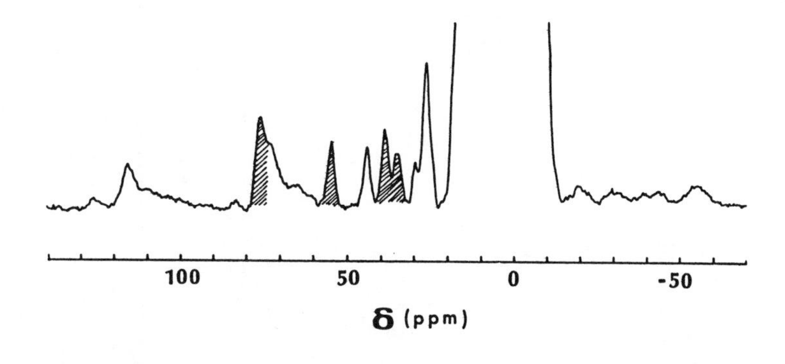

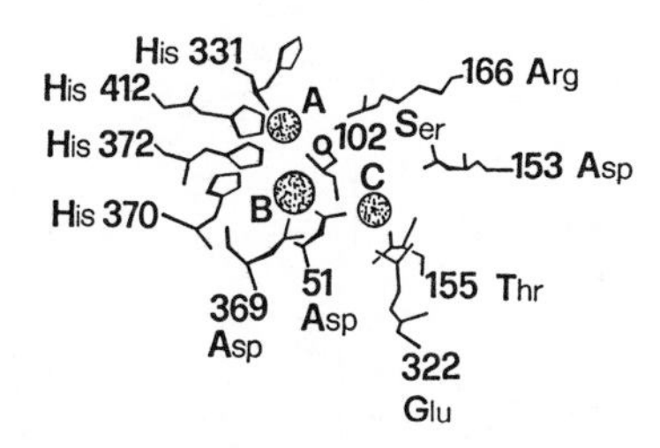

Figure 4 90 MHz ^{1}H NMR spectrum of $Cu_2Co_2Mg_2AP$ at 301 K at pH 6 (14). Shaded signals disappear in D_2O. The active–site scheme is also shown. (Reproduced with permission from Reference 45. Copyright 1986 Birkhäuser.)

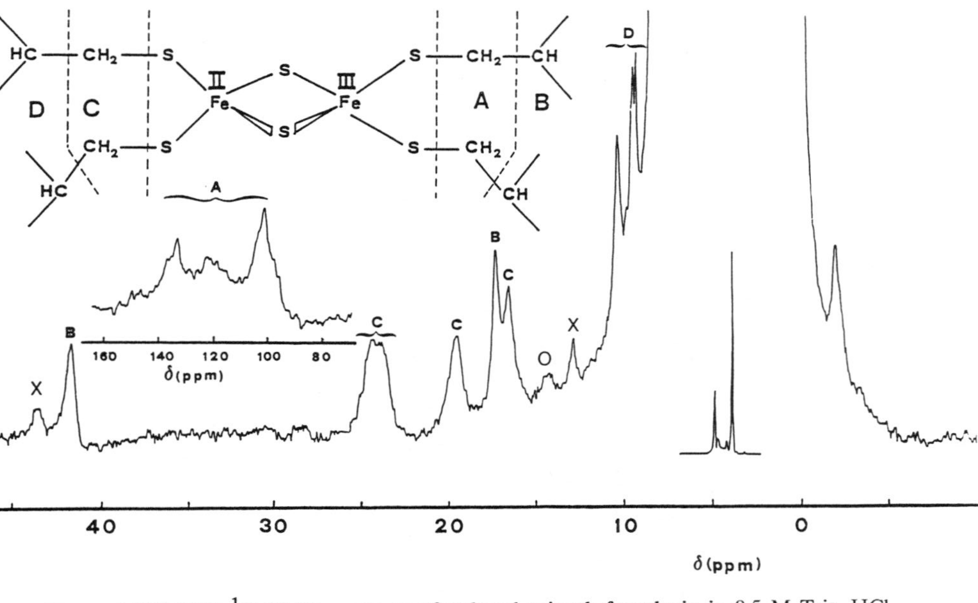

Figure 5 300 MHz ^{1}H NMR spectrum of reduced spinach ferredoxin in 0.5 M Tris–HCl buffer, pH* 7.4, 297 K: (o) signals from residual oxidized protein; (x) signals from impurities. A scheme of the dimetallic cluster in the Fe_2S_2 ferredoxin with the proposed assignment is also shown. (Reproduced from reference 39. Copyright 1984 American Chemical Society.)

be << J. Furthermore we have neglected the zero field splitting in the magnetic coupled levels which may in principle require different τ_S value for each electronic spin transition. These aspects, which are being taken into consideration in further refinements, dramatically complicate the problem when the aim is the full understanding of the electronic relaxation properties.

Literature Cited

1. Valentine, J.S.; Pantoliano, M.W. In *Copper Proteins*; Spiro, T.G., Ed.; Wiley: New York, 1981; Vol. 3, Ch. 8.
2. Fee, J.A. In *Metal Ions in Biological Systems*; Sigel, H., Ed.; Dekker: New York, 1981; Vol. 13, p 259.
3. Coleman, J.E.; Gettins, P. *Adv. Enzymol.* **1983**, *55*, 381; Wyckoff, H.W.; Handschumacher, M.; Murty, K.; Sowadski, J.M. *Adv. Enzymol.* 1983, *55*, 453.
4. Bertini, I.; Luchinat, C.; Viezzoli, M.S. In *Zinc Enzymes*; Bertini, I.; Luchinat, C; Maret, W.; Zeppezauer, M., Eds.; Birkhauser: Boston, 1986; Ch. 3.
5. *Iron–Sulfur Proteins*; Spiro, T.G., Ed.; Wiley: New York, 1982.
6. Antanaitis, B.C.; Aisen, P. *Adv. Inorg. Biochem.* **1983**, *5*, 111.
7. Que, L., Jr.; Maroney, M.J. In *Metal Ions in Biological Systems*; Sigel, H. Ed.; Dekker, New York, 1987; Vol. 21, p. 87.
8. Sanders–Loehr, J.; Loehr, T.M. *Adv. Inorg. Biochem.* **1979**, *1*, 235.
9. Klotz, I.M.; Kurtz, D.M. *Accts. Chem. Res.* **1984**, *17*, 16.
10. Bertini, I.; Luchinat, C. *NMR of Paramagnetic Species in Biological Systems*, Benjamin/Cummings: Boston, 1986.
11. Fee, J. A. *J. Biol. Chem.* **1973**, *248*, 4229.
12. Calabrese, L., Cocco, D.; Morpurgo, L.; Mondovi', B.; Rotilio, G. *Biochim. Biophys. Acta.* **1972**, *263*, 827.
13. Ming, L.–J.; Valentine, J.S. *J. Am. Chem. Soc.*, **1987**, *109*, 4426.
14. Banci, L.; Bertini, I.; Luchinat, C.; Viezzoli, M.S.; Wang, Y., in preparation.
15. Bertini, I.; Banci, L.; Brown, R.D. III; Luchinat, C.; Koenig, S.H., submitted for publication
16. Benelli, C.; Dei, A.; Gatteschi, D. *Inorg. Chem.* **1982**, *21*, 1284.
17. Owens, C.; Drago, R.S.; Bertini, I; Luchinat, C.; Banci, L. *J. Am. Chem. Soc.* **1986**, *108*, 3298.
18. Bertini, I.; Luchinat, C.; Owens, C.; Drago, R.S. *J. Am. Chem. Soc.* **1987**, *109*, 5208.
19. Lauffer, R.B.; Antanaitis, B.C.; Aisen, P.; Que, L., Jr. *J. Biol. Chem.* **1983**, *258*, 14212.
20. Fee, J.A.; Briggs, R.G. *Biochim. Biophys. Acta* **1975**, *400*, 439.
21. Bertini, I.; Lanini, G.; Luchinat, C.; Mancini, M.; Spina, G. *J. Magn. Reson.* **1985**, *63*, 56.
22. La Mar, G.N.; Horrocks, W. de W., Jr.; Holm, R.H., Eds. *NMR of Paramagnetic Molecules;* Academic: New York, 1973.
23. Solomon, I. *Phys. Rev.* **1955**, *99*, 559.
24. Bertini, I.; Luchinat, C.; Kowalewski, J. *J. Magn. Reson.* **1985**, *62*, 235.
25. Gueron, M. *J. Magn. Reson.* **1975**, *19*, 58.
26. Vega, A.J.; Fiat, D. *Mol. Phys.* **1976**, *31*, 347.
27. Tainer, J.A.; Getzoff, E.D.; Beem, K.M.; Richardson, J.S.; Richardson, D.C. *J. Mol. Biol.*, **1982**, *160*, 181.

28. Bertini, I.; Briganti, F.; Luchinat, C.; Mancini, M.; Spina, G. *J. Magn. Reson.* **1985**, *63*, 41.
29. Bertini, I.; Lanini, G.; Luchinat, C.; Messori, L.; Monnanni, R.; Scozzafava, A. *J. Am. Chem. Soc.*, **1985**, *107*, 4391.
30. Morgenstern–Badarau, I.; Cocco, D.; Desideri, A.; Rotilio, G.; Jordanov, J.; Dupre, N. *J. Am. Chem. Soc.* **1986**, *108*, 300.
31. Banci, L.; Bertini, I.; Luchinat, C.; Scozzafava, A. *J. Am. Chem. Soc.*, **1987**, *109*, 2328.
32. Ming, L.–J.; Banci, L.; Luchinat, C.; Bertini, I.; Valentine, J.S., in preparation.
33. Doddrell, D.M.; Healy, P.C.; Bendall, M.R. *J. Magn. Reson.* **1978**, *29*, 163.
34. Gottlieb, H.P.W.; Barfield, M.; Doddrell, D.M. *J. Chem. Phys.* **1977**, *67*, 3785.
35. Banci, L.; Bertini, I.; Luchinat, C.; Monnanni, R.; Scozzafava, A. *Inorg. Chem.* in press.
36. Mota de Freitas, D.; Luchinat, C.; Banci, L.; Bertini, I.; Valentine, J.S. *Inorg. Chem.* **1987**, *26*, 2788.
37. Banci, L.; Bertini, I.; Luchinat, C.; Hallewell, R.A., submitted for publication
38. Ming, L.–J; Banci, L.; Bertini, I.; Luchinat, C.; Valentine, J.S., in preparation
39. Bertini, I.; Lanini, G.; Luchinat, C. *Inorg. Chem.* **1984**, *23*, 2729, and references therein.
40. Unpublished results from our laboratory.
41. Dunham, W.R.; Palmer, G.; Sands, R.H.; Bearden, A.J. *Biochim. Biophys. Acta* **1971**, *253*, 373.
42. Maroney, M.J.; Kurtz, D.M.; Nocek, J.M.; Pearce, L.L.; Que, L., Jr. *J. Am. Chem. Soc.* **1986**, *108*, 6871.
43. Bertini, I.; Viezzoli, M.S.; Luchinat, C.; Stafford, E.; Cardin, A.D.; Behnke, W.D.; Bhattacharyya, L.; Brewer, C.F. *J. Biol. Chem.*, in press.
44. See for example: Morishima, I.; Ogawa, S. *Biochemistry*, **1978**, *17*, 4384; La Mar, G.N.; De Ropp, J.S.; Smith, K.M.; Langry, K.C. *J. Biol. Chem.* 1981, *256*, 237; Loew, G.H.; Herman, Z.S. *J. Am. Chem. Soc.*, **1980**, *102*, 6174
45. Coleman, J.E.; Gettins, P., In *Zinc Enzymes*; Bertini, I., Luchinat, C., Maret, W., Zeppezauer, M., Eds.; Birkhäuser; Boston, 1986; pp. 78–99.

RECEIVED February 18, 1988

Chapter 5

Models for Copper Proteins

Reversible Binding and Activation of Dioxygen and the Reactivity of Peroxo and Hydroperoxo Dicopper(II) Complexes

Zoltan Tyeklar, Phalguni Ghosh, Kenneth D. Karlin[1], Amjad Farooq, Brett I. Cohen, Richard W. Cruse, Yilma Gultneh, Michael S. Haka, Richard R. Jacobson, and Jon Zubieta

Department of Chemistry, State University of New York at Albany, Albany, NY 12222

Copper coordination complexes serving as models for copper-containing proteins which either reversibly bind or activate dioxygen are described. These include dinuclear copper(I) compounds which react with O_2 resulting in the hydroxylation of the arene-containing ligand, thus mimicking the action of tyrosinase. Dicopper(I) complexes which can duplicate the reversible dioxygen binding action of hemocyanin are also described. A related set of hydroperoxo and acylperoxo dicopper(II) complexes can also be generated. Comparisons of the complexes described to related model systems are made and the biological relevance is discussed.

Copper compounds have been established to be important and versatile catalysts for dioxygen-mediated reactions in both biological and synthetic systems (1-11). Studies on non-protein, low molecular weight compounds designed to mimic structural and spectral features as well as the chemistry of enzyme active sites have attracted a great deal of attention; these can and have contributed to the understanding of reversible binding and activation of molecular oxygen by copper-containing proteins (1-9,11,12). In addition, information gained in these investigations may lead to the discovery of synthetic systems capable of effecting mild, selective oxidations of organic substrates with dioxygen, and the improvement of existing catalysts of synthetic chemical or even industrial importance.

Our biomimetic investigations have focused on the metalloproteins hemocyanin (Hc) (11-17) and tyrosinase (11,12,14,16,18,29), which contain two copper ions in their active center. The function of hemocyanin is to bind and transport dioxygen in the hemolymph of molluscs and arthropods. Studies employing EXAFS spectroscopy have shown that in the deoxy form, two (19-21) or three (13,21) imidazole units from protein histidine residues coordinate to each cuprous ion. Upon addition of O_2 to give oxy-Hc, considerable changes take place in the coordination sphere giving rise to tetragonally coordinated Cu(II) ions

[1]Correspondence should be addressed to this author.

0097–6156/88/0372–0085$06.00/0

bridged by an exogenous peroxide (O_2^{2-}) ligand. An additional bridging ligand such as an oxygen atom from a tyrosine phenolate, serine hydroxyl, or water (i.e. as H_2O or OH^-) had been suggested; this could account for the presence of a 420 nm optical absorption band in oxy-Hc and the observed diamagnetism resulting from strongly antiferromagnetically coupled ("super-exchange" via the bridging ligands) Cu(II) ions (11,22,41). The recent crystallographic studies on a deoxy-Hc (23-25) indicate that three imidazole groups are ligated to each copper(I) ion. Along with amino acid sequence homology studies (23), phenolate and probably other protein-derived groups appear to be ruled out as possible bridging ligands.

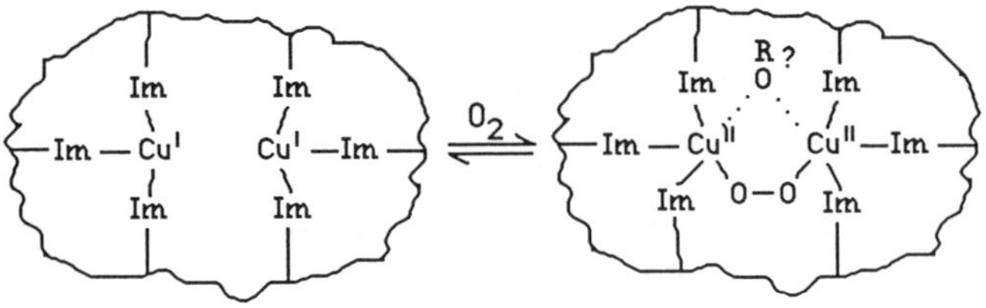

deoxy-Hc
Cu...Cu = 3.8 Å

oxy-Hc
Cu...Cu = 3.6 Å

Tyrosinase is a monooxygenase which catalyzes the incorporation of one oxygen atom from dioxygen into phenols and further oxidizes the catechols formed to *o*-quinones (oxidase action) . A comparison of spectral (EPR, electronic absorption, CD, and resonance Raman) properties of oxy-tyrosinase and its derivatives with those of oxy-Hc establishes a close similarity of the active site structures in these proteins (26-29). Thus, it seems likely that there is a close relationship between the binding of dioxygen and the ability to "activate" it for reaction and incorporation into organic substrates. Other important copper monooxygenases which are however of lesser relevance to the model studies discussed below include dopamine β-hydroxylase (16,30) and a recently described copper-dependent phenylalanine hydroxylase (31).

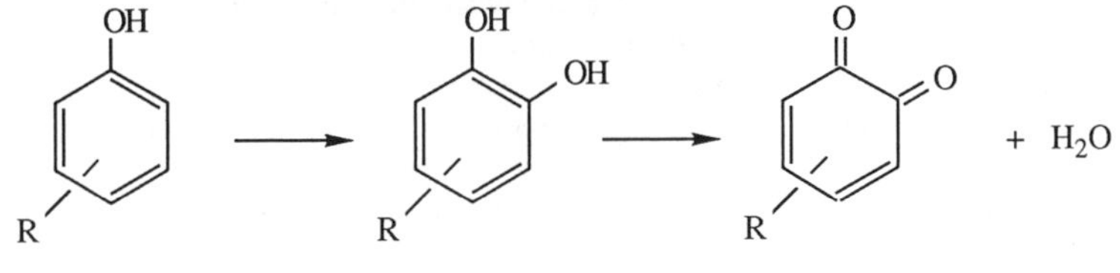

+ H_2O

A Copper Monooxygenase Model Reaction

In our studies with dinuclear copper(I) and copper(II) complexes (32), we have employed the dinucleating ligand, XYL (**1**), in which two tridentate donor units are connected by a *m*-xylyl group (32-35). Ligand **1** was fashioned following the approach of Martell (36) and Bulkowski, Osborn and coworkers (89), in which ligands capable of placing two metal ions in close proximity (i.e., dinucleating ligands), were formed by appending multidentate groups to a xylyl moiety. Here, pyridine donors are utilized to mimic the aromatic nitrogen ligands (imidazole from histidine) known to bind to copper in the protein active sites. Dinucleating ligand **1** also contains a tridentate unit for Cu(I), a feature which would be desirable in modeling a three-coordinate deoxy-Hc (reduced) active site (35).

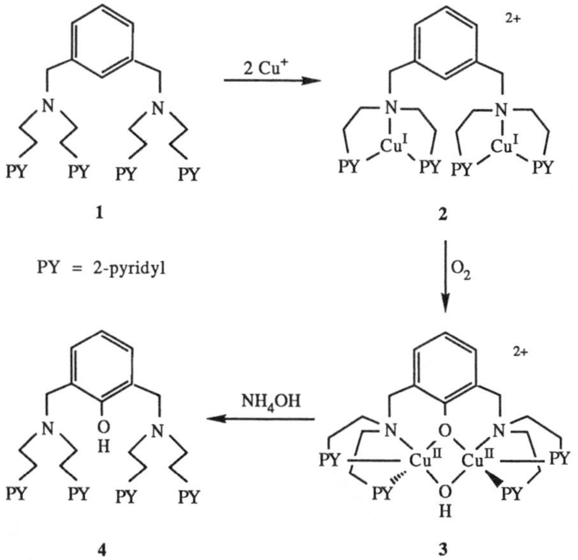

PY = 2-pyridyl

1 2 3 4

Ligand **1** does form a dinuclear copper(I) compound, **2**, with roughly planar three-coordinate moieties. Complex **2** reacts quantitatively with O_2 in the molar ratio 1:1 in dimethylformamide or dichloromethane, resulting in the hydroxylation of the *m*-xylyl connecting unit in **1** and the formation of a phenoxo and hydroxo-bridged dinuclear complex **3** (33). The structure of **3** consists of two crystallographically independent but very similar copper(II) coordination environments (Figure 1). The geometry around each copper is essentially undistorted from square-based pyramidal, and Cu...Cu = 3.1 Å. The copper(II) ions can be leached out of complex **3** to produce the free phenol **4**, completing the sequence involving the copper-mediated hydroxylation of an arene. The observed oxygen atom insertion into an aromatic C-H bond and the stoichiometry of the reaction **2**→**3** is reminiscent of the action of the copper monooxygenase tyrosinase; the reaction thus serves as a model system which may be of use in providing insights into possible mechanisms of copper-mediated dioxygen activation. While other copper complex systems have been reported to exhibit monooxygenase activity (11), the present system enjoys a number of advantages for further study, including the clean, high yield (> 90 %) reaction with O_2 and definitive structural characterization of both starting compound and the oxygenated product.

We have also found that reaction of hydrogen peroxide with the dinuclear copper(II) complex containing ligand **1** (prepared *in situ*) also results in the hydroxylation of **1** to give the same phenoxo-bridged dicopper(II) product (37). This observation suggests that a peroxo copper(II) species $[Cu(II)_2(O_2^{2-})]$ is a common intermediate in pathways developing either from $Cu(I)_2/O_2$ or $Cu(II)_2/H_2O_2$. Evidence for such an intermediate has also been obtained using copper(I) complexes of the ligands XYL-F (38) and N5PY2 (39). The ligand XYL-F is the 2-fluoro-substituted derivative of **1** (the 2-position is the site of hydroxylation). The ligand N5PY2 has a five-membered alkyl chain between the two tridentate donor units, instead of the *m*-xylyl group in **1**. The dinuclear

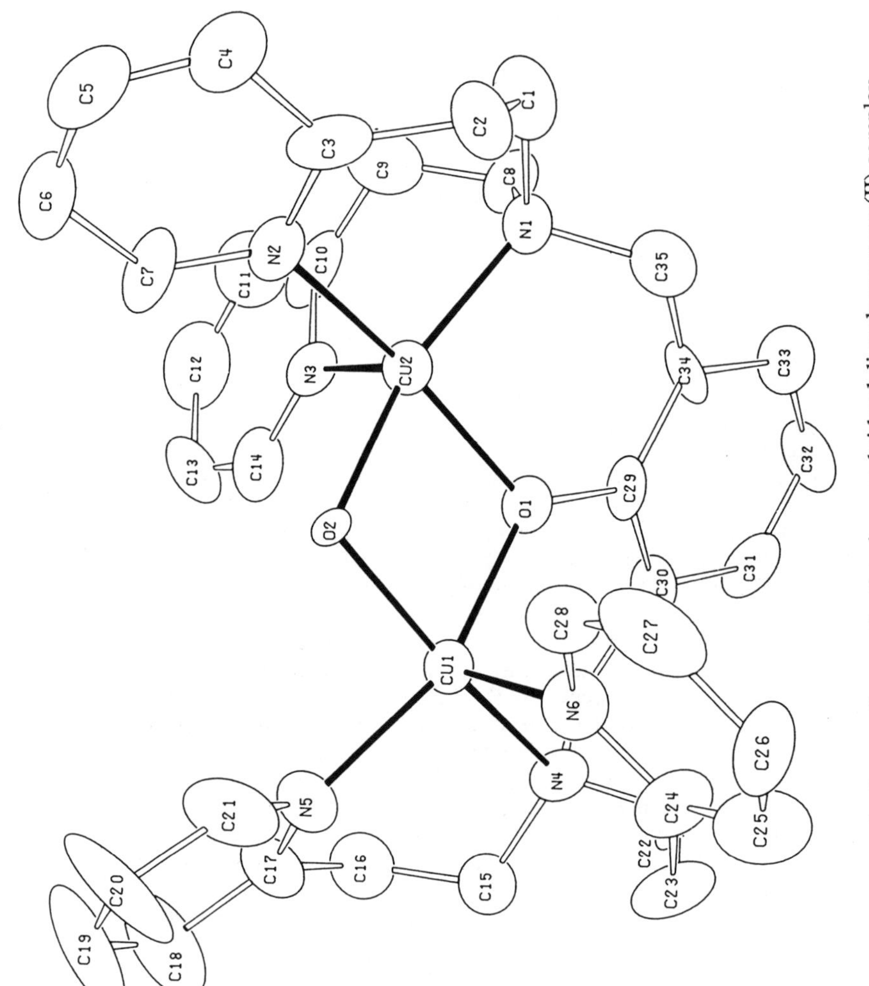

Figure 1. ORTEP diagram for the phenoxo-bridged dinuclear copper(II) complex $[Cu^{II}_2(XYL\text{-}O\text{-})(OH)]^{2+}$ (**3**), reference 33.

copper(I) complexes of these ligands react with O_2 in dichloromethane at -80 °C to give metastable species which we have characterized spectroscopically (39). The electronic spectra for both complexes are very similar and quite distinctive, and the stoichiometry of the reaction as determined manometrically is Cu:O_2 = 2:1. The combined data along with comparisons to other closely related systems (40,41) suggest that the low-temperature oxygenation products are peroxodicopper(II) complexes. Hydroxylation of the aromatic ring in XYL-F or of the aliphatic backbone in N5PY2 does not take place apparently because of the presence of the greater strength of the C-F bond, and the inert aliphatic C-H backbone, respectively.

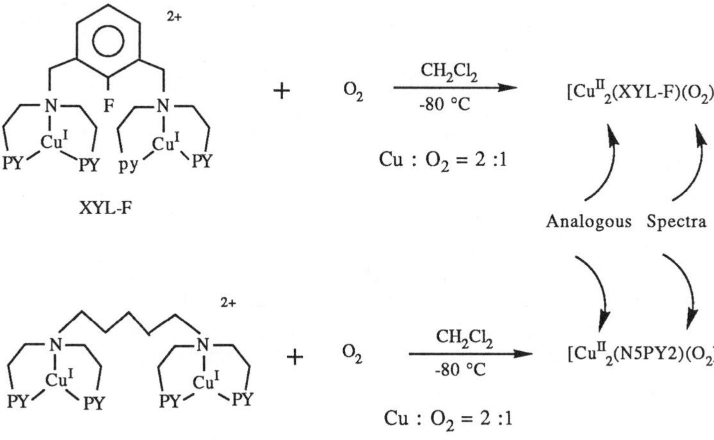

XYL-F + O_2 $\xrightarrow[-80\ °C]{CH_2Cl_2}$ $[Cu^{II}_2(XYL\text{-}F)(O_2)]^{2+}$

Cu : O_2 = 2 :1

Analogous Spectra

N5PY2 + O_2 $\xrightarrow[-80\ °C]{CH_2Cl_2}$ $[Cu^{II}_2(N5PY2)(O_2)]^{2+}$

Cu : O_2 = 2 :1

By contrast, when the dicopper(I) complex of XYL-CH_3 (**5**) is reacted with dioxygen, hydroxylation of the ligand does occur with concomitant migration of the methyl group in the 2-position on the *m*-xylyl group, loss of one PY2 tridentate unit, **7**, and formation of formaldehyde (42). This is the first example of copper-mediated hydroxylation-induced alkyl group migration (43,44). The process is reminiscent of the "N.I.H. shift" (45-47), where electrophilic attack results in hydroxylation-induced migrations in the iron-containing phenylalanine hydroxylase and cytochrome P-450 monooxygenases. In the present case, the observed behavior provides us an important new insight into the mechanism of hydroxylation observed in the conversions **5**→**6** and **2**→**3**, suggesting that these reactions proceed by ***electrophilic*** attack of a $Cu(I)_2/O_2$ derived species upon the aromatic ring.

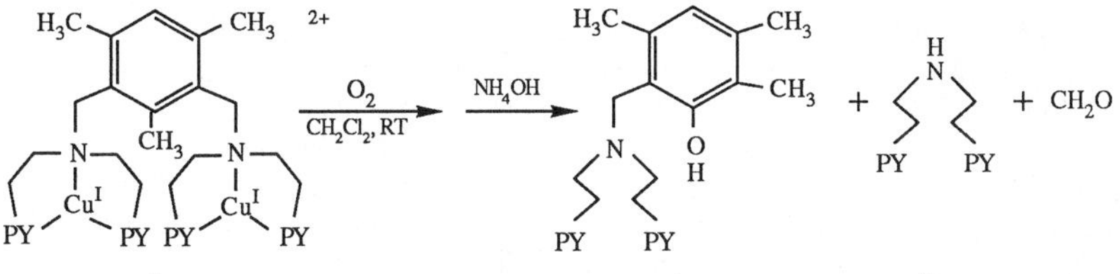

5 $\xrightarrow[CH_2Cl_2,\ RT]{O_2}$ $\xrightarrow{NH_4OH}$ **6** + **7** + CH_2O

We have recently prepared a number of 5-substituted (e.g., $-NO_2$ or -*t*-Bu *para* to the site of hydroxylation) XYL ligands, and their dicopper(I) complexes also give high yields of the corresponding hydroxylated dicopper(II) products (38). However, it seems that certain changes on the donor groups in **1** eliminate or drastically lower the monooxygenase activity. Thus, when a 6-methyl-2-pyridyl group is utilized instead of 2-pyridyl, the dicopper(I) complex is inert to O_2 (38). Also, when one of the four $-CH_2CH_2PY$ (PY = 2-pyridyl) groups in **1** is replaced by a $-CH_2PY$ unit, the yield of the hydroxylation is only *ca.* 40% and there is no hydroxylation at all when all four ligand arms consist of the $-CH_2PY$ unit.

Other researchers have also studied systems related to these *m*-xylyl dicopper compounds. Sorrell and coworkers (48) have described the dinuclear copper(I) complex of a ligand which is structurally analogous to **1** with 1-pryazolyl groups replacing the 2-pyridyl groups as donors for copper. No ligand hydroxylation occurs upon oxygenation of the pyrazolyl-containing compound and a dinuclear dihydroxo-bridged copper(II) complex is isolated instead (i.e. four-electron reduction of dioxygen took place). Casella and coworkers (49) have also described dicopper(I) complexes of analogous *m*-xylyl containing ligands having Schiff base nitrogen donors; their three-coordinate dicopper(I) complexes are completely inert to dioxygen. However, using *m*-xylyl dinucleating ligands containing two nitrogen donors per Cu(I) ion, Casella and Rigoni (50) observe hydroxylation in certain instances to give a phenoxo and hydroxo-bridged dicopper(II) compound. The course of the reaction depends critically upon the R group of the imidazole ligand donor; hydroxylation occurs with R = Me (monooxygenase stoichiometry $Cu:O_2 = 2:1$) but irreversible oxidation takes place with R = H ($Cu:O_2 = 4:1$).

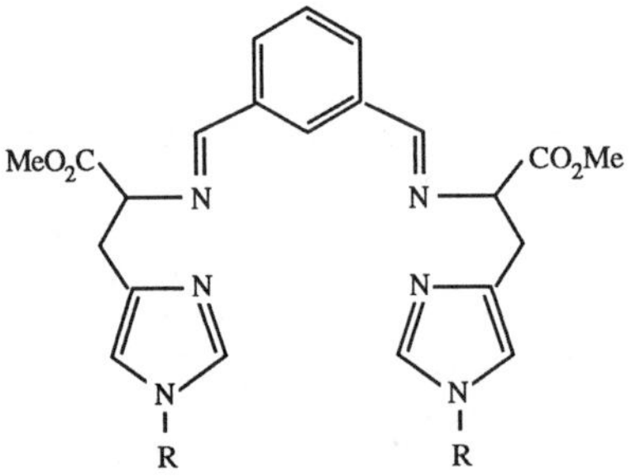

L1 R = H
L2 R = Me

$$[Cu^{I}_2(L1)]^{2+} + 1/2\ O_2 \longrightarrow [\quad\quad]$$ (no hydroxylation)

$$[Cu^{I}_2(L2)]^{2+} + O_2 \longrightarrow [Cu^{II}_2(L2\text{-}O^-)(HO^-)]^{2+}$$

At present, we do not completely understand why only some of these very similar *m*-xylyl dicopper(I) complexes systems described above undergo ligand oxygenation reactions. However, based on the results outlined above, we can speculate on a number of aspects of this O_2-activation process. Our studies implicate the presence of a copper-dioxygen (peroxo dicopper(II)) adduct as an intermediate in the oxygenation reaction and more recent kinetic studies (51) further support this conclusion. This adduct then either directly or via some further intermediate undergoes an electrophilic attack of the arene. The unique nature of this very fast reaction **2**→**3**, and the observed inability to intercept the active oxygenating species by the addition of external substrates (52) suggests that the active copper/oxy intermediate is formed in a proximity to the xylyl substrate suitable for reaction. Since it seems unlikely that replacing a pyridine by a pyrazole or Schiff base group would change the proximity of a copper/oxy intermediate to the xylyl ring, other explanations are required to account for the different courses of reactions in the these latter systems.

The answer perhaps lies in the variation in electronic environment provided by these different donor groups as reflected by the difference in the $E_{1/2}$ (cyclic voltammetry in CH_3CN), which is 0.15 V more negative for **2** than for Sorrell's pyrazolyl complex (48). Thus, peroxo dicopper(II) intermediates formed in these analogous systems will have different characteristics, since dioxygen binding to Cu(I) is a redox process involving at least some degree of electron transfer from copper(I) to dioxygen (11).

A related explanation has to do with the stability of the peroxo dicopper(II) intermediate, since it will either attack the substrate or 'decompose'; the kinetics of formation of the intermediate relative to those of the ensuing decomposition reactions will thus be important. Nelson (53) and Sorrell (90) have both described systems that undergo a $Cu:O_2 = 4:1$ reaction stoichiometry for dicopper(I) complexes where they propose that 'degradation' of the peroxo dicopper(II) intermediate proceeds by the fast bimolecular two-electron transfer from a second dicopper(I) molecule to the putative peroxo-dicopper(II) intermediate to give an aggregated oxo-copper(II) product. [The latter may form hydroxo-Cu(II) species in the presence of protic solvents].

$$\mathrm{Cu(I)...Cu(I)} \; + \; O_2 \rightleftharpoons \mathrm{Cu(II)\text{-}(O_2^{2-})\text{-}Cu(II)} \qquad k_1/k_{-1}$$

$$\mathrm{Cu(II)\text{-}(O_2^{2-})\text{-}Cu(II)} \;\; \mathrm{Cu(I)...Cu(I)} \rightarrow 2\,[\mathrm{Cu(II)\text{-}O\text{-}Cu(II)}]_n \qquad k_2$$

The stability of the intermediate will depend in part on the relative values of k_1 and k_2. The more negative Cu(II)/Cu(I) redox potential observed for the pyridyl complex **2** would probably result in a larger k_1 in the reaction of this compound with O_2, compared to the pyrazolyl analog. With the assumption that comparable k_2 values would be observed in the two systems, the peroxo dicopper(II) intermediate will be more 'stable' (i.e. longer lived) for the case of the pyridyl donors.

Further investigations with these and new synthetic model systems will be required to fully determine the nature of O_2-activation in these *m*-xylyl compounds and others. The systems described do provide a nice illustration of the model approach in bioinorganic chemistry. Although carried out by a number of different research groups having undoubtedly somewhat different objectives, studies of the structural and physical properties and reactivity studies of compounds which have been systematically varied with respect to parameters such as coordination number, chelate ring size and ligand donor type allow us to gain insights and draw conclusions concerning the possible mechanism of hydroxylation and to identify those special characteristics apparently present and/or required for O_2-activation.

Reactions of Dioxygen with Copper Complexes - Reversible Binding

While there has been a great deal of effort and success in the synthesis and characterization of dioxygen complexes of cobalt (54,55) and iron (54-57), attention to copper has come only more recently. This is in part due to the difficulty in handling of the kinetically labile copper(I) and copper(II) complexes, and the limited spectroscopic handles available for the d^{10} diamagnetic copper(I) ion (6,7). However, in trying to mimic the behavior of the biological copper active sites, there has been recent progress in finding ways of stabilizing appropriate copper(I) complexes and their dioxygen adducts (6,7,11,58,59).

In a recent review article (11), we have described systems either proposed or established to be dioxygen/copper adducts, including those systems which bind O_2 reversibly. Caution must be taken in this field, since oxygenation of a Cu(I) complex which is accompanied by color changes, even if reversible, does not guarantee that a dioxygen adduct was formed. A number of criteria for establishing the existence of Cu_n/O_2 adducts should be used (12,54), and often not all of these can be fulfilled for legitimate technical/experimental reasons.

Amongst the best characterized systems are those due to Thompson (60-62). He has described both a monomeric superoxo copper(II) complex (61) and a dinuclear peroxo dicopper(II) compound (60). $CuL(C_2H_4)$ [L = hydrotris(3,5-dimethyl-1-pyrazolyl)borate anion] reacts with O_2 to give $CuL(O_2)$ which can be isolated as a stable crystalline diethyl ether solvate. The observed EPR silence, normal ^1H NMR, and IR band at 1015 cm^{-1} (observed in the presence of $^{18}O_2$ but not $^{16}O_2$) firmly establish the complex as a superoxide species. The binding is reversible since ethylene can displace the bound O_2 ligand to again give $CuL(C_2H_4)$. When the copper(I) complex $[Cu(TEEN)(C_2H_4)]ClO_4$ (TEEN = N,N,N',N'-tetraethylethylenediamine) reacts with O_2 in wet methanol, the solid blue product which can be isolated is a peroxo-dicopper(II) complex [$\nu(^{16}O-^{16}O)$ = 825 cm^{-1}], based on a variety of analytical data (60,62). Reversible behavior is indicated by the observation that (TEEN)Cu(I)-ethylene or carbonyl complexes can be isolated by treating the Cu_2/O_2 complex with C_2H_4 or CO, respectively.

We have found that the dinuclear phenoxo-bridged copper(I) complex, $[Cu_2(XYL\text{-}O\text{-})]^+$ (**8**), can bind dioxygen reversibly (63,64). The x-ray structure of **8** (Figure 2) shows that it possesses some features which are strikingly similar to those of the proposed sites of either deoxy- and/or oxy-hemocyanin, including the Cu...Cu distance of 3.6-3.7 Å, an 'endogenous' phenoxo bridging group, and an empty "pocket" where a second small bridging group (X), such as OH^-, N_3^-, Cl^-, Br^-, and RCO_2^- (65) is known to coordinate in dicopper(II) complexes, $[Cu^{II}_2(XYL\text{-}O\text{-})(X)]^{2+}$.

When an orange dichloromethane solution of **8** is exposed to dioxygen below -50 °C, an intense purple color develops due to the formation of the dioxygen adduct **9**. Manometric measurements at -80 °C indicate that 1 mole of dioxygen is taken up per mole of **8** to give the product formulated as **9**, $[Cu^{II}_2(XYL\text{-}O\text{-})(O_2^{2-})]^+$. In resonance Raman spectra there are two enhanced vibrations, the copper-oxygen stretch at 488 cm^{-1} (464 cm^{-1} using $^{18}O_2$) and the O-O stretch at 803 cm^{-1} (750 cm^{-1} with $^{18}O_2$) (Figure 3). Oxygenation of **8** with mixed isotope dioxygen, $^{16}O\text{-}^{18}O$, reveals that the peroxide is asymmetrically bound, since the copper-peroxide stretch is split into two components at 465 and 486 cm^{-1} (66). Recent EXAFS results have indicated a 3.31 Å copper-copper separation for **9** (67). This finding rules out the μ-1,1-bridging peroxo geometry for $[Cu^{II}_2(XYL\text{-}O\text{-})(O_2^{2-})]^+$ (**9**) since structural data for the doubly bridged complexes $[Cu^{II}_2(XYL\text{-}O\text{-})(X)]^{2+}$ show that a μ-1,1-X (X = oxygen atom) bridging geometry is only compatible with

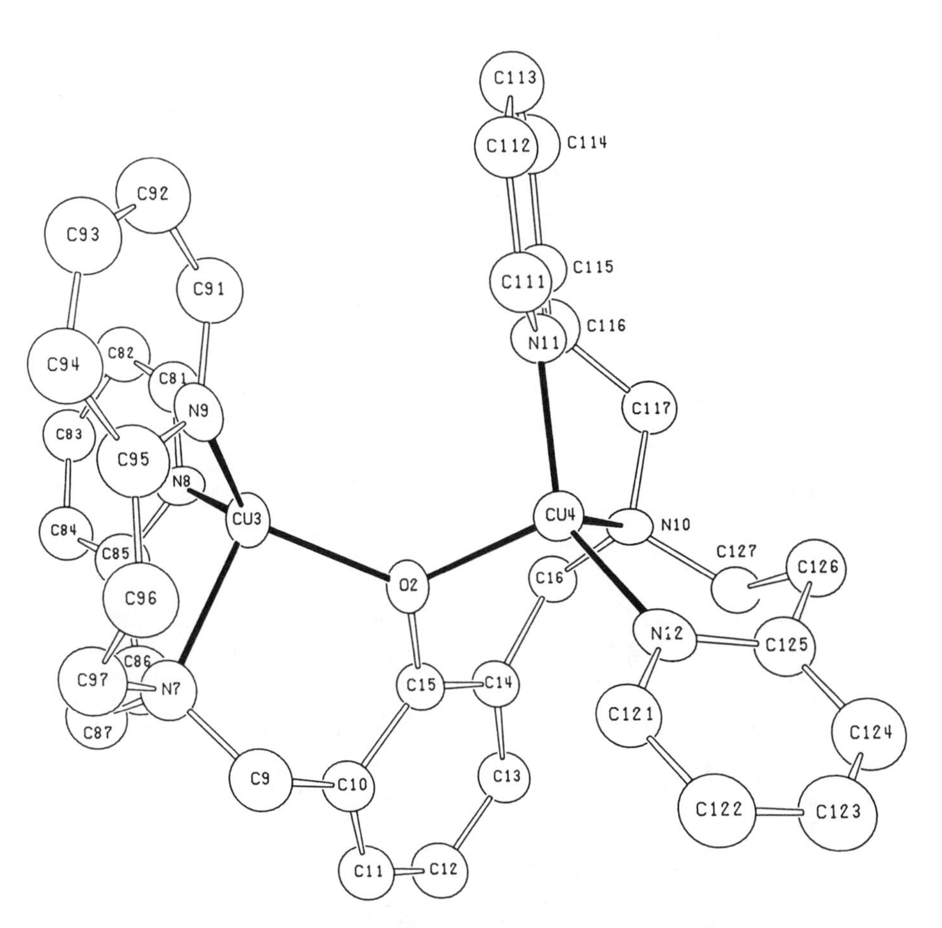

Figure 2. ORTEP diagram of the phenoxo-bridged dicopper(I) complex $[Cu^I_2(XYL\text{-}O\text{-})]^+$ (**8**), reference 64.

a Cu...Cu distance of < 3.15 Å (65,68,69). Consequently, the geometry of the peroxide moiety in $[Cu^{II}_2(XYL\text{-}O\text{-})(O_2^{\,2-})]^+$ (**9**) is either non-symmetrical μ-1,2-bridging (e.g. axial to one Cu, but equatorial to the other) or terminal in character (i.e. O-O-Cu...Cu).

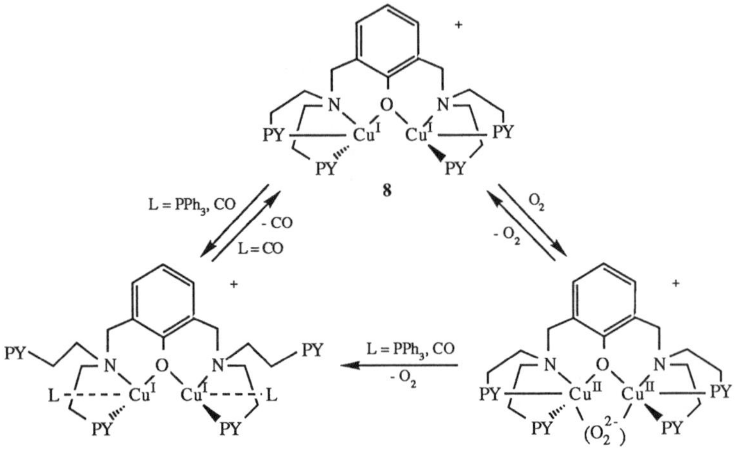

8

$L = PPh_3$, CO; - CO; L = CO

O_2; $- O_2$

$L = PPh_3$, CO; $- O_2$

10a L = CO
10b $L = PPh_3$

9

The application of a vacuum while rapidly warming solutions of **9** results in the removal of O_2 and regeneration of dicopper(I) complex **8** and quasi-reversible cycling between **8** and **9** using such vacuum-purge applications can be followed spectrophotometrically (64). Dioxygen can also be liberated by addition of carbon monoxide or triphenylphosphine to the purple solution of **9**; the purple color fades and the bis(carbonyl)dicopper(I) complex **10a**, or the bis(triphenylphosphine) adduct **10b** ($[Cu^I_2(XYL\text{-}O\text{-})(L)_2]^+$, **10**, L = CO, PPh3) is formed. Carbon monoxide can be removed from $[Cu^I_2(XYL\text{-}O\text{-})(CO)_2]^+$ (**10a**) by applying a vacuum to give back $[Cu^I_2(XYL\text{-}O\text{-})]^+$ (**8**) and several cycles of oxygenation, dioxygen displacement by CO and decarbonylation can be carried out without a severe amount of decomposition, Figure 4.

As previously discussed in another context, Borovik and Sorrell (90) have described a phenoxo-bridged dicopper(I) close analog of **8**, but with pyrazole donors instead of pyridine. In their case, oxygenation even at low temperature results in a reaction with a $4Cu{:}O_2$ stoichiometry, and no dioxygen complex is detected. As described above, they conclude that the kinetics of the ligand-complex oxygenation and ensuing decomposition reactions are unfavorable and that a metastable dioxygen adduct reacts rapidly with more dicopper(I) compound to give irreversible dioxygen reduction. Again, it seems likely that these kinetic differences are governed by the electronic environment around copper, as provided by the ligands, and that pyridine ligands favor a more rapid reaction of copper(I) with dioxygen compared to pyrazole, leading to a spectroscopically observable dioxygen-adduct (64).

As alluded to above, we have also recently characterized a related series of dioxygen-copper complexes utilizing ligands like N5PY2, shown above. The generalized ligand type, NnPY2, contains a variable length (n) methylene chain

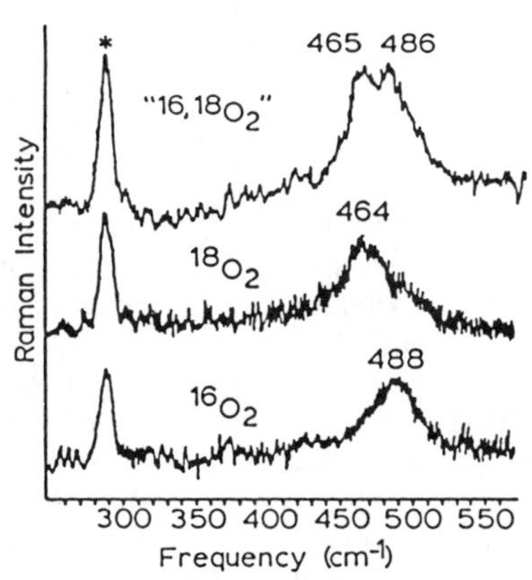

Figure 3. Resonance Raman spectra of the intraligand region of $[Cu^{II}_2(XYL\text{-}O)(O_2^{2-})]^+$ (**9**) prepared with $^{16}O_2$, $^{18}O_2$, and $^{16}O^{18}O$. (Reproduced from reference 66. Copyright 1987 American Chemical Society.)

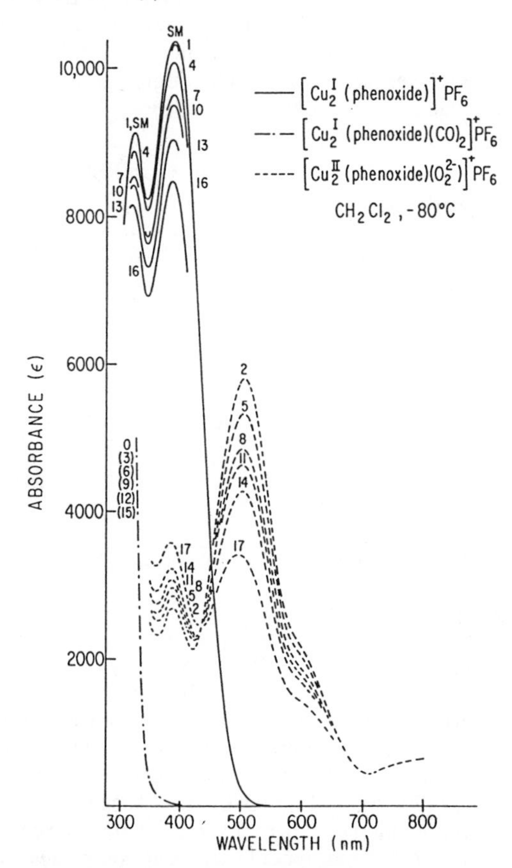

Figure 4. Absorption spectra showing the "carbonyl cycling" behavior of the dioxygen adduct $[Cu^{II}_2(XYL\text{-}O\text{-})(O_2{}^{2-})]^+$, **9**, reference 64.

which connects two N_3 tridentate groups, the same one used in the XYL and XYL-O- dinucleating ligands. Dicopper(I) complexes $[Cu_2(NnPY2)]^{2+}$, containing two three-coordinate Cu(I) ions, react reversibly at -80 °C in CH_2Cl_2 (Cu:O_2 = 2:1, manometry) to give O_2-adducts, $[Cu_2(NnPY2)(O_2)]^{2+}$. These are characterized by multiple and strong charge-transfer bands in the visible region, e.g., λ_{max} (ε, $M^{-1}cm^{-1}$) 360 (21400), 423 (3600), 520 (1200) nm, for the complex containing the N5PY2 ligand; this spectral pattern is unique in copper/dioxygen coordination chemistry (11) and also closely resembles that observed for oxy-Hc [λ_{max} (e); 345 (20000), 570 (1000), 485 (CD) nm] (40,41). The observation of d-d bands at > 650 nm, and results from x-ray absorption edge studies (41,80) suggest that Cu(II) is present, thus the O_2-adducts are best formulated as peroxo-dicopper(II) complexes. As in the XYL-O- containing system, carbon monoxide can displace the bound dioxygen (peroxo) ligand forming the bis-adduct, $[Cu_2(NnPY2)(CO)_2]^{2+}$, undoubtedly occuring by forcing the dioxygen binding equilibrium towards the deoxy form, $[Cu_2(NnPY2)]^{2+}$, with subsequent reaction with CO. Since the ligands NnPY2 have no potential Cu...Cu bridging group, the results indicate that a bridging ligand besides O_2^{2-} itself is *not* a prerequisite for systems capable of binding CO and O_2 reversibly, nor for generating spectral characteristics reminiscent of oxy-Hc. Recent experiments indicate that complexes $[Cu_2(NnPY2)(O_2)]^{2+}$ are diamagnetic (i.e. EPR silent, 'normal' NMR) suggesting that these dioxygen complexes possess a peroxo-bridging group capable of mediating strong magnetic coupling between Cu(II) ions.

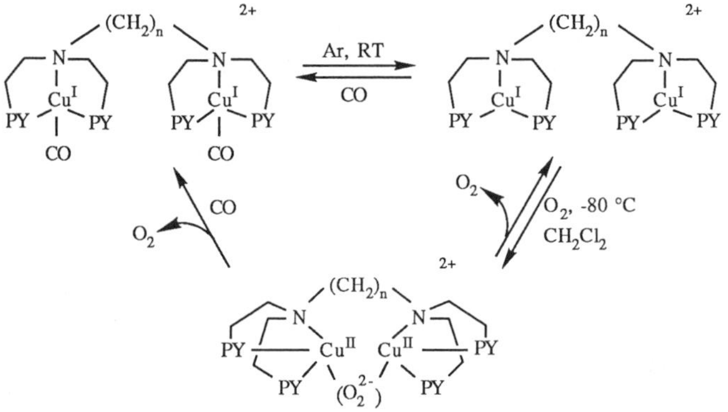

Hydroperoxo and Acylperoxo Dicopper(II) Complexes

Transition metal hydroperoxo species are well established as important intermediates in the oxidation of hydrocarbons (8,70,71). As they relate to the active oxygenating reagent in cytochrome P-450 monooxygenase, (porphyrin)M-OOR complexes have come under recent scrutiny because of their importance in the process of (porphyrin)M=O formation via O-O cleavage processes (72-74). In copper biochemistry, a hydroperoxo copper species has been hypothesized as an important intermediate in the catalytic reaction of the copper monooxygenase, dopamine β-hydroxylase (75,76). A Cu-OOH moiety has also been proposed to be involved in the disproportionation of superoxide mediated by the copper-zinc superoxide dismutase (77-78). Thus, model Cu_n-OOR complexes may be of relevance to active site intermediates in these and other copper proteins involved in dioxygen activation.

In our studies on reversible binding and activation of O_2 by dinuclear copper complexes, we have found that the dioxygen complex $[Cu^{II}_2(XYL\text{-}O\text{-})(O_2^{2-})]^+$ (**9**) reacts with acids to form a green hydroperoxo complex $[Cu^{II}_2(XYL\text{-}O\text{-})(OOH^-)]^{2+}$ (**11**) (Figure 5) (79). In fact, this species can be generated via three different routes: a) by protonation of **9** with one equivalent of $HBF_4 \cdot Et_2O$ in CH_2Cl_2 at -80 °C; here, spectrophotometric titration of $[Cu^{II}_2(XYL\text{-}O\text{-})(O_2^{2-})]^+$ (**9**) with the acid shows an isosbestic point indicating that only two species are involved in this reaction and thus **9** is straightforwardly converted to **11**, b) by decarbonylation of the bis(carbonyl)-dicopper(I) complex $[Cu^I_2(XYL\text{-}OH)(CO)_2]^{2+}$ (**12**) under reduced pressure at 0 °C, followed by oxygenation at -80 °C ($Cu:O_2 = 2:1$). Here, the uncoordinated and protonated phenol group serves as a stoichiometric source of H^+ to protonate a putative dioxygen adduct of the decarbonylated form of **12**, and c) by the addition of excess hydrogen peroxide to a dimethylformamide/CH_2Cl_2 solution of $[Cu^{II}_2(XYL\text{-}O\text{-})(OH)]^{2+}$ (**3**), Figure 5.

Attempts to isolate $[Cu^{II}_2(XYL\text{-}O\text{-})(OOH^-)]^{2+}$ (**11**) as a solid to determine its structure have not been successful. However, the close similarity of the UV-vis spectrum of **11** [λ_{max} = 395 nm, ε = 8000 $(M\text{-}cm)^{-1}$] with that of $[Cu^{II}_2(XYL\text{-}O\text{-})(OH)]^{2+}$ (**3**) and a solution EXAFS derived Cu...Cu distance of 3.05 Å in **11** (80) suggest that $[Cu^{II}_2(XYL\text{-}O\text{-})(OOH^-)]^{2+}$ possess a μ-1,1-bridged hydroperoxo moiety.

Recently, we have prepared a derivative of **11**, an acylperoxo dicopper(II) complex $[Cu^{II}_2(XYL\text{-}O\text{-})(RCO_3^-)]^{2+}$ (R = *m*-ClC_6H_4, **13**), and its structure has been determined by x-ray diffraction (Figure 6) which shows that indeed the peroxide moiety has a μ-1,1-bridging structure (81). Complex **13** can be prepared by two different pathways, a) by reacting the hydroxo-bridged complex **3** with *m*-chloroperbenzoic acid, or b) by acylation of the peroxo complex **9** with *m*-chlorobenzoyl chloride, followed by a metathesis reaction (Figure 5).

The reactivity of the dioxygen (peroxo) complex **9** is markedly different from that observed for both the hydroperoxo and acylperoxo complexes **11** and **13**. As described above, the addition of triphenylphosphine to **9** results in the quantitative displacement of the dioxygen ligand with the concomitant production of the bis(triphenylphosphine) adduct, $[Cu^I_2(XYL\text{-}O\text{-})(PPh_3)_2]^+$ (**10b**). By contrast, the Cu_2-OOR complexes **11** and **13** react with one equivalent of triphenylphosphine to give essentially quantitative yields of triphenylphosphine oxide (79,81), Figure 5. These results are in accord with observations on other transition metal peroxide complexes where the oxidation of organic substrates is enhanced by the presence of electrophiles such as H^+ or RCO^+ (82-86). In the present case, protonation (or acylation) of the dioxygen-copper complex appears to result in activation via formation of the Cu_n-OOR species which is capable of transferring an oxygen atom to a substrate while leaving behind a stable hydroxo, **3**, (or carboxylato, **14**)-copper(II)$_n$ moiety. Further mechanistic work is required to distinguish between a metal-based reaction or a pathway involving displacement of the coordinated RO_2^- ligand followed by its direct reaction with a substrate (87-88).

As suggested above, we can speculate that the biological relevance of the observations described above pertain to the activation of dioxygen in copper monooxygenases via protonation of a peroxo-Cu_n (derived from Cu(I) and O_2)

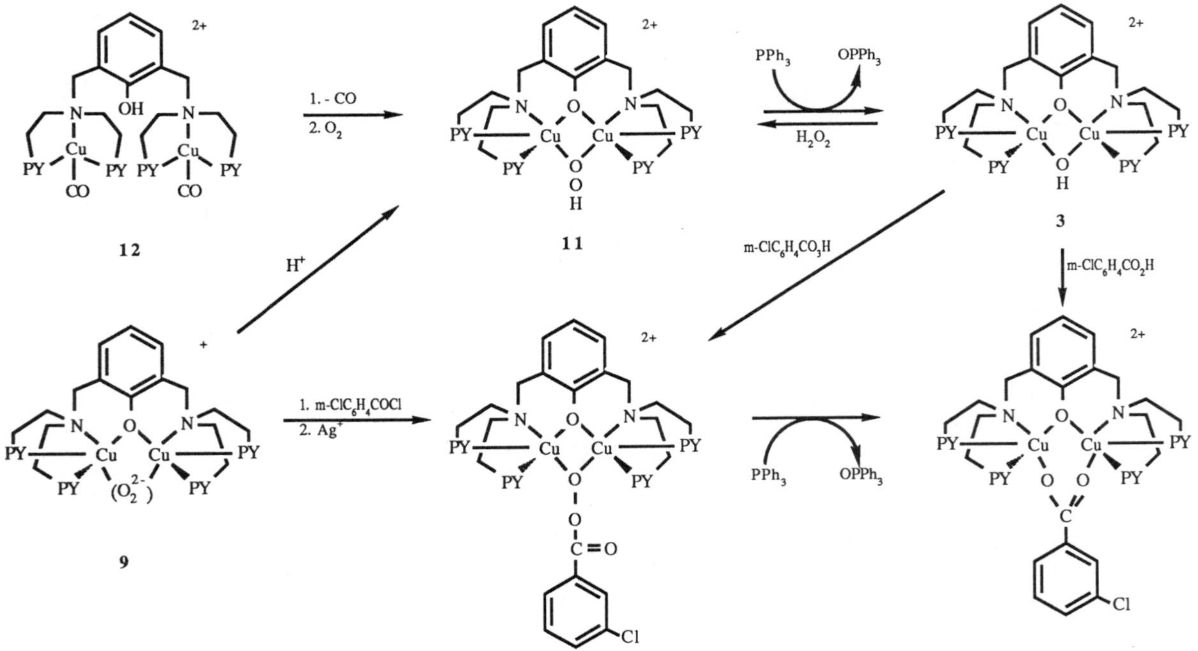

Figure 5. Scheme describing the reactions of the dioxygen-complex $[Cu^{II}_2(XYL\text{-}O\text{-})(O_2^{2-})]^+$ (**9**), and the transformations involving the $[Cu^{II}_2(XYL\text{-}O\text{-})(OOR^-)]^{2+}$ compounds (R = H-, **11**; R = m-$ClC_6H_4C(O)$-, **13**).

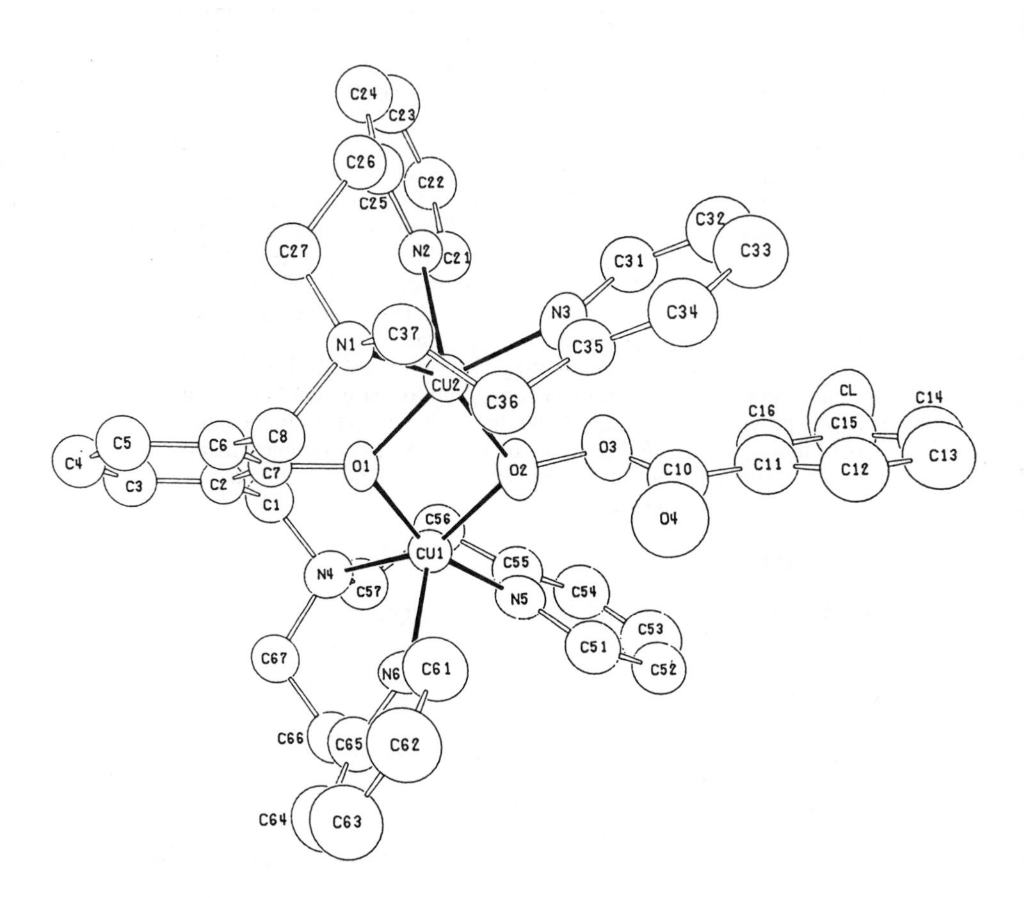

Figure 6. ORTEP diagram of the dicationic portion of $[Cu^{II}_2(XYL\text{-}O\text{-})(m\text{-}ClC_6H_4C(O)O_2\text{-})](ClO_4)_2{\cdot}CH_3CN$ (**13**), reference 81.

species which through O-O 'degradation' processes in the presence of substrates can lead to oxygen atom transfer reactions.

Conclusions

The purpose of model bioinorganic studies is to establish the relevant coordination chemistry, providing evidence for the reasonable existence of hypothesized biological structures or reaction intermediates, physical characterization of these species and determination of the competence of such moieties towards reactivity patterns found in the biochemical systems.

Here, we have described model systems for copper metalloproteins which mimic the reversible binding of O_2 in hemocyanin. While the reversible binding of O_2 to **8** to give **9** mimics the action of Hc, the UV-vis spectral characteristics of **9** do not closely match the unique features observed for oxy-hemocyanin. However, oxygenated dicopper(I) complexes containing NnPY2 ligating systems do have UV-vis spectral features that closely resemble those of oxy-Hc (40,41) and these may presently serve as closer biomimics for Hc. From our work and that of others, it is clear that a number of Cu_n/O_2 structural types can exist.

The hydroxylation reactions on *m*-xylyl systems described mimic certain aspects of the aromatic hydroxylation occuring in tyrosinase. Further investigations will hopefully provide greater insights into the mechanism(s) of O_2 activation processes of this type. The studies with Cu_2-OOR complexes suggest that these and/or reactive metal-oxo reagents derived from these are important in oxygen atom transfer (i.e. oxygenation) processes.

Acknowledgments

We thank the National Institutes of Health for support of the research described by the authors of this article.

Literature Cited

1. *Copper Proteins and Copper Enzymes*; Lontie, R., Ed.; CRC: Boca Raton, FL, 1984; Vol. 1-3.
2. Solomon, E. I. In *Metal Ions in Biology*; Spiro, T. G., Ed.; Wiley-Interscience: New York, 1981; Vol. 3, pp 41-108.
3. *Metal Ions in Biological Systems*; Sigel, H., Ed.; Marcel Dekker: NY, 1981; Vol. 13.
4. Owen, C. A., Jr. *Biochemical Aspects of Copper*; Noyes: Park Ridge, NJ, 1982.
5. *Biological Roles of Copper*, CIBA Foundation Symposium 79; Excerpta Medica: New York, 1980.
6. *Copper Coordination Chemistry: Biochemical and Inorganic Perspectives*; Karlin, K. D.; Zubieta, J., Eds.; Adenine Press: Guilderland, N.Y., 1983.
7. *Biological & Inorganic Copper Chemistry*; Karlin, K. D.; Zubieta, J., Eds.; Adenine Press: Guilderland, N.Y., 1986; Vol. 1-2.
8. Sheldon, R. A.; Kochi, J. K. *Metal-Catalyzed Oxidations of Organic Compounds*; Academic Press: New York, 1981.
9. Gampp, H.; Zuberbuhler, A. D. *Met. Ions Biol. Syst.* **1981**, *12*, 133-190.

10. Nigh, W. G. In *Oxidation in Organic Chemistry*; Trahanovsky, W. S., Ed.; Academic: New York, 1973; Part B, pp 1-96.
11. Karlin, K. D.; Gultneh, Y. *Prog. Inorg. Chem.* **1987**, *35*, 219-327 and refs. cited therein.
12 Karlin, K. D.; Gultneh, Y. *J. Chem. Educ.* **1985**, *62* (11), 983-990.
13. Brown, J. M.; Powers, L.; Kincaid, B.; Larrabee, J. A.; Spiro, T. G. *J. Am. Chem. Soc.* **1980**, *102*, 4210-4216.
14. Solomon, E. I.; Penfield, K. W.; Wilcox, D. E. *Struct. Bonding (Berlin)* **1983**, *53*, 1-57.
15. Lontie, R.; Witters, R. *Met. Ions Biol. Syst.* **1981**, *13*, 229-258.
16. Lerch, K. *Met. Ions Biol. Syst.* **1981**, *13*, 143-186.
17. Solomon, E. I. In *Copper Coordination Chemistry: Biochemical & Inorganic Perspectives*; Karlin, K. D.; Zubieta, J., Eds.; Adenine: Guilderland, New York, 1983; pp 1-22.
18. Robb, D. A. In *Copper Proteins and Copper Enzymes*; Lontie, R., Ed.; CRC: Boca Raton, FL, 1984; Vol. 2, pp 207-241.
19. Co, M. S.; Hodgson, K. O.; Eccles, T. K.; Lontie, R. *J. Am. Chem. Soc.* **1981**, *103*, 984-986.
20. Co, M. S.; Hodgson, K. O. *J. Am. Chem. Soc.* **1981**, *103*, 3200-3201.
21. Spiro, T. G.; Wollery, G. L.; Brown, J. M.; Powers, L.; Winkler, M. E.; Solomon, E. I. In *Copper Coordination Chemistry: Biochemical & Inorganic Perspectives*; Karlin, K. D.; Zubieta, J., Eds.; Adenine Press: Guilderland, NY, 1983; pp 23-42.
22. Wilcox, D. E.; Long, J. R.; Solomon, E. I. *J. Am. Chem. Soc.* **1984**, *106*, 2194-2196 and references cited therein.
23. Linzen, B.; Soeter, N. M.; Riggs, A. F.; Schneider, H.-J.; Schartau, W.; Moore, M. D.; Yokota, E.; Behrens, P. Q.; Nakashima, H.; Takagi, T.; Nemoto, T.; Vereijken, J. M.; Bak, H. J.; Beintema, J. J.; Volbeda, A.; Gaykema, W. P. J.; Hol, W. G. J. *Science* **1985**, *229*, 519-524.
24. Gaykema, W. P. J.; Volbeda, A.; Hol, W. G. J. *J. Mol. Biol.* **1985**, *187*, 255-275.
25. Gaykema, W. P. J.; Hol, W. G. J.; Vereijken, J. M.; Soeter, N. M.; Bak, H. J.; Beintema, J. J. *Nature(London)* **1984**, *309*, 23-29.
26. Himmelwright, R. S.; Eickman, N. C.; LuBien, C. D.; Lerch, K.; Solomon, E. I. *J. Am. Chem. Soc.* **1980**, *102*, 7339.
27. Woolery, G. L.; Powers, L.; Winkler, M.; Solomon, E. I.; Lerch, K.; Spiro, T. G. *Biochim. Biophys. Acta* **1984**, *788*, 155-161.
28. Jolly, R. L.; Evans, L. H.; Makino, N.; Mason, H. S. *J. Biol. Chem.* **1974**, *249*, 335-345.
29. Wilcox, D. E.; Porras, A. G.; Hwang, Y. T.; Kerch, K.; Winkler, M. E.; Solomon, E. I. *J. Am. Chem. Soc.* **1985**, *107*, 4015-4027.
30. Villafranca, J. J. In *Metal Ions in Biology* Spiro, T. G. Ed.; Wiley-Interscience: New York, 1981: Vol. 3, pp. 263-290.
31. Pember, S. O.; Villafranca, J. J.; Benkovic, S. J. *Biochem.* **1986**, *25*, 6611-6619.
32. Zubieta, J.; Karlin, K. D.; Hayes, J. C. In *Copper Coordination Chemistry: Biochemical and Inorganic Perspectives*; Karlin, K. D.; Zubieta, J., Eds.; Adenine Press: Albany, N.Y., 1983; pp 97-108; Karlin, K. D.; Hayes, J. C.; Zubieta, J. In *Copper Coordination Chemistry: Biochemical and*

Inorganic Perspectives; Karlin, K. D.; Zubieta, J., Eds.; Adenine Press: Albany, N.Y., 1983; pp 457-472; Karlin, K. D.; Cruse, R. W.; Gultneh, Y.; Hayes, J. C.; McKown, J. W.; Zubieta, J. In *Biological & Inorganic Copper Chemistry*; Karlin, K. D.; Zubieta, J., Eds.; Adenine: Guilderland, N. Y., 1986; Vol. 2, pp 101-114.

33. Karlin, K. D.; Hayes, J. C.; Gultneh, Y.; Cruse, R. W.; McKown, J. W.; Hutchinson, J. P.; Zubieta, J. *J. Am. Chem. Soc.* **1984**, *106*, 2121-2128.
34. Karlin, K. D.; Dahlstrom, P. L.; Cozzette, S. N.; Scensny, P. M.; Zubieta, J. *J. Chem. Soc. Chem. Commun.* **1981**, 881.
35. Karlin, K. D.; Gultneh, Y.; Hutchinson, J. P.; Zubieta, J. *J. Am. Chem. Soc.* **1982**, *104*, 5240-5242.
36. Taqui-Khan, M. M.; Martell, A. E. *Inorg. Chem.* **1975**, *14*, 676-680.
37. Blackburn, N. J.; Karlin, K. D.; Concannon, M.; Hayes, J. C.; Gultneh, Y.; Zubieta, J. *J. Chem. Soc. Chem. Commun.* **1984**, 939-940.
38. Karlin, K. D. et al., unpublished results.
39. Karlin, K. D.; Cruse, R. W.; Haka, M. S.; Gultneh, Y.; Cohen, B. I. *Inorg. Chim. Acta* **1986**, *125*, L43-L44.
40. Karlin, K. D.; Haka, M. S.; Cruse, R.W.; Gultneh, Y. *J. Am. Chem. Soc.* **1985**, *107*, 5828-5829.
41. Karlin, K. D.; Haka, M. S.; Cruse, R. W.; Meyer, G. J.; Farooq, A.; Gultneh, Y.; Hayes, J. C.; Zubieta, J. *J. Am. Chem. Soc* **1988**, *110*, 0000.
42. Karlin, K. D.; Cohen, B. I.; Jacobson, R. R.; Zubieta, J. *J. Am. Chem. Soc.* **1987**, *109*, 6194-6196.
43. The migration of hydrogen atoms has been observed in the copper ion catalyzed oxidative coupling of 2,6-xylenol; Butte, W. A., Jr., Price, C. C. *J. Am. Chem. Soc.* **1962**, *84*, 3567-3570.
44. Systems containing metals other than copper which exhibit the "N.I.H. shift" include: Sharpless, K. B.; Flood, T. C. *J. Am. Chem. Soc.* **1971**, *93*, 2316-2318; Castle, L.; Lindsay-Smith, J. R.; Buxton, G. V. *J. Mol. Cat.* **1980**, *7*, 235-243; Sakurai, H.; Hatayama, E.; Fumitani, K.; Kato, H. *Biochem. Biophys. Res. Commun.* **1982**, *108*, 1649-1654.
45. Guroff, G.; Daly, J. W.; Jerina, D. M.; Renson, J.; Witkop, B.; Udenfriend, S. *Science*, **1967**, *158*, 1524.
46. Daly, J.; Guroff, G.; Jerina, D. M.; Udenfriend, S.; Witkop, B. *Adv. Chem. Ser.* **1968**, *77*, 270.
47. Jerina, D. M.; *Chemtech.* **1973**, *3*, 120-127.
48. Sorrell, T. N.; Malachowski, M. R.; Jameson, D. L. *Inorg. Chem.* **1982**, *21*, 3250-3252.
49. Casella, L.; Ghelli, S. *Inorg. Chem.* **1983**, *22*, 2458-2463.
50. Casella, L.; Rigoni, L. *J. Chem. Soc. Chem. Commun.* **1985**, 1668-1669.
51. Karlin, K. D.; Zuberbuhler, A. D., unpublished results.
52. Karlin, K. D., and co-workers, unpublished observations.
53. Nelson, S. M.; Esho, F. S.; Lavery, A.; Drew, M. G. B. *J. Am. Chem. Soc.* **1983**, *105*, 5693.
54. Niederhoffer, E. C.; Timmons, J. H.; Martell, A. E. *Chem. Rev.* **1984**, *84*, 137.
55. Jones, R. D.; Summerville, D. A.; Basolo, F. *Chem. Rev.* **1979**, *79*, 139.
56. Collman, J. P.; Halpert, T. R.; Suslick, K. S. In *Metal Ion Activation of Dioxygen*; Spiro, T. G., Ed.; Wiley: New York, 1980; pp 1-72.

57. Suslick, K. S.; Reinert, T. J. *J. Chem. Educ.* **1985**, *62*, 974-983.
58. Urbach, F. L., in reference 3, pp. 73-116.
59. Fenton, D.E. In *Advances in Inorganic and Bioinorganic Reaction Mechanisms*; Sykes, A.G., Ed.; Academic Press: New York, 1983; Volume 2, pp. 187-257.
60. Thompson, J. S. *J. Am. Chem. Soc.* **1984**, *106*, 8308-8309.
61. Thompson, J. S. *J. Am. Chem. Soc.* **1984**, *106*, 4057-4059.
62. Thompson, J. S. In reference 7, Volume 2, pp. 1-10.
63. Karlin, K. D.; Cruse, R. W.; Gultneh, Y.; Hayes, J. C.; Zubieta, J. *J. Am. Chem. Soc.* **1984**, *106*, 3372-3374.
64. Karlin, K. D.; Cruse, R. W.; Gultneh, Y.; Farooq, A.; Hayes, J. C.; Zubieta, J. *J. Am. Chem. Soc.* **1987**, *109*, 2668-2679.
65. Karlin, K. D.; Farooq, A.; Hayes, J. C.; Cohen, B. I.; Zubieta, J. *Inorg. Chem.* **1987**, *26*, 1271-1280.
66. Pate, J. E.; Cruse, R. W.; Karlin, K. D.; Solomon, E. I. *J. Am. Chem. Soc.* **1987**, *109*, 2624-2630.
67. Blackburn, N. J.; Strange, R. W.; Cruse, R. W.; Karlin, K. D. *J. Am. Chem. Soc.* **1987**, *109*, 1235-1237.
68. Sorrell, T. M.; Jameson, D. L.; O'Connor, C. J. *Inorg. Chem.* **1984**, *23*, 190-195.
69. Maloney, J. J.; Glogowski, M.; Rohrbach, D. F.; Urbach, F. L. *Inorg. Chim. Acta*, **1987**, *127*, L33-L35.
70. Parshall, G. W. *Homogeneous Catalysis. The Application of Catalysis by Soluble Transition Metal Complexes*; Wiley: New York, 1980.
71. See, for example, Saussine, L.; Brazi, E.; Robina, A.; Mimoun, H.; Fischer, J.; Weiss, R. *J. Am. Chem. Soc.* **1985**, *107*, 3534; Sugimoto, H.; Sawyer, D. T. *J. Am. Chem. Soc.* **1985**, *107*, 5712; Durand, R. R., Jr.; Bencosme, C. S.; Collman, J. P.; Anson, F. C. *J. Am. Chem. Soc.* **1983**, *105*, 2710.
72. Balasubramanian, P. N.; Bruice, T. C. *Proc. Natl. Acad. Sci. USA* **1987**, *84*, 1734-1738.
73. Groves, J. T.; Watanabe, Y. *Inorg. Chem.* **1987**, *26*, 785-786; Groves, J. T.; Watanabe, Y. *J. Am. Chem. Soc.* **1986**, *108*, 7834-7836.
74. Traylor, T. G.; Lee, W. A.; Stynes, D. *J. Am. Chem. Soc.* **1984**, *106*, 755-764.
75. Miller, S. M.; Klinman, J. P. *Biochem.* **1985**, *24*, 2114-2127.
76. Wimalasena, K.; May, S. W. *J. Am. Chem. Soc.* **1987**, *109*, 4036-4046.
77. Osman, R.; Basch, H. *J. Am. Chem. Soc.* **1984**, *106*, 5710-5714.
78. Rosi, M.; Sgamellotti, A.; Tarantelli, F.; Bertini, I.; Luchinat, C. *Inorg. Chem.* **1986**, *25*, 1005-1008.
79. Karlin, K. D.; Cruse, R. W.; Gultneh, Y. *J. Chem. Soc. Chem. Commun.* **1987**, 599-600.
80. Karlin, K. D.; Blackburn, N. J. and co-workers, submitted for publication.
81. Ghosh, P.; Tyeklar, Z.; Karlin, K. D.; Jacobson, R. R.; Zubieta, J. *J. Am. Chem. Soc.*, **1987**, *109*, 6889-6891.
82. Chen, M. J. Y.; Kochi, J. K.; *J. Chem. Soc. Chem. Commun.* **1977**, 204-205.
83. Tatsuno, Y.; Otsuka, S.; *J. Am. Chem. Soc.* **1981**, *103*, 5832-5839.
84. Strukul, G.; Ros, R.; Michelin, R. A.; *Inorg. Chem.* **1982**, *21*, 495-500.
85. Fukuzumi, S.; Ishikawa, K.; Tanaka, T. *Chem. Lett.* **1986**, 1-4.

86. Mimoun, H.; Saussine, L.; Daire, E.; Postel, M.; Fischer, J.; Weiss, R. *J. Am. Chem. Soc.* **1983**, *105*, 3101-3110.
87. Sen, A.; Halpern, J. *J. Am. Chem. Soc.* **1977**, *99*, 8337-8339.
88. Read, G.; Urgelles, M. *J. Chem. Soc. Dalton Trans.* **1986**, 1383-1387.
89. Bulkowski, J. E.; Burk, P. L.; Ludmann, J.-L.; Osborn, J. A. *J. Chem. Soc. Chem. Commun.* **1977**, 498.
90. Sorrell, T. N.; Borovik, A. S. *J. Chem. Soc. Chem. Commun.* **1984**, 1489-1490.

RECEIVED January 4, 1988

Chapter 6

Oxygen Atom Transfer Reactions at Binuclear Copper Sites

Lawrence D. Margerum, Kazuko I. Liao, and Joan Selverstone Valentine

Department of Chemistry and Biochemistry, University of California, Los Angeles, CA 90024

A brief account is given of the reactions of copper-containing mono-oxygenase enzymes. Theories concerning the detailed steps in the dioxygen activation process are presented and recent studies of the reactivity of binuclear copper complexes with the single oxygen atom donor iodosylbenzene, OIPh, are described. The binuclear complexes were found to be significantly better catalysts for the oxidation of cyclohexene than their mononuclear analogues. Some speculations on the mechanism of these reactions are discussed.

Monooxygenase enzymes catalyze reactions in which one atom of oxygen derived from dioxygen is incorporated into an organic substrate while the other atom is reduced by two electrons to form water (1,2).

$$R\text{-}H + O_2 + 2\,e^- + 2\,H^+ \longrightarrow R\text{-}O\text{-}H + H_2O$$

The enzymes of this type that have been characterized contain some type of redox-active cofactor, such as a flavin (3), or a metal ion (heme iron, non-heme iron, or copper), or both (4-6). Our understanding of the mechanism of these enzymes is most advanced in the case of the heme-containing enzyme cytochrome P450. But in spite of the availability of a crystal structure of an enzyme-substrate complex (7) and extensive information about related reactions of low molecular weight synthetic analogues of cytochrome P450 (8), a detailed picture of the molecular events that are referred to as "dioxygen activation" continues to elude us.

The purpose of this paper is to give a brief account of the properties of two of the better characterized copper-containing monooxygenase enzymes, to present theories concerning the detailed steps in the dioxygen activation process for each enzyme, and to describe the results of our recent studies of the reactivity of binuclear copper complexes with the single oxygen atom donor iodosylbenzene, OIPh, which may be relevant to the mechanisms of copper-containing monooxygenase enzymes.

0097–6156/88/0372–0105$06.00/0

Dioxygen Complexes as Models

Studies of metal-dioxygen complexes (including metal-superoxide and metal-peroxide complexes) and the reactions of dioxygen, superoxide, and peroxides with metal complexes (9,10) suggest possibilities for the mode in which dioxygen may be reacting with the metal centers in monooxygenase enzymes. If we consider the reaction of single metal ions or complexes and binuclear metal complexes with dioxygen, the likely intermediates in these reactions are those shown in Scheme 1. It is unknown what role is played by ligand environments in proteins and in synthetic analogues in stabilizing different species as it is also unknown which species represent active oxidants capable of transfering oxygen atoms to substrate in the enzyme systems. Moreover, it is not known how a binuclear metal active site might differ from a mononuclear active site or if there is one type of reaction mechanism that operates in all or most of the monooxygenase enzymes or if each type of enzyme follows a different mechanism.

Monooxygenase Enzymes

Tyrosinase. This enzyme is found in numerous organisms ranging from bacteria and fungi to plants and animals. The best characterized tyrosinases are those from mushrooms and from *N. crassa* (11-14). The reactions catalyzed by tyrosinase are the hydroxylation of phenols to give catechols and the subsequent oxidation of catechols to give *o*-quinones.

$$\text{PhOH} + O_2 + 2e^- + 2H^+ \longrightarrow \text{C}_6\text{H}_4(\text{OH})_2 + H_2O$$

$$\text{C}_6\text{H}_4(\text{OH})_2 + 1/2\ O_2 \longrightarrow \text{C}_6\text{H}_4\text{O}_2 + H_2O$$

Evidence from a variety of sources indicates that the active site of tyrosinase is very similar to that of hemocyanin, a dioxygen-binding protein found in molluscs and arthropods (15,16). This type of active site contains two copper ions, which are cuprous in the deoxy state, and which reversibly bind dioxygen, forming the oxy form of the enzyme or protein in which a peroxy ligand bridges between two cupric ions.

$$\text{Cu(I)} \quad \text{Cu(I)} + O_2 \longrightarrow \text{Cu(II)-}(O_2)^{2-}\text{-Cu(II)}$$

The very different reactivities of hemocyanin and tyrosinase toward oxidizable substrates seem to be due to the presence of a substrate binding site in the latter. It appears, therefore, that the oxygenation of the phenol substrate occurs either by reaction of copper-bound peroxide or hydroperoxide with the ortho position of the copper-bound phenol (16) or that the O-O bond of the peroxide ligand is cleaved

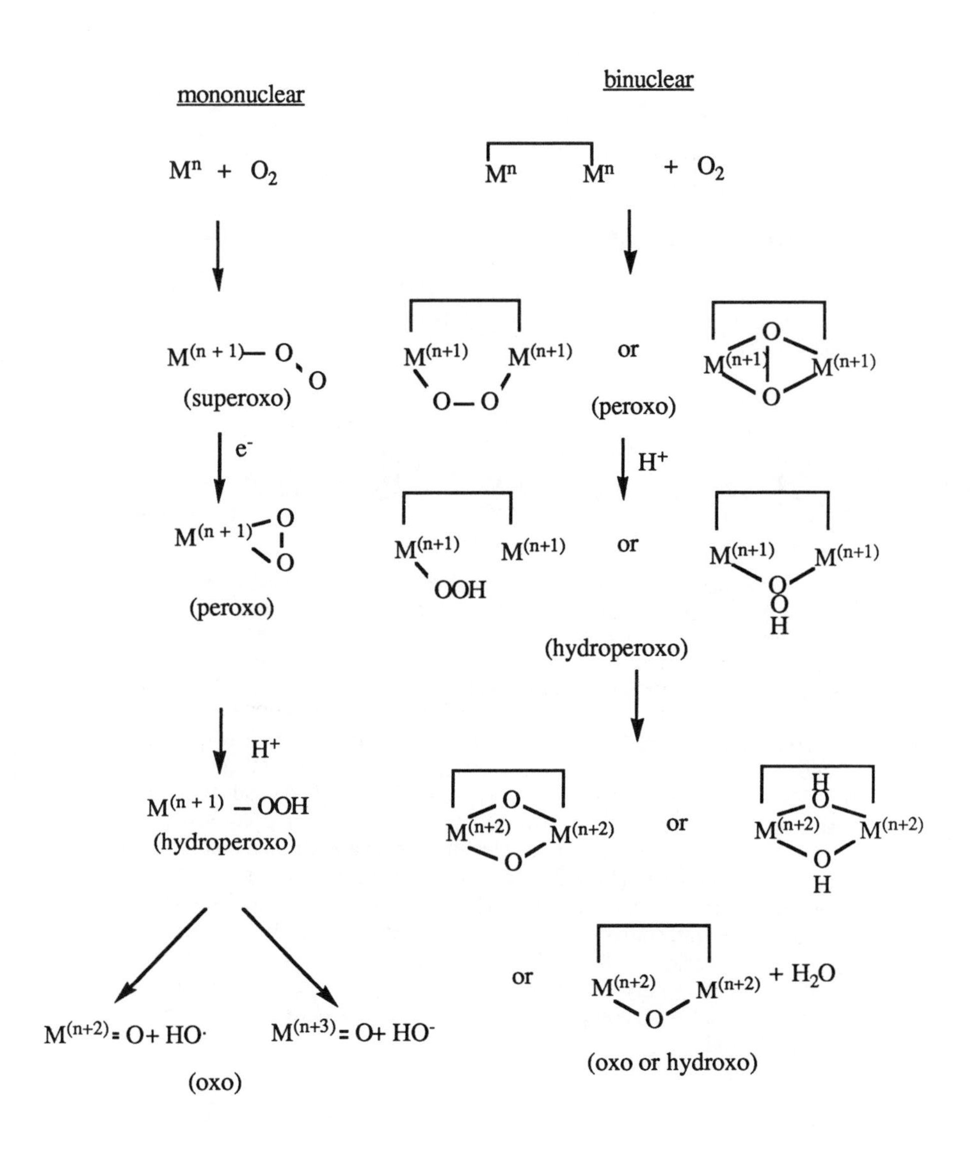
mononuclear
binuclear
Mn + O2
Mn Mn + O2
M(n + 1)— O O
(superoxo)
e-
M(n + 1) O O
(peroxo)
H+
M(n + 1) – OOH
(hydroperoxo)
M(n+2) = O+ HO·
M(n+3) = O+ HO-
(oxo)
M(n+1) M(n+1) O—O
or
(peroxo)
M(n+1) M(n+1)
H+
M(n+1) M(n+1) OOH
or
(hydroperoxo)
M(n+2) M(n+2)
or
H O H
or
M(n+2) M(n+2) O + H2O
(oxo or hydroxo)

Scheme 1

and that an oxygen atom is then transferred to the copper bound phenol (see Figure 1).

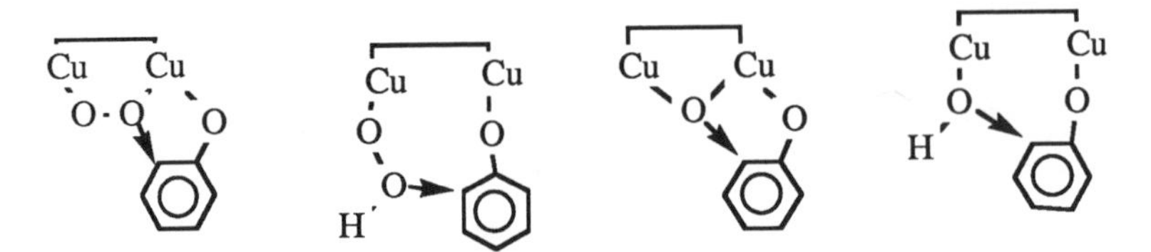

Figure 1. Examples of hypothetical modes of reaction for the enzyme tyrosinase

Dopamine β-monooxygenase This enzyme (DBM) catalyzes the hydroxylation of dopamine to the neurotransmitter norepinephrine and is found in a number of higher organisms (17).

$$\text{HO, OH-}C_6H_3\text{-}CH_2CH_2NH_2 + O_2 + \text{Ascorbate} \xrightarrow{\text{DBM}} \text{HO, OH-}C_6H_3\text{-}CH(OH)CH_2NH_2 + H_2O + \text{Dehydroascorbate}$$

The enzyme requires two copper ions per subunit for full expression of activity (18), but, unlike tyrosinase and hemocyanin, there is an absence of magnetic coupling between the two Cu(II) sites and both appear to be separate, isolated mononuclear copper sites (17). The process of dioxygen binding and activation appears to involve interaction of the dioxygen molecule with only one copper ion, and it is also found that a proton is required for the hydroxylation of substrate (19).

Oxidation by DBM has been proposed by Klinman and coworkers (18,20) to proceed through the following steps. First, the catalytic Cu(II) center is reduced, followed by binding of substrate and dioxygen. General acid catalysis of dioxygen reduction is required and may be provided by the presence of a protonated base. Partial O-O bond homolysis generates electrophilic character on an oxygen atom which, in turn, initiates C-H bond homolysis. This process leads to a transition state, in which both C-H bond breaking (H atom abstraction) and O-H bond making are important, consistent with the observation of significant primary and secondary isotope effects (18,20). Finally, the transition state yields a radical intermediate which rapidly recombines with an oxygen atom bound to copper. The proposed alkoxide-Cu(II) complex then undergoes slow dissociation to yield the hydroxylated product.

$$\text{Enz-Cu}^{II}\text{-O-O-H} + \text{H-C}{\lessgtr} \longrightarrow \text{Enz-Cu}^{III}\text{-O}^{2-}\ \text{O-H (H)} + \cdot\text{C}{\lessgtr} \longrightarrow \text{Enz-Cu}^{II}\text{-O-C}{\lessgtr} + H_2O$$

General Considerations Concerning Mechanism. A key unanswered question in our current picture of the mechanism of dioxygen activation in monooxygenase enzymes concerns the sequence of events involved in O-O bond cleavage and bond formation between oxygen atoms and the enzyme-bound substrate. There is strong evidence indicating that the catalytic sequence of dioxygen activation commences with binding of dioxygen to the reduced metal center (9,21), presumably forming a peroxide or a hydroperoxide complex. In each case where selectivity of attack on the substrate is observed, it is believed that the site to be oxygenated is directed toward the metal center by specific substrate binding to the enzyme. At issue is the nature of the oxygen-containing species that reacts with the enzyme-bound substrate.

In the case of cytochrome P450, it has been proposed that O-O bond cleavage precedes attack on substrate (22). It is generally believed that this cleavage is heterolytic, although homolytic cleavage of peroxide derivatives by cytochrome P450 has also been observed (23).

$$\text{Homolytic: } H_2O_2 \longrightarrow 2\cdot OH$$

$$\text{Heterolytic: } H_2O_2 + e^- \longrightarrow \cdot OH + OH^-$$

In the case of heme-containing systems, it is believed that the activation barrier for O-O bond cleavage can be lowered by the complexation of the resulting oxygen atom by the iron porphyrin center, i.e.:

$$\text{Homolytic: } Fe^{III}(P)\text{-}OOH \longrightarrow Fe^{IV}(P){=}O + \cdot OH$$

$$\text{Heterolytic: } Fe^{III}(P)\text{-}OOH \longrightarrow [Fe^{IV}(P^+){=}O]^+ + OH^-$$

The activation barrier for homolytic cleavage in DBM is estimated to be lowered due to binding of hydroperoxide to Cu (18). The activation barrier for heterolytic cleavage is apparently lowered in peroxidases (24) by interaction of the hydroperoxide ligand with residues in the active site which help to stabilize the charge-separated transition state. The crystal structure of cytochrome $P450_{cam}$ provides no clues as to comparable effects on the mechanism of O-O bond cleavage; i.e. there are no amino acid residues situated in such a way that they might either protonate the dioxygen ligand or stabilize charge separation as it occurs during the formation of hydroxide or water by heterolytic cleavage of the hydroperoxide ligand (7). In the case of the copper-containing monooxygenase enzymes, we have even less information about the nature of the enzymatic reaction since we have neither crystal structures of the enzymes nor synthetic analogues of candidates for active oxygenating intemediates.

The role of the metal center in either heterolytic or homolytic O-O bond cleavage is that of a reducing agent.

$$M^{n+}\text{-}OOH^- \longrightarrow M^{(n+1)+}\text{-}O^{2-} + \cdot OH$$

In the case of the heme systems, the accessibility of high oxidation states of the iron plus porphyrin ligand is well-documented by model studies (8). In the case of the non-heme systems, however, comparable high-valent metal oxo species have not been characterized and therefore the analogous mechanism in such systems is not as appealing. While high oxidation states of iron and copper are known to exist in certain complexes, they generally are supported by ligands such as oxide and fluoride that are particularly resistant to oxidation (25) and not by the non-heme ligands typically found in metalloenzyme active sites, such as imidazole or phenol. One possible explanation is that in the non-heme systems (and possibly in the heme systems as well (26)), the oxygen is not present as an oxide ligand but has been inserted in a nitrogen metal bond, forming an N-oxide complex, i.e.,

$$\text{HOO-M}^{n+}\text{-N(ligand)} \longrightarrow \text{HO-M}^{n+}\text{-O-N(ligand)}$$

$$\text{HO-M}^{n+}\text{-O-N(ligand)} + \text{substrate} \longrightarrow \text{HO-M}^{n+}\text{-N(ligand)} + \text{substrate(O)}$$

Such a species cannot be ruled out in reactions of iron-EDTA complexes with hydroperoxides recently described by Bruice and coworkers (27). On the other hand, a hydroperoxide complex that reacts with the substrate such that bond formation from O to substrate is concerted with O-O bond breaking, as proposed by Klinman for dopamine β-monooxygenase (18), could provide compensation for the cost of O-O bond cleavage in the transition state. In fact, it is interesting to speculate that for each of these enzymes, the mechanism by which the substrate is oxidized may be dependent on the reactivity of the substrate. One could envision certain substrates that would react with the metal-bound hydroperoxide ligand prior to or concerted with O-O bond cleavage. This possibility is difficult to assess because of our lack of information concerning the reactivity of HO_2^- when complexed to different metal ions.

Reactions of Copper Complexes

Recent studies by Karlin and co-workers on reactions of binuclear cuprous complexes with dioxygen have been dramatically successful both in mimicking the reaction of monooxygenase enzymes by causing the hydroxylation of a copper bound ligand by dioxygen and in the synthesis of analogues of the peroxo-bridged binuclear cupric species of oxyhemocyanin (28,29). However, these synthetic binuclear copper systems have not yet been shown to effect the oxygenation of externally added substrates by dioxygen. Moreover, they have not been observed to catalyze more than one turnover in oxygenations of added substrates by either dioxygen or hydrogen peroxide.

Iodosylbenzene, OIPh, has been useful as a reagent for the study of oxygen atom transfer from metalloporphyrins to organic substrates (8). Such studies have provided insight into the mechanisms of catalytic oxygenation catalyzed by heme-containing monooxygenase enzymes such as cytochrome P450 (8). Species generated by reaction of metalloporphyrins and single oxygen atom donors such as iodosylbenzene have been characterized as high-valent metal oxo complexes and represent likely synthetic analogues to intermediates in the cytochrome P450 pathway. We have been studying the reactions of iodosylbenzene with binuclear copper complexes in order to gain some understanding of the process of oxygen atom transfer occurring in copper-containing monooxygenase enzymes.

Our initial studies were of the epoxidation of olefins by OIPh catalyzed by simple copper salts such as cupric nitrate (30). These reactions showed a marked dependence upon the ratio of copper to OIPh in a manner that suggested that the reaction required two cupric ions per OIPh molecule. For this reason, we initiated a study of the reactivity of binuclear copper complexes, and in some cases their mononuclear analogues, with OIPh in the presence of externally added olefins (31). A series of copper complexes have been tested as catalysts for the epoxidation of cyclohexene by OIPh . These complexes are shown in Table I. The following conclusions have been drawn from this study: (a) the primary product derived from cyclohexene is cyclohexene oxide and yields of allylic oxidation products are quite low, (b) binuclear complexes are considerably more reactive than analogous mononuclear complexes or cupric triflate, (c) the binuclear copper complexes, in addition to being catalysts for cyclohexene epoxidation, are also good catalysts for the conversion of OIPh into IPh, and (d) the reactivity of these binuclear cupric complexes is not greatly influenced by small changes in the nature of the ligands bound to each copper. This last conclusion is in sharp contrast to results observed for the reactions of cuprous complexes of ligands **1** and **4** with dioxygen where the former undergoes ligand oxygenation (32) and the latter does not (33). We have also examined the free ligand isolated from the complex of ligand **1** after it had been used to catalyze epoxidation of olefins by OIPh and found it to be unchanged. For

Table I. Copper complexes used as catalysts for epoxidation of cyclohexene by iodosylbenzene in acetonitrile.

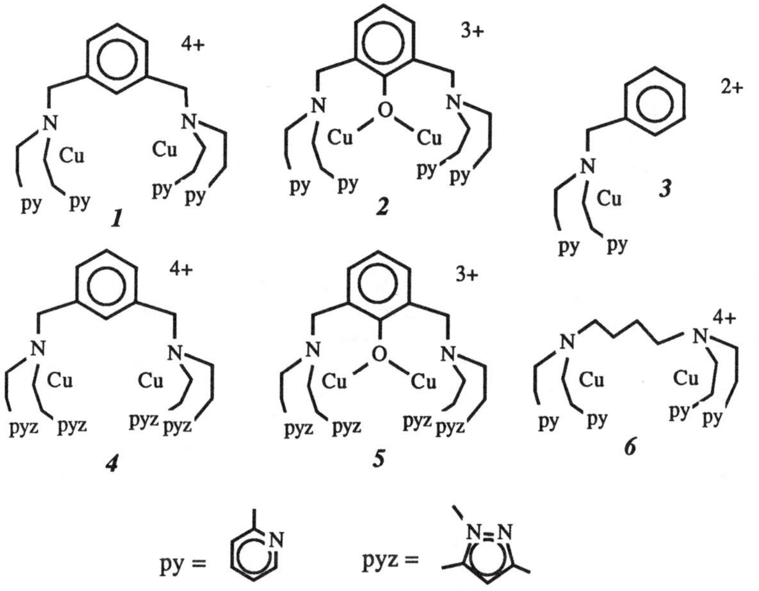

this reason, we conclude that the reaction does not occur via the same intermediate generated in reactions of the binuclear cuprous complex of this ligand with dioxygen, since in that case the ligand is found to be hydroxylated (32).

In an effort to elucidate the nature of the intermediates in the epoxidation reactions, a Hammett σ^+ reactivity plot was obtained by carrying out competition reactions between styrene and substituted styrenes. These results are summarized in Table II along with the results of several other metal complex-catalyzed styrene epoxidation reactions using OIPh. The large negative ρ value found for the binuclear copper system is similar to values found with other metal complex-catalyzed epoxidations by OIPh (with the exception of the reaction with $Mn^{III}(salen)^+$ which is thought to be radical in character (36)) and indicates that the transition state is markedly electrophilic in character.

We also observe a pronounced influence of the nature of the anion in these copper catalyzed epoxidation reactions in acetonitrile. We have found that copper complexes isolated as salts of oxygen-containing anions, especially triflate ($CF_3SO_3^-$) and perchlorate (ClO_4^-), show the highest reactivity compared to those of nitrate (NO_3^-) complexes which are less reactive and tetrafluoroborate (BF_4^-) complexes which show almost no reactivity in acetonitrile. We have shown previously that reactions of blue solutions of complex **1** with one equiv of OIPh lead to unstable blue solutions which, upon addition of cyclohexene, give high conversions to epoxide (31). Taken together, these results suggest to us that the active oxygenating intermediate in this reaction may be a complex of an I^{III} ligand derived from OIPh and that the anion may be an integral part of this intermediate

Table II. Comparison of σ^+ Hammett correlations for epoxidation of para-substituted styrenes.

Epoxidizing System	ρ	reference
$Fe^{III}TPPCl$ + OIPh	-0.9	34
$Cr^{III}(salen)^+$ + OIPh	-1.9	35
$Mn^{III}(salen)^+$ + OIPh	-0.3	36
$O{=}Fe^{IV}TMP^{+\cdot}$	-1.9	37
$(1)(OSO_2CF_3)_4$ + OIPh	-1.5	this work
$Fe^{II}(OSO_2CF_3)_2$ + OIPh	-1.1	38
Perbenzoic Acid	-1.3	39

Mechanistic Speculations. It is doubtful that the iodosylbenzene monomer, 0IPh, is a stable entity, even when associated with a metal center. Solid iodosylbenzene is an insoluble polymer linked by -0-I-0- bonds (40). Typical nonredox reactions of iodosylbenzene give products with three bonds to iodine, e.g. $(Me0)_2IPh$ or the series of μ-oxo dimer complexes which have recently been isolated and characterized (41).

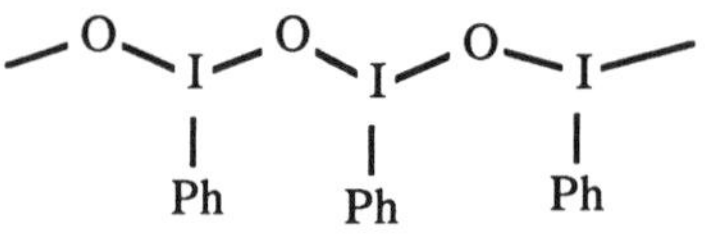

iodosylbenzene

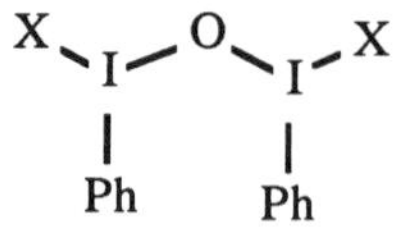

μ-oxo dimer, X = anion

In these latter compounds, the hypervalent iodine may be associated with oxyanions such as perchlorate, trifluoroacetate, and triflate, as verified by the crystal structure of the trifluoroacetate derivative (41). It seems likely that the mode of binding of iodosylbenzene in metal complexes is similar, as has previously been suggested by Hill and coworkers (42).

$$M^{n+}—O\diagdown I \diagup X$$
$$|$$
$$Ph$$

Iodosylbenzene is sufficiently reactive on its own to epoxidize electron-deficient olefins such as tetracyanoethylene (43). It is possible that coordinated monomeric iodosylbenzene is substantially more reactive than polymeric iodosylbenzene and that complexation of a monomeric form is sufficient to provide the requisite reactivity with normal olefins.

Alternate candidates for active oxygenating species are high-valent metal oxo complexes that might be monomeric or polymeric with μ-oxo bridges, similar to complexes proposed as intermediates in reactions of porphyrin and salen metal complexes. Any solvated metal oxide produced in this manner is also likely to be a strong oxidant easily capable of organic substrate oxidation. In the case of the copper-catalyzed reactions, the enhanced reactivity of the binuclear cupric complexes is not yet understood. While it is possible that the active oxidant is a μ-oxo bis-Cu(III) complex, it is also possible that the binuclear copper catalyst forms a iodosylbenzene-bridged complex with enhanced reactivity toward olefinic substrates or enhanced stability toward catalyst deactivation. The mononuclear copper complexes are also observed to catalyze the disproportionation of OIPh to O_2IPh and IPh suggesting another possibility, i.e. that the binuclear complexes may form iodosylbenzene complexes which are relatively stable toward disproportionation and thus more avaailable for reaction with olefin. Future work in our laboratory will address these possibilities.

Conclusions

There is a rich diversity of reaction types observed in reactions of copper ions or copper coordination complexes with dioxygen, reduced derivatives of dioxygen, and single oxygen atom donors. Results to date clearly demonstate differences in reactivities of mononuclear and binuclear copper complexes in reactions with dioxygen and with iodosylbenzene. These differences may be of importance in future efforts to understand how binuclear copper centers occurring in biological systems react with and activate dioxygen.

Acknowledgments

We thank Professors M. Malachowski and T. Sorrell for providing us with samples of binucleating ligands. Support of this work by the National Science Foundation is gratefully acknowledged.

Literature Cited

1. White, R.E.; Coon, M. J. *Ann. Rev. Biochem.* **1980**, *49*, 315-356 .
2. Hayaishi, O. In "Molecular Mechanisms of Oxygen Activation"; Hayaishi, O., Ed.; Academic Press: New York, 1974; pp 1-28.

3. Guengerich, F.P.; Macdonald, T.L. *Acc. Chem. Res.*, **1984**, *17*, 9-16.
4. Kaufman, S.; Fisher, D. B. In "Molecular Mechanisms of Oxygen Activation"; Hayaishi, O., Ed.; Academic Press: New York, 1974; pp 285-369.
5. Hamilton, G. A., In "Metal Ions in Biology: Copper Proteins"; Spiro, T. G., Ed.; John Wiley and Sons: New York, 1981; Vol. 3, pp 193-218.
6. "Cytochrome P-450: Structure, Mechanism, and Biochemistry"; Ortiz de Montellano, P. R., Ed.; Plenum Press: New York, 1986.
7. Poulos, T. L.; Finzel, B. C.; Gunsalus, I. C.; Wagner, G. C.; Kraut, J. *J. Biol. Chem.* **1985**, *260*, 16122-16130
8. McMurry, T. J.; Groves, J. T. In "Cytochrome P-450: Structure, Mechanism, and Biochemistry"; Ortiz de Montellano, P. R., Ed.; Plenum Press: New York, 1986; pp 1-28.
9. "Metal Ions in Biology: Metal Ion Activation of Dioxygen"; Spiro, T. G., Ed.; John Wiley and Sons: New York, 1980; Vol. 2.
10. Karlin, K. D.; Gultneh, Y. *Prog. Inorg. Chem.*, **1987**, *35*, 219-327 .
11. Mason, H. S. *Adv. Exp. Med. Biol.*, **1976**, *74*, 464-469 .
12. Lerch, K. *Met. Ions Biol. Sys.,* **1981**, *13*, 143-186.
13. Soloman, E. T. In "Metal Ions in Biology: Copper Proteins"; Spiro, T. G., Ed.; John Wiley and Sons: New York, 1981; Vol. 3, pp 41-108.
14. Robb, D. A. In "Copper Proteins and Copper Enzymes"; Lontie, R., Ed.; CRC: Boca Raton, FL, 1984; Vol. 2, pp 207-240.
15. Huber, M; Hintermann, G.; Lerch, K. *Biochemistry*, **1985**, *24*, 6038-6044 .
16. Wilcox, D. E.; Porras, A. G.; Hwang, Y. T.; Lerch, K.; Winkler, M. E.; Solomon, E. I. *J. Am. Chem. Soc.*, **1985**, *107*, 4015-4027.
17. Villafranca, J. J. In "Metal Ions in Biology: Copper Proteins"; Spiro, T. G., Ed.; John Wiley and Sons: New York, 1981; Vol. 3, pp 263-289.
18. Miller, S. M.; Klinman, J. P. *Biochemistry*, **1985**, *24*, 2114-2127 .
19. Klinman, J. P.; Krueger, M.; Brenner, M.; Edmondson, D. E. *J. Biol. Chem.*, **1984**, *259*, 3399-3402.
20. Ahn, N.; Klinman, J.P. *Biochemistry*, **1983**, *22*, 3096-3106.
21. Ochiai, E.I., "Bioinorganic Chemistry: An Introduction"; Allyn and Bacon: Boston, 1977.
22. White, R.E.; Sligar, S.G.; Coon, M.J. *J. Biol. Chem.*, **1980**, *255*, 11108-11111.
23. Ortiz de Montellano, P. R., In "Cytochrome P-450: Structure, Mechanism, and Biochemistry"; Ortiz de Montellano, P. R., Ed.; Plenum Press: New York, 1986; pp 217-271.
24. Poulos, T. L.; Kraut, J. *J. Biol. Chem.*, **1980**, *255*, 8199-8205.
25. Cotton, F. A.; Wilkinson, G. "Advanced Inorganic Chemistry: A Comprehensive Text"; 4th Ed, Wiley: New York, 1980; pp 765, 818.
26. Groves, J. T.; Watanabe, Y. *J. Am. Chem. Soc.*, **1986**, *108*, 7836-7837 .
27. Balasubramanian, P.N.; Bruice, T. C. *J. Am. Chem. Soc.*, **1986**, *108*, 5495-5503.
28. Karlin, K.D.; Cruse, R. W.; Gultneh, Y.; Amjad, F.; Hayes, J.C.; .; Zubieta, J. *J. Am. Chem. Soc.*, **1987**, *109*, 2668-2679 .
29. Karlin, K. D.; Cruse, R. W.; Gultneh, Y.; Hayes, J. C.; McKown, J. W.; Zubieta, J. In "Biological and Inorganic Copper Chemistry"; Karlin, K. D. and Zubieta, J., Eds., Adenine: Guilderland, New York, 1986; Vol. 2, pp 101-114.
30. Franklin, C. C.; VanAtta, R. B.; Tai, A. F.; Valentine, J. S. *J. Am. Chem. Soc.*, **1984**, *106*, 814-816.
31. Tai, A. F.; Margerum, L. D.; Valentine, J. S. *J. Am. Chem. Soc.*, **1986**, *108*, 5006-5008.

32. Karlin, K. D.; Hayes, J. C.; Gultneh, Y.; Cruse, R. W.; McKown, J. W.; Hutchinson, J. P.; Zubieta, J. *J. Am. Chem. Soc.*, **1984**, *106*, 2121-2128.
33. Sorrell, T. N., Malachowski, M. R.; Jameson, D. L. *Inorg. Chem.*, **1982**, *21*, 3250-3252.
34. Smith, J. R. L.; Sleath, P. R. *J. Chem. Soc., Perkin Trans. II*, **1982**, 1009-1015.
35. Samsel, E. G.; Srinivasan, K.; Kochi, J. K. *J. Am. Chem. Soc.*, **1985**, *107*, 7606-7617.
36. Srinivasan, K.; Michaud, P.; Kochi, J. K. *J. Am. Chem. Soc.*, **1986**, *108,* 2309-2320.
37. Value given is for the complexation reaction of styrene derivatives with O=FeTMP$^{\cdot+}$. Groves, J. T.; Watanabe, Y. *J. Am. Chem. Soc.*, **1986**, *108*, 507-508.
38. Yang, Y.; Valentine, J. S.; in preparation.
39. Ishii, Y.; Inamato, Y.; *Kogyo Kagaku, Zasshi*, **1960**, *63*, 765-768.
40. Schardt, B. C.; Hill, C. L. *Inorg. Chem.*, **1983**, *22*, 1563-1565.
41. Gallos, J.; Varvoglis, A.; Alcock, N. W. *J. Chem. Soc., Perkin Trans. I*, **1985**, 757-763.
42. Smegal, J. A.; Schardt, B. C.; Hill, C. L. *J. Am. Chem. Soc.*, **1983**, *105*, 3510-3515.
43. Moriarty, R. M.; Gupta, S. C.; Hu, H.; Berenschot, D. R.; White, K. B. *J. Am. Chem. Soc.*, **1983**, *103*, 686-688.

RECEIVED December 30, 1987

Chapter 7

Coupled Binuclear Copper Active Sites

Edward I. Solomon

Stanford University, Department of Chemistry, Stanford, CA 94305

A coupled binuclear copper active site is found in a variety of different metalloproteins involved in dioxygen reactions. These include hemocyanin (reversible O_2 binding), tyrosinase (O_2 activation and hydroxylation) and laccase (O_2 reduction to water). Unique excited state spectral features of active site derivatives and binuclear copper model complexes are used to define exogenous ligand bridging between the two coppers at the active sites of hemocyanin and tyrosinase. A chemical and spectroscopic comparison of the coupled site in hemocyanin to the binuclear center (Type 3, T3) in laccase indicates that the sites are similar with respect to an endogenous bridge responsible for antiferromagnetic coupling but differ in that exogenous ligands bind to only one copper at the T3 site. In addition to the T3 coupled binuclear copper site, laccase contains two additional coppers, the Type 1 and Type 2 centers. X-ray absorption edge spectral studies of the 1s->4p transition at 8984 eV are used to demonstrate that in the absence of the T2 copper, the T3 site is reduced and will not react with O_2. The role of the T2 and T3 coppers in exogenous ligand binding has been probed through low temperature magnetic circular dichroism (MCD) which allows a correlation between the excited state and ground state spectral features. Through these low temperature MCD studies it is demonstrated that one azide produces charge transfer transitions to both the paramagnetic T2 and the antiferromagnetic T3 center defining a new trinuclear copper cluster active site which appears to be important in the irreversible multi-electron reduction of dioxygen to water.

The Coupled Binuclear Copper Active Site

A coupled binuclear active site is found in a wide variety of proteins and enzymes which are involved in different biological

0097–6156/88/0372–0116$09.75/0

functions. All reactions involve dioxygen and it has been the major focus of research of this class of metalloproteins to establish correlations between differences in active site geometric and electronic structure and variation in protein function. Proteins containing a coupled binuclear copper active site are summarized in Table 1.

Table 1. **Proteins Containing Coupled Binuclear Copper Active Sites and Their Functions**

HEMOCYANIN:	Arthropod	deoxy + $O_2 \rightleftarrows$ oxy
	Mollusc	deoxy + $O_2 \rightleftarrows$ oxy
		$2H_2O_2 \rightleftarrows O_2 + 2H_2O$
TYROSINASE:		deoxy + $O_2 \rightleftarrows$ oxy
		$2H_2O_2 \rightleftarrows O_2 + 2H_2O$
		phenol + $2e^-$ + O_2 + $2H^+ \rightleftarrows$ *o*-diphenol+H_2O
		2*o*-diphenol + $O_2 \rightleftarrows$ 2*o*-quinone + $2H_2O$
MULTICOPPER OXIDASES:		
	Laccase	
	Ceruloplasmin	$4AH + O_2 \rightarrow 4A + 2H_2O$
	Ascorbate Oxidase	

The hemocyanins which cooperatively bind dioxygen are found in two invertebrate phyla; arthropod and mollusc. The mollusc hemocyanins additionally exhibit catalase activity. Tyrosinase, which also reversibly binds dioxygen and dismutates peroxide, is a monooxygenase, using the dioxygen to hydroxylate monophenols to ortho-diphenols and to further oxidize this product to the quinone. Finally, the multicopper oxidases (laccase, ceruloplasmin and ascorbate oxidase) also contain coupled binuclear copper sites in combination with other copper centers and these catalyze the four electron reduction of dioxygen to water.

Our earlier research on the coupled binuclear copper proteins generated a series of protein derivatives in which the active site was systematically varied and subjected to a variety of spectroscopic probes. These studies developed a Spectroscopically Effective Model for the oxyhemocyanin active site.(1) The coupled binuclear copper active site in tyrosinase was further shown to be extremely similar to that of the hemocyanins with differences in reactivity correlating to active site accessibility, and to the monophenol coordinating directly to the copper(II) of the oxytyrosinase site.(2) These studies have been presented in a number of reviews.(3) In the first part of this chapter, we summarize some of our more recent results related to the unique spectral features of oxyhemocyanin, and use these to obtain insight into the dioxygen-copper bond. In the second part, our chemical and spectroscopic studies of the coupled binuclear site are extended to the simplest of the multicopper oxidases, laccase. The coupled binuclear copper site in this enzyme is found to be significantly different from that in hemocyanin with respect to exogenous ligand binding; in particular, low temperature magnetic circular dichroism (MCD) is used to determine that an additional copper is involved in this binding, thus defining a new trinuclear copper cluster active site involved in the reduction of dioxygen to water.

Electronic Structure of Oxyhemocyanin

Resonance Raman spectroscopy has shown that in oxyhemocyanin (and oxytyrosinase) bound dioxygen has been formally reduced to peroxide and thus the binuclear site exists in the [Cu(II)Cu(II)] oxidation state.(4) However, the spectroscopic features of oxyhemocyanin shown in Figure 1 are quite unusual when compared to those of normal, monomeric, tetragonal cupric complexes. Instead of the normal weak d-d transitions at 600-700 nm, oxyhemocyanin exhibits two intense absorptions bands in the visible and near-UV. Further, in contrast to the usual tetragonal Cu(II) EPR spectra, oxyhemocyanin is EPR nondetectable. Since both coppers in the active site are known to be d^9 cupric from X-ray absorption edge studies,(5,6) they must be strongly antiferromagnetically coupled. SQUID magnetic susceptibility studies have only been able to put a lower limit on the magnitude of this coupling of $-2J > 600\ cm^{-1}$ ($H_{ex} = -2JS_1 \cdot S_2$).(7)

Since the ground state is spectroscopically inaccessible, the excited state features must be employed to probe the interactions of dioxygen with the coppers in the oxyhemocyanin active site. Figure 2 presents the spectroscopic changes observed by displacement of peroxide from the oxyhemocyanin site, producing the met derivative [Cu(II)Cu(II)].(1) The two intense bands in the oxyhemocyanin absorption spectrum at 350 nm ($\varepsilon = 20{,}000\ M^{-1}\ cm^{-1}$) and 600 nm ($\varepsilon = 1000\ M^{-1}\ cm^{-1}$) are eliminated as is a feature in the circular dichroism (CD) spectrum at 486 nm ($\Delta\varepsilon = +2.53\ M^{-1}\ cm^{-1}$). These three features are therefore assigned as O_2^{2-} --> Cu(II) charge transfer transitions. Weak d-d transitions which are sensitive to the average ligand field at the coppers remain at $\sim$ 630 nm ($\varepsilon \sim 250\ M^{-1}\ cm^{-1}$) and reflect an approximately tetragonal coordination geometry.

It is thus important to define the charge transfer spectral features associated with an O_2^{2-}-Cu(II) bond and to use these features to obtain insight into the charge transfer spectrum of oxyhemocyanin. Karlin and coworkers have synthesized a binuclear copper(I) model complex which reacts with dioxygen to give the absorption spectrum changes illustrated in Figure 3a.(8) Resonance Raman excitation into these features(9) produces a metal ligand stretch at 488 cm^{-1} and an intraligand stretch at 803 cm^{-1} which shifts to 750 cm^{-1} with $^{18}O_2$ indicating that dioxygen is bound as peroxide (Figure 3b). The mode of peroxide coordination was probed through a mixed isotope perturbation of the resonance Raman spectrum combined with normal coordinate analysis. In Figure 3c is presented the resonance Raman spectrum of the oxygenated complex in the $Cu-O_2$

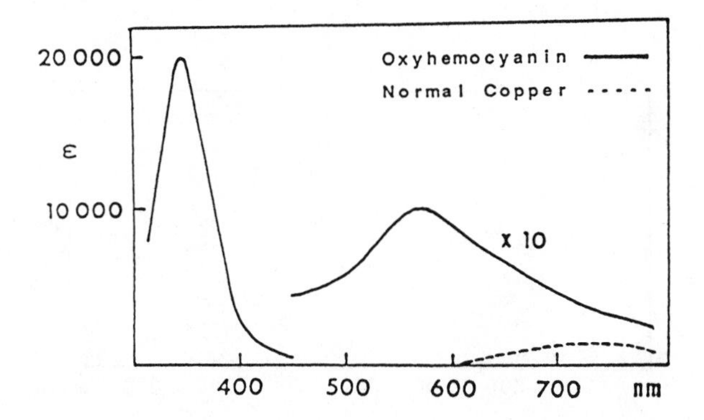

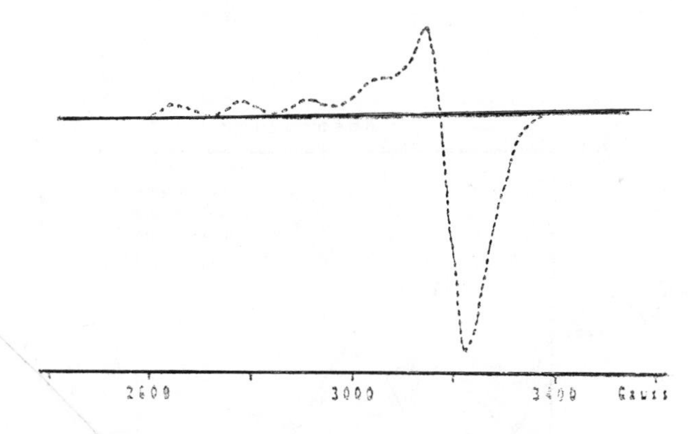

Figure 1. Comparison of absorption (top) and EPR (bottom) spectra of normal tetragonal copper(II) and oxyhemocyanin.

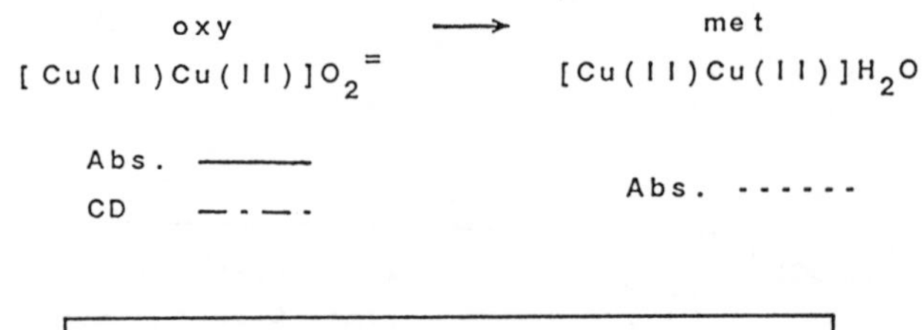

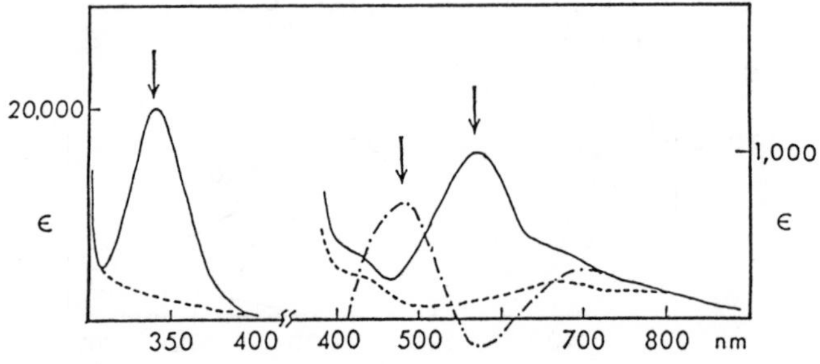

Figure 2. Absorption and circular dichroism spectra of oxyhemocyanin compared with absorption spectrum of methemocyanin.

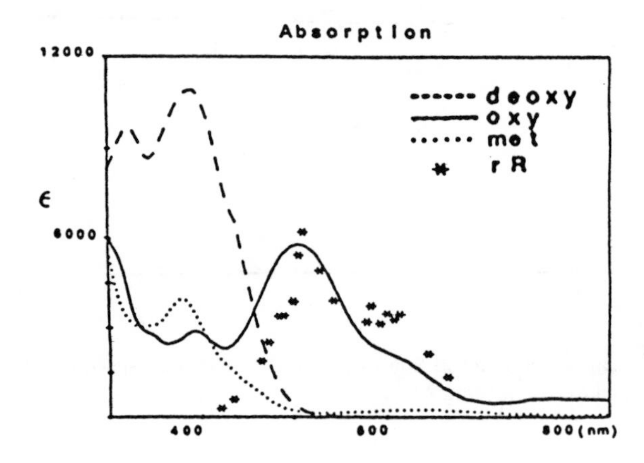

Figure 3a. Absorption spectra of deoxy, oxy, and met forms of O_2 binding binuclear copper model complex. Also included is the resonance Raman profile for the O-O stretch in the oxy complex.

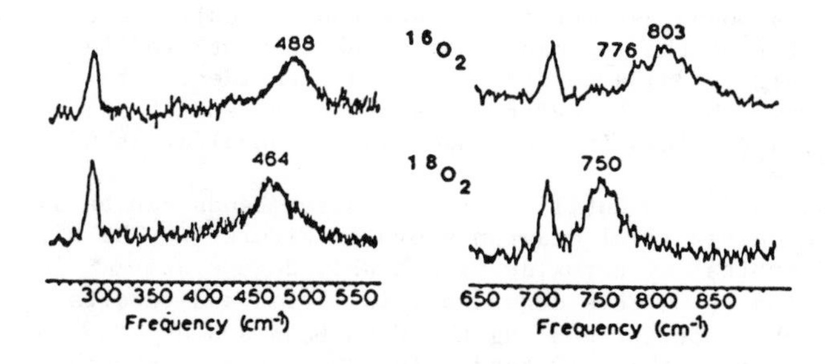

Figure 3b. Resonance Raman spectra in $^{16}O_2$ and $^{18}O_2$ of the oxy complex. The unlabelled peaks are due to the CH_2Cl_2 solvent.

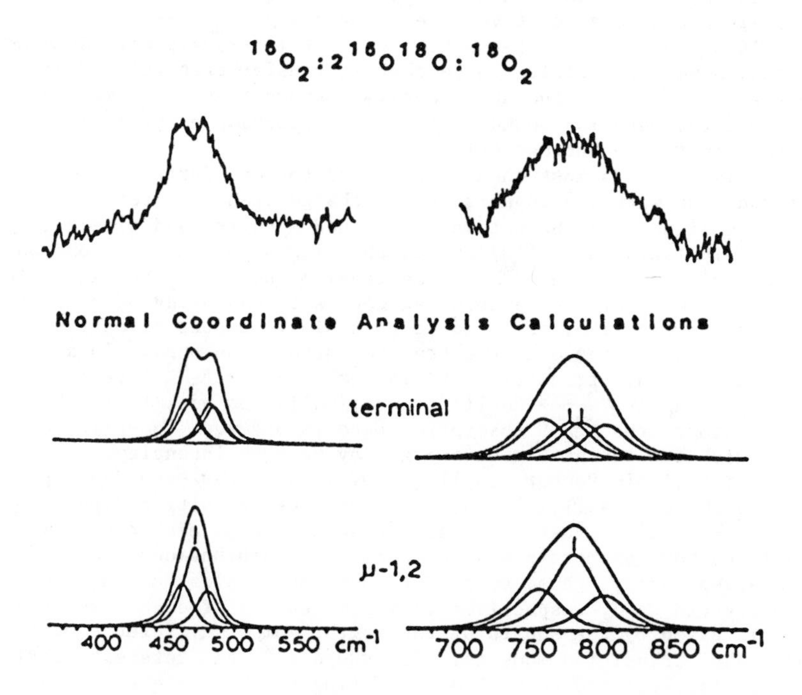

Figure 3c. Comparison of observed resonance Raman spectra for isotope mixture with predicted spectra for two possible coordination geometries.

and O-O stretch regions using a statistical mixture of oxygen isotopes ($^{16}O_2$: $2^{16}O^{18}O$: $^{18}O_2$). Also shown is the vibrational spectrum calculated using experimental lineshapes for both terminal and bridging peroxide. The metal-ligand stretch clearly allows these possible binding geometries to be distinguished and indicates that peroxide is bound asymmetrically and thus to only one copper. Therefore, from the resonance Raman enhancement profile of the O-O stretch (Figure 3a) a terminally bound peroxide-cupric complex exhibits two charge transfer transitions, a band at 503 nm (ε = 6300 M^{-1} cm^{-1}) and a less intense lower energy shoulder at 625 nm (ε = 1100 M^{-1} cm^{-1}).

These O_2^{2-} --> Cu(II) charge transfer bands can be assigned using the energy level diagram shown in Figure 4. The highest energy occupied orbital of peroxide is a doubly degenerate π^* level. Upon coordination to Cu this degeneracy is lifted and results in a π_σ^* level which is oriented along the Cu-O bond and a π_v^* which is perpendicular to the Cu-O bond. The π_σ^* is more strongly stabilized by the bonding interaction with the copper. Further, the highest energy, half-occupied orbital of tetragonal Cu(II) is $d_{x^2-y^2}$ which has a lobe directed at the peroxide ligand in the equatorial plane. Since the intensity of charge transfer transitions is proportional to overlap (S) of the donor and acceptor orbitals ($I_{ct} \propto (SR)^2$), this analysis predicts a relatively weak low energy π_v^* --> Cu(II) transition and a more intense π_σ^* --> Cu(II) transition. Oxyhemocyanin, however, exhibits three charge transfer transitions (arrows in Figure 2). Since the above analysis shows that only two bands are possible for peroxide bound to a single copper, the three band pattern must have another origin.

This analysis must then be extended to consider the effects of interactions with two coppers on the charge transfer transitions of a bridging ligand.(1) Structurally defined monomers and dimers of Cu(II) with azide exist(10) and as this ligand has a π^{nb} homo which is very similar to the π^* valence level of peroxide, these complexes are suitable models for probing exogenous ligand bridging effects. As shown in Figure 5, azide bound to a single copper gives rise to one relatively intense charge transfer absorption band. In analogy to peroxide, this transition originates from the π_σ^{nb} level; the corresponding π_v^{nb} --> Cu(II) transition is too weak to be observed. In the dimer, the π_σ^{nb} absorption band is observed to split into two bands with the one at higher energy having more intensity. A Transition Dipole Vector Coupling (TDVC) Model has been developed to analyze these data.(1,11) In the dimer, the transition dipoles of the N_3^- --> Cu(II) charge transfer transitions to each copper will couple to form symmetric and antisymmetric combinations (Table 2). From group theory, these coupled transition dipoles have A_1 and B_1 symmetry (in C_{2v}), respectively. The intensity of each dimer transition is given by the square of the vector sum or difference of the individual transition moments. For the dimer, the intensity ratio I_{A1}/I_{B1} is predicted to be $\sim$ 8 indicating that the higher energy intense band is the A_1 component of π_σ^{nb} --> Cu(II) transfer. Experimentally, the intensity ratio is somewhat lower which derives from vibronic mixing of A_1 intensity into the B_1 transition. Further, the coulomb interaction between the two transition moments can be estimated for the μ-1,3 geometry to obtain an approximate

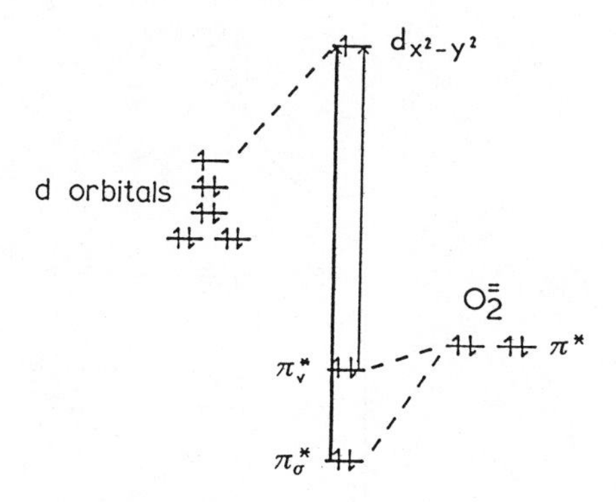

Figure 4. Molecular orbital diagram for terminally bound peroxide-Cu(II) complex. The two possible charge transfer transitions are indicated with relative intensity.

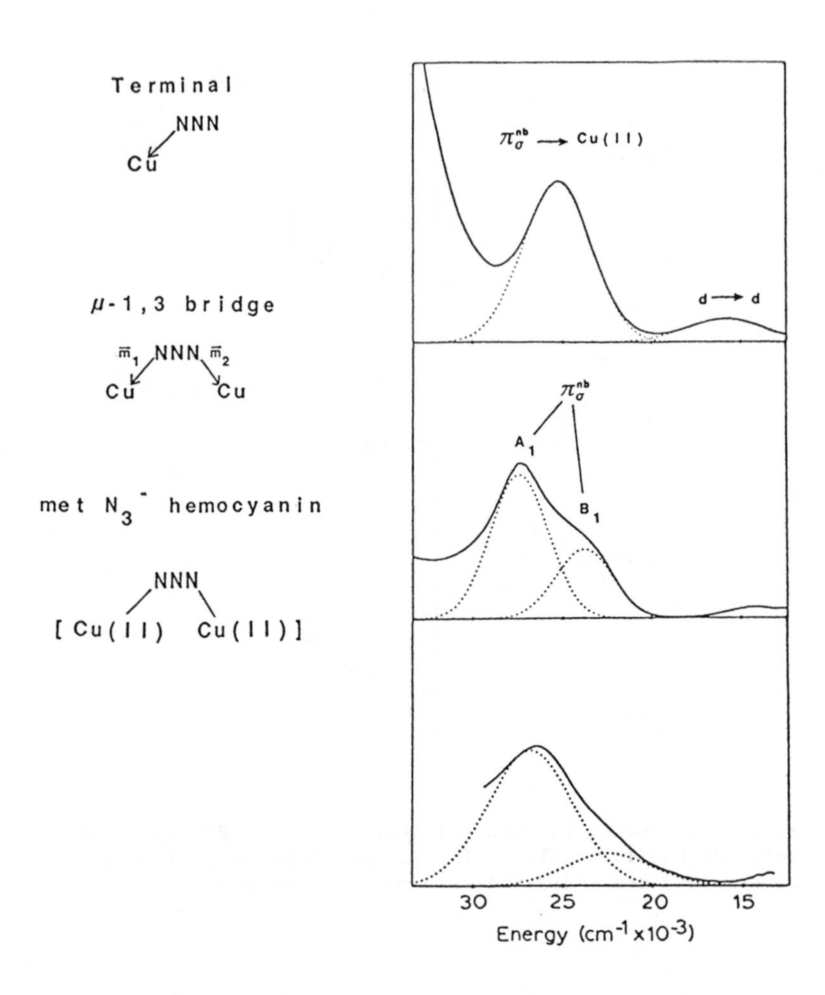

Figure 5. Absorption spectra with Gaussian resolution for terminally bound and bridging azide model complexes and for met-azide hemocyanin.

Table 2. Transition Dipole Vector Coupling

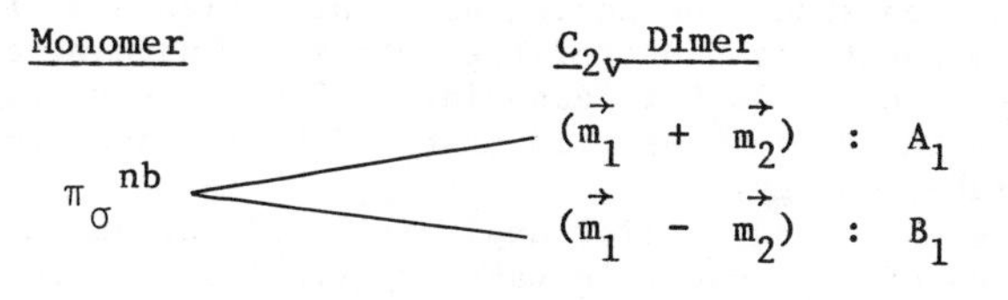

	calc.	exp.
I_{A_1}/I_{B_1}	~ 8	~ 3
$E_{A_1}-E_{B_1}$	350 cm^{-1}	3500 cm^{-1}

value for the energy difference between the A_1 and B_1 dimer states. The calculated value gives the correct energy ordering with the A_1 component at higher energy but is much smaller than the observed splitting, which indicates that excited state exchange terms contribute significantly to the splitting (see reference 12). Finally, addition of azide to methemocyanin, Figure 5 (Bottom), yields a charge transfer spectrum which shows an A_1, B_1 pattern very similar to the spectrum of the μ-1,3 azide model complexes. Thus azide binds in an analogous μ-1,3 bridging geometry to the binuclear site in methemocyanin.

This approach can now be applied to interpret the charge transfer spectrum of oxyhemocyanin with peroxide bridging in a μ-1,2 fashion.(1) In the dimer, the π_V^* and π_σ^* charge transfer transitions of the monomer each split into two bands. The symmetry types, selection rules, intensity ratios, and coulomb splittings have been calculated and are given in Table 3. The π_V^* --> Cu(II) transition will split into two components, a lower energy B_2 band which should contain most of the absorption intensity and a higher energy A_2 transition having little absorption intensity. This higher energy transition is, however, magnetic dipole allowed and is therefore predicted to be relatively intense in the CD spectrum. The predicted relative energies and absorption and CD intensities correlate with the bands at 600 nm and 486 nm (Figure 6), allowing

Table 3. Transition Dipole Vector Coupling-Oxyhemocyanin

Monomer	C_{2v} Dimer		
π_V^*	A_2	R_z	I_{A_2}/I_{B_2} = 0.0
	B_2	y, R_x	$E_{A_2} - E_{b_2}$ = 840 cm^{-1}
π_σ^*	A_1	z	I_{A_1}/I_{B_1} = 2.6
	B_1	x, R_y	$E_{A_1} - E_{B_1}$ = -5000 cm^{-1}

their assignment as the B_2 and A_2 components, respectively. The π^*_σ --> Cu(II) transition should be analogously split (Table 3) into a low energy A_1 component with most of the intensity and a B_1 component at higher energy. The A_1 band is associated with the very intense 350 nm band while the B_1 component is obscured by intense protein absorption at higher energy.

From the preceding and earlier analysis(1) of the unique spectroscopic features of oxyhemocyanin (and oxytyrosinase) a spectroscopically effective model of the active site has been generated which is illustrated in Figure 7. In the site, two tetragonal cupric ions are bridged by both an endogenous ligand (OR^-) and the exogenous μ-1,2 peroxide. The endogenous bridge is responsible for the strong antiferromagnetic coupling between the coppers and hence the lack of an EPR signal, and the bridging peroxide gives rise to the intense charge transfer features in the electronic absorption spectrum of oxyhemocyanin. While, unfortunately, no high resolution crystallographic information is presently available on oxyhemocyanin(13a), the structure of deoxy hemocyanin has recently been determined to 3.2 A resolution(13b). For deoxy, the two Cu(I) are each found to have two short Cu-N bonds (His at $\sim$ 2.0 A) and one long, axial Cu-His bond (at $\sim$ 2.6 A) consistent with the tetragonal effective Cu(II) symmetry of oxyhemocyanin. These crystallographic results indicate that the endogenous bridge is not a protein residue but is derived from water, most likely in the form of hydroxide.

Chemical and Spectroscopic Comparison With Laccase

Characterization of the Type 2 Depleted Derivative of Laccase. The model for the coupled binuclear copper site in hemocyanin and tyrosinase (Figure 7) may now be compared to the parallel site in laccase which contains a blue copper (denoted Type 1 or T1), a normal copper (Type 2, T2), and a coupled binuclear copper (Type 3, T3) center. As shown in Figures 8a and b, native laccase has contributions from both the T1 and T2 copper centers in the EPR spectrum (the T3 cupric ions are coupled and hence EPR nondetectable as in hemocyanin), and an intense absorption band at $\sim$600 nm (ε=5700 M^{-1} cm^{-1}) associated with the T1 center (a thiolate --> Cu(II) CT transition).(14) The only feature in the native laccase spectra believed to be associated with the T3 center was the absorption band at 330 nm ($\varepsilon \sim$ 3200 M^{-1} cm^{-1}) which reduced with two electrons, independent of the EPR signals.(15) Initial studies have focussed on the simplified Type 2 depleted (T2D) derivative(16) in which the T2 center has been reversibly removed. From Figure 8 the T2 contribution is clearly eliminated from the EPR spectrum of T2D and the T1 contribution to both the EPR and absorption spectrum remains. However, upon removal of the T2 copper the 330 nm absorption is eliminated, raising concern with respect to the involvement of the T2 copper in this spectral feature.

When T2D laccase is reacted with peroxide the 330 nm band reappears (Figure 8a).(17) This observation suggested the possibility that the T3 site might be reduced in the T2D derivative but could be oxidized by H_2O_2. Alternatively, other researchers assigned the 330 nm band as an O_2^{2-} --> Cu(II) charge transfer

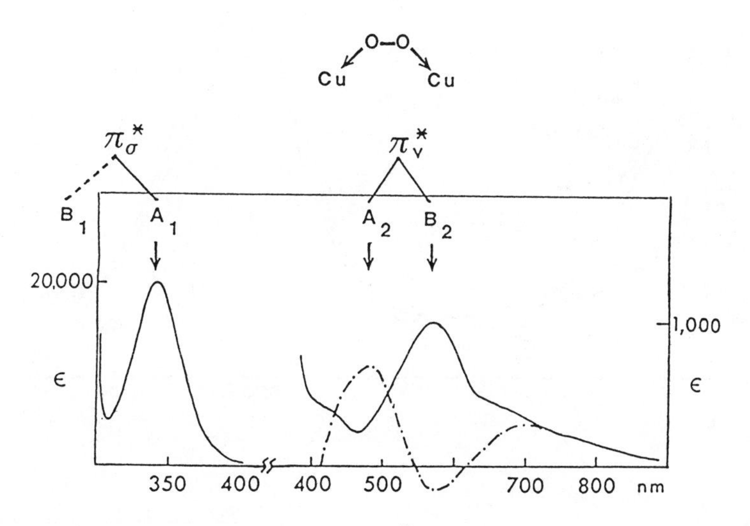

Figure 6. Assignment of the O_2^{2-} --> Cu(II) charge transfer spectrum of oxyhemocyanin using the TDVC model.

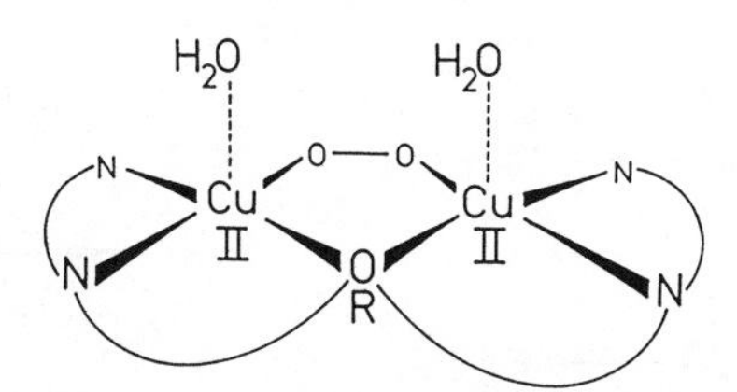

Figure 7. The Spectroscopically Effective Active Site of hemocyanin and tyrosinase.

a

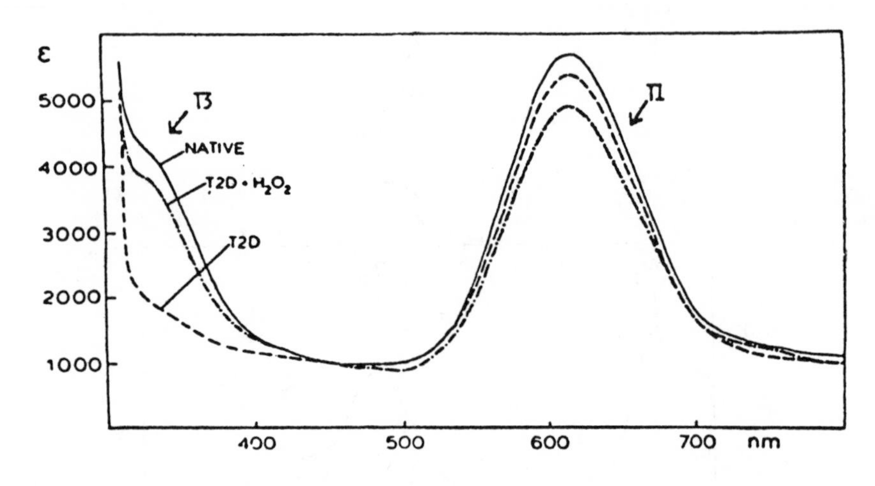

c

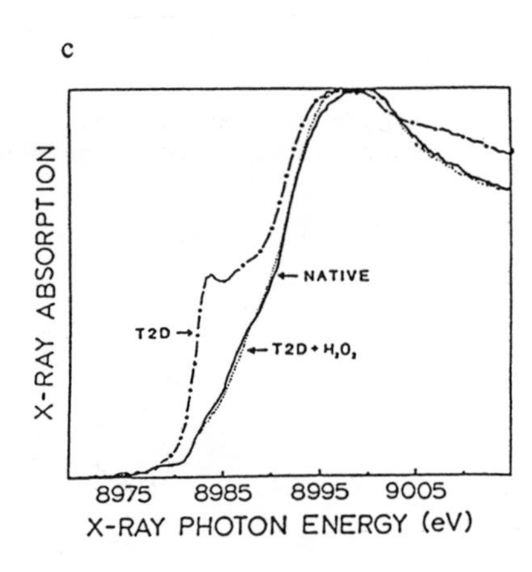

b

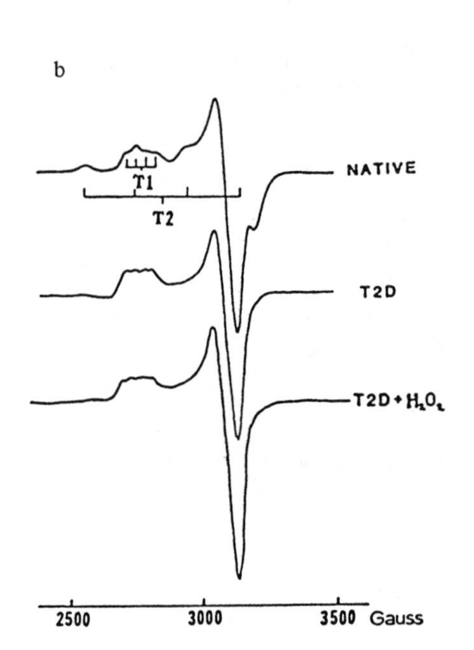

Figure 8. (a) Optical absorption, (b) EPR, and (c) X-ray absorption edge spectra for native and T2D laccase and after reaction of T2D with peroxide.

transition in analogy to oxyhemocyanin and therefore associated this spectral feature with peroxide binding to the T3 site.(18) An absorption feature at 330 nm, however, cannot directly distinguish between these possibilites. Also, the EPR spectrum is not diagnostic of the T3 oxidation state since both a reduced site and an antiferromagnetically coupled oxidized binuclear cupric site would be EPR nondetectable. A direct probe of oxidation state is, however, available through X-ray absorption edge spectroscopy. The T2D form exhibits a peak at 8984 eV which is eliminated by reaction with peroxide (Figure 8c), qualitatively indicating that Cu(I) is being oxidized to Cu(II).(17) Such conclusions must be drawn cautiously however, since covalent cupric complexes can also exhibit X-ray absorption edge intensity at low energy.(19)

The oxidation state dependence of the X-ray absorption edge feature was defined and quantitated by a systematic study of a series of twenty Cu(I) and forty Cu(II) model complexes in collaboration with Professor Keith Hodgson.(6) The Cu K edge band energies and shapes depend on oxidation state, ligation, and coordination geometry. Figure 9 (Top) presents representative edges for two, three, and four coordinate Cu(I), and tetragonal Cu(II) complexes with a variety of ligand sets. Two or three coordinate Cu(I) complexes always exhibit a peak at energies below 8985 eV. In addition, the three coordinate complexes display a double peak with lower intensity. No Cu(II) complex studied shows a peak below 8986 eV. There is however a low energy tail in the 8984 eV region which is associated with a peak at higher energy.

These spectral differences can be understood through ligand field theory.(6) For two coordinate Cu(I) with an approximately linear geometry, the electric dipole allowed 1s --> 4p transition is predicted to split into 1s --> $4p_z$ and 1s --> $4p_{x,y}$ components. Further, the 1s --> $4p_z$ transition is expected to be highest in energy due to repulsive interactions with the axial ligands. Polarized single crystal X-ray edge spectroscopy(20) confirms the assignment of the lowest energy peak as 1s --> $4p_{x,y}$. As shown at the bottom of Figure 9, increasing the ligand field strength by adding a third ligand along the y axis leads to an additional repulsive interaction which raises the energy of the $4p_y$ level relative to the $4p_x$. This accounts for the experimentally observed edge splitting in the three coordinate Cu(I) complexes. For a tetrahedral geometry, all p orbitals are equally destabilized by the ligand field and the 1s --> 4p transitions are all above 8985 eV.

For the tetragonal Cu(II) complexes, both the low energy tail and its associated higher energy peak shift to lower energy as the covalent interaction with the equatorial ligand set increases. Since this band is z polarized(20) with an energy shift dependent on equatorial ligand ionization energy, this transition can be assigned as the 1s --> $4p_z$ combined with a ligand --> Cu(II) charge transfer shake-up, which is also supported by final state calculations.(21)

Turning to the X-ray absorption edge spectrum of the T3 site in T2D laccase (Figure 10a), a peak is observed below 8984 eV. This energy is characteristic of Cu(I) while the shape of the peak suggests that it is due to three coordinate copper. The amount of reduced copper present can then be quantitated from the normalized edge intensities of copper model complexes with the appropriate

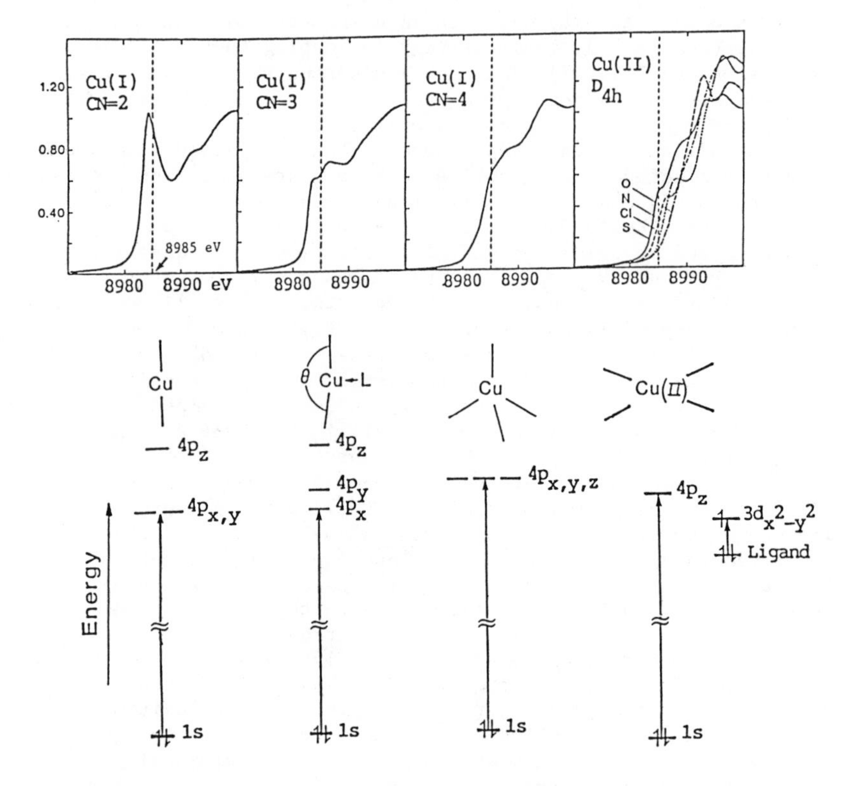

Figure 9. Representative X-ray absorption edge data for Cu(I) and Cu(II) complexes in a variety of coordination environments. Also shown (bottom) are ligand field splitting diagrams for these complexes.

coordination number. For T2D laccase, the 8984 eV peak indicates that 98 ± 22% of the T3 sites are reduced.(22) Peroxide reacted T2D laccase shows no 8984 eV feature and corresponds quantitatively to a fully oxidized T3 site.

The possibility of the 330 nm absorption band being a O_2^{2-} --> Cu(II) charge transfer transition was investigated through a correlation of the 330 nm absorption intensity increase with the loss of the 8984 eV Cu(I) peak as T2D was titrated with H_2O_2. The linear correlation in Figure 10c demonstrates that all change at 330 nm is associated with oxidation of the T3 site with no evidence for additional peroxide binding to the oxidized Cu(II). Thus, as indicated in Scheme I, the T3 site in T2D laccase is in a deoxy [Cu(I)Cu(I)] form which, in strong contrast to deoxyhemocyanin and deoxytyrosinase, does not react with dioxygen. It is however oxidized by peroxide to give the binuclear cupric met derivative, [Cu(II)Cu(II)].(23)

Scheme I.

$$[Cu(I)Cu(I)] \text{ deoxy T2D} \xrightarrow{O_2} \text{N.R.}$$

$$[Cu(I)Cu(I)] \text{ deoxy T2D} \xrightarrow{H_2O_2} [Cu(II)Cu(II)] \text{ Met T2D}$$

Comparison of the T3 Site in T2D Laccase with the Coupled Binuclear Site in Hemocyanin. In comparing the spectroscopic properties of coupled binuclear copper site in met T2D laccase with that of methemocyanin, it is noted that both have no EPR signal. Since both sites contain two Cu(II)'s, the sites must be antiferromagnetically coupled. SQUID magnetic susceptibility measurements presently place a lower limit on the singlet-triplet splitting of $-2J > 400\ cm^{-1}$ for methemocyanin and $-2J > 200\ cm^{-1}$ for met T2D laccase.(24) By lowering the pH and adding azide a new broad EPR signal is produced in both met derivatives (Figure 11).(23,25) This signal accounts for <15% of the T3 sites and can be simulated (dashed lines in Figure 11) based on two dipolar interacting cupric ions. In met T2D laccase this signal clearly shows a high field component at $g_{eff} = 1.86$ associated with a larger zero field splitting and thus shorter Cu-Cu distance relative to methemocyanin. These results are interpreted in terms of an endogenous bridge, RO^-, being present in both sites, which provides the superexchange pathway for strong antiferromagnetic coupling and the lack of an EPR signal. Protonation of this bridge uncouples the sites and a quantitative study of the pH dependence of the dipolar EPR signal indicates an intrinsic pK_a of > 7.0 for the endogenous bridge.(23,25) Thus, met T2D laccase and methemocyanin are similar with respect to exchange coupling through an endogenous bridging ligand.

Reduction of the met derivatives by one electron produces the mixed-valent 1/2-met [Cu(II)Cu(I)] derivatives. A comparison of the chemical and spectroscopic properties of 1/2-met hemocyanin and 1/2-met T2D laccase reveals a number of important differences (Figure 12).

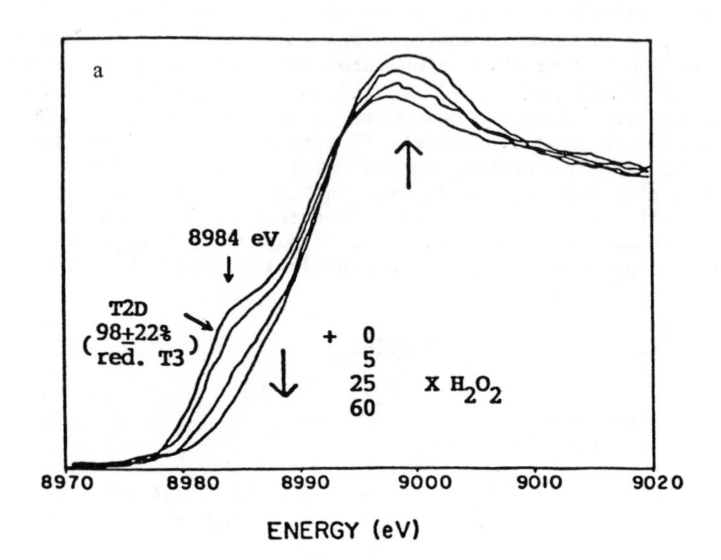

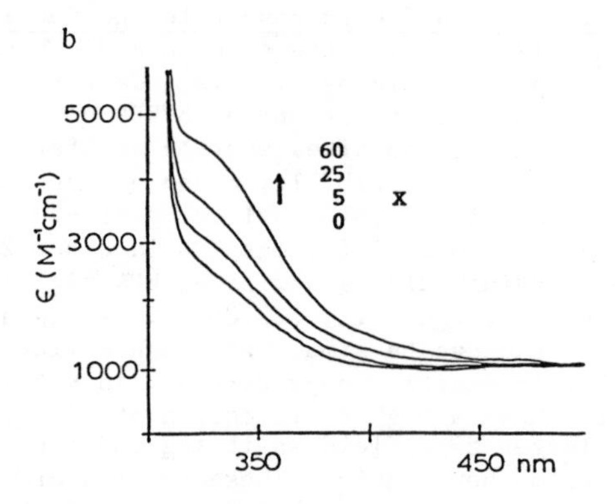

Figure 10. Correlation of X-ray edge data and optical absorption for the reaction of T2D laccase with peroxide.

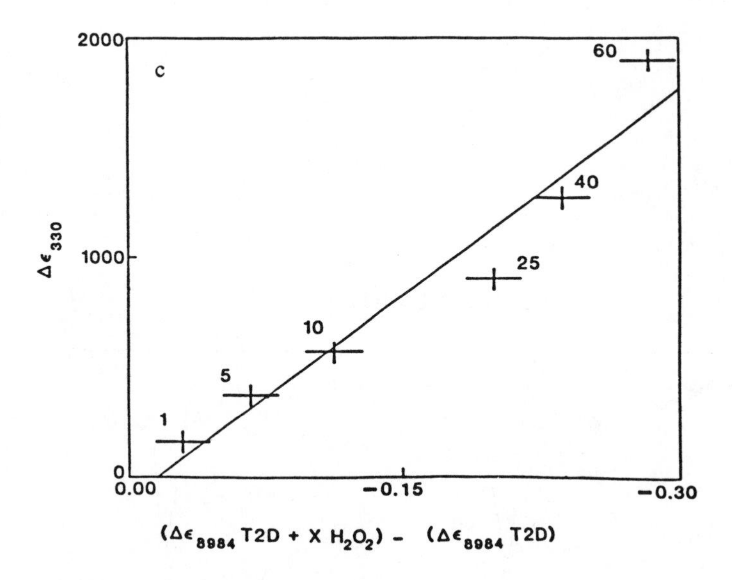

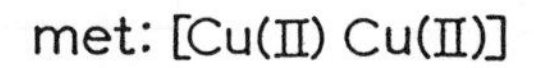
Figure 10. Continued.

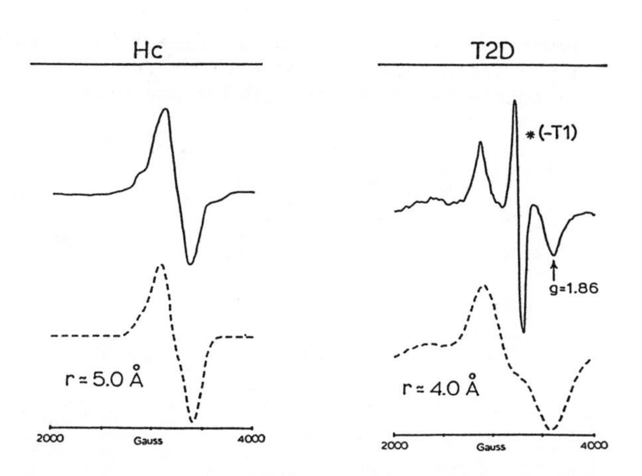

Figure 11. Comparison of EPR spectra of the uncoupled met derivatives of hemocyanin and T2D laccase with simulations for dipolar interacting copper(II)'s at distances indicated. The peak indicated by the asterisk results from subtraction of the T1 signal.

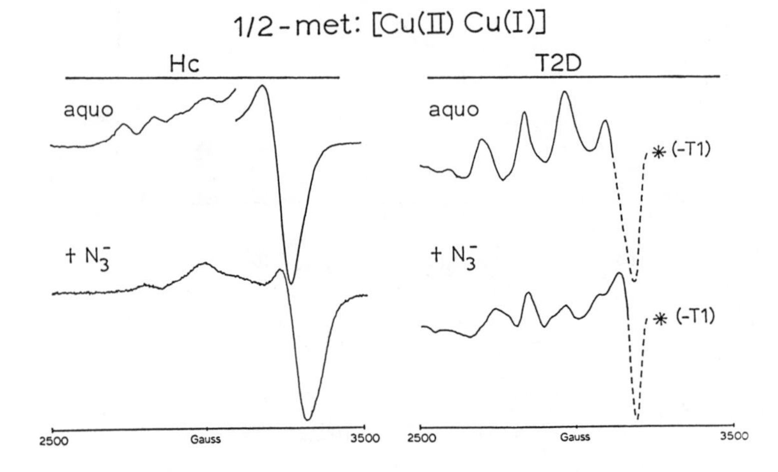

Figure 12. Comparison of EPR spectra of 1/2-met hemocyanin and T2D laccase in the absence and presence of azide. The T1 copper signal has been subtracted from the T2D laccase data.

In particular, the 1/2-met T3 site in T2D has EPR spectra and equilibrium binding constants ($K_{N_3^-} = 10^2\ M^{-1}$) for exogenous ligands which are similar to those observed for normal tetragonal Cu(II) complexes. In 1/2-met hemocyanin, however, binding constants of exogenous ligands are, in general, two orders of magnitude greater ($K_{N_3^-} > 10^4\ M^{-1}$) and the spectroscopic features are very unusual relative to those of normal cupric complexes.

The unusual chemical and spectroscopic features of 1/2-met hemocyanin have been studied in some detail. Specifically, a met apo [Cu(II) -] derivative of hemocyanin has been prepared which contains a single Cu(II) in the active site.(26) The strength of exogenous ligand binding in this derivative very closely parallels aqueous Cu(II) chemistry. For example, as shown in Figure 13a, addition of azide causes changes in the Cu(II) EPR spectrum which indicates azide binding. Dialysis to remove azide rapidly restores the original met apo EPR signal. In contrast, addition of azide to the 1/2-met hemocyanin derivative followed by extensive dialysis does not produce the original EPR signal, but instead the unusual spectrum at the bottom of Figure 13b which corresponds to one extremely tightly bound azide at the active site. Further, reversible coordination of CO to the Cu(I) changes this Cu(II) EPR signal and the azide is now rapidly removed from the active site by dialysis. Thus, the presence of the Cu(I) greatly increases the affinity of the 1/2-met site for azide while coordination of CO to the Cu(I) greatly reduces this affinity. Both observations indicate that in addition to the Cu(II), Cu(I) must also bind the exogenous ligand at the active site. This bridging mode is supported by the unusual spectroscopic features of the 1/2-met hemocyanins.

Figure 14 presents the absorption and EPR spectra of a series of 1/2-met hemocyanin derivatives with variation in tightly bound exogenous ligand, L.(26) The 1/2-met NO_2^- spectral features are representative of normal tetragonal Cu(II) complexes with d --> d transitions in the 600-700 mm region and four hyperfine lines in the $g_\parallel$ region of the EPR spectrum. In contrast, the 1/2-met L derivatives with L = Cl^-, Br^-, I^-, and N_3^- exhibit an increasingly more complex EPR spectrum, and a new absorption band at low energy which increases in intensity over this series. These unusual spectral features have now been studied in some detail.(27) S, X, and Q-band EPR spectra have been obtained and simulated (see Figure 15A for 1/2 met Br^-). In particular, the Q-band data show that only one paramagnetic specie is present and has axial g values. The resolution of the additional hyperfine splittings is improved in the S-band data which demonstrate that the unpaired electron interacts with both copper nuclei. The fitted g values and parallel hyperfine coupling constants are listed in the Table 4. Further, a correlation of the low temperature MCD spectra with the absorption data allows assignment of the new low energy absorption band of the 1/2 met L derivatives. Half met NO_2^- exhibits absorption and low temperature MCD features associated with the d-->d transitions of a tetragonal Cu(II) site (Figure 15B, top). Comparison to the 1/2 met Br^- spectrum (Figure 15B, bottom) indicates that while the d-->d transitions shift in energy consistent with some change in the ligand field at the Cu(II), there is no significant additional low

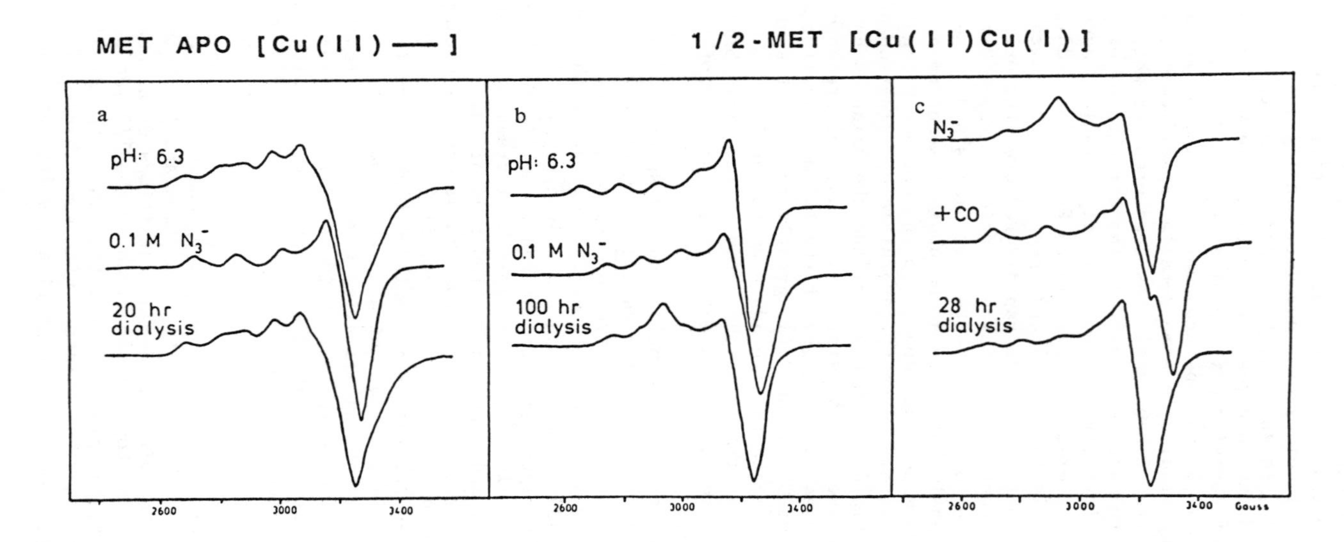

Figure 13. (a) EPR data indicating reversible azide binding to met apo hemocyanin. (b) EPR data indicating irreversible azide binding to 1/2-met hemocyanin. (c) Effect of CO binding on high affinity azide binding in 1/2-met hemocyanin.

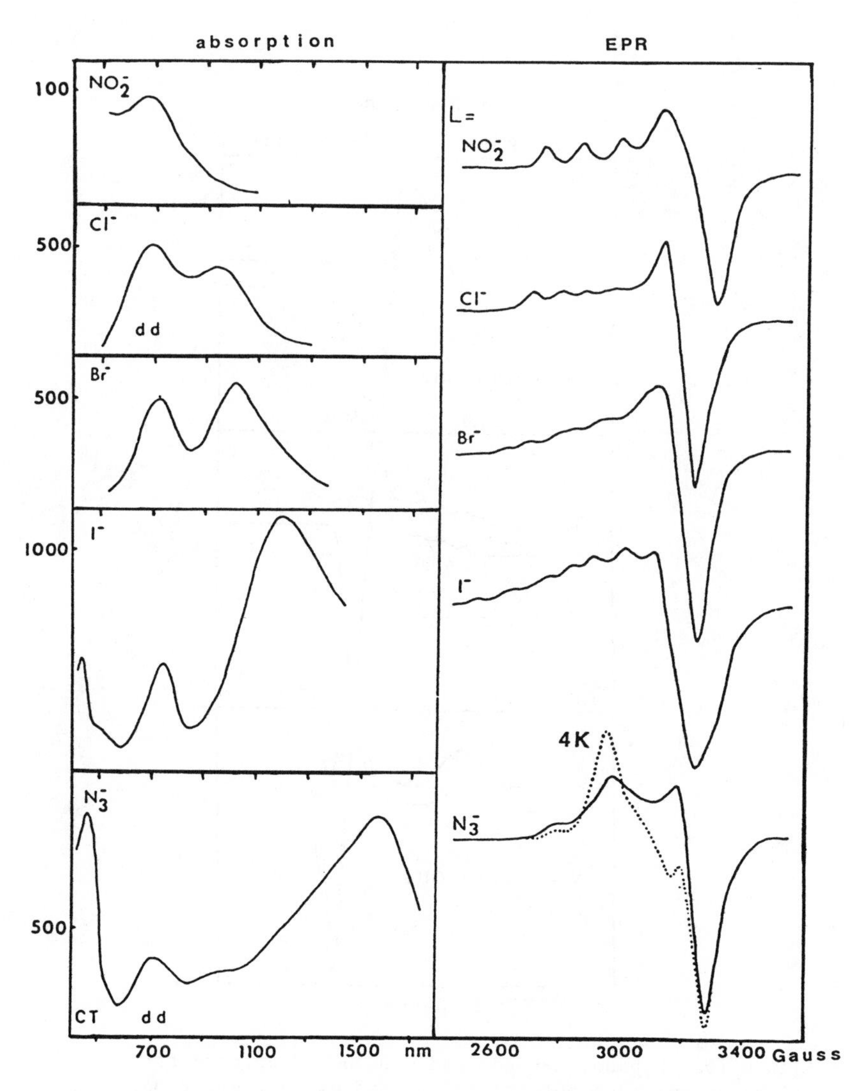

Figure 14. Low temperature absorption spectra (7K) and EPR spectra (77K) for a series of 1/2-met hemocyanin derivatives.

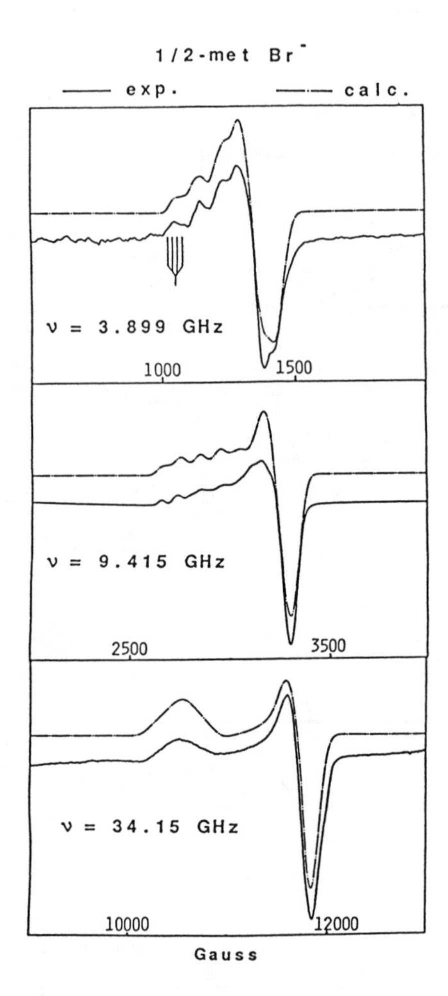

Figure 15a. S-band, X-band, and Q-band EPR spectra for 1/2-met Br^- hemocyanin and spectra calculated using the parameters given in Table 4.

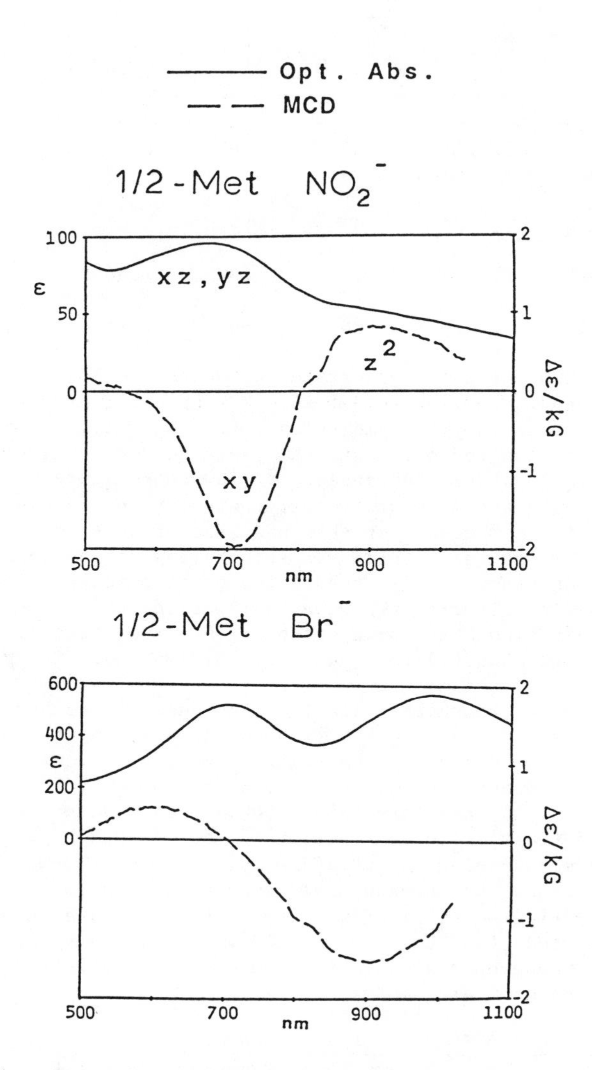

Figure 15b. Comparison of the low-temperature MCD and absorption spectra of 1/2-met NO_2^- and 1/2-met-Br^- hemocyanin.

Table 4. EPR and Intervalence Transfer Transition Parameters for Half Met L Hemocyanins

L	$g_{\parallel}$	$g_{\perp}$	$A^{A}_{\parallel}$ ($\times 10^{-4}$ cm^{-1})	$A^{B}_{\parallel}$ ($\times 10^{-4}$ cm^{-1})	f_{IT} ($\times 10^{-3}$)	%Cu_A	%Cu_B
NO_2^-	2.3022	2.1058 2.0414	120	<16	0.00	>90	<10
Cl^-	2.3470	2.0780	93	<5	5.67	>94	<6
Br^-	2.3115	2.0774	92	38	7.86	71	29
I^-	2.2610	2.0710	98	84	16.78	54	46

temperature MCD intensity associated with the new low energy absorption band. the low temperature MCD spectrum of a paramagnetic center is dominated by C terms (vida infra, Sec. 3). From the general expression for orientationally averaged C terms ($I_c = g_z m_x m_y + g_x m_y m_z + g_y m_x m_z$)(28), the absence of low temperature MCD intensity must result from the lack of two perpendicular non-zero electric dipole transition moments for this new absorption band. The unidirectional nature of this transition allows it to be assigned as an intervalent transfer (IT) transition corresponding to optical excitation of an electron from the Cu(I) to the Cu(II) and thus polarized only along the Cu-Cu vector. The experimentally observed oscillator strengths (f) for the IT transitions are also given in Table 4.

From Table 4, both the relative magnitude of the hyperfine coupling to each copper and the IT absorption intensity correlate with the covalent nature of the tightly bound exogenous ligand. Thus, this exogenous ligand must provide the pathway for electron delocalization and therefore bridge between the two coppers, as shown in Figure 16 (left). In contrast, the lack of both high affinity binding and mixed-valent spectral features in the 1/2-met T2D laccase strongly suggests that exogenous ligands bind to only one copper at the T3 site (Figure 16, right). Since both sites are accessible to exogenous ligands,(23) the lack of dioxygen reactivity of deoxy T2D relative to hemocyanin appears to relate to this difference in exogenous ligand binding mode.

Low Temperature MCD Studies of Exogenous Ligand Binding to Native Laccase. Native laccase reduces dioxygen to water but the reduced T3 site in T2D laccase does not react with dioxygen. Thus, the T2 copper must play an important role in exogenous ligand interactions with the T3 site. We have employed low temperature MCD spectroscopy to probe exogenous ligand interactions with the T2 and T3 centers, since the optical features associated with the paramagnetic T2 site can be distinguished from those associated with the antiferromagnetic T3 centers.(29) As illustrated in Figure 17 (left) application of a magnetic field to the T2 (S=1/2) center splits both the ground and the excited state energies by $g\beta H$. The selection rules for MCD

L
Cu(II) Cu(I)
O
R

L
Cu(II) Cu(I)
O
R'

Coupled Binuclear Site T3 Center in T2D

Figure 16. Comparison of exogenous ligand binding modes in 1/2-met hemocyanin (left) and 1/2-met laccase (right).

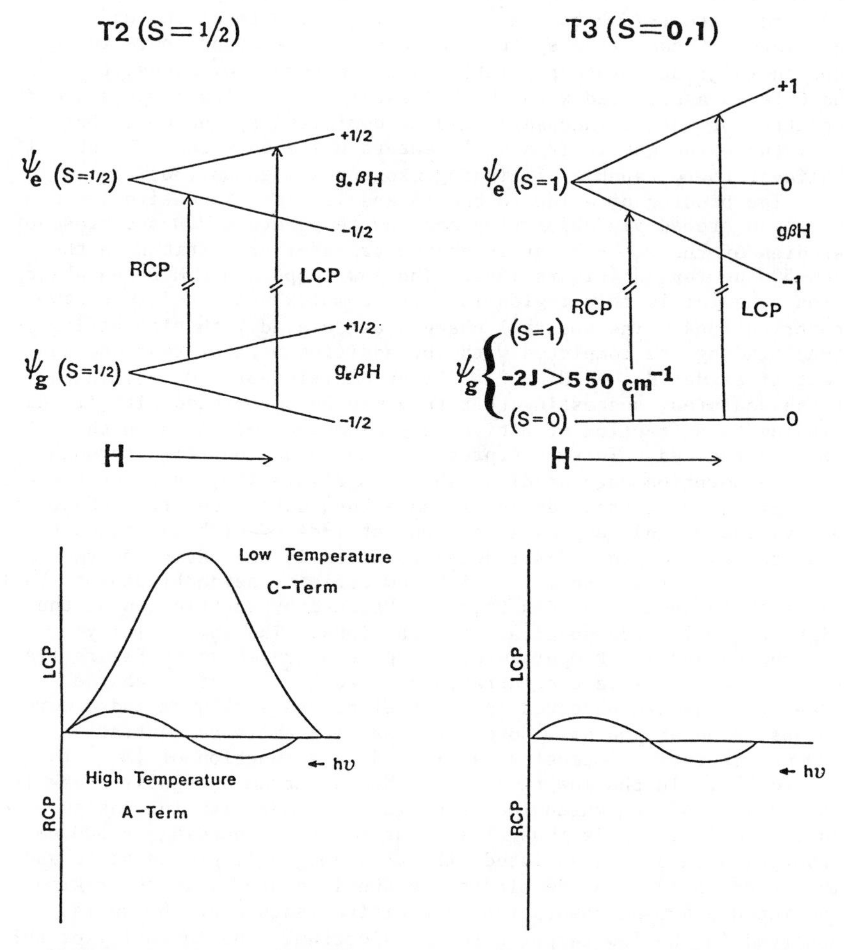

Figure 17. Electronic structure origin of low temperature MCD features for the T2 and T3 centers in native laccase.

transitions are $\Delta m = +1$ for left circularly polarized (LCP) light and $\Delta m = -1$ for right circularly polarized (RCP) light. Thus, one predicts for the T2 center two transitions with equal intensity and opposite sign. Since the Zeeman splitting is on the order of 10 cm^{-1} and the absorption bands are several thousand cm^{-1} wide, the transitions will mostly cancel except for a broad, weak derivative-shaped signal at high temperatures (an A term). However, as the temperature is lowered the Boltzmann population of the m_s=+1/2 level is reduced relative to that of the m_s=-1/2 level. This leads to a large increase in absorption of LCP light and results in a temperature dependent C term. At very low temperatures ($\leq$4K), these C terms will be two to three orders of magnitude more intense than the A terms. In contrast, the strong antiferromagnetic coupling in the T3 center leads to an S=0 ground state and a thermally inaccessible S=1 excited state (-2J > 550 cm^{-1}). Since the S=0 level is nondegenerate, it cannot be split by a magnetic field and there is no change in ground state population with temperature. Thus, there are no C terms associated with the T3 center and the low temperature MCD spectrum of native laccase should be dominated by the contributions from the paramagnetic T2 and T1 centers where only the T2 center is believed to be capable of binding exogenous ligands.(30)

The binding of azide to the T3 and T2 sites in native laccase has been probed via absorption and low temperature MCD spectroscopic studies of the N_3^- --> Cu(II) charge transfer transitions in the 400-500 nm region (Figure 18a). The low temperature MCD and absorption behavior in this region is quite complex and consists of two contributions. The spectral changes associated with high affinity (HA) binding are completed with the addition of less than one equivalent of azide ($K_{HA} \geq 10^4$ M^{-1}). Prior addition of H_2O_2 eliminates these features, suggesting that they may be associated with ligand binding to a fraction of native laccase molecules in which the T3 site is reduced. This interpretation has been confirmed through X-ray absorption edge studies, shown in Figure 18b, in which the edge features of H_2O_2 treated laccase have been subtracted from those of native enzyme.(6) A peak is present at 8984 eV with an intensity that corresponds to ∿22% reduced T3 sites in the native enzyme.

The features associated with low affinity azide binding to fully oxidized laccase sites can then be obtained by subtraction of the high affinity azide spectral contributions. The low affinity absorption and low temperature MCD spectra are shown in Figure 19a for increasing azide concentrations. Two N_3^- -> Cu(II) charge tranfer bands are observed in absorption, one at 500 nm and a more intense band at 400 nm. Both increase with azide concentration; the 400 nm absorption intensity is plotted as a function of [N_3^-] in Figure 19b. In the low temperature MCD spectrum a negative peak is observed at 485 nm which increases in magnitude with increasing azide concentration. It is thus clear that the corresponding ∿ 500 nm absorption band is associated with a paramagnetic ground state and must correspond to azide binding to the T2 center. In the region of the intense 400 nm absorption, a positive feature at 385 nm is observed in the low temperature MCD spectrum. The intensity of this feature first increases, then decreases with increasing azide concentration. As shown in Figure 19b, the intensity of this low temperature MCD feature does not correlate with the 400 nm absorption

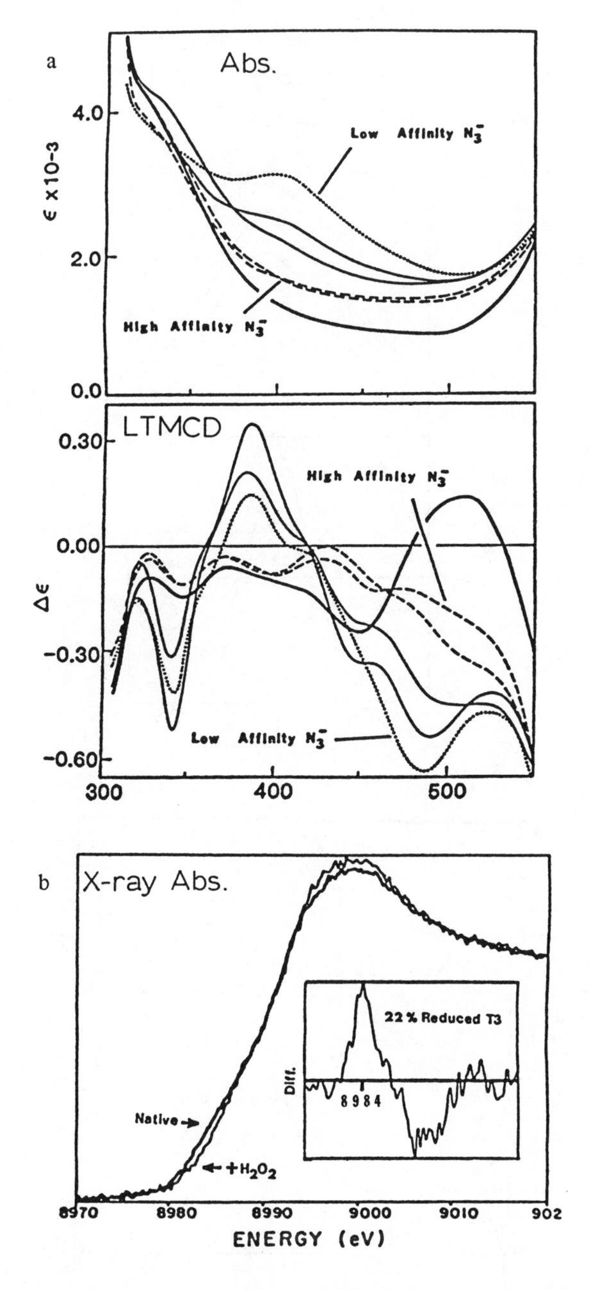

Figure 18. (a) Absorption and low temperature MCD spectra of azide binding to native laccase. (b) X-ray absorption edge data for reaction of native laccase with peroxide (insert presents normalized difference edge).

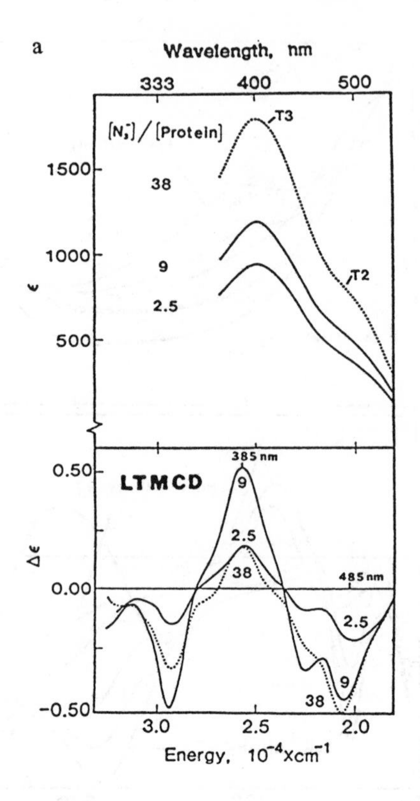

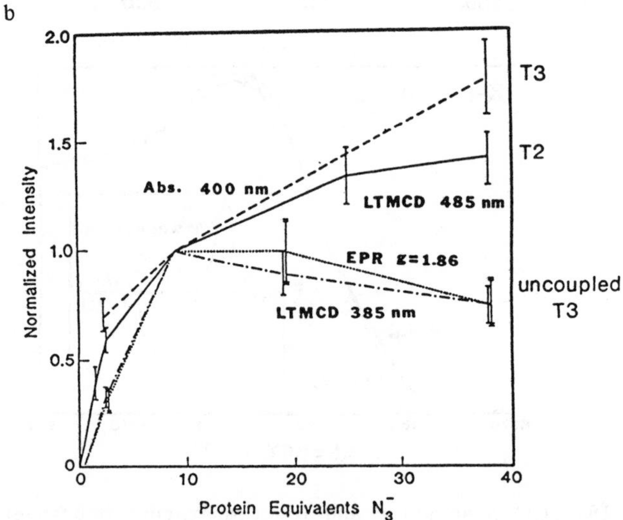

Figure 19. (a) Absorption and low temperature MCD spectra for binding of low affinity azide to native laccase. (b) Correlation of the LT MCD signals at 485 and 385 nm to the 400 nm absorption and the g = 1.86 EPR signal with increasing $[N_3^-]$.

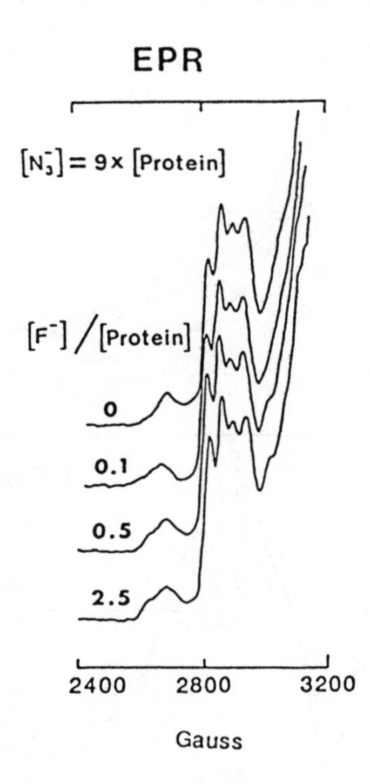

Figure 20. Summary of N_3^-/F^- binding competition experiments for native laccase. Continued on next page.

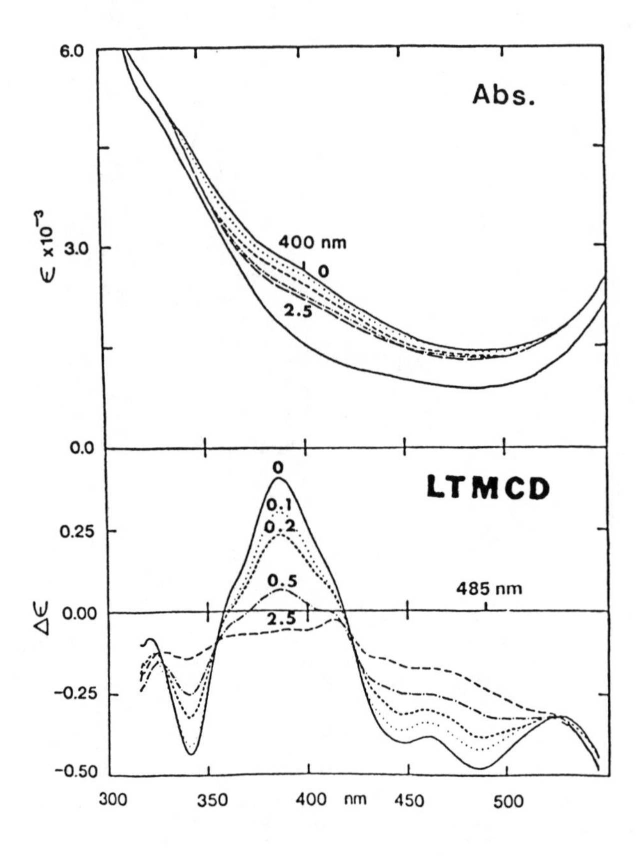

Figure 20. Continued. Summary of N_3^-/F^- binding competition experiments for native laccase.

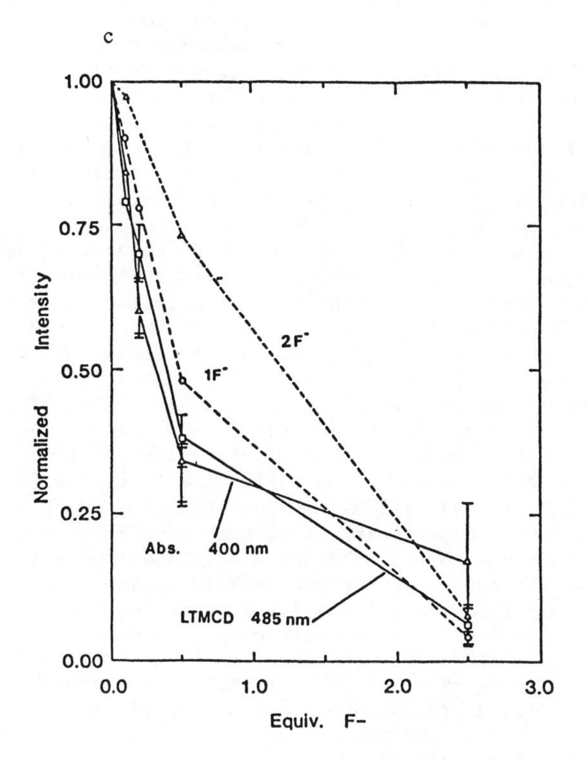

Figure 20. Continued.

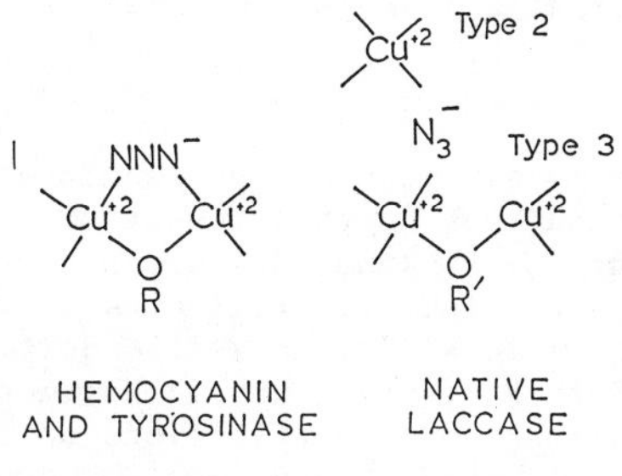

Type 2

N_3^- Type 3

HEMOCYANIN AND TYROSINASE

NATIVE LACCASE

Figure 21. Comparison of the spectroscopically effective models for azide binding at the binuclear copper active site in hemocyanin and the trinuclear copper cluster site in laccase.

intensity, but in fact parallels the appearance of the g=1.86 EPR signal which is associated with the small fraction of uncoupled T3 sites (Figure 11, right). Thus, no low temperature MCD intensity is associated with the intense 400 nm absorption band and it must be associated with azide binding to the antiferromagnetically coupled T3 site.

From the low temperature MCD data it is thus found that azide binds to both the T2 and T3 centers with similar binding constants. The stoichiometry of this binding has been determined through a series of ligand competition experiments, one of which is summarized in Figure 20. From Figure 20a, addition of F^- to native laccase containing 9-fold excess N_3^- (corresponding to the presence of both HA and LA forms) produces a superhyperfine splitting (I_n = 1/2 for ^{19}F) of the T2 hyperfine line, indicating that one F^- is binding to the T2 center. In the absorption and low temperature MCD spectra (Figure 20b), the low affinity N_3^- -> Cu(II) charge transfer transitions associated with the T2 (485 nm MCD) and the T3 (400 nm absorption) centers are eliminated with the addition of increasing concentrations of fluoride. The quantitative loss of these features parallel each other as shown by the solid line in Figure 20c. Fitting these data to models(29b) for competitive binding of one or two fluorides clearly demonstrates that only one fluoride binds to the T2 center and is involved in displacing azide from both the T2 and T3 centers. Therefore one exogenous azide ligand must bridge between the T2 and T3 centers in native laccase defining a trinuclear copper cluster active site in the multicopper oxidases(29).

Figure 21 summarizes the differences between exogenous azide binding to the coupled binuclear copper site in hemocyanin (and tyrosinase) as compared to laccase. For hemocyanin and tyrosinase exogenous ligands bridge the two coppers at the active site while in laccase azide binds to only one copper of the T3 center and bridges to the T2 center forming a trinuclear copper cluster. It is the focus of current research to define in detail the contribution of the μ-1,2 peroxy bridge in reversible O_2 binding and activation and the T2-T3 trinuclear copper cluster with respect to its role in the multi-electron reduction of dioxygen to water.

Acknowledgments

I would like to thank my students and postdoctoral associates who have made major contributions to this research: Dr. Richard S. Himmelwright, Dr. Nancy C. Eickman, Dr. Cynthia D. LuBien, Dr. Yeong T. Hwang, Dr. Marjorie E. Winkler, Dr. Thomas J. Thamann, Dr. Arturo G. Porras, Dr. Dean E. Wilcox, Dr. Mark Allendorf, Dr. Darlene Spira-Solomon, Lung-Shan Kau, Dr. T. David Westmoreland, and Dr. James Pate. I also acknowledge contributions of the following collaborators: Prof. K. Lerch, Prof. K.O. Hodgson, Prof. T.G. Spiro, Prof. K.D. Karlin, Prof. C.A. Reed, Prof. T.N. Sorrell, Dr. W.B. Mims, and Dr. M.S. Crowder. NIH grant #DK31450 is gratefully acknowledged for supporting this work.

Literature Cited

1. Eickman, N.C.; Himmelwright, R.S.; Solomon, E.I. Proc. Natl. Acad. Sci. USA 1979, 76, 2094.

2. (a) Himmelwright, R.S.; Eickman, N.C.; LuBien, C.D.; Lerch, K.; Solomon, E.I. J. Am. Chem. Soc. 1980, 102, 7339. (b) Winkler, M.E.; Lerch, K.; Solomon, E.I. J. Am. Chem. Soc. 1981, 103, 7001. (c) Wilcox, D.E.; Porras, A.G.; Hwang, Y.T.; Lerch, K.; Winkler, M.E.; Solomon, E.I. J. Am. Chem. Soc. 1985, 107, 4015.
3. (a) Solomon, E.I. in Copper Proteins, ed. T.G. Spiro, Wiley, NY, 1981, pp. 41-108. (b) Solomon, E.I.; Penfield, K.W.; Wilcox, D.E. in Structure and Bonding, Springer-Verlag, 1983, Vol. 53, pp. 1-57. (c) Solomon, E.I. Pure and Applied Chemistry 1983, 55, 1069.
4. (a) Freedman, T.B.; Loehr, J.S.; Loehr, T.M. J. Am. Chem. Soc. 1976, 98, 2809. (b) Eickman, N.C.; Solomon, E.I.; Larrabee, J.A.; Spiro, T.G.; Lerch, K. J. Am. Chem. Soc. 1978, 100, 6529.
5. Brown, J.M.; Powers, L.; Kincaid, B.; Larrabee, J.A.; Spiro, T.G. J. Am. Chem. Soc. 1980, 102, 4210.
6. (a) Kau, L.S.; Spira, D.J.; Penner-Hahn, J.E.; Hodgson, K.O.; Solomon, E.I. J. Am. Chem. Soc. 1987, 109, 6433. (b) Kau, L.S.; Solomon, E.I.; Hodgson, K.O. J. Physique 1986, 47, 289 (1986).
7. (a) Solomon, E.I.; Dooley, D.M.; Wang, R.H.; Gray, M.; Cerdonio, M.; Mogno, F.; Romani, G.L. J. Am. Chem. Soc. 1976, 98, 1029. (b) Dooley, D.; Scott, R.A.; Ellinghaus, J.; Solomon, E.I.; Gray, H.B. Proc. Nat. Acad. Sci. USA 1978, 75, 3019.
8. Karlin, K.D.; Cruse, R.W.; Gultneh, Y.; Hayes, J.C.; Zubieta, J. J. Am. Chem. Soc. 1984, 106, 3372.
9. Pate, J.E.; Cruse, R.W.; Karlin, K.D.; Solomon, E.I. J. Am. Chem. Soc. 1987, 109, 2624.
10. (a) McKee, V.; Dagdigian, J.V.; Bau, R.; Reed, C.A. J. Am. Chem. Soc. 1981, 103, 7000. (b) McKee, V.; Zvagulis, M.; Dagdigian, J.V.; Patch, M.G.; Reed, C.A. J. Am. Chem. Soc. 1984, 106, 4765. (c) Karlin, K.D.; Cohen, B.I.; Hayes, J.C.; Farooq, A.; Zubieta, J. Inorg. Chem. 1987, 26, 147.
11. (a) Pate, J.E.; Thamann, T.J.; Solomon, E.I. Spectrochimica Acta 1986, 42A, 313. (b) Pate, J.; Karlin, K.D.; Reed, C.A.; Sorrell, T.; Solomon, E.I., manuscript in preparation.
12. Desjardins, S.R.; Wilcox, D.E.; Musselman, R.L.; Solomon, E.I. Inorg. Chem. 1987, 26, 288.
13. (a) Magnus, K.A. and Love, W.G. J. Mol. Biol. 1977, 116, 171. (b) Volbeda, A. and Hol, W.G.J., in Invertebrate Oxygen Carriers, ed. B. Linzen, 1986, pp. 135-47.
14. (a) Solomon, E.I.; Hare, J.W.; Gray, H.B. Proc. Nat. Acad. Sci. USA 1976, 73, 1389. (b) Penfield, K.W.; Gay, R.R.; Himmelwright, R.S.; Eickman, N.C.; Norris, V.A.; Freeman, H.C.; Solomon, E.I. J. Am. Chem. Soc. 1981, 103, 4382. (c) Penfield, K.W.; Gewirth, A.A.; Solomon, E.I. J. Am. Chem. Soc. 1985, 107, 4519.
15. Reinhammar, B. Biochim. Biophys. Acta 1972, 275, 245.
16. Graziani, M.T.; Morpurgo, L.; Rotilio, G.; Mondovi, B. FEBS Lett. 1976, 70, 87.
17. LuBien, C.D.; Winkler, M.E.; Thamann, T.J.; Scott, R.A.; Co, M.S.; Hodgson, K.O.; Solomon, E.I. J. Am. Chem. Soc. 1981, 103, 7014.
18. (a) Farver, O.; Frank, P.; Pecht, I. Biochem. Biophys. Res.

Comm. 1982, 108, 273. (b) Frank, P.; Farver, O.; Pecht, I. J. Biol. Chem. 1983, 258, 11112. (c) Frank, P.; Farver, O.; Pecht, I. Inorg. Chim. Acta 1984, 91, 81.

19. Powers, L.; Blumberg, W.E.; Chance, B.; Barlow, C.; Leight, J.S. Jr.; Smith, J.C.; Yonetani, T.; Vik, S.; Peisach, J. Biochim. Biophys. Acta 1979, 546, 520.
20. Smith, T.A.G., Ph.D. Thesis, Stanford University, 1985.
21. (a) Blair, R.A.; Goddard, W.A. Phys. Rev. B. 1980, 22, 2767. (b) Kosugi, N.; Yokoyama, T.; Asakuna, K.; Kuroda, H. Springer Proc. Phys. 1984, 2, 55 (1984). (c) Kosugi, N.; Yokoyama, T.; Asakuna, K.; Kuroda, H. Chem. Phys. 1984, 91, 249. (d) Gewirth, A.A.; Cohen, S.L.; Schugar, H.J.; Solomon, E.I. Inorg. Chem. 1987, 26, 1133.
22. It should be noted that there has been some confusion in the literature with respect to the apparent oxidation state of the T3 site in different T2D laccase preps. This derives from the use of indirect methods to define oxidation state. We have run x-ray edge spectra of T2D laccase prepared by four of the research groups strongly involved in this field, and in all cases the Type 3 site is found to be fully reduced.
23. Spira D.J.; Solomon, E.I., J. Am. Chem. Soc. 1987, 109, 6421.
24. Wilcox, D.E.; Westmoreland, T.D.; Sandusky, P.O.; Solomon, E.I., unpublished results.
25. Wilcox, D.E.; Long, J.R.; Solomon, E.I. J. Am. Chem. Soc. 1984, 106, 2186.
26. Himmelwright, R.S.; Eickman, N.C.; Solomon, E.I. J. Am. Chem. Soc. 1979, 101, 1576.
27. Westmoreland, T.D.; Solomon, E.I., to be published.
28. (a) Stephens, P.J. Ann. Rev. Phys. Chem. 1974, 25, 201. (b) Piepho S.B.; Schatz, P.N. Group Theory in Spectroscopy, Wiley-Interscience: New York, 1983.
29. (a) Allendorf, M.D.; Spira, D.J.; Solomon, E.I. Proc. Natl. Acad. Sci. USA 1985, 82, 3063 (1985). (b) D.J. Spira, M.D. Allendorf, and E.I. Solomon, J. Am. Chem. Soc. 1986, 108, 5318.
30. The oxidized T1 site in laccase is spectroscopically similar to that of plastocyanin and azurin, indicating that exogenous ligands can coordinate only to the T2 and T3 coppers at the native active site. (see Spira, D.J.; Co, M.S.; Solomon, E.I.; Hodgson, K.O. Biochem. Biophys. Res. Commun. 1983, 112, 746.)

RECEIVED December 18, 1987

METAL–OXO CENTERS

Chapter 8

Active Sites of Binuclear Iron–Oxo Proteins

Lawrence Que, Jr., and Robert C. Scarrow

Department of Chemistry, University of Minnesota, Minneapolis, MN 55455

The binuclear iron unit consisting of a (μ-oxo(or hydroxo))bis(μ-carboxylato)diiron core is a potential common structural feature of the active sites of hemerythrin, ribonucleotide reductase, and the purple acid phosphatases. Synthetic complexes having such a binuclear core have recently been prepared; their characterization has greatly facilitated the comparison of the active sites of the various proteins. The extent of structural analogy among the different forms of the proteins is discussed in light of their spectroscopic and magnetic properties. It is clear that this binuclear core represents yet another structural motif with the versatility to participate in different protein functions.

A binuclear iron unit has recently emerged as a potential common structural feature of the active sites of several binuclear iron proteins, namely, hemerythrin (1), ribonucleotide reductase (2), and the purple acid phosphatases (3). Such proteins are characterized by strong antiferromagnetic coupling ($-J > 100\ cm^{-1}$ for $H = -2JS_1 \cdot S_2$) between the iron centers in their Fe(III)-Fe(III) forms and EPR signals with g_{av} values ≈ 1.7-1.8 in the mixed-valence Fe(II)-Fe(III) forms (Figure 1). The binuclear iron unit consists of a (μ-oxo(or hydroxo))bis(μ-carboxylato)diiron core and can be distinguished from other iron prosthetic groups such as hemes and iron-sulfur clusters. The binuclear unit is readily recognized as a wedge of the basic ferric acetate structure (4) which consists of a μ_3-oxo-triiron core with carboxylates bridging between pairs of iron atoms.

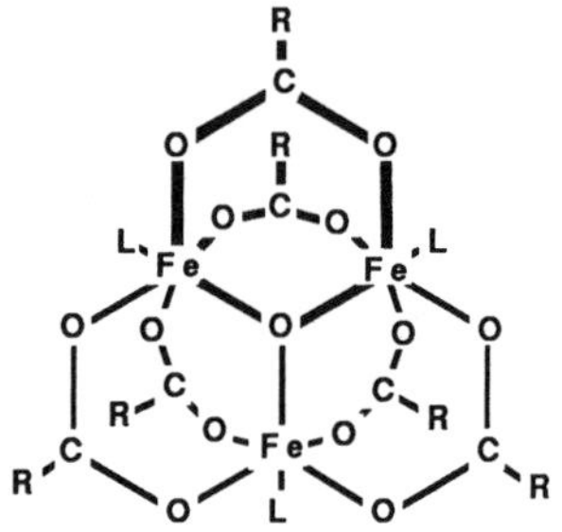

0097–6156/88/0372–0152$07.75/0

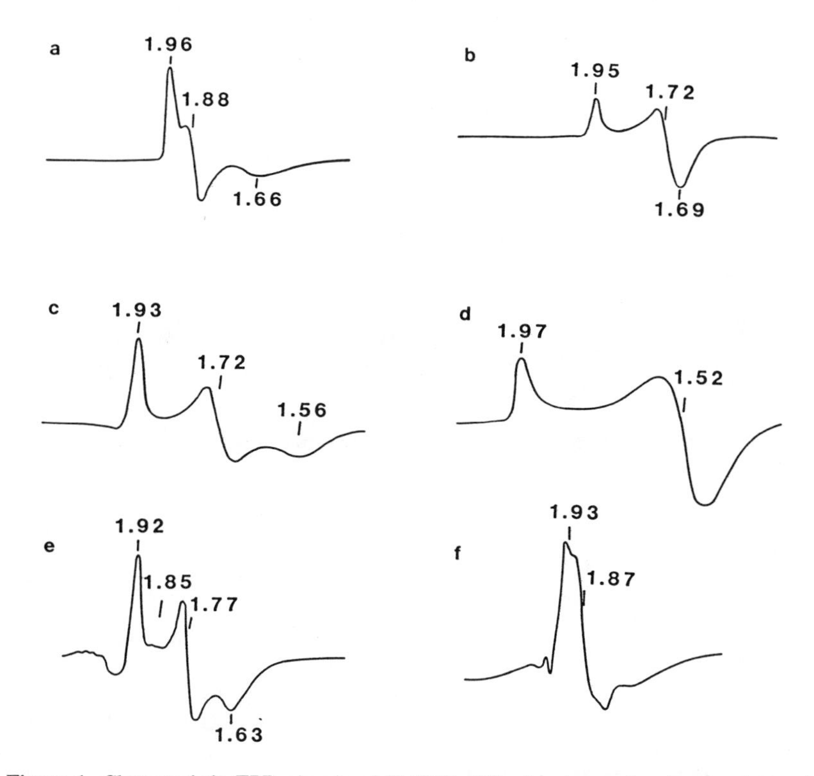

Figure 1. Characteristic EPR signals of Fe(II)Fe(III) sites in semimethemerythrin$_R$ (a), semimethemerythrin$_O$ (b), reduced uteroferrin (c), reduced uteroferrin-molybdate complex (d), reduced bovine spleen purple acid phosphatase (e), reduced component A of methane monooxygenase (f). (Reproduced with permission from ref. 26. Copyright 1987 Elsevier.)

Synthetic Analogues

The thermodynamic stability of such a binuclear unit has been recently demonstrated by the spontaneous self assembly of the core structure with a number of trigonal face-capping ligands. Complexes of the type $[(LFe^{III})_2O(OAc)_2]$ have been crystallized with ligands such as $HBpz_3$ (**1**) (5), tacn (**2**) (6), and Me_3tacn (**3**) (7,8). With a hexadentate ligand such as **4**, a tetranuclear complex is formed with $Fe_2O(OAc)_2$ units connecting N_3 faces of different ligands (9). Relevant structural parameters are compared in Table I.

1, $HBpz_3$ **2, tacn (R=H)** **3, Me_3tacn (R=Me)** **4, tpbn**

Table I. Comparison of Structural and Spectroscopic Parameters of Binuclear Iron Complexes

Complex	$r_{Fe\text{-}\mu\text{-}O(R)}$ (Å)	$r_{Fe\cdots Fe}$ (Å)	ΔE_Q (mm/s)	J^a (cm^{-1})	Ref
Fe(III)-Fe(III)					
$[(HBpz_3Fe)_2O(OAc)_2]$	1.784	3.146	1.60	-121	5
$(HBpz_3Fe)_2O(O_2CH)_2]$	1.781	3.168			5
$[(tacnFe)_2O(OAc)_2]^{++}$	1.781	3.063	1.72		6
$[(Me_3tacnFe)_2O(OAc)_2]^{++}$	1.800	3.12	1.50	-119	8
$[tbpnFe_2O(OAc)_2]_2^{4+}$	1.794	3.129	1.27	-120[b]	9
$[(HBpz_3Fe)_2O(O_2P(OPh)_2)_2]$	1.808	3.337		-98	13
$[(HBpz_3Fe)_2OH(OAc)_2]^+$	1.956	3.439	0.25	-17	18
$[Fe_2(5\text{-Me-HXTA})(OAc)_2]^-$	2.008	3.442	0.58	-10[c]	19
Fe(II)-Fe(II)					
$[(Me_3tacnFe)_2OH(OAc)_2]^+$	1.987	3.32	2.83	-13	8
Fe(II)-Fe(III)					
$[Fe_2(5\text{-Me-HXTA})(OAc)_2]^{2-}$				-5[c]	23
$[Fe_2(bbmp)(OBz)_2]^{2+}$				-5	24
$[Fe_2(bpmp)(OPr)_2]^{2+}$	1.943 2.090	3.365			22

[a]Determined from fits of variable temperature solid state magnetic susceptibility data, unless otherwise noted.
[b]J not determined for this complex but for a related complex.
[c]Estimated from temperature dependence of NMR isotropic shifts

The presence of an oxo bridge gives rise to short Fe-O_{oxo} bonds of 1.78-1.80 Å, while the carboxylate bridges restrict the Fe-O-Fe angle to ca. $120^o \pm 10^o$ and the Fe-Fe distance to 3.15 ± 0.1 Å. The variation of the Fe-Fe distance and the Fe-O-Fe angle in the oxo-bridged complexes appears to be a function of the cone angle defined by the iron and the three ligating atoms on the trigonal face. The tacn ligand has a more restricted cone angle (trigonal "bite") and gives rise to a larger trigonal bite for the O_3 face. The necessity to maintain reasonable Fe-O bond lengths results in the closer approach of the other iron atom (Table I).

The binuclear unit also confers characteristic physical and spectroscopic properties to the complexes. Electronic spectra reveal a strong ($\varepsilon \sim 3000$-4000 $M^{-1}cm^{-1}$) absorption band at 300-400 nm with a characteristic low energy shoulder, both associated with the presence of an oxo bridge; weaker ($\varepsilon \sim 300$-600 $M^{-1}cm^{-1}$) features in the 400-550 nm region associated with the (μ-oxo)bis(μ-carboxylato)-diiron(III) core; and d-d bands in the near IR region (Figure 2a) (5,10). Laser excitation into the visible and near UV bands results in the observation of enhanced Raman features due to Fe-O-Fe vibrations (5,11) (see Chapter 3 for a more detailed discussion into the Raman spectra of these complexes).

The short Fe-O bond results in a strong antiferromagnetic interaction between the high spin ferric centers. This interaction, described by the general isotropic exchange Hamiltonian ($H = -2JS_1 \cdot S_2$) (12), couples the two $S = 5/2$ centers to a ladder of spin states $S_T = 0, 1, 2, 3, 4, 5$, separated by $2S_T(S_T+1)J$ from the next higher state, with the ground state being the $S_T = 0$ state. The J value can be derived from a fit of the temperature dependence of the magnetic susceptibility of the complex to the following equation which reflects the Boltzmann population of the various S_T states,

$$\chi = \frac{Ng^2\mu_B{}^2\,(2e^{2x} + 10e^{6x} + 28e^{12x} + 60e^{20x} + 110e^{30x})}{kT\,(1 + 3e^{2x} + 5e^{6x} + 7e^{12x} + 9e^{20x} + 11e^{30x})}$$

where $x = J/kT$. The J values found from the (μ-oxo)bis(μ-carboxylato)diiron(III) complexes are ca. -100 cm^{-1} (5,8,13), similar to values observed for other oxo-bridged complexes (14).

The strong coupling is also reflected in the relatively small isotropic shifts observed for these complexes in their NMR spectra. Since the contact shift is directly proportional to the electron-nuclear hyperfine splitting constant, A, and the magnetic susceptibility of the metal center (15), the shift in a coupled system would be expected to decrease as the magnetic susceptibility, assuming that A remains relatively constant in the various S_T states. As an example, the bridging acetate resonance near 11 ppm found for $Fe_2O(OAc)_2$ complexes of **1** and **2** can be compared to that of mononuclear Fe(salen)OAc at ca. 140 ppm (5,16). The isotropic shift of the binuclear acetate is greatly diminished relative to that of the mononuclear acetate. For $[(HBpz_3Fe)_2O(OAc)_2]$, which exhibits a J value of -121 cm^{-1} (5), the above equation calculates that the complex will exhibit a susceptibility that is 8% of the expected uncoupled high spin ferric value, which is in reasonable agreement with the observed decrease in the acetate shift. Thus, NMR shifts serve as a convenient probe of the antiferromagnetic interaction of these complexes in solution.

The short Fe-O bond also exerts a characteristic effect on the Mössbauer spectra of these complexes. The Fe-oxo bond is usually significantly shorter than the other metal-ligand bonds (1.8 vs 2.0-2.2 Å), and this large difference in the

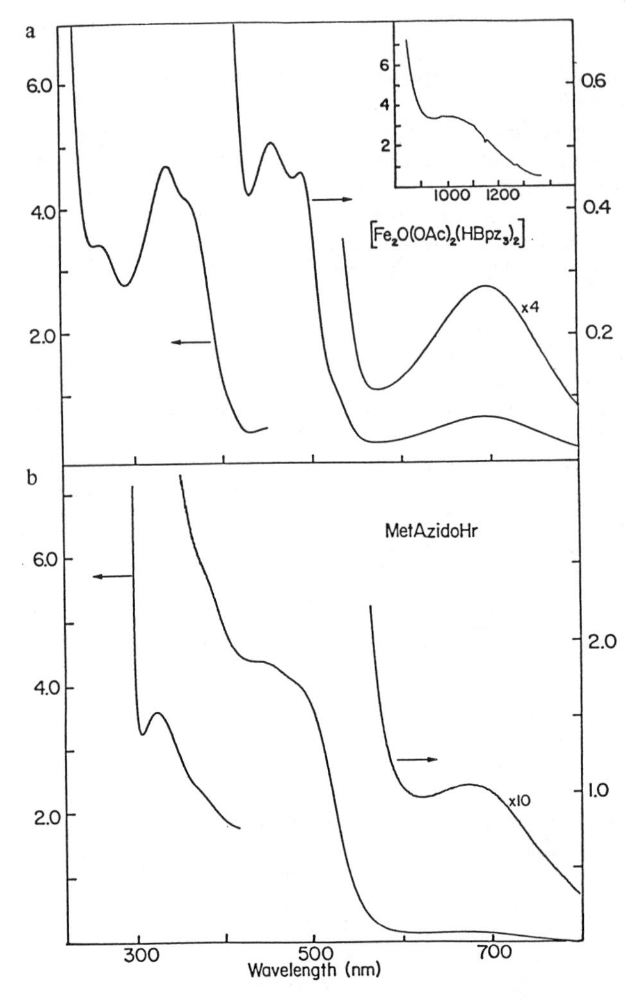

Figure 2. Electronic spectra of $[(HBpz_3Fe)_2O(OAc)_2]$ (a) and methemerythrin azide (b). The y-axis units are for ϵ_{Fe} in $cm^{-1}mM^{-1}/Fe$. (Reproduced from ref. 5. Copyright 1984 American Chemical Society.)

iron-ligand bond lengths should give rise to a rather anisotropic electric field gradient about the iron nucleus and result in large values for ΔE_Q (> 1.25 mm/s). This large value for ΔE_Q is observed in a variety of oxo-bridged complexes (5,14). The important exceptions are the binuclear complexes with salicylideneamine ligands, which have ΔE_Q values of 0.9-1.0 mm/s. The smaller values reflect the intermediate Fe-O bond lengths for the phenolates (1.9 Å) (17), which diminish the anisotropy of the electric field gradient about the iron nucleus.

Complexes with structural variations in the triply bridged core have also been synthesized. The acetato groups of $[(HBpz_3Fe)_2O(OAc)_2]$ undergo facile ligand exchange reactions under acidic conditions with other carboxylates. More interestingly, the acetates can be also replaced by diphenylphosphates (13). The larger bite of the bridging phosphate increases the Fe-O-Fe angle and the Fe···Fe separation; the phosphate substitution slightly lengthens the Fe-oxo bond and weakens the magnetic interaction between the metal centers (Table I).

The oxo group of $[(HBpz_3Fe)_2O(OAc)_2]$ can be protonated by HBF_4 to yield $[(HBpz_3Fe)_2OH(OAc)_2]^+$ (18). The protonation of the bridge results in the lengthening of the Fe-μ-O bonds to 1.96 Å, an increase in the Fe···Fe separation, and a decrease in the Fe-O-Fe angle. The antiferromagnetic coupling between the ferric centers is diminished to $J = -17\ cm^{-1}$, which is also reflected by the NMR spectrum of the complex showing substantially larger isotropic shifts for all resonances. In particular, the acetato methyl resonance appears at 70 ppm, a value corresponding to ca. 50% of the shift expected of a high spin ferric complex, in agreement with the J value observed.

A triply bridged complex with a heptadentate binucleating ligand, **5** (5-Me-HXTA), has a phenoxo oxygen in place of the oxo bridge (19). Such a complex, $[Fe_2(5\text{-Me-HXTA})(OAc)_2]^-$, exhibits even longer Fe-μ-O bonds (ca. 2.01 Å) and an Fe···Fe separation of 3.44 Å. Unlike the previously discussed structures, which have an actual or pseudo-plane of symmetry bisecting the triply bridged core, the latter complex has a twofold axis through the phenolate C-O bond relating the two halves of the molecule, because the phenolate ring plane is rotated 45° relative to the Fe-Fe axis. This structure of this complex also differs from the previously discussed structures in that the acetates are unsymmetrically bridged, i. e. each iron atom has a short Fe-OAc and a long Fe-OAc bond. The J value for this complex is estimated at $-10\ cm^{-1}$ from variable temperature NMR measurements and its acetate shift, 89 ppm, reflects this (19).

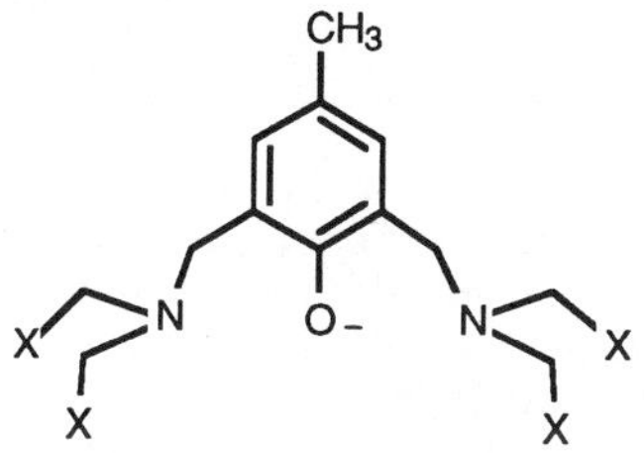

CH_3

N O- N
X X X X

5: 5-Me-HXTA (X = COO^-)

6: bbmp (X = 2-benzimidazolyl)

7: bpmp (X = 2-pyridyl)

This collection of triply bridged diferric complexes, together with other oxo- and hydroxo-bridged complexes affords a range of structural and spectroscopic properties that should serve as useful models for comparison with the proteins; in particular, the magnetic properties correlate with the length of the shortest Fe-μ-O bond (20) (see Chapter 10 for an in-depth discussion).

There is currently a paucity of structural data for binuclear iron complexes in other oxidation states. Wieghardt et al. have reported the structure of the only triply bridged diferrous complex thus far, $[(Me_3tacnFe)_2OH(OAc)_2]^+$ (7,8). Its structure is closely related to that of the corresponding diferric complex, with Fe-μ-OH bond lengths of 1.99 Å and an Fe···Fe separation of 3.4 Å. The two metal centers are antiferromagnetically coupled with a J value of $-13\ cm^{-1}$.

Mixed-valence forms of some of these complexes have also been reported. Early attempts to obtain mixed-valence species by one-electron reduction of $[(HBpz_3Fe)_2O(OAc)_2]$ and $[(tacnFe)_2O(OAc)_2]^{++}$ resulted in the breakdown of the binuclear unit. The Me_3tacn ligand, however, has permitted the generation of mixed-valence forms either by reduction of the Fe(III)Fe(III) form or oxidation of the Fe(II)Fe(II) complex (8). The steric hindrance of the Me_3tacn ligand presumably prevents the formation of the mononuclear bis(ligand) complex, depriving the decomposition pathway of its driving force. The mixed-valence complex has been characterized only as the majority species in solution, because of its propensity to disproprotionate. It exhibits EPR signals at 1.95, 1.89, and 1.82 (8), characteristic of antiferromagnetically coupled Fe(II)-Fe(III) units (21). Other properties have not yet been determined.

Another strategy for enhancing the stability of the binuclear unit in mixed-valence complexes is the use of binucleating ligands **5-7**. Complexes of the type $[Fe_2L(O_2CR)_2]^{2+/-}$ have been isolated (22,23,24), and a crystal structure has been obtained for the complex of **7** with R = propyl (22). The mixed-valence nature of the complex is indicated by the difference in average Fe-O bond lengths for the two iron atoms (2.09 vs. 1.96 Å); these values are typical of average Fe(II)- and Fe(III)-O bond lengths in similar complexes, respectively. The mixed-valence complexes exhibit near-IR transitions which are tentatively assigned as intervalence charge transfer bands. For the complexes with **5** and **7**, two quadrupole doublets are observed in the Mössbauer spectra at 55 K with parameters characteristic of high spin Fe(II) and Fe(III), respectively. At 4.2 K, the doublets broaden to reveal magnetic hyperfine structure expected of a half-integer spin state (22,23). Variable temperature studies show that complexes of **5** and **6** exhibit J values of $-5\ cm^{-1}$ (23,24). No EPR signal with g values less than 2 has been observed for the mixed-valence complex of **5**; the rationale for this intriguing observation awaits further work.

Hemerythrin

The best characterized and thus prototypical of the binuclear iron-oxo proteins is hemerythrin (Hr), an oxygen carrier from some classes of marine invertebrates, such as sipunculids. The protein consists of several identical 13 kDa subunits which contain two iron atoms and reversibly bind one molecule of O_2. Myohemerythrins contain a single 13 kDa subunit (25,26,27).

Electrons from ferrous ions in deoxyhemerythrin are transferred to O_2 during formation of oxyhemerythrin, so that the latter is a diferric hydroperoxide complex. The electron transfer is reversed upon oxygen release. Oxidation of the metal centers by other processes yields inactive methemerythrin or one of its complexes with small anions. Our discussion of the various states of hemerythrin shall start with the well-characterized diferric *met* and *oxy* states and proceed to the *deoxy* and mixed-valence *semimet* states, whose structures are less well understood.

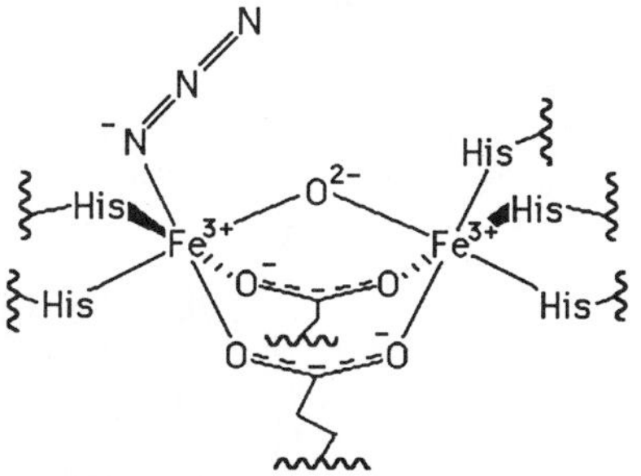

metHrN3

Met- and oxyhemerythrins. The view of the active site shown above is taken from the X-ray crystal structure of $metHrN_3$ (28). The iron centers are bridged by an oxo group (with Fe-O bondlengths averaging 1.76 Å) and carboxylates from one aspartate and one glutamate. The remaining coordination sites are occupied by five histidines and the azide. Comparison of electron density maps for oxyHr and $metHrN_3$ shows that oxygen and azide bind end-on to the same site (29). In metHr (formerly called "aquomethemerythrin") the coordination site normally occupied by the anion is vacant, so that one iron is approximately trigonal bipyramidal (28). A comparison of the metrical parameters of the binuclear site in metHr and $metHrN_3$ with those of synthetic complexes shows some unusually long iron bond distances and an asymmetric μ-oxo bridge; these may be due to systematic errors in the crystallographic data or analysis (28). A recently published high-resolution structure of $metmyoHrN_3$ (30) shows that the binuclear site has a nearly symmetric μ-oxo bridge and iron bond lengths very similar to those of the $(LFe)_2O(OAc)_2$ complexes of ligands **1-3** (5,6,8,9). As in these model complexes, the longest Fe-N bonds of $metmyoHrN_3$ are *trans* to the oxo ligand. Also, the carboxylate ligands appear to bridge asymmetrically, as observed for $[Fe_2(5\text{-MeHXTA})(OAc)_2]^-$ (19).

EXAFS studies of met and oxyhemerythrins confirm the presence of the triply bridged diiron core with Fe-μ-O bond lengths of ca. 1.8 Å and Fe-Fe distances of ca. 3.2 Å (Table II) (31,32,33). The Fe-μ-O bondlengths determined by EXAFS for oxyHr and metHrOH are marginally longer than those for $metHrN_3$ and the synthetic complexes (32), the lengthening may be due to the effect of hydrogen bonding of the bound hydroperoxide or hydroxide to the oxo bridge (vide infra). The 3.20 - 3.26 Å Fe-Fe distance determined by EXAFS and crystallography in methemerythrins is longer than in the model complexes; this could be due to the larger trigonal "bite" offered by the terminal ligands to the binuclear core in the protein relative to the those of tridentate ligands in the synthetic complexes.

The presence of the oxo bridge in oxyHr and metHrX is reflected in their magnetic susceptibility, NMR, and Mössbauer properties. The complexes exhibit strong antiferromagnetic coupling, small NMR isotropic shifts, and large Mössbauer quadrupole splittings. J values obtained for metHr and oxyHr from fits to variable temperature susceptibility data are -134 and -77 cm^{-1}, respectively (34). The J for metHr compares favorably with those observed for the synthetic

Table II. Properties of Binuclear Iron-Oxo Proteins

Protein	ΔE_Q (mm/s)	J^a (cm^{-1})	Fe-O,N^b (Å)	Fe···Fe^b (Å)
Hemerythrin				
metHrN$_3$ (37,38,28)	1.91		1.78, 2.24^c	3.25^c
(31)			1.76, 2.10	3.20
(33)			1.80, 2.13	3.19
(32)			1.80, 2.15	(3.49)
metmyoHrN$_3$ (30)			1.79, 2.16^c	3.22^c
metHr (37,38,34)	1.57	-134(χ)		
oxyHr (37,38,34,32)	1.92, 1.00	-77(χ)	1.83, 2.16	(3.57)
metHrCl (38)	2.04			
metHrOH (32)			1.82, 2.15	(3.54)
semimetHrN$_3$ (36,33)		-20(NMR)	1.87, 2.14	3.44
deoxyHr (38,36,47)	2.76	-13±2(χN)	2.02, 2.14^d	(3.13)
(48)		-25±13(MCD)		
Ribonucleotide reductase				
native (69,60,33,62,69)	1.66, 2.45	-108(χ)	1.79, 2.06	3.22
(63)			1.78, 2.04	3.26^e
radical free (69,33,62)	1.66, 2.45		1.78, 2.06	3.19
Purple acid phosphatases				
Uf$_o$ (92)	1.65, 2.12			
Uf$_o$-P$_i$ (90,82)	1.02, 1.38		1.96, 2.10^d	3.2±0.1
Uf$_r$ (92,81,82)	1.78, 2.63	-10(NMR) -7(EPR)	2.02, 2.13^d	3.18
Uf$_r$-P$_i$ (90)	0.78, 2.76			
BSPAP$_o$-P$_i$ (73,92)		<-150(χ)	1.98, 2.13^d	3.01
BSPAP$_r$ (73)		-5(EPR)		

[a]J values obtained by fits to variable temperature data obtained by technique indicated. χN denotes a room temperature Evans susceptibility measurement.
[b]Unless noted otherwise, structural parameters are obtained from EXAFS measurements; the shorter distance corresponds to the shortest Fe-O bond and the longer value to the average of 5 longer Fe-O and Fe-N distances. Values in parentheses are considered questionable.
[c]X-ray crystallography. The listed bond lengths were refined without restraints.
[d]The distances corresponds to 3 Fe-O and 3 Fe-N distances per iron.
[e]See text and footnote (64).

analogues, but that for oxyHr is somewhat small. The smaller value observed for oxyHr may be a consequence of the hydrogen bonding of the hydroperoxide to the oxo bridge, which would be expected to weaken the Fe-O-Fe interaction. The NMR spectra of metHr, metHrX and oxyHr exhibit solvent-exchangable resonances between 13 and 25 ppm, assigned to imidazole N-H protons of the five coordinated histidines (35,36). These shifts contrast those found for imidazoles in mononuclear high spin ferric complexes at ca. 100 ppm and demonstrate the differences in room temperature magnetic susceptibilities of two sets of complexes. The iron centers exhibit large Mössbauer quadrupole splittings because of the anisotropic electric field gradient imposed by the short Fe-μ-O bond (37,38) The anisotropy is somewhat mitigated by the binding of anions *cis* to the oxo ligand on one iron. This results in the two iron atoms exhibiting distinct quadrupole splittings, which is most apparent in oxyHr (37,38) and metHrN_3 from *Phascolosoma lurco* (39).

The visible spectra of oxyHr and metHrN_3 are dominated by ligand-to-metal charge transfer bands from the hydroperoxide or azide anions, but otherwise they are similar to those of the synthetic complexes (Figure 2) (38). The d-d transitions observed at 700 and 1000 nm are more intense than usually observed for high-spin iron(III) complexes, probably due to the strong antiferromagnetic coupling interaction (38,40).

The resonance Raman spectrum of oxyHr includes peaks at 844 and 504 cm^{-1} assigned to O-O and Fe-O_2 stretching frequencies by isotopic substitution experiments; the data obtained with $^{18}O_2$ and $^{16}O^{18}O$ also show that oxygen binds as a monodentate ligand (25,41,42). The resonance Raman frequencies unique to oxyHr change in D_2O, indicating that oxygen is bound as hydroperoxide (43). The resonance Raman spectra of most forms of methemerythrin show peaks at 506-514 and 750-780 cm^{-1} which have been assigned by ^{18}O substitution to ν_s and ν_{as} of the Fe-O-Fe unit (44,45). OxyHr and one form of metHrOH show lower ν_s frequencies (486 and 492 cm^{-1}, respectively) which are unique in shifting to higher energy in D_2O. The lowered energy is ascribed to hydrogen bonding to the oxo bridge, which weakens the Fe-O bond; deuterium bonds are weaker and hence do not lower ν_s as much (45).

Deoxyhemerythrin. Hemerythrin appears to retain its triply bridged core structure in the deoxy form. A low (3.9 Å) resolution difference electron density map of deoxyHr vs. metHr from X-ray diffraction suggests that the iron atoms move slightly further apart in deoxyHr, but remain five and six coordinate, respectively (29). Confirmation of the iron coordination comes from near-IR absorption and circular dichroism spectra (40,46,47). Based on model high-spin ferrous complexes, the six coordinate iron is expected to give two of the three observed transitions near 10000 cm^{-1} while the five-coordinate iron accounts for the d-d transition at ca. 5000 cm^{-1}.

The magnetic properties of deoxyHr indicate that a bridge between the iron centers persists but may be altered from that in oxyHr. The NMR spectrum of deoxyHr exhibits large isotropic shifts for the N-H's on histidines coordinated to the binuclear unit. Mössbauer parameters of deoxyHr (Table II) are typical of high spin iron(II) centers which are at best weakly coupled (37,38,39). Room temperature magnetic susceptibility measurements using the Evans NMR technique referenced to oxyHr indicate deoxyHr has lower susceptibility than expected for isolated high spin iron(II), indicating $J \approx -15 \pm 3$ cm^{-1}. The variable-temperature near-IR MCD spectra also indicate that deoxyHr is antiferromagnetically coupled with J between -12 and -38 cm^{-1}. Analysis of the MCD spectra was complicated by the effects of zero-field splitting, which for high spin iron(II) can be as high as D = 15 cm^{-1}. The extent of antiferromagnetic coupling observed for deoxyHr compares favorably with that observed for $[(Me_3tacnFe)_2OH(OAc)_2]^+$ ($J = -13 \pm 1$ cm^{-1}) and thus strongly suggests that the oxo group in metHr has become protonated in deoxyHr.

The EXAFS analysis on deoxyHr differs in part from the results from the crystallographic and magnetochemical studies. In agreement with the other studies, no resolvable short Fe-O bonds are found, consistent with the lack of an oxo bridge (Table II). However, a peak in the 80 K spectrum was analyzed to give an Fe-Fe distance (Table II) which is *shorter* by 0.1 Å than those found in methemerythrins (48). The same peak is not observed at 300 K presumably due to thermal motions. Such a short Fe-Fe distance is unlikely, given the absence of an oxo bridge and the likely involvement of multiple bridging. Perhaps the second sphere EXAFS oscillations are due to histidine carbons ca. 3.1 Å from the ferrous ions, a possibility not considered by the original authors.

The binding of anions to deoxyHr alters the properties of the binuclear unit. Dioxygen binding is retarded by the presence of azide, cyanate, or fluoride. Evidence for the coordination of the anions to the binuclear center has been found in the near-IR CD, MCD, and NMR spectra of these complexes. And it has been demonstrated from kinetic studies that the anions bind to the protein as the conjugate acid. When azide, cyanate or fluoride bind deoxyHr, only the 10000 cm^{-1} transition found in deoxyHr is retained, and the 5000 cm^{-1} transition is no longer observed, suggesting both irons are six coordinate in deoxyHrX (47). Anion binding also affects the magnetic properties of the binuclear unit, implying possible changes in the bridging groups. Evans susceptibilities of deoxyHr(OCN) and deoxyHrN_3 complexes are significantly higher than that for deoxyHr and are reasonable for uncoupled iron(II) pairs (36), while MCD measurements suggest that J is small and positive (i.e. weak ferromagnetic coupling) (47). An unusual EPR signal at $g_{eff} \approx 13$ observed below 30 K for deoxyHrN_3 may arise from an S=4 ground state (46,47).

Taken together, the physical evidence suggests deoxyHr contains a (μ-hydroxo)bis(μ-carboxylato)diferrous core. Upon binding anions, the hydroxo group may either be protonated or become non-bridging, either of which would account for the loss of antiferromagnetic coupling in deoxyHrX.

Semimethemerythrins. Mixed valence (semimet) forms of hemerythrin are thermodynamically unstable intermediates in the oxidation of deoxyHr or reduction of metHr. The kinetic stability of semimethemerythrins is apparently due to a conformational change which must occur between the deoxy and met states of the protein. Two forms of semimetHr are generated depending on whether deoxyHr is oxidized (to generate semimetHr_O) or metHr is reduced (to generate semimetHr_R) (49,50). They exhibit distinct EPR signals with $g_{av} \approx 1.8$ (50), the g_{av} value of < 2 being indicative of antiferomagnetic coupling between the Fe(II) and Fe(III) centers (21). This coupling is likely to be weak, because the EPR signals disappear above 30 K presumably due to the accessibility of higher spin states.

Both forms bind small anions; the anion bound forms are more stable and appear to be the same when prepared from either the O or R forms of semimetHr (49). The spectroscopic evidence favors a μ-hydroxo bridge for semimetHrX, with X^- bound to the ferric ion, as shown below. Less evidence is available for the structures of semimetHr_O and semimetHr_R.

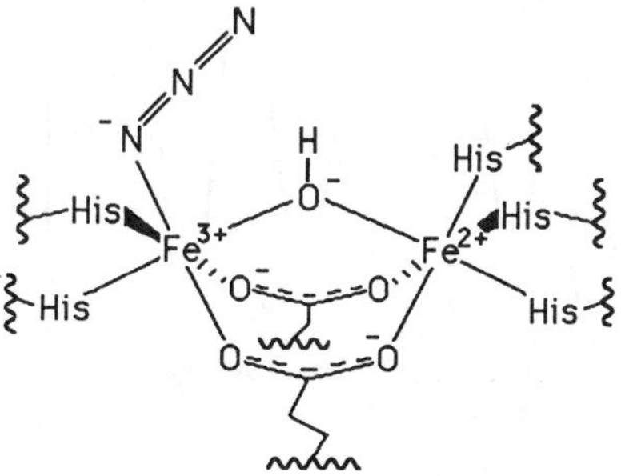

semimetHrN_3

The visible spectrum of semimetHrN_3 includes an charge transfer band from the azide at 470 nm (51). The similarity of this band to the 446 nm band in metHrN_3 indicates the azide is bound to the ferric ion of the mixed-valence form. The resonance Raman spectrum confirms the role of azide in this transition and isotopic substitution indicates the same end-on mode of binding found in metHrN_3 (51). The *decrease* in energy for the azide to iron(III) charge transfer band suggests the ferric ion has become a stronger Lewis acid upon reduction. Since the reduction of one of the ferric centers in a μ-oxodiferric unit would render the oxo bridge more Lewis basic and diminish the Lewis acidity of the remaining ferric center, the most likely explanation for the red shift of the N_3-to-Fe(III) band is the protonation of the oxo bridge.

Further evidence for the protonation of the oxo bridge is derived from the EXAFS spectrum of semimetHrN_3 (Table II) (33). A comparison of the first shell EXAFS spectra of metHrN_3 and semimetHrN_3 shows that the latter complex lacks the significant destructive interference at low k values due to a short Fe-O bond as observed in metHrN_3 (Figure 3). The Fe-μ-O bond has thus lengthened in semimetHrN_3. In addition, the Fe-Fe distance derived from a fit of the second shell data is considerably longer than in metHrN_3 and in the range found for model diferric and diferrous complexes with the (μ-hydroxo)bis(μ-carboxylato) bridges. No crystallization of a semimet form of hemerythrin has been reported.

The NMR spectra of semimetHrX show exchangeable resonances between 40 and 70 ppm (36). The spectrum of semimetHrN_3 is particularly simple, with

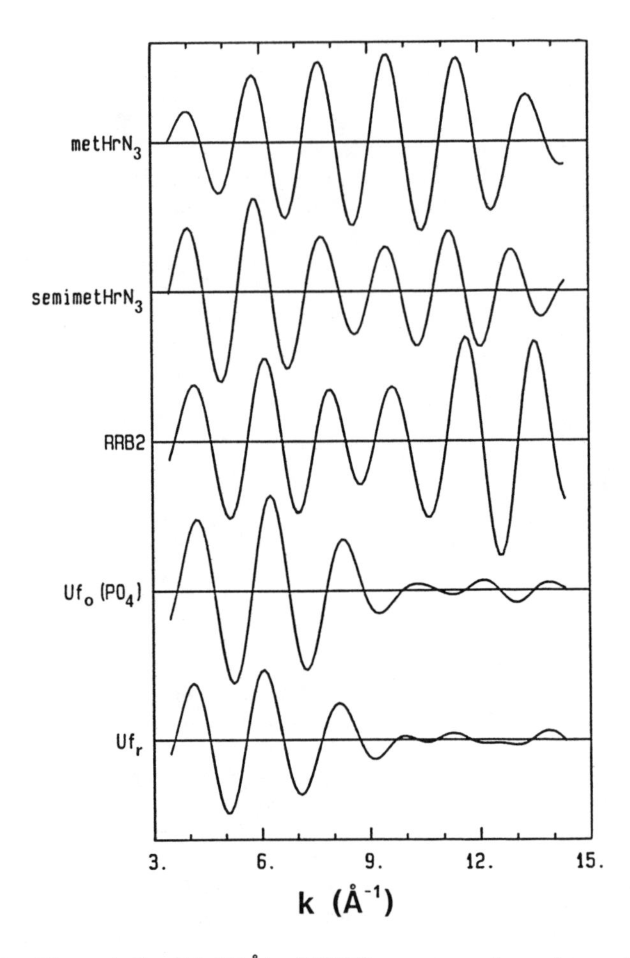

Figure 3. First shell (1.1–2.3Å) EXAFS spectra of methemerythrin azide, semimethemerythrin azide, ribonucleotide reductase B2 subunit, oxidized uteroferrin-phosphate complex, and reduced uteroferrin.

resonances of intensity 2:3 at 72 and 54 ppm assigned to N-H protons of histidines bound to iron(III) and iron(II) respectively (35). The smaller contact shift of the 72 ppm peak relative to those of mononuclear ferric imidazole complexes was used to estimate $J \approx -18\ cm^{-1}$. The temperature dependence of the Fe(III)-N-H shifts also indicates $J \approx -20\ cm^{-1}$ for semimetHrX, where $X^- = N_3^-$, Cl^-, and F^- (36). This J is similar to that observed in diferric and diferrous model compounds with a hydroxo bridge (see Table I).

Mechanism of oxygen binding by hemerythrin. The oxygenation of deoxyHr involves a one-electron oxidation of each of the iron atoms and reduction of O_2 to the peroxide oxidation level. The facile reversibility of this process (in contrast to the slow multistep interconversion of deoxyHr and metHr by chemical oxidants or reductants) requires minimal structural rearrangement and no high energy intermediates. Comparisons of the structures of deoxy and oxyhemerythrin have led to the following proposal for oxygen binding by hemerythrin: (30)

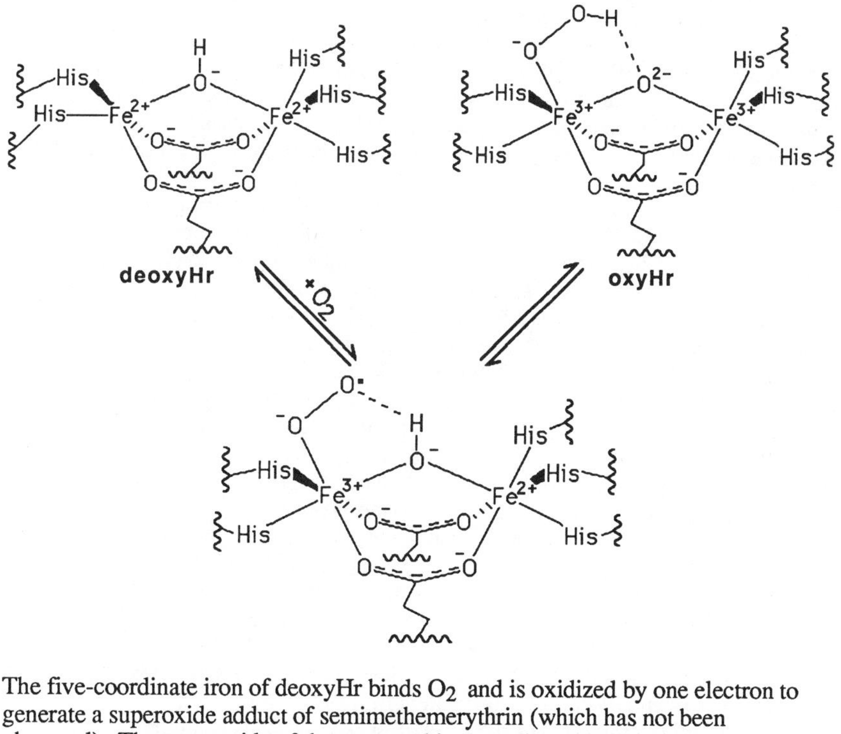

The five-coordinate iron of deoxyHr binds O_2 and is oxidized by one electron to generate a superoxide adduct of semimethemerythrin (which has not been observed). The superoxide of the proposed intermediate rapidly abstracts the nearby hydrogen atom of the hydroxo bridge and becomes hydroperoxide. At the same time the remaining ferrous ion donates an electron to the oxygen of the bridge to complete the two-electron oxidation of the binuclear iron center.

Ribonucleotide reductase

The reduction of ribonucleoside di- or tri-phosphates to give deoxyribonucleotides is a key step in DNA biosynthesis. The enzymes involved in this transformation, the ribonucleotide reductases, generally have a subunit containing either a cobalt (B12) or binuclear iron center which is involved in generation of a catalytically important radical within the same subunit (52,53). This radical abstracts the 3' proton from the ribonucleotide substrate as the first step of its reduction and is regenerated during turnover (54,55). The reducing equivalents come from cysteine residues on another subunit. Binuclear iron centers occur in ribonucleotide reductases from humans and other mammals and in a variety of other species, including *Escherichia coli* bacteria. The enzyme from *E. coli* has been studied far more extensively than any other of this class, and we will focus on its iron-containing B2 subunit (abbreviated "RRB2"). Studies of RRB2 have been facilitated by genetically altered cells capable of its overproduction (56,57,58).

Although diffracting crystals of RRB2 have been grown, no structural information is yet available from this technique (59). Similarities between RRB2 and the diferric forms of hemerythrin have been noted in the visible, EXAFS, resonance Raman, NMR and Mössbauer spectra and in the magnetic susceptibility of these proteins. For this reason, it is likely that RRB2 contains a (μ-oxo)bis(μ-carboxylato)diferric core as found in met- and oxyHr. The presence of an oxo bridge has been established, but no definitive evidence for the carboxylate bridges has been presented.

Both the binuclear iron center and the tyrosyl radical contribute to the visible spectrum of active RRB2. By reducing the radical with hydroxyurea, the contributions of the two chromophores can be separated (Figure 4) (60). The radical has sharp absorption peaks at 390 and 410 nm and a broad absorbance at 600 nm which are also found in 2,4,6-tri(*tert*-butyl)phenoxy radical (61). The iron chromophore has a spectrum very similar to that of metHr(OH).

Analyses of the EXAFS spectra of RRB2 (both native and radical-free) clearly show the 1.80 Å Fe-O bond length characteristic of an oxo bridge (Table II). Figure 3 compares the first shell EXAFS of RRB2 with metHrN_3, showing the similarity of the amplitude modulation in the two complexes. The low amplitude at low k is symptomatic of the destructive interference of a short Fe-O bond with longer Fe-(O,N) bonds. These (4-5) longer bonds average ca. 2.05 Å in length (33,62,63), and are significantly shorter than the average bond length found by EXAFS in metHrN_3 or oxyHr. The shorter average bond length in RRB2 relative to hemerythrin is attributed to more anionic oxygen ligands and fewer histidine ligands (which have relatively long bond lengths to iron) (33,62). The small magnitude in the EXAFS Fourier transform of a peak near r' = 4 Å, generally associated with multiple scattering from histidine ligands, is consistent with at most three histidines bound to the two iron centers (63).

The Fourier transforms of the EXAFS spectra of RRB2 and metHrN_3 are very similar near r' = 3 Å, suggesting similar second sphere coordination; analyses using single scattering theory of the EXAFS spectrum of RRB2 indicate an Fe-Fe separation of 3.22 (33,62) or 3.26 Å (63). These distances are considerably less than the values expected from multiple scattering pathways, and should be considered accurate estimates of the true interiron distance (64,65). The similarity of the Fe-Fe distance determined for RRB2 with those found in metHrN_3 and the

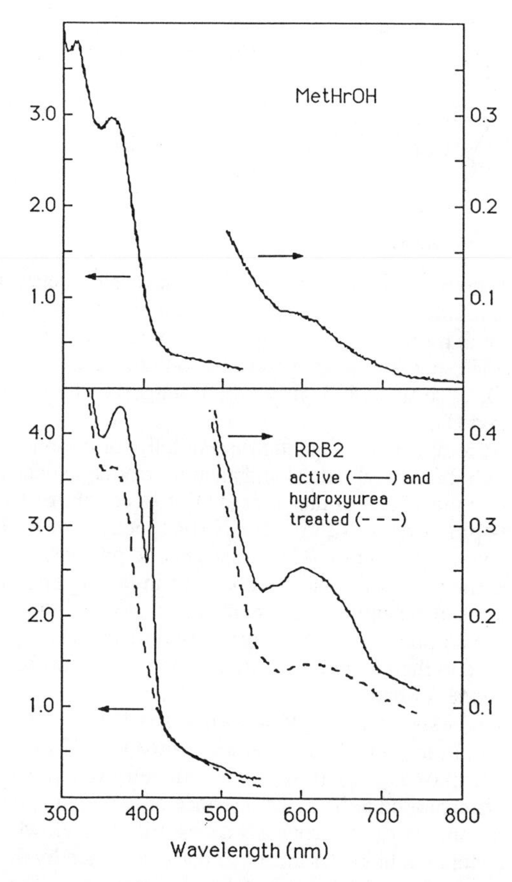

Figure 4. Electronic spectra of methemerythrin hydroxide and native and radical free ribonucleotide reductase B2 subunit. The y-axis units are for ϵ_{Fe} in $cm^{-1}mM^{-1}/Fe$. (Data adapted from refs. 38 and 60.)

synthetic complexes with (μ-oxo)bis(μ-carboxylato)diiron(III) cores argues strongly for the presence of similar carboxylate bridges in RRB2 which limit the Fe-Fe separation.

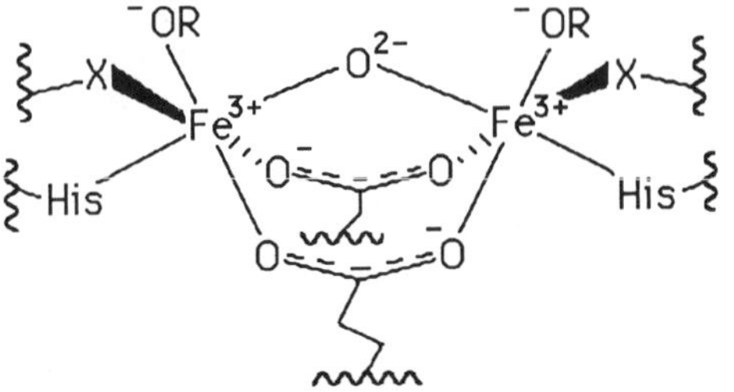

a plausible structure for binuclear iron site of RRB2

The resonance Raman spectrum of RRB2 shows symmetric and asymmetric Fe-O-Fe stretching modes which have been assigned by isotopic substitution (66,67). In solution at room temperature these resonances are observed at 492 cm^{-1} and ca. 756 cm^{-1}. These values are very similar to the values for oxyHr (486 and 753 cm^{-1}). In D_2O, ν_s shifts to 495 cm^{-1}, which suggests a hydrogen bond to the oxo bridge (as in oxyHr).

The variable temperature magnetic susceptibility of ribonucleotide reductase shows effects of both the tyrosyl radical and an antiferromagnetically coupled binuclear iron center; the effects of the latter are fit to a coupling of $J = -108$ cm^{-1} between two high-spin ferric ions (66). The NMR spectrum of RRB2 shows an exchangeable resonance at 25 ppm. This is assigned to protons of coordinated histidines based on the similar resonances observed in met and oxy forms of hemerythrin. There is also a non-exchangeable resonance at 19 ppm with three times the area of the 25 ppm peak (68). This peak remains unassigned. However, its magnitude relative to the 25 ppm peak argues for a small number of bound histidines, perhaps one for each iron.

The Mössbauer spectrum of RRB2 shows two distinct high-spin iron(III) environments ($\delta = 0.5$ mm/s) with ΔE_Q values (1.66 and 2.45 mm/s) which are only consistent with an oxo bridge (69). The asymmetry is suprising since the binuclear iron site is proposed to lie at the interface of two identical polypeptides of RRB2 (52). The symmetry of the protein is necessarily broken when the single tyrosyl radical is generated; however the asymmetry indicated by the Mössbauer spectrum persists when the radical is reduced by hydroxyurea. The difference in ΔE_Q values is similar to that in oxyHr; however, the visible and resonance Raman spectra offer no evidence of peroxide binding in RRB2.

The iron in ribonucleotide reductase can be removed by dialysis against a solution containing reductant (hydroxylamine) and an iron(II) chelator (69). In the process the tyrosyl radical is reduced. Active enzyme can be regenerated from apoenzyme by addition of iron(II) in the presence of O_2 (60). In contrast, hydroxyurea-treated ribonucleotide reductase, which is diferric but radical-free, cannot regenerate the tyrosyl radical; apparently Fe(II) is important for tyrosyl formation. Cell extracts from *E. coli* are able to activate RRB2 previously treated with hydroxyurea, and probably do so by first reducing the diiron center, although iron is not exchanged during the reactivation (70,71).

The spectroscopic similarities of hemerythrin and ribonucleotide reductase suggest that regeneration of the latter from apoenzyme or its reactivation by the cell extracts proceeds through a diferrous form similar to deoxyhemerythrin. Oxygen, which is necessary for reactivation (60,70), could bind to make an intermediate resembling oxyhemerythrin, which then causes oxidation of the tyrosine. It is unclear why such an intermediate would be unstable for ribonucleotide reductase while oxyhemerythrin is stable; the proximity of the tyrosine and an alternate H-bond donor in ribonucleotide reductase may be factors. Further experiments are required to elucidate the mechanism of oxidation of the tyrosine in ribonucleotide reductase.

Purple Acid Phosphatases

A third member of this class is the mammalian purple acid phosphatases, which catalyze the hydrolysis of phosphate esters under acidic conditions (pH optimum of 4.9) (3). The best characterized of these are the enzymes from porcine uterus (also called uteroferrin) (72) and bovine spleen (73). Similar enzymes have been isolated from rat spleen (74) and cell cultures of hairy cell leukemia cells from human spleen (75). The porcine and bovine enzymes have molecular weights of 35-40 kD and exhibit substantial sequence homologies (76). The enzymes contain a binuclear iron unit which can be found in two oxidation states, with only the reduced form exhibiting catalytic activity. Two properties suggest that the purple acid phosphatases belong to this class of iron-oxo proteins: the strong antiferromagnetic coupling observed in the [Fe(III),Fe(III)] forms (73,77) and the EPR signals observed for the [Fe(III),Fe(II)] forms with g_{av} of 1.7-1.8 (72,78), similar to those found for semimetHr complexes (Figure 1). There are also purple acid phosphatases from plant sources, which are related to the mammalian enzymes but are likely to have different active sites (79). These plant enzymes will not be discussed further.

To what extent are the binuclear units in the purple acid phosphatases analogous to those found in hemerythrin and ribonucleotide reductase? There are similarities and differences. The oxidized form is purple and EPR silent with strong antiferromagnetic coupling between the two Fe(III) centers; the reduced form is pink and EPR active (g_{av} = 1.7-1.8) with weak antiferromagnetic coupling between the Fe(III)-Fe(II) centers (3,72,73).

The spectroscopic property that readily distinguishes these enzymes from the other two already discussed is their purple/pink chromophore, which contrasts with the greenish color of methemerythrin and ribonucleotide reductase. The visible transitions of the phosphatases are ligand-to-metal charge transfer in nature, as indicated by their extinction coefficients of 4000 $cm^{-1}M^{-1}$ and confirmed by resonance Raman studies to be tyrosinate-to-iron(III) charge transfer (73,80) The persistence of the transition in the reduced form and the almost identical extinction coefficients in the two forms strongly suggest that tyrosine is coordinated solely to the iron center that remains ferric in the two states.

Further insights into the coordination chemistry of the binuclear unit have come principally from magnetic resonance studies on the reduced form of uteroferrin. The mixed valence form of uteroferrin exhibits an NMR spectrum with paramagnetically shifted features (Figure 5) that are reasonably sharp (300-500 Hz FWHM at 300 MHz) because of the fast relaxing Fe(II) center (81). The shifted features span a range of ±100 ppm, indicating a substantial amount of paramagnetism at ambient temperatures, and, therefore, weak antiferromagnetic

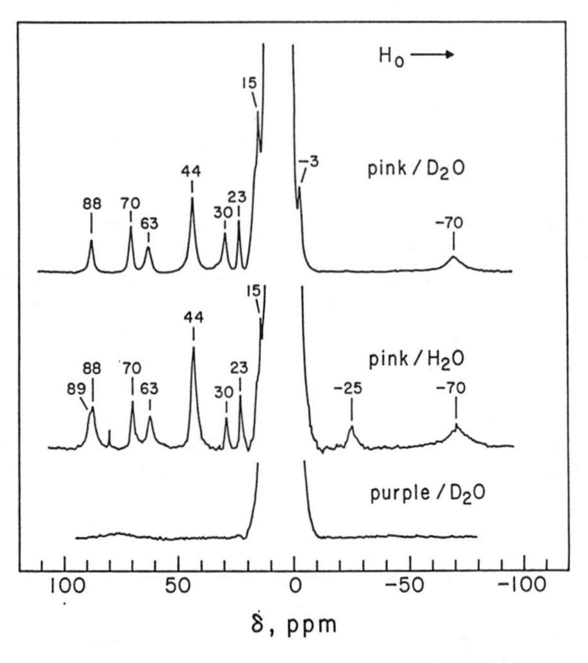

Figure 5. ^{1}H-NMR spectrum of reduced (pink) and oxidized (purple) uteroferrin. (Reproduced with permission from ref. 81. Copyright 1983 the authors.)

coupling (vide infra). By comparison of the shifts observed with model compounds, the spectral features have been assigned to tyrosine [Fe(III)], histidine [Fe(III)], and histidine [Fe(II)] in a 1:1:1 ratio. Absolute integration results indicate that there is only one of each ligand present (82). The presence of two distinct histidines is also suggested by the observation of two different ^{14}N components in the spin echo EPR experiments (83). Two other NMR features yet unassigned suggest the presence of at least one more endogenous protein ligand to the binuclear unit.

Oxyanions also affect the coordination chemistry of the metal center (84). Molybdate and tungstate are tightly bound noncompetitive inhibitors (K_I's of ca. 4 μM) (85). These anions bind to the reduced form of the enzyme, changing the rhombic EPR spectrum of the native enzyme to axial (Figure 1) and affecting the NMR shifts observed (84,85). Comparisons of the ENDOR spectra of reduced uteroferrin and its molybdate complex show that molybdate binding causes the loss of ^1H features which are also lost when the reduced enzyme is placed in deuterated solvent (86). These observations suggest that molybdate displaces a bound water upon complexation.

Phosphate and arsenate, not surprisingly, are competitive inhibitors of the enzymes with K_I's of 1-10 mM (85,87). Binding of phosphate to reduced uteroferrin results in a red shift of the λ_{max} and loss of most of the intensity of the characteristic EPR spectrum (84,87). Upon standing in air, the oxidized uteroferrin-phosphate complex is formed, in which the phosphate is tightly bound (87,88,89).

The effects of phosphate binding to the purple acid phosphatases are a matter of current disagreement (84,87,89,90,91). Antanaitis and Aisen have concluded that the oxidized uteroferrin-phosphate complex ($Uf_o{\cdot}P_i$) is formed immediately upon addition of phosphate to reduced uteroferrin on the basis of the observed color change and loss of the EPR signal (84). However, these spectral changes occur anaerobically as well (84). The immediate formation of $Uf_o{\cdot}P_i$, in which the phosphate is tighly bound (87,88), conflicts with the observation that phosphate is a competitive inhibitor of the phosphatases (87). In addition, Zerner et al. have reported that phosphatase activity is lost with a $t_{1/2}$ of 51 minutes during preincubation with phosphate (87). The apparently conflicting observations may be rationalized by the following scheme:

$$Uf_r + P_i \rightleftharpoons Uf_r{\cdot}P_i \rightarrow Uf_o{\cdot}P_i$$

The reduced uteroferrin-phosphate complex ($Uf_r{\cdot}P_i$) forms immediately and reversibly and becomes oxidized over a period of hours (90). The reversibility of its formation would be consistent with the competitive inhibition, and a rate-determining oxidation to $Uf_o{\cdot}P_i$ would be consistent with the slow loss of enzyme activity.

The postulated $Uf_r{\cdot}P_i$ would thus be purple and EPR silent. Though the purple color is characteristic of the oxidized form, Antanaitis and Aisen have shown that color and oxidation state in uteroferrin are not necessarily coupled (84). The loss of the EPR signal is more surprising, since a mixed valence state is normally expected to be EPR active. The Mössbauer spectrum of $Uf_r{\cdot}P_i$ at 55 K (Figure 6) exhibits two quadrupole doublets with parameters consistent with a high spin Fe(III) and high spin Fe(II) formulation (90). The 4.2 K spectrum shows magnetic hyperfine broadening characteristic of a half-integer spin state. These observations

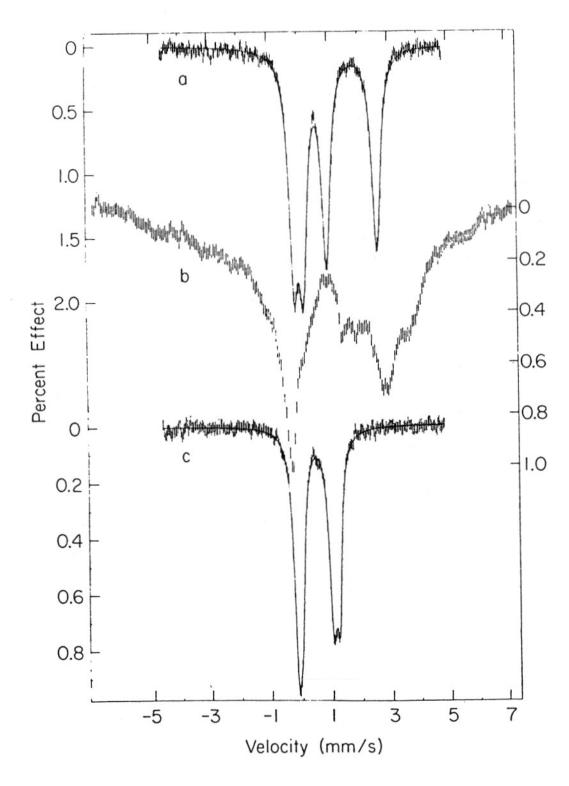

Figure 6. Mössbauer spectra of the reduced uteroferrin phosphate complex at 55K (a) and 4.2K (b) and the oxidized uteroferrin-phosphate complex at 4.2K (c). (Reproduced with permission from ref. 90. Copyright 1986 American Society for Biological Chemists.)

demonstrate that the purple and EPR silent complex obtained immediately after addition of phosphate to reduced uteroferrin is indeed $Uf_r \cdot P_i$. The EPR silence of this mixed valence complex is still not understood, but a recently synthesized mixed valence complex has also been found to be EPR silent (22,23). Observations on the interaction of phosphate with the bovine enzyme are somewhat different (89). In the bovine case, the oxidation of the enzyme in the presence of phosphate is slow like that observed for uteroferrin ($k = 0.07\ min^{-1}$); however, the color change and loss of EPR also follow kinetics similar to the loss of enzyme actvity.

A comparison of the Mössbauer parameters of reduced uteroferrin and its phosphate complex shows that phosphate binding perturbs the ferrous center only slightly but significantly affects the ferric center, suggesting that it coordinates to the ferric site (90,92). However, the potentiation of phosphatase oxidation by phosphate has led to the suggestion that phosphate can bind at the Fe(II) site to alter the Fe(III)/Fe(II) redox potential (73,89). In $Uf_o \cdot P_i$, the Mössbauer parameters of both iron sites differ significantly from those of oxidized uteroferrin, suggesting that the phosphate may bridge the iron centers in this form. A scheme which accommodates these observations is as follows:

Uf
Fe(II) Fe(III) + O⁻–P(=O)(O⁻)(OH) ⇌ (rapid) Fe(II)–Uf–Fe(III)(–O–P(=O)(O⁻)(OH)) ⇌ (slow) Fe(II)–Uf–Fe(III) (O–P(=O)(OH)–O bridging) —oxdn→ Fe(III)–Uf–Fe(III) (O–P(=O)(OH)–O bridging)

It is clear from the foregoing discussion that the terminal ligands in the phosphatases differ somewhat from those found in hemerythrin, but to what extent is the triply bridged diiron unit retained? The analyses of the second-shell EXAFS spectra of bovine (93) and porcine (94) oxidized phosphate complexes have identified Fe-Fe and Fe-P components (Table II). For the bovine enzyme, these scatterers are found at ca. 3 Å; for the porcine enzyme, they are found at 3.2±0.1 Å. The Fe-P distances of 3.1 - 3.2 Å are consistent with a bridging phosphate ligand. The short Fe-Fe distances observed indicate the participation of at least one single atom bridge. Indeed the distance found for the porcine enzyme is comparable to those observed for $metHrN_3$ and RRB2, suggesting the presence of a triply bridged core structure.

Variable temperature magnetic susceptibility experiments on the oxidized phosphate complex from bovine spleen confirm the antiferromagnetic coupling between the iron centers and -J is estimated to be > 150 cm^{-1} (73). Studies on oxidized uteroferrin suggest that -J is at least 40 cm^{-1} (77,81). The strong antiferromagnetic coupling found in the oxidized enzymes would argue for an oxo bridge between the metal centers, since only single atom bridges thus far have been shown to give rise to such a strong interaction between iron(III) centers. (A sulfido bridge could mediate such strong coupling, as in μ-sulfidomethemerythrin (35,95), but, unlike μ-sulfidomethemerythrin, the purple acid phosphatases exhibit neither multiple $S^{2-} \rightarrow Fe(III)$ charge transfer transitions in their visible spectra nor Fe-S-Fe vibrations in their Raman spectra that would be expected of a sulfide bridge.)

For hemerythrin and ribonucleotide reductase, resonance Raman and EXAFS spectroscopy provide clear evidence for an oxo bridge (31-33,43,45, 62,63,65-67). Surprisingly, such evidence is lacking for the oxidized acid phosphatases. The $\nu_{Fe\text{-}O\text{-}Fe}$ feature expected at ca. 500 cm^{-1} has not been observed by resonance Raman experiments, despite efforts to find it (73). Its absence can be rationalized by the results from model compound studies which show that the intensity of the $\nu_{Fe-O-Fe}$ feature depends on the nature of the terminal ligands (see Chapter 3 for an in depth discussion). The terminal ligand arrangement in the purple acid phosphatases could conceivably be such as to render only weak enhancement of the $\nu_{Fe-O-Fe}$ feature. EXAFS studies of the bovine (93) and porcine (94) oxidized phosphate complexes also do not provide definitive evidence for an oxo bridge (Table II). The first shell EXAFS spectrum of $Uf_o{\cdot}P_i$ (Figure 3) clearly differs from those of $metHrN_3$ and RRB2 in the lack of significant destructive interference at low k values due to a short Fe-O bond (33). Two-component fits of the first shell EXAFS of the bovine and the porcine complexes show a shell at 1.95-1.98 Å and another shell at 2.10-2.13 Å. The presence of short (1.9 Å) Fe-O(tyrosine) bonds may cause sufficient interference to obscure the even shorter Fe-oxo bond (93); perhaps more likely is the participation of hydrogen bonding to the postulated oxo bridge which may lengthen the Fe-oxo bond (94). The resolution of these inconsistencies await further data.

The magnetic interactions in the reduced porcine and bovine enzymes are considerably weaker than in the oxidized forms. The temperature dependences of the intensity of the EPR signals afford an estimate of -7 and -5.5 ± 1 cm^{-1} for the J values of the reduced porcine and bovine enzymes, respectively (73,81). The large shifts observed in the NMR spectra of both enzymes also indicate that the coupling is weak; from the temperature dependence of the isotropic shifts, J is estimated to be ca. -10 cm^{-1} for reduced uteroferrin (81). By analogy to the proposed structure for $semimetHrN_3$, the reduced enzymes would have a hydroxo bridge.

Overview

Hemerythrin in its met and oxy forms has been characterized crystallographically to have a (μ-oxo)bis(μ-carboxylato)diiron(III) core. From a comparison of spectroscopic and magnetic data, it is clear that the B2 subunit of ribonucleotide reductase has an active site which closely resembles that found in methemerythrin. The purple acid phosphatases may have a similar active site core structure, but one that is modified in some respects. Their active site picture will evolve with further experiments. Other proteins that are likely to contain such an active site include the recently isolated rubrerythrin (96) and methane monooxygenase (97). Rubrerythrin has been recently purified from *Desulfovibrio gigas* and found to have two mononuclear iron sites resembling that of rubredoxin and a binuclear site similar to that of methemerythrin. The binuclear site in rubrerythrin consists of antiferromagnetically coupled high spin ferric centers with large quadrupole splittings. Its biological functional has yet to be determined. The hydroxylase component of methane monooxygenase from *Methanobacterium capsulatas* Bath is a nearly colorless protein which contains two iron atoms (97). As isolated, it is EPR silent, but yields signals with g_{av} of 1.7-1.8 upon treatment with NADH (98). The resemblance of these signals to those found for semimethemerythrins and reduced purple acid phosphatases suggests the possibility of a similar triply-bridged diiron unit in methane monooxygenase. With an appropriate modification of the active site, methane monooxygenase may utilize the dioxygen binding chemistry of hemerythrin to effect hydroxylation, by analogy to hemocyanin and tyrosinase which have similar active sites but differing functions (99).

Acknowledgments

This work has been supported by grants from the National Institutes of Health and the National Science Foundation. Robert C. Scarrow is grateful for a postdoctoral fellowship from the American Cancer Society.

References and Notes

1. Klotz, I. M.; Kurtz, D. M., Jr. *Accts. Chem. Res.* **1984**, *17,* 16-22 and references therein.
2. Reichard, P.; Ehrenberg, A. *Science (Washington, D. C.)* **1983**, *221,* 514-519.
3. Antanaitis, B. C.; Aisen, P. *Adv. Inorg. Biochem.* **1983**, *5,* 111-136.
4. Cotton, F. A.; Wilkinson, G. *Advanced Inorganic Chemistry (4th ed)* Wiley:1980; pp. 154-155.
5. Armstrong, W. H.; Spool, A.; Papaefthymiou, G. C.; Frankel, R. B.; Lippard, S. J. *J. Am Chem. Soc.* **1984**, *106,* 3653-3667.
6. Wieghardt, K.; Pohl, K.; Gebert, W. *Angew. Chem. Intl. Ed. Engl.* **1983**, *22,* 727.
7. Chaudhuri, P.; Wieghardt, K.; Nuber, B.; Weiss, J. *Angew. Chem. Intl. Ed. Engl.* **1985**, *24,* 778-779.
8. Hartman, J. R.; Rardin, R. L.; Chaudhuri, P.; Pohl, K.; Wieghardt, K.; Nuber, B.; Weiss, J.; Papaefthymiou, G. C.; Frankel, R. B.; Lippard, S. J. *J. Am Chem. Soc.* **1987**.
9. Toftlund, H.; Murray, K. S.; Zwack, P. R.; Taylor, L. F.; Anderson, O. P. *J. Chem. Soc. Chem. Commun.* **1986**, 191-193.
10. Spool, A.; Williams, I. D.; Lippard, S. J. *Inorg. Chem.* **1985**, *24,* 2156-2162.
11. Czernuszewicz, R. S.; Sheats, J. E.; Spiro, T. G. *Inorg. Chem.* **1987**, *26,* 2063-2067.
12. O'Connor, C. J. *Prog. Inorg. Chem.* **1982**, *29,* 204-283.
13. Armstrong, W. H.; Lippard, S. J. *J. Am Chem. Soc.* **1985**, *107,* 3730-3731.
14. Murray, K. S. *Coord. Chem. Rev.* **1974**, *12,* 1-35.
15. Bertini, I.; Luchinat, C. *NMR of Paramagnetic Molecules in Biological Systems*, Benjamin/Cummings: Menlo Park, 1986; Chapter 2.
16. Que, L., Jr.; Maroney, M. J. *Metal Ions Biol. Syst.* **1987**, *21,* 87-120.
17. Gerloch, M.; McKenzie, E. D.; Towl, A. D. C. *J. Chem. Soc. A* **1969**, 71-.
18. Armstrong, W. H.; Lippard, S. J. *J. Am. Chem. Soc.* **1984**, *106,* 4632-4633.
19. Murch, B. P.; Bradley, F. C.; Que, L., Jr. *J. Am. Chem. Soc.* **1986**, *108,* 5027-5028.
20. Gorun, S. M.; Lippard, S. J. *Rec. Trav. Chim. Pays-Bas* **1987**, *106*, 417.
21. Gibson, J. F.; Hall, D. O.; Thornby, J. H. M.; Whatley, F. R. *Proc. Natl. Acad. Sci. U. S. A.* **1966**, *56*, 987-.
22 Borovik, A. S.; Que, L., Jr. submitted.
23 Borovik, A. S.; Murch, B. P.; Que, L., Jr.; Papaefthymiou, V.; Münck, E. *J. Am. Chem. Soc.* **1987**, *109*, 7190-7191.

24. Suzuki, M.; Uehara, A. *Inorg. Chim. Acta* **1986**, *123*, L9-L10.
25. Klotz, I. M.; Kurtz, D. M., Jr. *Acc. Chem. Res.* **1984**, *17*, 16-22.
26. Wilkins, P. C.; Wilkins, R. G.; *Coord. Chem. Rev.* **1987**, *79*, 195-214.
27. Sanders-Loehr, J.; Loehr, T. M. *Adv. Inorg. Biochem.* **1979**, *1*, 235-252.
28. a) Stenkamp, R. E.; Sieker, L. C.; Jensen, L. H. *J. Am. Chem. Soc* **1984**, *106*, 618-622. b) Stenkamp, R. E.; Sieker, L. C.; Jensen, L. H. *Acta Crystallogr. Sect. B* **1983**, *B39,* 697-703.
29. Stenkamp, R. E.; Sieker, L. C.; Jensen, L. H.; McCallum, J. D.; Sanders-Loehr, J. *Proc. Natl. Acad. Sci. U.S.A.* **1985**, *82*, 713-716.
30 Sheriff, S.; Hendrickson, W. A.; Smith, J. L. *J. Mol. Biol.,***1987**, *197,* 273-296.
31. Hendrickson, W. A.; Co, M. S.; Smith, J. L.; Hodgson, K. O.; Klippenstein, G. L. *Proc. Natl. Acad. Sci. U.S.A.* **1982**, *79*, 6255-6259. Hedman, B.; Co, M. S.; Armstrong, W. H.; Hodgson, K. O.; Lippard, S. J. *Inorg. Chem.* **1986**, *25*, 3708-3711.
32. Elam, W. T.; Stern, E. A.; McCallum, J. D.; Sanders-Loehr, J. *J. Am. Chem. Soc* **1982**, *104*, 6369-6373.
33. Scarrow, R. C.; Maroney, M. J.; Palmer, S. M.; Que, L., Jr.; Roe, A. L.; Salowe, S. P.; Stubbe, J. *J. Am. Chem. Soc* **1987**, *109*, 7857-7864.
34. Dawson, J. W.; Gray, H. B.; Hoenig, H. E.; Rossman, G. R.; Shredder, J. M.; Wang, R.-H. *Biochemistry* **1972**, *11*, 461-465.
35. Maroney, M. J.; Lauffer, R. B.; Que, L., Jr.; Kurtz, D. M., Jr. *J. Am. Chem. Soc* **1984**, *106*, 6445-6446.
36. Maroney, M. J.; Kurtz, D. M., Jr.; Nocek, J. M.; Pearce, L. L.; Que, L., Jr. *J. Am. Chem. Soc* **1986**, *108*, 6871-6879.
37. Okamura, M. Y.; Klotz, I. M.; Johnson, C. E.; Winter, M. R. C.; Williams, R. J. P. *Biochemistry* **1969**, *8*, 1951-1959.
38. Garbett, K.; Darnall, D. W.; Klotz, I. M.; Williams, R. J. P. *Arch. Biochem. Biophys.* **1969**, *135*, 419-434.
39. Clark, P. E.; Webb, J *Biochemistry* **1981**, *20*, 4628-4632.
40. Loehr, J. S.; Loehr, T. M.; Mauk, A. G.; Gray, H. B. *J. Am. Chem. Soc* **1980**, *102*, 6992-6996.
41. Dunn, J. B. R.; Shriver, D. F.; Klotz, I. M. *Biochemistry* **1975**, *14*, 2689-2694.
42. Kurtz, D. M. Jr.; Shriver, D. F.; Klotz, I. M. *J. Am. Chem. Soc* **1976**, *98*, 5033-5035.
43. Shiemke, A. K.; Loehr, T. M.; Sanders-Loehr, J. *J. Am. Chem. Soc* **1984** *106*, 4951-4956.
44. Freier, S. M.; Duff, L. L; Shriver, D. F.; Klotz, I. M. *Arch. Biochem. Biophys.* **1980**, *205*, 449-463.
45. Shiemke, A. K.; Loehr, T. M.; Sanders-Loehr, J. *J. Am. Chem. Soc* **1986** *108*, 2437-2443.
46. Reem, R. C.; Solomon, E. I. *J. Am. Chem. Soc* **1984**, *106*, 8323-8325.
47. Reem, R. C.; Solomon, E. I. *J. Am. Chem. Soc* **1987**, *109*, 1216-1226.
48. Elam, W. T.; Stern, E. A.; McCallum, J. D.; Sanders-Loehr, J. *J. Am. Chem. Soc* **1983**, *105*, 1919-1923.
49. Babcock, L. M.; Bradic, Z.; Harrington, P. C.; Wilkins, R. G.; Yoneda, G. S. *J. Am. Chem. Soc* **1980**, *102*, 2849-2850.
50. Muhoberac, B. B.; Wharton, D. C.; Babcock, L. M.; Harrington, P. C.; Wilkins, R. G. *Biochem. Biophys. Acta* **1980**, *626*, 337-345.

51. Wilkins, R. G.; Harrington, P. C. *Adv. Inorg. Biochem.* **1983**, *5*, 52-85.
52. Sjöberg, B.-M.; Gräslund, A. *Adv. Inorg. Biochem.* **1983**, *5*, 87-110.
53. Lammers, M.; Follmann, H. *Struct. Bonding (Berlin)* **1983**, *54*, 27-91.
54. Stubbe, J.; Ator, M.; Krenitsky, T. *J. Biol. Chem.* **1983**, *258*, 1625-1630.
55. Ator, M. A.; Stubbe, J.; Spector, T. *J. Biol. Chem.* **1986**, *261*, 3595-3599.
56. Sjöberg, B.-M.; *J. Biol. Chem.* **1986**, *261*, 5658-5662.
57. Sjöberg, B.-M.; Hahne, S.; Karlsson, M.; Jörnvall, H.; Göransson, M.; Uhlin, B.-E. *J. Biol. Chem.* **1986**, *261*, 5658-5662.
58. Salowe, S. P.; Stubbe, J. *J. Bacteriol.* **1986**, *165*, 363-366.
59. Joelson, T.; Uhlin, U.; Eklund, H.; Sjöberg, B.-M.; Hahne, S.; Karlsson, M *J. Biol. Chem.* **1984**, *259*, 9076-9077.
60. Peterson, L.; Gräslund, A.; Ehrenberg, A.; Sjöberg, B.-M.; Reichard, P. *J. Biol. Chem.* **1980**, *255*, 6706-6712.
61. Land, E. J.; Porter, G.; Strachan, E. *Trans. Faraday Soc.* **1961**, *57*, 1885-1893.
62. Scarrow, R. C.; Maroney, M. J.; Palmer, S. M.; Que, L., Jr.; Salowe, S. P.; Stubbe, J. *J. Am. Chem. Soc.* **1986**, *108*, 6832-6834.
63. Bunker, G.; Petersson, L.; Sjöberg, B.-M.; Sahlin, M.; Chance, M.; Chance, B.; Ehrenberg, A. *Biochemistry* **1987**, *26*, 4708-4716.
64. In μ-oxo dimers where the Fe-O-Fe angle is greater than 150°, multiple scattering along the Fe-O-Fe bonds contributes more to the EXAFS spectrum than does single scattering along the Fe-Fe vector. In such cases, the observed Fe-Fe distance (neglecting multiple scattering) is between 3.34 and 3.36 Å, which is twice the Fe-O bond length minus a phase shift factor of 0.22 Å due to the scattering from the oxygen atoms.[65] Thus we dispute the claim that the determined Fe-Fe distance of 3.26 Å could be an artifact of multiple scattering.[63]
65. Co, M. S.; Hendrickson, W. A.; Hodgson, K. O.; Doniach, S. *J. Am. Chem. Soc* **1983**, *105*, 1144-1150.
66. Sjöberg, B.-M.; Loehr, T. M.; Sanders-Loehr, J. *Biochemistry* **1982**, *21*, 96-102.
67. Sjöberg, B.-M.; Sanders-Loehr, J.; Loehr, T. M. *Biochemistry* **1987**, *26*, 4242-4247.
68. Sahlin, M.; Ehrenberg, A.; Gräslund, A.; Sjöberg, B.-M. *J. Biol. Chem.* **1986**, *261*, 2778-2780.
69. Atkin, C. L.; Thelander, L.; Reichard, P.. Lang, G. *J. Biol. Chem.* **1973**, *248*, 7464-7472.
70. Barlow, T.; Eliasson, R.; Platz, A.; Reichard, P.; Sjöberg, B.-M. *Proc. Natl. Acad. Sci. U.S.A.* **1983**, *80*, 1492-1495.
71. Eliasson, R.; Jörnvall, H.; Reichard, P. *Proc. Natl. Acad. Sci. U.S.A.* **1986**, *83*, 2373-2377.
72. Antanaitis, B. C.; Aisen, P.; Lilienthal, H. R. *J. Biol. Chem.* **1983**, *258*, 3166-3172.
73. Averill, B. A.; Davis, J. C.; Burman, S.; Zirino, T.; Sanders-Loehr, J.; Loehr, T. M.; Sage, J. T., Debrunner, P. G. *J. Am Chem. Soc.* **1987**, *109*, 3760-3767.
74. Hara, A.; Sawada, H.; Kato, T.; Nakayama, T.; Yamamoto, H.; Matsumoto, Y. *J. Biochem. (Tokyo)* **1984**, *95*, 67-74.

75. Ketcham, C. M.; Baumbach, G. A.; Bazer, F. W.; Roberts, R. M. *J. Biol. Chem.* **1985**, *260,* 5768-5776.
76. Hunt, D. F.; Yates, J. R., III; Shabanowitz, J.; Zhu, N.-Z.; Zirino, T.; Averill, B. A.; Daurat-Larroque, S. T.; Shewale, J. G.; Roberts, R. M.; Brew, K. *Biochem. Biophys. Res. Comm.* **1987**, *144,* 1154-1160.
77. Sinn, E.; O'Connor, C. J.; de Jersey, J.; Zerner, B. *Inorg. Chim. Acta* **1983**, *78,* L13-L15.
78. Antanaitis, B. C.; Aisen, P. *J. Biol. Chem.* **1982**, *257,* 5330-5332.
79. For example: Hefler, S. K.; Averill, B. A. *Biochem. Biophys. Res. Comm.* **1987**, *146,* 1173-1177. Sugiura, Y.; Kawabe, H.; Tanaka, H.; Fujimoto, S.; Ohara, A. *J. Biol. Chem.* **1981**, *256,* 10664-10670. Uehara, K.; Fujimoto, S.; Taniguchi, T. *J. Biochem. (Tokyo)* **75**, 627-638. Uehara, K.; Fujimoto, S.; Taniguchi, T.; Nakai, K. *J. Biochem. (Tokyo)* **1974**, *75,* 639-649. Fujimoto, S.; Nakagawa, T.; Ohara, A. *Chem. Pharm. Bull.* **1977**, *25,* 3282-3288.
80. Gaber, B. P.; Sheridan, J. P.; Bazer, F. W.; Roberts, R. M. *J. Biol. Chem.* **1979**, *254,* 8340-8342. Antanaitis, B. C.; Strekas, T.; Aisen, P. *J. Biol. Chem.* **1982**, *257,* 3766-3770.
81. Lauffer, R. B.; Antanaitis, B. C.; Aisen, P.; Que, L., Jr. *J. Biol. Chem.* **1983**, *258,* 14212-14218.
82. Scarrow, R. C.; Que, L., Jr., unpublished observations
83. Antanaitis, B. C.; Peisach, J.; Mims, W. B.; Aisen, P. *J. Biol. Chem.* **1985**, *260,* 4572-4574.
84. Antanaitis, B. C.; Aisen, P. *J. Biol. Chem.* **1985**, *260,* 751-756.
85. Pyrz, J. W. Ph. D. thesis, Cornell University, 1986.
86. Doi, K.; Gupta, R.; Aisen, P. *Rec. Trav. Chim. Pays-Bas* 1987, *106,* 251.
87. Keough, D. T.; Dionysius, D. A.; de Jersey, J.; Zerner, B. *Biochem. Biophys. Res. Comm.* **1982**, *108,* 1643-1648.
88. Antanaitis, B. C.; Aisen, P. *J. Biol. Chem.* **1984**, *259,* 2066-2069.
89. Burman, S.; Davis, J. C.; Weber, M. J.; Averill, B. A. *Biochem. Biophys. Res. Comm.* **1986**, *136,* 490-497.
90. Pyrz, J. W.; Sage, J. T.; Debrunner, P. G.; Que, L., Jr. *J. Biol. Chem.* **1986**, *261,* 11015-11020.
91. Doi, K.; Gupta, R.; Aisen, P. *J. Biol. Chem.* **1987**, *262,* 6892-6895.
92. Debrunner, P. G.; Hendrich, M. P.; DeJersey, J.; Keough, D. T.; Sage, J. T.; Zerner, B. *Biochim. Biophys. Acta* **1983**, *745*, 103-106.
93. Kauzlarich, S. M.; Teo, B. K.; Zirino, T.; Burman, S.; Davis, J. C.; Averill, B. A. *Inorg. Chem.* **1986**, *25,* 2781-2785.
94. Scarrow, R. C.; Que, L. , Jr., manuscript in preparation
95. Lukat, G. S.; Kurtz, D. M., Jr.; Shiemke, A. K.; Loehr, T. M.; Sanders-Loehr, J. *Biochemistry* **1984**, *23,* 6416-6422.
96. Huynh, B. H.; Prickril, B., C.; Moura, I.; Moura, J. J. G.; LeGall, J. *Rec. Trav. Chim. Pays-Bas* **1987**, *106,* 233.
97. Woodland, M. P.; Dalton, H. *J. Biol. Chem.* **1984**, *259,* 53-59.
98. Woodland, M. P.; Patil, D. S.; Cammack, R.; Dalton, H *Biochim. Biophys. Acta* **1986**, *873,* 237-242.
99. Himmelwright, R. S.; Eickman, N. C.; Solomon, E. I. *J. Am. Chem. Soc.* **1979**, *101,* 1576-1586. Himmelwright, R. S.; Eickman, N. C.; LuBien, C. D.; Lerch, K.; Solomon, E. I. *J. Am. Chem. Soc.* **1980**, *102,* 7339-7344.

RECEIVED December 18, 1987

Chapter 9

Fe(III) Clusters on Ferritin Protein Coats and Other Aspects of Iron Core Formation

Elizabeth C. Theil

Department of Biochemistry, North Carolina State University, Raleigh, NC 27695–7622

Ferritin, found in plants, animals, and some bacteria, serves as a reserve of iron for iron-proteins, such as those of respiration, photosynthesis, and DNA synthesis, as well as providing a safe site for detoxification of excess iron. The structure of ferritin, unique among proteins, is a protein coat of multiple, highly conserved polypeptides around a core of hydrous ferric oxide with variable amounts of phosphate. Variations in ferritin protein coats coincide with variations in iron metabolism and gene expression, suggesting an interdependence. Iron core formation from protein coats requires Fe(II), at least experimentally, which follows a complex path of oxidation and hydrolytic polymerization; the roles of the protein and the electron acceptor are only partly understood. It is known that mononuclear and small polynuclear Fe clusters bind to the protein early in core formation. However, variability in the stoichiometry of Fe/oxidant and the apparent sequestration and stabilization of Fe(II) in the protein for long periods of time indicate a complex microenvironment maintained by the protein coats. Full understanding of the relation of the protein to core formation, particularly at intermediate stages, requires a systematic analysis using defined or engineered protein coats.

The Structure and Function of Ferritin

Iron is required by essentially all living organisms, participating in respiration, photosynthesis, and DNA synthesis. An apparent exception is certain members of the bacterial genus *Lactobacillus* (1). In general, the reactions mediated by iron proteins are electron transfer or activation of dioxygen or nitrogen. However, the very properties of iron which render it so

0097–6156/88/0372–0179$06.00/0

useful are hazardous when uncontrolled. For example, the transfer of a single electron from Fe(II) to dioxygen or to organic compounds produces dangerous radicals which can derange the delicate structures required for normal life. Moreover, the oxidation product, hydrated Fe(III), is extraordinarily insoluble [K_s = 10^{-18}M (2) at the pH of biological materials], forming rust-like precipitates. Stabilizing Fe(III) as a complex with, e.g., citrate creates additional problems. For example, the amount of citrate required to stabilize the iron released from senescent red cells of a human in a single day could only be supplied by drinking 5 liters of orange juice (3), an amount likely to disrupt the normal acid/base balance in the body and to derange the metabolism of other metal ions. The problem of using iron is solved in biological systems by the use of ferritin (3), a protein which sequesters the iron in a safe, available form for plants, animals, and bacteria. Such a widespread distribution of an iron storage protein and the high conservation of the protein structure suggests that ferritin is an old protein. Certainly the need for ferritin is old and may even have preceded the appearance of dioxygen in the atmosphere, since even the most primitive organisms appear to have depended upon iron.

Ferritin Structure. Ferritin is a large complex protein composed of a protein coat which surrounds a core of polynuclear hydrous ferric oxide ($FeO \cdot OH$ or $Fe_2O_3 \cdot nH_2O$). Points of contact which have been observed (4) between the inner surface of the protein coat and the iron core may reflect sites of cluster formation and core nucleation.

The Protein Coat. Twenty-four polypeptides assemble into a hollow sphere, of ca. 100-120 Å in outer diameter, to form the protein coat of ferritin. The diameter of the interior, which becomes filled with hydrous ferric oxide, is ca. 70-80 Å. Subunit assembly appears to be spontaneous; the coat remains assembled even without the iron core. Subunit biosynthesis is actually controlled by the amount of iron to be stored by a cell; the subunit templates (mRNAs) are stored in the cytoplasm of a cell in a repressed form and are recruited for biosynthesis when the concentration of iron increases (3).

Many of the current ideas about the shape of the ferritin molecule are derived from the high resolution x-ray crystallographic studies of Harrison and coworkers (5,6) (Figure 1) on the protein coat of ferritin from the spleen of horses, in which essentially all (> 90%) of the polypeptide subunits are identical. However, protein coats of ferritin from other animals, and indeed from different cells and tissues in the same animal, can be composed of assemblages of similar, but distinct, subunits (3). The maximum number of different ferritin subunits possible in a given cell, tissue, or organism is not known, but three distinct subunits have been characterized so far (7). Although as many as 16 different DNA fragments encoding ferritin-like sequences have been identified (8-11) in the genome of a number of animals, some of the sequences may not be functional genes; the exact number of real and pseudoferritin genes is the subject of intense

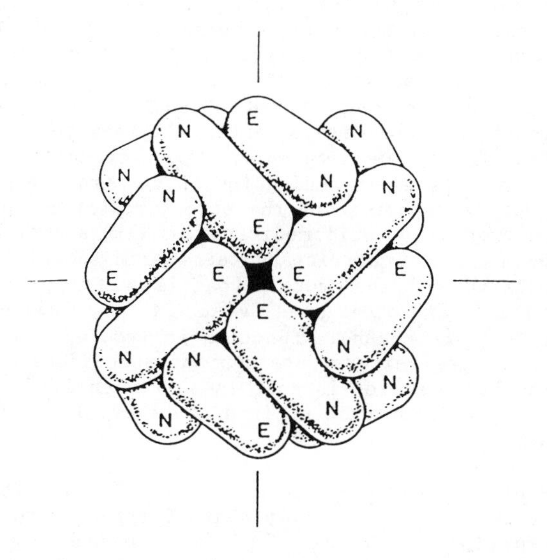

Figure 1. The protein coat of ferritin from horse spleen. N refers to the N-terminus and E refers to the location of the E-helix, a short helix with an axis perpendicular to the long axis of the subunit and which lines the channels formed at the four-fold axes. (Reproduced with permission from Ref. 5. Copyright 1983 Elsevier.)

investigation. Overall, the morphology of the assembled protein coat appears to be conserved not only within an organism but also among plants, animals, and bacteria (3,12), in spite of variations in the ratio of ferritin subunit types, the number of ferritin subunit types, and, in the case of ferritin from bacteria, the presence of heme (13-15).

Dimeric, trimeric, and tetrameric interactions of ferritin subunits each have distinctive features which may relate to the function of the protein coat. Among vertebrates, at least, the amino acid sequence in such regions is highly conserved, and forms structures which may have functional roles.

The structure of the dimer interface in the crystals from the protein coats of horse spleen ferritin (5,6), for example, is a densely packed mass of amino acid sidechains on the outer surface, where the protein would be in contact with the cytoplasm, but forms a flexible groove with potential ligands for iron on the inner surface facing the iron core. Speculation that the dimer interface participates in nucleation of the iron core has been frequent, not only because of the structure (6) but also because of the effect of amino acid modification (16) and of cross-linking subunit pairs (17). Trimeric and tetrameric subunit interactions form channels through the protein coat large enough to permit the movement of iron and other small molecules and leading to the hypothesis that the channels function in iron uptake and release from ferritin; preliminary data support such ideas (18,19) but the possibility that the protein coat is sufficiently flexible to permit the entrance or exit of iron at other sites cannot be discounted.

The Iron/Protein Interface. Interactions of iron with the protein coat of ferritin are most easily characterized in the early stages of core formation when most, if not all, of the iron present is in contact with the protein coat. In the complete core, bulk iron is inorganic. To date, the protein coat has been little examined early in iron core formation except in terms of effects on the iron environment. Studies of the iron early in core formation will be discussed later.

The Iron Core. All ferritin molecules store iron inside the protein coat as a polynuclear oxo-bridged complex. Experimentally, ferritin can only be formed from protein coats using Fe(II). Two reactions are involved: oxidation of Fe(II) and hydrolysis of Fe(III) to form oxo-bridged complexes. [Some Fe(II) atoms might form oxo or hydroxo bridges before oxidation, depending upon local microenvironments in the growing core.] The reverse reaction, i.e., release of iron from ferritin as Fe(II), requires addition of electrons and hydration of the oxo bridges (or ligand exchange). Whether or not the two reactions are temporally coupled is not known, although it is sometimes assumed. However, Watt and Frankel observed the apparent uncoupling of the two reactions when coulometric reduction of iron in ferritin cores produced a ferritin with 50% Fe(II) but with essentially all the iron retained inside the protein coat (20).

Variations in ferritin iron cores include the number of iron atoms, composition, and the degree of order (3,6,21-23). Size variations of the iron core range from 1-4500 Fe atoms and appear to be under biological control (e.g. Ref. 15). The distribution of iron core sizes in a particular ferritin preparation can be easily observed after sedimentation of ferritin through a gradient of sucrose.

Known compositional variations of ferritin iron cores only involve phosphate, which can range from as much as 80% (21) to as little as 5% of the iron (21); in normal mammalian liver or spleen, the amount of phosphate in the ferritin iron core is ca. 12% of the iron (24). When the phosphate content is high, the distribution of phosphate is clearly throughout the core rather than on the surface. However, interior locations for phosphate are also suggested when the phosphate content is lower, by data on an Fe(III)ATP model complex (P:Fe = 1:4) (25) or by phosphate accessibility studies in horse spleen ferritin (P:Fe = 1:8) (24). Based on model studies, other possible variations in core composition could include H_2O or sulfate (26).

The degree of order in ferritin iron cores has been shown to vary depending upon the organism and the amount of phosphate. For example, high resolution electron microscopy shows that the iron cores of many mammalian ferritins are microcrystalline and are composed of several small or one large crystal (27,28). Such crystals are similar to ferrihydrite, a mineral (29) which can also be formed experimentally by heating solutions of ferric nitrate (30); ferrihydrite has no phosphate. In contrast, the cores of ferritin from bacteria (21) and some invertebrates (22) are highly disordered. While the disorder is related to the amount of phosphate in some cases, in others the disorder appears to depend on the environment in which the core forms (the protein coat? other anions?). In mammalian ferritins, it is not known if phosphate might be responsible for regions of disorder that were not detected by the analysis of the diffraction pattern. In contrast to phosphate, sulfate might increase the order of ferritin iron cores since, in model systems, the amount of order in soluble hydrous ferric oxides appears to be increased by sulfate (26). However, the effect of sulfate on ferritin iron cores has not yet been examined.

Ferritin iron cores, or polynuclear iron complexes in lipid vesicles or in matrices of protein and complex carbohydrates, appear to be the precursors of minerals such as hematite and magnetite that form in certain bacteria (31), marine invertebrates (22), insects, and birds. The conversion from ferritin-like iron cores requires partial changes in the oxidation state of and/or ordering of the iron atoms, and may depend on some of the natural variations in ferritin core structure.

Clearly, the final structure of the iron core of ferritin is the resultant of effects of the protein coat and the environment on the formation of the polynuclear oxo-bridged iron complex. Although the important interplay among iron, various anions, and different ferritin protein coats (or other organic surfaces) has only begun to be examined in a systematic way, the early results promise new insights to understanding both the significance of variations in ferritin protein coats and core structures as well as the mechanisms of biomineralization.

Ferritin Function

The function of all ferritin molecules is to store iron. However, the mechanisms by which iron enters the core or is released from the core *in vivo* is poorly understood. Experimentally, Fe(II), but not Fe(III), mixed with ferritin protein coats forms normal iron cores. Moreover, reductants such as thioglycollate or reduced flavins can reverse the process of core formation and release Fe(II) from the core. Since such reductants occur *in vivo*, reduction of ferritin cores may also occur *in vivo*.

In contrast to the dearth of information about physiological uptake and release mechanisms for iron stored in ferritin, there is a great deal of information about the physiological roles of iron stored in ferritin. There are at least three purposes for which the stored iron can be used. First, ferritin can store iron for ordinary cellular metabolism such as the synthesis of iron proteins, e.g., the cytochromes or ribonucleotide reductase; all cells store iron for such housekeeping purposes, but the amount of iron (and of ferritin) will vary depending upon the quantity of iron proteins needed, cf. hemoglobin in red blood cells to cytochromes in an adipocyte. Second, ferritin can store excess iron for detoxification. Although in normal cells the incorporation of iron by a cell is tightly coupled to cellular needs, the safeguards can be breached in abnormal conditions that produce iron overload. For example, excess iron accumulates in the genetic disease hemochromatosis or in copper poisoning, where excess iron in the spleen is stored in a ferritin with a modified protein coat that allows more iron to be sequestered/molecule (17). Finally, some cells have the specialized role of storing extra iron (beyond the need for housekeeping or detoxification) for other cells. There are several types of specialized iron storage cells and each appears to release iron under different circumstances depending on the need of the animal, e.g., daily recycling of iron, long term iron storage for emergencies such as hemorrhaging, or iron storage for large and rapid demands during development and growth (3). It is tempting to think that the different types of ferritin protein coats in a specific cell type or in a particular physiological condition reflect the different iron storage roles of ferritin and different modes of gene regulation. While some observations support such an idea (7,17), it is far from being proven.

Iron Clusters and the Early Stages of Ferritin Iron Core Formation

The sequence of steps in the biosynthesis of the ferritin iron core has been studied by analyzing the incorporation of ^{59}Fe into ferritin during synthesis of the protein *in vivo*. Ferritin, collected at various intervals after the induction of synthesis, was fractionated according to iron core size by sedimentation through gradients of sucrose (32). ^{59}Fe appeared first in ferritin with small amounts of Fe, and later, the ^{59}Fe appeared in fractions further down the gradient as the core size and the ratio of Fe to protein increased. The data indicate the assembly of the polypeptide subunits into protein coats with no or small amounts of iron, followed by accretion of the iron core. Iron core formation in ferritin has been studied experimentally either by adding small amounts ($\leq$ 10-12 Fe atoms) of Fe(II) or other metal ions to ferritin protein coats or by adding enough Fe(II) to form relatively complete cores (400-2000 Fe atoms/protein coat). Note that cores do not form experimentally with Fe(III) and protein coats, although small amounts of Fe(III) can be added to existing ferritin iron cores (33). Each experimental approach has particular virtues. For example, at low Fe/protein ratios the contributions of Fe bound to the protein are emphasized. At high Fe(II)/protein ratios [and Fe(II) concentrations], measurement of oxygen uptake or Fe(II) oxidation is facilitated. Each set of experimental conditions has a biological counterpart. However, experimental conditions which correspond to the intermediate states of core formation, i.e. the repeated addition of small amounts of Fe(II), have been little used as yet.

Low Fe/Protein (Fe <10-12/Molecule). Ferritin protein coats have multiple (8-12) binding sites for a variety of metals, including Fe(II), Fe(III), V(IV), Mn(II), Tb(III), Cd(II), Zn(II), and Cu(II) (e.g., 5,34-36, and reviewed in Ref. 37). At least some of the metals bind at the three-fold channels. The location of the nucleation sites is presently unknown. However, if the three-fold channels are the nucleation sites for core formation, core growth could block the channels, thus inhibiting further accretion INSIDE the protein coat and could lead to the addition of Fe OUTSIDE the protein coat. Such an effect would obviate the sequestering function of the protein. Three forms of Fe have been observed bound to ferritin protein coats (apoferritin): mononuclear, dinuclear, and multinuclear clusters.

Mononuclear Fe-apoferritin has mainly been characterized in the Fe(II) state. The EPR signal characteristic of mononuclear Fe(III), $g' = 4.3$, which has been observed at very low Fe/protein ratios (38), appears to be damped by Fe(III) cluster formation even before all the metal-binding sites on the protein are saturated, suggesting that it is a minor form relative to multinuclear species. Mössbauer spectroscopy has been used to characterize Fe(II)-apoferritin directly (39), and EPR spectroscopy to characterize the site indirectly by competition with V(IV) and Mn(II) (35,36). Fe(II) bound to apoferritin appears to be an octahedrally coordinated, high-spin complex in which the ligands are most likely carboxylate and water; the EPR spectra of reporter metal ions collected at varying Fe(II)/protein ratios indicate more than one mode of Fe(II) binding (35,36). Fe(II) and Tb(III) binding have also been examined by difference absorption spectroscopy in the UV-visible region of the spectrum (34); a significant change in the spectrum upon Fe(II) binding occurs at 290-310 nm, where tyrosine and tryptophan absorb. Such results suggest that the protein undergoes conformational changes when metals bind but confound the examination of the metal environment itself.

Binuclear iron has been observed on ferritin coats, using EPR spectroscopy, by saturating the protein with Fe(II) (12 Fe/

molecule), converting to Fe(III) and adding a large excess of Fe(II) (120/ molecule) (40). A putative Fe(II)-O-Fe(III) binuclear species is only a small fraction of the total iron (39,40) and may be a transient precursor of the Fe(III) clusters.

Multinuclear clusters of Fe(III) occur bound to the protein when Fe(II), at amounts less than or equal to that required to saturate the protein, is allowed to oxidize _in situ_. The clusters, predicted from the results of EPR spectroscopy (35,38) and UV-difference spectroscopy (34), were observed and characterized by x-ray absorption (EXAFS) and Mössbauer spectroscopy (Figure 2; Ref. 39). Measurements were made with a complex of Fe(III) and the protein coats of apoferritin after the binding of 10 Fe(II) atoms/molecule, the admission of air and equilibration for 24 hours.

Fe(III) clusters on apoferritin are coordinated to six oxy ligands at 1.94 Å, one to two of which are contributed by -C-O (or -C-N), with the second low Z atom at 2.79 Å; models with carboxylate ligands fit the data well. Note that the low Z atom at 2.79 Å is not detectable in the environment of iron in ferritin with large iron cores (n = 2000) (39), presumably because the number of iron atoms attached directly to the protein is small relative to the number of iron atoms in the bulk phase. An average Fe(III) atom in the small cluster attached to the protein is also surrounded by 2-3 Fe atoms at $\sim$3.5 Å, suggesting a spherical cluster size of 3-4; however, if the cluster were elongated, the cluster size could be larger, with the average coordination number remaining the same. The Fe(III) clusters also show apparent blocking of superparamagnetic fluctuations at low temperatures, i.e. the Mössbauer spectrum is transformed from a quadrupole doublet to a magnetic sextet. The absence of mononuclear high spin Fe(III), for which a similar spectral change occurs at low temperatures, is indicated by the extraordinarily weak EPR signal at $g' = 4.3$ that only accounts for a small fraction of the Fe in the sample (39, see also Ref. 38). A reliable estimate of the cluster size was not possible from the Mössbauer data, either because the anisotropy constant differs from that for full ferritin cores or because the clusters are more disordered than full ferritin cores.

The presence of multinuclear Fe(III) complexes attached to the protein coats of ferritin indicates that the protein participates in nucleation of the iron core rather than merely serving as a surface for binding (or oxidizing) individual Fe atoms that then leave the protein surface to polymerize in a purely inorganic phase. Nucleation of the ferritin iron core on the protein surface has several ramifications. First, variations in the protein coat could alter the rate of nucleation or the structure of the nucleus, leading to possible protein-dependent variations in ferritin function. Second, the protein surface could influence the binding of anions and their incorporation into the core nucleus, leading again to protein-dependent variations in iron core structure (and function?). Finally, nucleation on the protein allows the orientation of core growth, defining the role of the protein coat in terms of structure rather than catalysis.

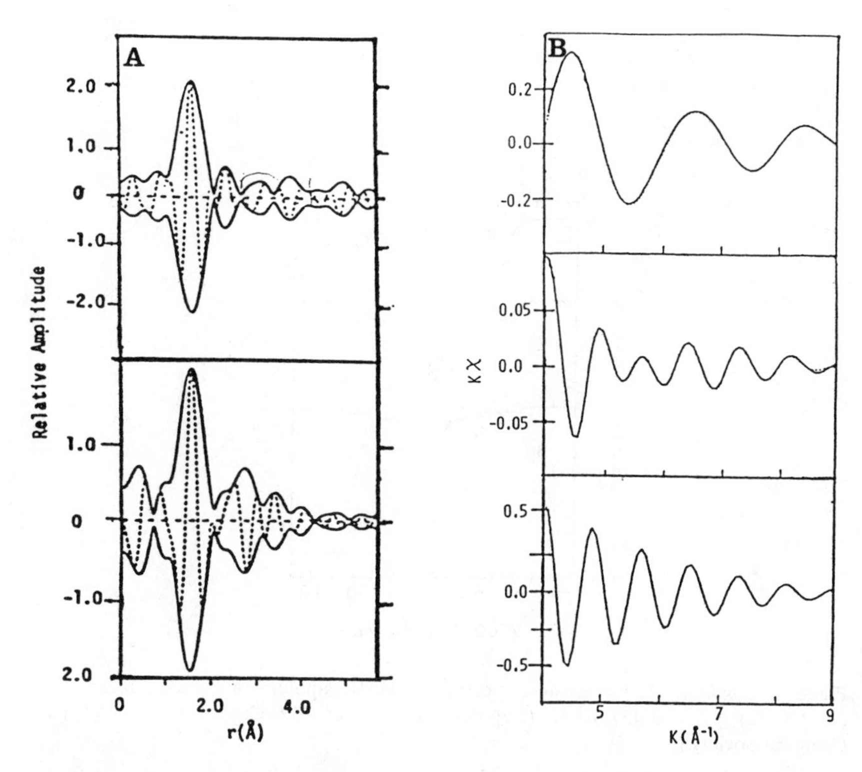

Figure 2. Fe(III) clusters on protein coats from horse spleen ferritin. (A) Fourier transform (3.0–10.2 $k\text{Å}^{-1}$) of the Fe(III)-apoferritin complex (10 Fe/molecule). Ferritin (ca 2000 Fe/molecule) is included for comparison. Upper panel, Fe(III)-apoferritin complex; lower panel, ferritin; k^3 weighting, not corrected for the phase shift. The combinations of ranges used in modeling the data were 0.52–2.06, 0.52–2.64, 0.52–3.38, 0.52–4.19, 2.64–3.38, and 2.64–4.19 Å. (B) Comparison of $k\chi$ for filtered experimental data and numerical fits for selected ranges [viz 0.52–2.64 Å (top), 2.64–4.19 Å (middle), and 3.38–4.19 Å (bottom)] of Fourier transform of Fe(III)-apoferritin complex [see (2)]. The ranges were chosen to show the nearest Fe–O and Fe–C interactions (top), the Fe–Fe and distant Fe–O interactions (middle), and the distant Fe–O interaction alone (bottom). Note that the bottom panel is displayed after analysis with k^3 weighting to emphasize the absence of the high Z (Fe) component at the longer distances. Solid line, experimental data; dashed line, modeling fits. The difference can only be observed at $k \geq 8$ because the variance is so small. The variance is 3.05×10^{-5} for Fe–O and Fe–C interactions, 1.60×10^{-5} for the Fe–Fe and distant Fe–O interactions, and 8.56×10^{-10} for the distant Fe–O interaction. (Reproduced from ref. 39. Copyright 1987 American Chemical Society.) *Continued on next page.*

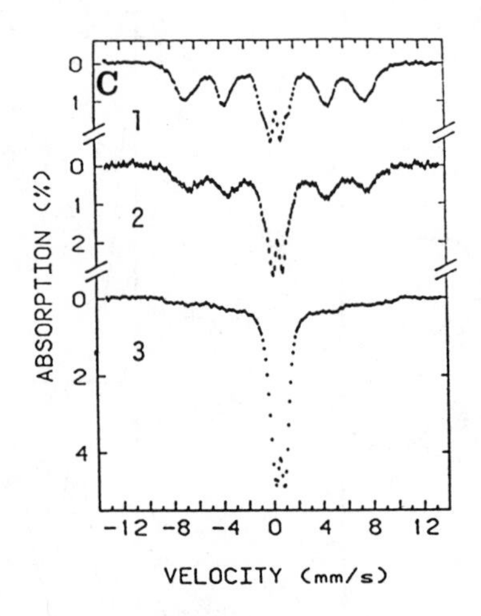

Figure 2. *Continued.* (C) Mössbauer spectra of ^{57}Fe(III)-apoferritin complex in zero field: (1) 1.5, (2) 4.2, and (3) 10 K. (Reproduced from ref. 39. Copyright 1987 American Chemical Society.)

Where in the protein coat the Fe(III) clusters form is not yet known. Nor is it known how many Fe(III) atoms can be accommodated by the protein coat; eventually, during iron core formation, Fe atoms added to the cluster will be in the environment of bulk core iron. Each conundrum has as a possible solution a testable hypothesis. In the case of the location of the sites of Fe(III) cluster formation, inspection of conserved carboxylate residues in the sequences of ferritins from mammals and frogs (3) shows a cluster of three at the trimer interface and a line of three pairs on the inner surface, the dimer interface; no patches of carboxylate ligands appear to exist on the outer surface of the protein coat of ferritin, making it unlikely that the Fe(III) cluster is on the surface. The trimer interface is not a likely site for core nucleation, as indicated above, because the growing core would block the channels and because polymerization of iron could extend to the outside of the protein coat. Moreover, the trimer interaction appears to be quite inflexible in the crystals and unlikely to expand readily to admit an Fe(III) cluster. In contrast, the dimer interface has the following features: 1. more potential Fe ligands on the inner surface than the three-fold channel (3); 2. a structure which can orient core growth toward the hollow center of the protein coat, i.e. a flexible groove on the inner surface toward the hollow center of the protein, and a boundary for the groove at the outer surface of the protein coat, which is a seemingly impenetrable mass of amino acid side chains (6); and 3. altered iron uptake when the dimer interaction is altered by covalent cross-links (17). Analysis of iron core formation in a recently prepared form of ferritin in which carboxylate residues at the dimer interface have been replaced using site-directed mutagenesis (S. Sreedharan and E. Theil, unpublished results) will test the hypothesis. In the case of the number of Fe(III) atoms that can be accommodated by the protein coat of ferritin, the number of binding sites (8-12) x the number of Fe(III) atoms/cluster (3-5) defines the lower limit of the number of clustered Fe(III) bound (24-60). Analysis of x-ray absorption spectra in complexes with a variety of numbers of Fe atoms added can provide limits to the numbers of iron atoms the protein can accommodate in the clusters.

High Fe/Protein (Fe = 400-2000/Molecule). The reconstitution of large ferritin iron cores from Fe(II) and protein coats involves the oxidation and polymerization of Fe. Usually, dioxygen is the electron acceptor in the experiments. A number of investigators have examined the stoichiometry of Fe(II) oxidation (Table I) (41-43), with results varying from 1.5-4 Fe/O_2; a stoichiometry of 2 Fe/O_2 was the value most frequently obtained. The amounts of iron and the buffers used were similar, although the method of measuring dioxygen consumption varied from manometry (41), to incorporation of ^{18}O from $^{18}O_2$ (42), to the use of an oxygen electrode (43). No evidence for oxygen radical intermediates was obtained. The lack of a truly satisfying explanation for the stoichiometry of oxygen consumption and Fe oxidation emphasizes the complexity of the path of ferritin iron core formation and the amount of knowledge still to be acquired. Recent evidence

Table I. Stoichiometry Fe/O_2 in Ferritin Formation [Horse Spleen Apoferritin, $Fe(II)NH_4SO_4$]

Fe(II) oxidized / O_2 consumed	Fe/Molecule	Buffer	Analysis	Reference
3.8-4.1	2300-1500	Imidazole, pH 6.4	Manometric	(41)
1.78-2.00	240-1900	Hepes-Mes-Imidazole, $H_2{}^{18}O$, pH 7.0	Mass Spectrometry	(42)
2.5-3.2*	222-2220	Hepes, Tris Imidazole, pH 7.4	Oxygen Electrode	(43)

*At 22 Fe/molecule, the ratio : Fe/O_2 consumed = 1.5.

emphasizes how much still needs to be learned (Figure 3; 44). For example, when changes in the oxidation state of bulk Fe are measured directly, e.g. by x-ray absorption spectroscopy, the changes continue for up to 24 hours. Such observations contrast with the results obtained when the changes in the Fe environment are measured indirectly. For example, when colorless solutions of Fe(II) and apoferritin are mixed, the solution quickly becomes the amber color of solutions of ferritin. The absorption spectrum of both ferritin and Fe(II) mixed with apoferritin is broad, with a series of transitions in the visible range. Measurements of color formation can be made at wavelengths between 310 and 500 nm; extinction coefficients vary with the amount of iron/molecule (34). The color change of the solutions, which previously has been used as a measure of Fe(II) oxidation during ferritin iron core formation, is complete long before the x-ray absorption spectrum changes from that of Fe(II) to that of Fe(III) (Figure 3). Note that the x-ray absorption spectrum of mixtures of solutions of Fe(II) and Fe(III) salts in 0.1 N HNO_3 can be reproduced very precisely by mathematically varying the weighting of averages of the spectra of solutions of pure Fe(II) or Fe(III) salts equivalent to those in the actual mixtures. Such results indicate that x-ray absorption spectra can be used to estimate reliably the relative amounts of Fe(II) and Fe(III) in a mixture. Other analyses of the rate of oxidation of Fe(II) in ferritin iron core formation have used the decline in reactivity of Fe(II) with *o*-phenanthroline or bipyridyl as an index. Almost immediately, upon mixing Fe(II) with solutions of ferritin protein coats, the availability to *o*-phenanthroline diminishes [the aliquots were mixed with the chelator in 0.01 N HCl (Figure 3) to allow complex formation before adjusting the pH]; all of the Fe(II) in solutions of Fe(II) without protein reacted with *o*-phenanthroline. By 2 hours after Fe(II) solutions were mixed with solutions of the protein, little of the Fe(II) in the protein solution was available to complex with *o*-phenanthroline, although the x-ray absorption spectrum was essentially unchanged and was indistinguishable

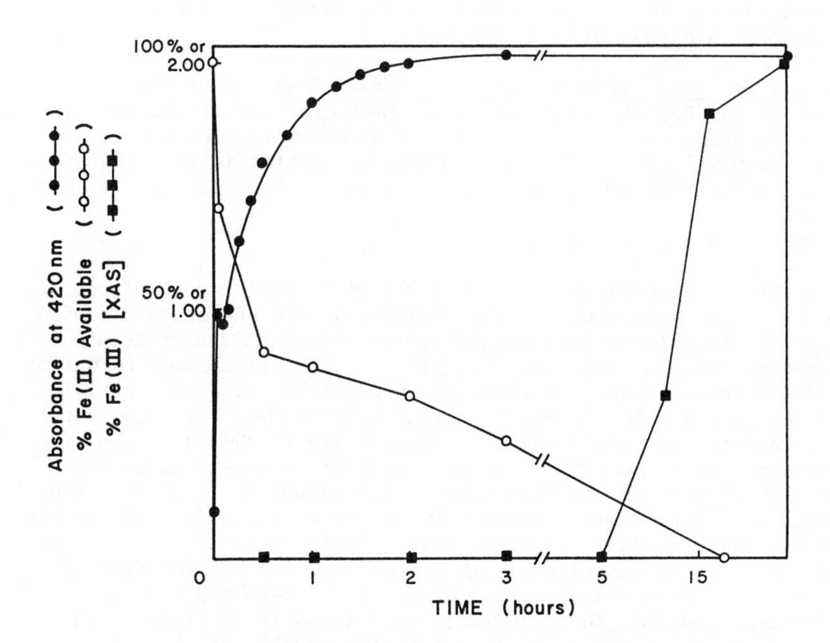

Figure 3. Comparison of the rate of oxidation of Fe(II) when mixed with apoferritin coats (480 Fe/molecule) in 0.15 M Hepes·Na, pH 7.0, using absorbance at 420 nm (●—●—●), availability to react with o-phenanthroline (o—o—o), change in the x-ray absorption near edge structure (XANES) (■—■—■). All three types of measurements were made under the same experimental conditions, including the sample holder. (Data are taken from Ref. 44.) Fe(II) is released to react with o-phenanthroline after boiling the protein in 1 N HCl.

from that for solutions of Fe(II). Boiling the protein solution in acid to denature the protein coats rendered Fe(II) available to complex with o-phenanthroline (J. Rohrer and E. Theil, unpublished results). Apparently, the protein rapidly incorporated Fe(II) inside the coats in an environment unavailable to the complexing agent, where the oxidation and/or the transition to polynuclear hydrous ferric oxide occurred slowly. What causes the apparent stabilization of Fe(II) inside the protein coats is unknown, but a microenvironment rich in protons or deficient in electron acceptors are possible explanations.

Models for Nucleation of Ferritin Cores. A number of small polynuclear Fe(III) complexes of defined size and stabilized by organic coats have been crystallized and characterized (e.g. 45-48). A complete description of their properties is the subject of Chapter 10 in this volume by S. M. Gorun.

Summary and Conclusions

Ferritin is a complex composed of a protein coat and a hydrous ferric oxide core with variable amounts of phosphate; points of contact which occur between the inner surface of the protein and the iron core may represent the sites of Fe(III) cluster formation and core nucleation. Animals, plants, and bacteria use ferritin to maintain a safe, available source of iron that overcomes both the toxicity of Fe(II) and the insolubility of Fe(III). The protein coat of ferritin is assembled from 24 polypeptide chains (subunits) to form a hollow sphere with channels at sites of the trimeric and tetrameric subunit interactions and a flexible groove on the inner surface of the dimeric subunit contacts. Variations in the types of subunits ($n \geq 3$) and the numbers of each type of subunit occur in the ferritins of different cell types of an organism, possibly to accommodate variations in the role of the stored iron such as normal intracellular metabolism (housekeeping), detoxification, long-term storage in specialized cells, or rapid recycling of iron in specialized cells which process iron from old cells. Variations in subunits of the protein coat are superimposed on a highly conserved sequence, which is found not only in all the ferritin of different cells in an animal but also in different animals. Variations also occur in the structure of the iron core of ferritin and include differences in size, composition (e.g. phosphate and, possibly, other anions), and degree of order. The functional significance of differences in core structure and the relation to difference in protein structure are only beginning to be explored.

Ferritin forms by addition of iron to assembled protein coats, which are stable even with no iron. The protein coats of ferritin can bind a variety of metal ions, e.g. Fe, Cd, Mn, V, Tb, and Zn, but to date, only Fe has been observed to form a core. *In vitro*, Fe(II) is required to initiate core formation. The steps include binding to the protein coat, oxidation and migration to form an Fe(III) cluster on the protein (or migration and oxidation on a cluster already formed on the protein), followed by the addition and oxidation of hundreds to thousands of Fe atoms; a subfraction of the iron, added to protein coats in small amounts to restrict core size, has been detected as a putative Fe(II)-O-Fe(III) binuclear species indicative of a possible intermediate step. The location of the Fe(III) cluster in the protein coat is not yet known, but consideration of the structure of the groove on the inner surface of the subunit dimer interface, the effect of covalent linking of subunits into dimers, and the ability of the dimer interface to orient iron core growth toward the hollow center of the protein coat suggest it is a reasonable site to consider for the cluster and core nucleation. Complete conversion of bulk Fe(II) to polynuclear Fe(III) appears to be relatively slow (16-24 hours at room temperature and pH = 7.0), although sequestration of the Fe(II) is relatively rapid. The detection both of Fe(III) clusters on the protein and the apparent control by the protein of Fe(II) sequestration and oxidation emphasize the important contribution of the ferritin protein coat to iron core formation, as well as providing indices for assessing the effect of variations in the structure of the protein coats. In addition, the study of the interrelationship between the ferritin protein coat and the structure and formation of the iron core should not only clarify the significance of the structure of ferritin related to function, but should also provide lessons for understanding the formation of iron biominerals and mineralization on organic surfaces in general.

Acknowledgments

The author has been fortunate to collaborate on much of this work with Dale E. Sayers, N. Dennis Chasteen, Boi Hanh Huynh, and Alain Fontaine, and to have received support from the North Carolina Agricultural Research Service and the National Institutes of Health (grants DK20251 and GM34675).

Literature Cited

1. Neilands, J.B. *Microbial Iron Metabolism*; Academic Press: New York, 1974; p 25.
2. Biedermann, G.; Schindler, P. *Acta Chem. Scand.* **1957**, *11*, 731-740.
3. Theil, E.C. *Ann. Rev. Biochem.* **1987**, *56*, 289-315.
4. Massover, W.H. *J. Mol. Biol.* **1978**, *123*, 721-726.
5. Rice, D.W.; Ford, G.C.; White, J.L.; Smith, J.M.A.; Harrison, P.M. In *Advances in Inorganic Biochemistry*; Theil, E.C.; Eichhorn, G.L.; Marzilli, L.G., Eds.; Elsevier: New York, **1983**; Vol. 5, pp 39-50.
6. Ford, G.C.; Harrison, P.M.; Rice, D.W.; Smith, J.M.A.; Treffry, A. *Philos. Trans. R. Soc. Lond. B Biol. Sci.* **1984**, *304*, 561-565.
7. Dickey, L.F.; Sreedharan, S.; Theil, E.C.; Didsbury, J.R.; Wang, Y.-H.; Kaufman, R.E. *J. Biol. Chem.* **1987**, *262*, 7901-7907.
8. Didsbury, J.R.; Theil, E.C.; Kaufman, R.E.; Dickey, L.F. *J. Biol. Chem.* **1986**, *261*, 949-955.

9. Leibold, E.A.; Munro, H.N. *J. Biol. Chem.* 1987, *262*, 7335-7341.
10. Santoro, C.; Marone, M.; Ferrone, M.; Constanzo, F.; Colombo, M.; Minganti, C.; Cortese, R.; Silengo, L. *Nucleic Acids Res.* **1986**, *14*, 2863-2876.
11. Constanzo, F.; Colombo, M.; Staempfli, S.; Santoro, C.; Marone, M.; Rainer, F.; Hajo, D.; Cortese, R. *Nucleic Acids Res.* **1986**, *14*, 721-736.
12. Clegg, G.A.; Fitton, J.E.; Harrison, P.M.; Treffry, A. *Prog. Biophys. Mol. Biol.* **1981**, *36*, 53-86.
13. Stiefel, E.F.; Watt, G.D. *Nature* 1979, *279*, 81-83.
14. Yariv, J.; Kalb, J.; Sperling, R.; Banninger, E.R.; Cohen, S. G.; Ofer, S. *Biochem. J.* **1981**, *197*, 171-175.
15. Moore, G.R.; Mann, S.; Bannister, J.V. *J. Inorg. Biochem.* **1987**, *28*, 329-336.
16. Wetz, K.; Crichton, R.R. *Eur. J. Biochem.* **1976**, *61*, 545-550.
17. Mertz, J.R.; Theil, E.C. *J. Biol. Chem.* **1983**, *258*, 11719-11726.
18. Treffry, A.; Harrison, P.M. *Abstracts, Proc. 8th Int. Conf. on Proteins of Iron Transport and Storage*, **1987**, p 8.
19. Levi, S.; Luzzago, A.; Ruggeri, G.; Cozzi, A.; Campanini, S.; Arioso, P.; Cesarini, G. *Abstracts, Proc. 8th Int. Conf. on Proteins of Iron Transport and Storage*, **1987**, p 53.
20. Watt, G.D.; Frankel, R.B.; Papaefthymiou, G.C. *Proc. Natl. Acad. Sci. U.S.A.* 1985, *82*, 3640-3643.
21. Mann, S.; Bannister, J.V.; Williams; R.J.P. *J. Mol. Biol.* **1986**, *188*, 225-232.
22. St. Pierre, T.G.; Bell, S.H.; Dickson, D.P.K.; Mann, S.; Webb, J.; Moore, G.R.; Williams, R.J.P. *Biochim. Biophys. Acta* **1986**, *870*, 127-134.
23. Bell, S.; Weir, M.P.; Dickson, D.P.K., Gibson, J.F.; Sharp, G.A.; Peters, T.J. *Biochim. Biophys. Acta* **1984**, *787*, 227-236.
24. Granick, S.; Hahn, P.F. *J. Biol. Chem.* **1944**, *155*, 661-669.
25. Mansour, A.N.; Thompson, C.; Theil, E.C.; Chasteen, N.D.; Sayers, D.E. *J. Biol. Chem.* **1985**, *260*, 7975-7979.
26. Yang, C.-Y.; Bryan, A.M.; Theil, E.C.; Sayers, D.E.; Bowen, L.H. *J. Inorg. Biochem.* **1986**, *28*, 393-405.
27. Harrison, P.M.; Fischbach, F.A.; Hoy, T.G.; Haggis, G.A. *Nature* **1967**, *216*, 1188-1190.
28. Massover, W.H.; Crowley, J.M. *Proc. Natl. Acad. Sci. U.S.A.* **1973**, *70*, 3847-3851.
29. Towe, K.M.; Bradley, W.F. *J. Colloid. Interface Sci.* **1967**, *24*, 384-392.
30. Schwertmen, V.; Schulze, D.G.; Murad, E. *Soil Sci. Soc. Amer. J.* **1982**, *46*, 869-875.
31. Mann, S.; Frankel, R.B.; Blakemore, R.P. *Nature* **1984**, *310*, 405-407.
32. Hoy, T.G.; Harrison, P.M. *Brit. J. Haem.* **1976**, *33*, 497-504.
33. Treffry, A.; Harrison, P.M. *Biochem. J.* **1979**, *181*, 709-716.
34. Treffry, A.; Harrison, P.M. *J. Inorg. Biochem.* **1984**, *21*, 9-20.
35. Chasteen, N.D.; Theil, E.C. *J. Biol. Chem.* **1982**, *257*, 7672-7677.

36. Wardeska, J.G.; Viglione, B.; Chasteen, N.D. *J. Biol. Chem.* **1986**, *261*, 6677-6683.
37. Theil, E.C. In *Advances in Inorganic Biochemistry*; Theil, E.C.; Eichhorn, G.L.; Marzilli, L.G., Eds.; Elsevier: New York, **1983**; Vol. 5, pp 1-38.
38. Rosenberg, L.P.; Chasteen, N.D. In *The Biochemistry and Physiology of Iron*; Saltman, P.; Hegenauer, J., Eds.; Elsevier Biomedical: New York, **1982**; pp 405-407.
39. Yang, C.-Y.; Meagher, A.; Huynh, B.H.; Sayers, D.E.; Theil, E.C. *Biochemistry* **1987**, *26*, 497-503.
40. Chasteen, N.D.; Aisen, P.; Antanaitis, B.C. *J. Biol. Chem.* **1985**, *260*, 2926-2929.
41. Melino, G.; Stefanini, S.; Chiancone, E.; Antonini, E. *FEBS Lett.* **1978**, *86*, 136-138.
42. Mayer, D.E.; Rohrer, J.S.; Schoeller, D.A.; Harris, D.C. *Biochemistry* **1983**, *22*, 876-880.
43. Treffry, A.; Sowerby, J.M.; Harrison, P.M. *FEBS Lett.* **1978**, *95*, 221-224.
44. Rohrer, J.S.; Joo, M.-S.; Dartyge, E.; Sayers, D.E.; Fontaine, A.; Theil, E.C. *J. Biol. Chem.* **1987**, *262*, 13385-13387.
45. Murch, B.P.; Boyle, P.D.; Que, L., Jr. *J. Am. Chem. Soc.* **1985**, *107*, 6728-6729.
46. Gorun, S.M.; Lippard, S.J. *J. Am. Chem. Soc.* **1985**, *107*, 4568-4570.
47. Gorun, S.M.; Lippard, S.J. *Nature* **1986**, *319*, 666-668.
48. Wieghardt, K.; Pohl, K.; Jebril, I.; Huttner, G. *Angew. Chem.*, Int. ed., **1984**, *230*, 77-78.

RECEIVED December 18, 1987

Chapter 10

Oxo–Iron(III) Aggregates

A Current Perspective

Sergiu M. Gorun

Exxon Research and Engineering Company, Route 22 East, Annandale, NJ 08801

Polynuclear oxo-iron(III) aggregates form a new class of inorganic complexes of the type found at the active site of iron storage proteins like ferritin. Crystallographically characterized, synthetic members of this class range in size from three to eleven oxo-hydroxo- or alkoxo-bridged metal atoms. Preparative methods include hydrolytic polymerization of common iron salts and stepwise controlled oligomerization of mono- and oxo-bridged binuclear Fe(III) complexes. The position and intensity of the bands found in the electronic and vibrational spectra of the aggregates are related to their nuclearity. Within the Heisenberg-Dirac-Van Vleck formalism, small size antiferromagnetically coupled aggregates can be magnetically decomposed into a sum of interacting mono- and binuclear metal centers of the type employed in their synthesis. Magnetic and Mössbauer data suggest that the largest of the aggregates, which contains eleven metal atoms, is at the borderline between a molecule and a solid. Formation of various size oxo-aggregates parallels the accretion process by which the core of metal storage proteins forms *in vivo*.

A new class of metalloproteins containing polynuclear, non-heme oxo-bridged iron complexes has emerged recently. Dinuclear centers are present in hemerythrin (Hr), ribonucleotide reductase (RR), purple acid phosphatases (PAP) and, possibly, methane monooxygenase (MMO); these centers as well as model compounds are reviewed in Chapter 8.

Higher nuclearity oxo-iron centers are present in iron storage proteins like ferritin (Ft) and hemosiderin (Hs)(1) as well as in magnetotactic organisms and in the teeth of chitons(2). In the latter two cases magnetite is the predominant phase. In the case of the storage proteins (see Chapter 9), the metallic cores are structurally poorly characterized and numerous questions concerning their formation, structure and functionality remain to

0097–6156/88/0372–0196$07.25/0

be answered. Modelling the structure and the formation of the ferritin core represent a challenge for the inorganic chemist;(3a) recent efforts will be discussed here. In order to facilitate a better understanding of the aggregates, this review compares the same type of properties (e.g. spectroscopy, magnetism) for aggregates of various sizes.

We define "oxo-iron aggregates" as a collection of three or more iron ions linked continuously by bridging oxo-, hydroxo- or alkoxo groups. The aggregates are labelled according to their nuclearity. Undoubtedly the molecular structures of the aggregates represent the key to understanding their properties and therefore they will be discussed first.

Structural Chemistry

The structural chemistry of oxo-iron aggregates is dominated by the presence of μ- and μ_3-oxo groups which, in conjunction with other ligands, confer stability to the polynuclear core. For structurally characterized synthetic complexes, the degree of aggregation varies to date from three to eleven, but neither a decanuclear nor any odd numbered intermediate size is known.

Trinuclear Aggregates. The basic iron salts presented schematically in Figure 1 constitute the vast majority of this class. Their general formula can be written $[Fe_3OL_6L'_3]$, **3**, where L is a bidentate bridging ligand, usually a carboxylate or sulphate anion, and L' is a monodentate ligand such as water or pyridine(3,4). For all trinuclear aggregates, the metals are octahedrally coordinated, high spin and their oxidation states can vary from 2+ to 3+. The mixed valence complexes will not be discussed here.

Depending upon the geometry of the triangle formed by the metal ions, two types could be distinguished:

Type A: The metals form an equilateral triangle with the μ_3-oxo oxygen at its geometric center but not necessarily in the plane. The μ_3-oxo-Fe bond lengths are typically 1.92 ± 0.02 Å, significantly longer than the 1.79 ± 0.02 Å values found for dinuclear μ-oxo-Fe centers, but closer to the 1.97 ± 0.02 Å values found in their hydroxo bridged counterparts. The Fe-O (terminal) bonds are at least 2.01 Å long. The Fe-μ_3-O-Fe angles are all equal, 120° when the μ_3-oxo atom is in the plane of the metals(3) but, when L = $(CH_3)_3CCOO^-$, the μ_3-oxo group is above the metal plane(3e).

Type B: The metals form an isosceles triangle, the μ_3-oxo group moves off center toward the base of the triangle. Only one compound of this type is known(5), $[Fe_3O(TIEO)_2(O_2CPh)_2Cl_3]$, **3B**, shown in Figure 2. The two metal ions that form the base of the triangle, Fe(1) and Fe(2), are 3.667 Å apart and 3.022 Å from the apical Fe(3). The Fe(1)-μ_3-O-Fe(2), angle is 159°, more toward the ~ 180° values found in single bridged μ-oxo diiron complexes than the 120° value found for type A. Undoubtedly, the biologically relevant TIEOH ligand dictates the geometry of the $\{Fe_3O\}$ core. As in type A, two benzoate residues bridge Fe(1)-Fe(3) and Fe(2)-Fe(3)

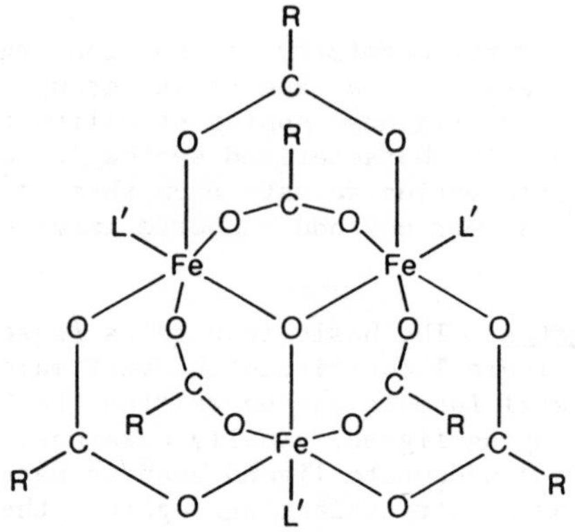

Fig. 1. The schematic structure of basic iron carboxylates. R is an alkyl or aryl group. (Reproduced from Ref. 5b. Copyright 1987 American Chemical Society.)

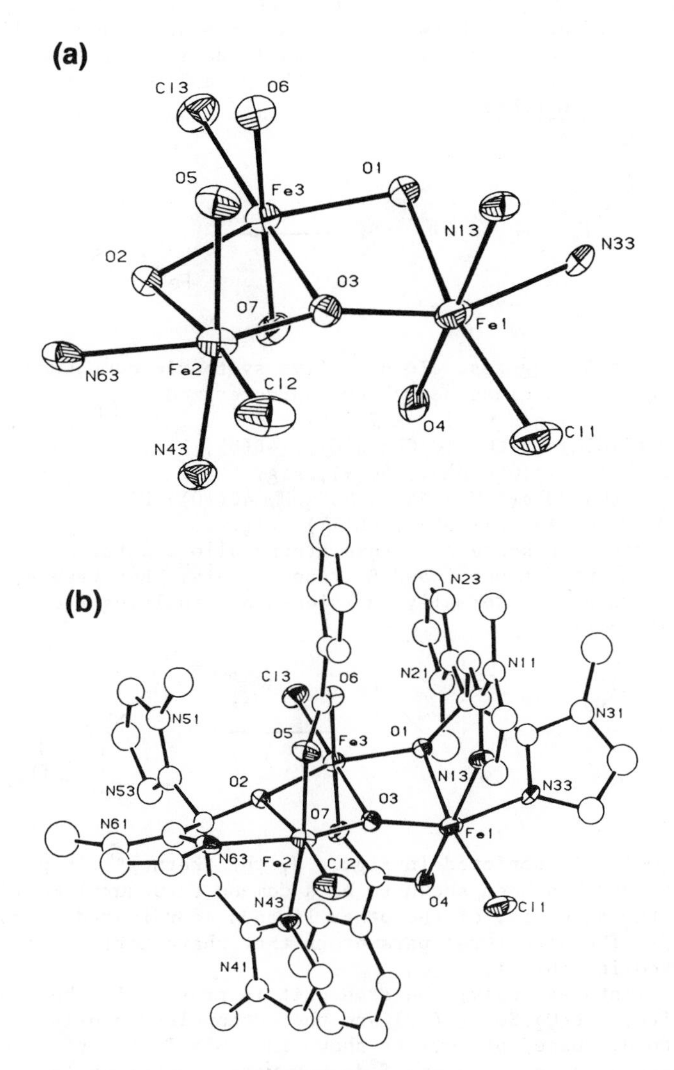

Fig. 2. The structure of $[Fe_3O(TIEO)_2(O_2CPh)_2Cl_3]$, 3B, showing the 40% probability thermal ellipsoids.
(a) the iron coordination spheres.
(b) the full molecule; the carbon atoms are not labelled. TIEOH stands for 1,1,2-tris(N-methylimidazol-2-yl)-1-hydroxyethane. TIEO is the alkoxide. (Reproduced from Ref. 5a. Copyright 1985 American Chemical Society.)

and each metal has a terminal L′ ligand, in this case chloride. The non-equivalent metals are also linked by alkoxy bridges. The structural differences between types A and B are dramatically reflected in their physical properties (vide infra). Formally, the trinuclear centers can be viewed(5a,8b) as a 2 + 1 addition of the type depicted in scheme 1:

$$Fe{-}O{-}Fe \; + \; Fe \longrightarrow (\mu_3\text{-}O)Fe_3 \qquad (1)$$

Tetranuclear Aggregates. To date five synthetic compounds belonging to this class have been characterized:

$[Fe_4O_2(O_2CCF_3)_8(H_2O)_6]$, **4A**(7), Fig. 3,
$[Fe_4O_2(BICOH)_2(BICO)_2(O_2CPh)_4]Cl_2$, **4B**(8), Fig. 4,
$[Fe_4O_2(H_2Bpz_2)_2(O_2CPh)_7]$, **4C**(9), Fig. 5,
$(PH)_4[Fe_4O_2(OH)_2(5\text{-}MeHXTA)_2]\cdot 2CH_3OH$, **4D**(10), Fig. 6,
$Na_6[Fe_4O_2(CO_3)_2L_2'']\cdot 20H_2O$, **4E**(11), Fig. 7.

The presence or absence of μ_3-oxo groups allows a further subdivision into types A and B, respectively. For type A, the $\{Fe_4O_2\}$ core can be formally envisioned as resulting from a 2 + 2 condensation:(8)

$$Fe{-}O{-}Fe \; + \; Fe{-}O{-}Fe \longrightarrow Fe_4(\mu_3\text{-}O)_2 \qquad (2)$$

The $\{Fe_4O_2\}$ unit depicted in Figure 8, the "kernel", is planar in **4A** and **4B** and has been shown to be a common structural motif despite the diversity of the other ligands coordinated to the metal atoms(8). The structural parameters that characterize this unit are listed in Table I.

Interestingly, the same unit is present in the mineral amarantite, $[Fe_4O_2(SO_4)_4]$(12) and the octanuclear oxo-iron aggregate discussed below. As shown in Table I, the structural parameters a, b, c, f, β and δ do not vary significantly. Any change in the planarity of the $\{Fe_4O_2\}$ core, however, will result in variations within the kernel as in **4C** and for the mineral leucophosphite(13), $K_2[Fe_4(OH)_2(H_2O)_2(PO_4)_4]\cdot 2H_2O$. In the latter case the μ_3-oxo groups are protonated and no longer in the plane defined by the four iron atoms.

Type B is represented by **4D** and **4E** which contain only μ-oxo groups. In **4D** the metal atoms form a tetrahedron and are bridged by two μ-oxo, two μ-hydroxo and two μ-phenoxo groups(10). The Fe-μ-oxo bond lengths, 1.791(3) Å, are normal, similar to those found in μ-oxo dinuclear complexes. The same type of bonds are 1.829(4) Å long for the centrosymmetric **4E** which comprises two

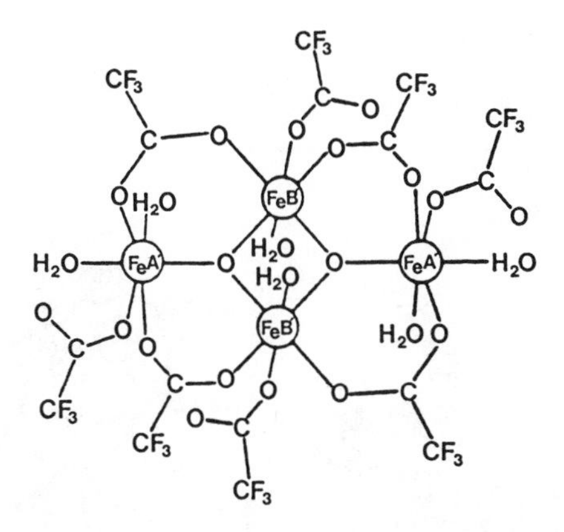

Fig. 3. The schematic structure of 4A, $[Fe_4O_2(O_2CCF_3)_8(H_2O)_6]$. (Data taken from Ref. 24.)

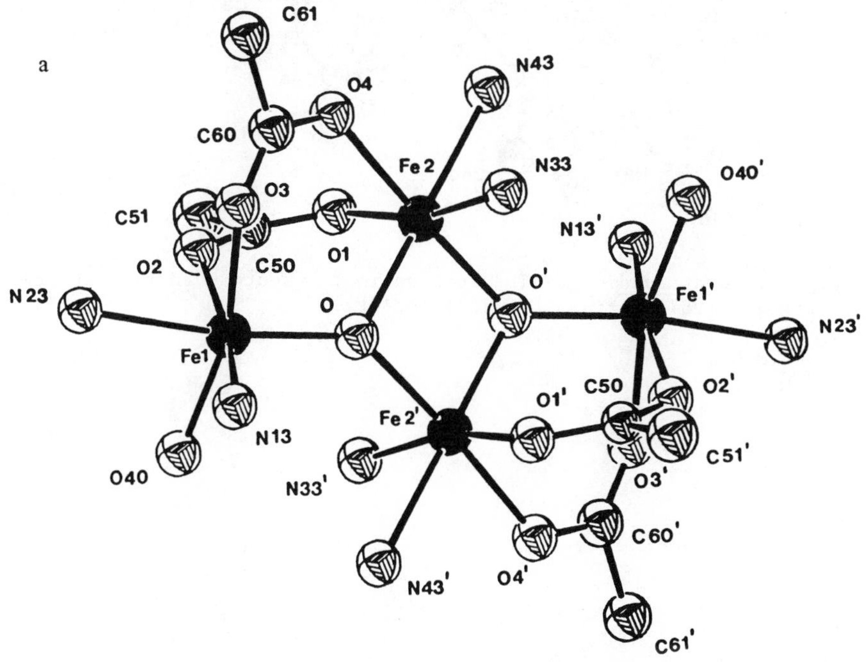

Fig. 4. The structure of the cation of 4B, $[Fe_4O_2(BICOH)_2(BICO)_2(O_2CPh)_4]^{2+}$ showing the 40% probability thermal ellipsoids. Only the first atom of the phenyl rings is shown. BICOH stands for bis(N-methylimidazol-2-yl)carbinol. BICO is the alkoxide anion. (a) The Fe_4O_2 core: iron coordination spheres and bridging benzoate groups. Continued on next page.

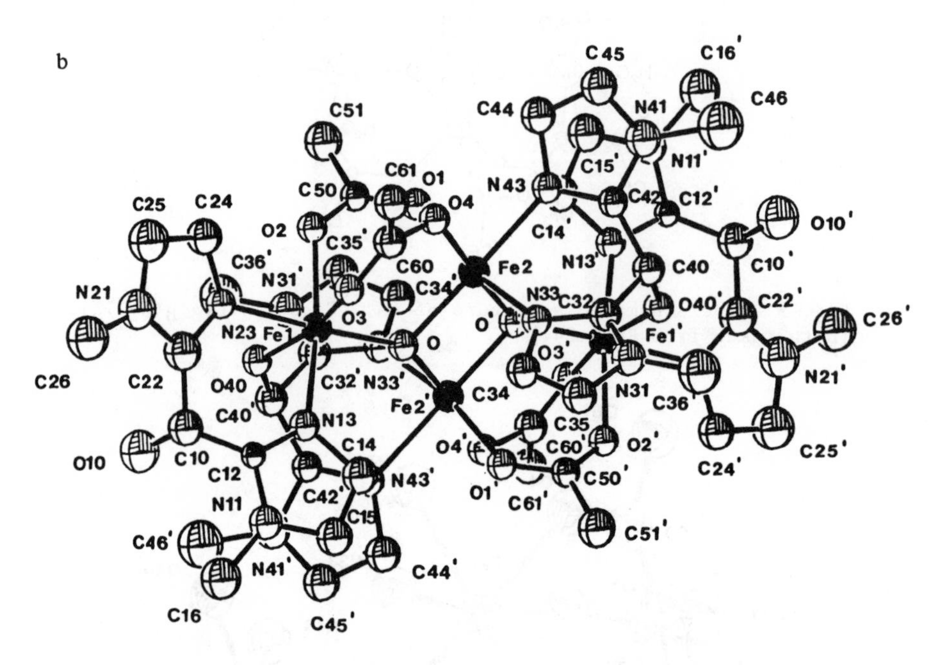

Fig. 4. Continued. (b) The full cation. (Reproduced from Ref. 8a. Copyright 1987 American Chemical Society.)

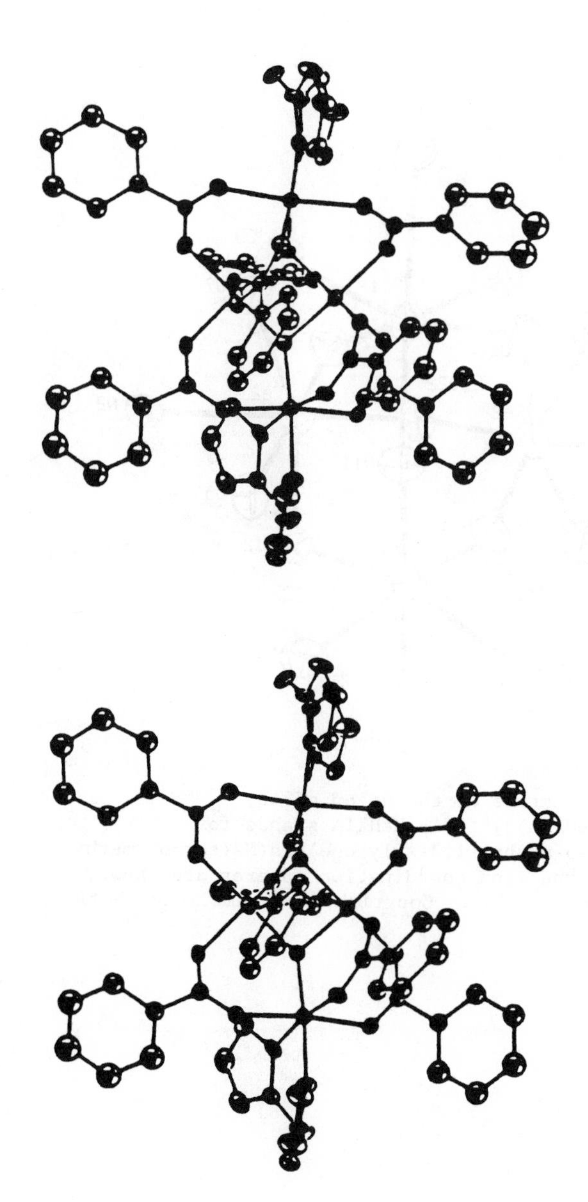

Fig. 5. Stereo view of the structure of 4C $[Fe_4O_2(H_2Bpz_2)_2(O_2CPh)_7]$. H_2Bpz_2 stands for the dihydrobis(1-pyrazolyl)borate anion. (Reproduced from Ref. 9. Copyright 1987 American Chemical Society).

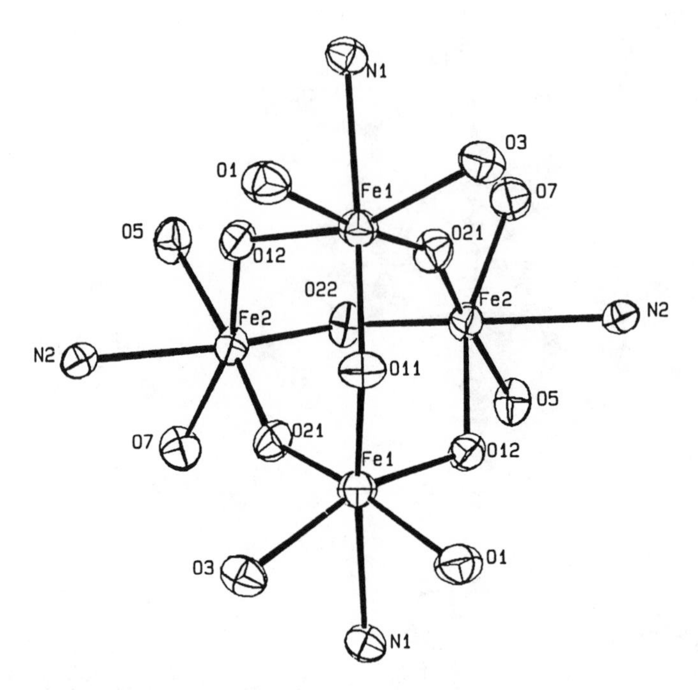

Fig. 6. The structure of the anion of **4D**, $[Fe_4O_2(OH)_2(5\text{-MeHXTA})_2]^{4-}$. 5-MeHXTA stands for N,N'-(2-hydroxy-5-methyl-1,3-xylylene)bis(N-(carboxymethyl) glycine). Only the iron coordination spheres are shown. (Reproduced from Ref. 10a. Copyright 1988 American Chemical Society.)

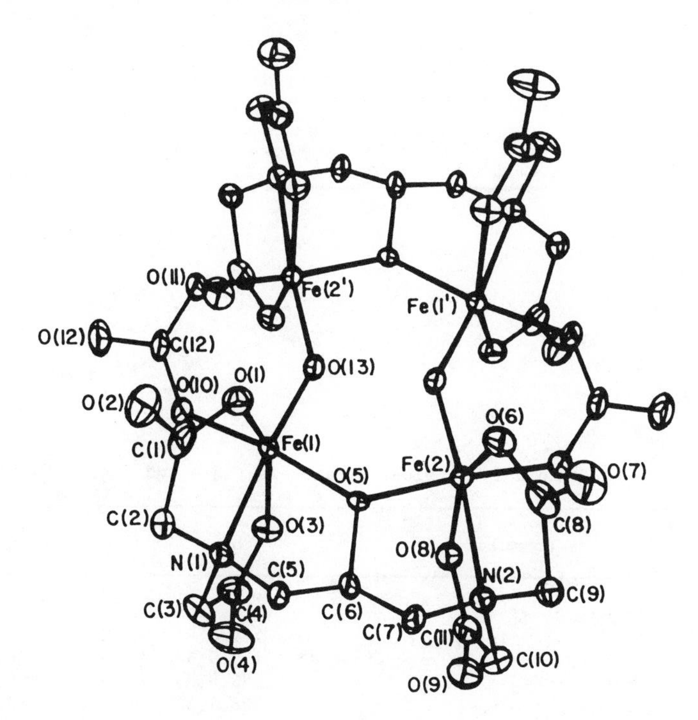

Fig. 7. The structure of the anions of **4E**, $[Fe_4O_2(CO_3)_2L_2'']^{6-}$. L′′ stands for the pentaanionic form of the [(2-hydroxy-1,3-propanediyl)diimino]tetraacetic acid. (Reproduced from Ref. 11. Copyright 1987 American Chemical Society.)

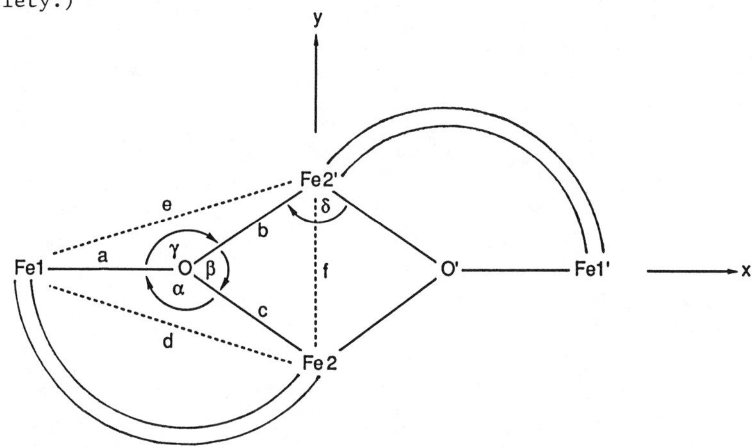

Fig. 8. Diagram of the $\{Fe_4O_2\}$ kernel showing the definition of its metrical parameters. Primed and unprimed atoms are related by an inversion center. (Reproduced from Ref. 8a. Copyright 1987 American Chemical Society.)

Table I. Metrical Parameters (in A and deg) of Selected

Compound	Core Formula	α	β	γ
amarantite	$[Fe_4O_2(SO_4)_4]$	131.9	96.3	131.9
leucophosphite	$[Fe_4(OH)_2(PO_4)_4(H_2O)_2]$	124.05[b]	91.82[b]	122.9[b]
4B	$[Fe_4O_2(BICOH)_2(BICO)_2(O_2CPh)_4]^{2+}$	119.0(6)	95.7(5)	136.8(6)
4A	$[Fe_4O_2(O_2CCF_3)_8(H_2O)_6]$	133.9(2)	96.5(2)	129.5(2)
8	$\{Fe_4O_2\}$ core in $[Fe_8O_2(OH)_{12}(TACN)_6]^{8+}$	128.7(3)	96.8(4)	128.7(1)

[a] See Figure 8 for labels used to identify bond lengths and angles.

[b] Calculated using the data presented in the original references.

$\{Fe_4O_2\}^{n+}$ Type Cores with μ_3-O or μ_3-OH Bridges

δ	a	b	c	d	e	f
83.7	1.892	1.969	1.929	3.440[b]	3.526[b]	2.93
88.2[b]	2.150	2.175	2.152[b]	3.820[b]	3.778[b]	3.108[b]
84.3(5)	1.884(9)	1.98(1)	1.94(1)	3.29(1)	3.59(1)	2.90(1)
82.9(2)	1.842(4)	1.961(3)	1.936(3)	3.476(2)	3.436(3)	2.915(3)
83.2(5)	1.859(9)	1.968(18)	1.961(5)	3.444(12)	3.425(6)	2.939(4)

$[Fe_2O(CO_3)L'']$ units bridged by alkoxide groups(11). Partial protonation of the μ-oxo bridges could explain the long Fe-μ-oxo bond and is suggested by the unusually short μ-oxo-μ-oxo' non-bonding contact of only 2.408(9) Å. Compounds **4D** and **4E** can still be envisioned as the result of 2 + 2 condensations:

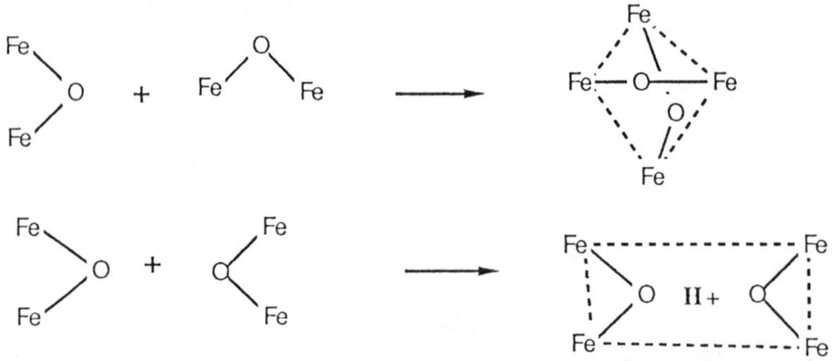

(3)

In all the tetranuclear aggregates described here the metal-ligand distances are normal and typical for high-spin complexes.

The tetranuclear complex $[Fe_2O(O_2CCH_3)_2(tpbn)_2]_2^{4+}$ recently reported(14) is not an oxo-iron aggregate according to our definition but comprises two quasi-independent dinuclear μ-oxo diiron units. Its spectroscopic and magnetic properties are consistent with this view.

Hexanuclear Agregates. The only such aggregate structurally characterized so far,(15) $[Fe_3O(OH)(O_2CC(CH_3)_3)_6]_2$, **6** is presented in Figure 9. The hexanuclear complex can be viewed as two μ_3-oxo centered centers crosslinked by hydroxyl and carboxylate groups. Interestingly, $[Fe_3O(O_2CC(CH_3)_3)_6(H_2O)_3]^+$ was the starting material used to prepare **6**. The (μ-oxo)bis(μ-carboxylato) subset which is also present at the active site of hemerythrins, is found in **6**. The two dinuclear units are linked by μ_3-oxo and μ-carboxylato groups. The third position in the "coordination sphere" of the μ_3-oxo groups is occupied by an unusual 5 coordinate, trigonal bipyramidal iron atom. This is the only case known so far in oxo-iron aggregates. The μ_3-oxo bridge is of type B since the surrounding metal atoms form an isosceles triangle.

Octanuclear Aggregates. $\{[(C_6H_{15}N_3)_6Fe_8(\mu_3\text{-}O)_2(\mu_3\text{-}OH)_{12}]Br_7(H_2O)\}Br\cdot 8H_2O$, **8**, is the only member of this class known so far.(16) The cation, shown in Fig. 10, has idealized C_i symmetry. It contains as a subset the planar $\{Fe_4O_2\}$ kernel discussed above (see Table I). Four other metal atoms are bound at the periphery of the kernel through bridging hydroxy groups. Bond lengths and angles are typical for high-spin, octahedrally coordinated Fe(III).

Undecanuclear Aggregates. The largest member of the series of discrete oxo-iron aggregates is $[Fe_{11}O_6(OH)_6(O_2CC_6H_5)_{15}]$, **11**. Two different crystalline forms, $\mathbf{11}\cdot H_2O\cdot 8CH_3CN$, **11A**, space group P1 and $\mathbf{11}\cdot 6THF$, **11B**, space group R32 have been described(17). In both

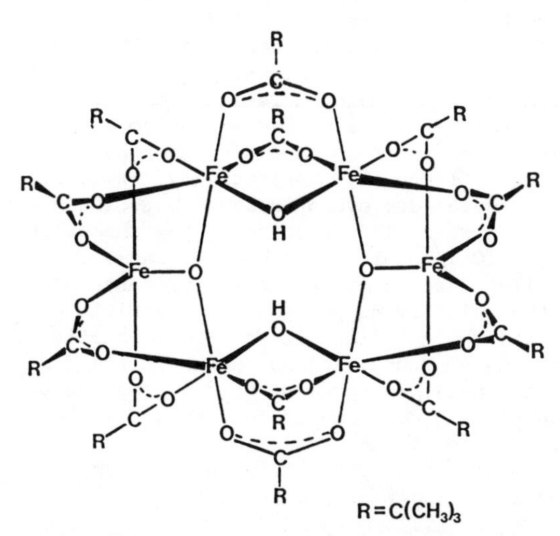

Fig. 9. Schematic structure of **6**, $[Fe_3O(OH)(O_2CC(CH_3)_3)_6]_2$. For the octahedral metals, the average Fe-μ_3-O and Fe-μ-OH distances are 1.946(2) and 1.962(9)Å, respectively. For the trigonal bipyramidal metals, Fe-μ_3-O is 1.848(5)Å. (Data taken from Ref. 15.)

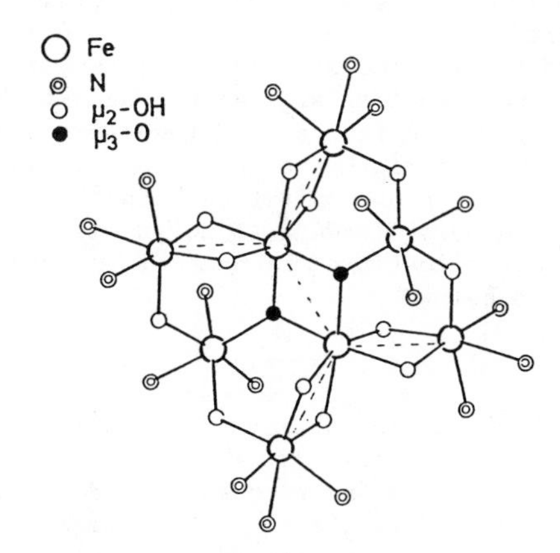

Fig. 10. The structure of the cation **8**, $[(C_6H_{15}N_3)_6Fe_8(\mu_3\text{-}O)_2(\mu_3\text{-}OH)_{12}]^{8+}$. $C_6H_{15}N_3$ is 1,4,7-triazacyclononane. Only the iron coordination spheres are shown. (Reproduced with permission from Ref. 16. Copyright 1984 Verlag Chemie GmbH).

forms the eleven, distorted, octahedrally coordinated, iron atoms form a pentacapped trigonal prism which is twisted 17.4° for **11B** as shown in Figure 11. The metals are linked by six planar μ_3-oxo groups and six pyramidal μ_3-hydroxo groups. The fifteen bidentate benzoate groups provide additional bridges. The twelve triply bridging oxygen atoms define an irregular icosahedron, with the μ_3-oxo subset forming a close-packed antiprism at the core as shown in Figure 12. The oxo-hydroxo iron core and the full molecule are depicted in Figures 13 and 14, respectively. Water and acetonitrile molecules are hydrogen bounded to the μ_3-OH groups in the triclinic form while THF is bound in the rhombohedral one. Since the $\{Fe_{11}O_6(OH)_6\}$ unit is stable, it has been speculated(8b,17b) that it might also be present in the ferritin core. Since the majority of phosphate in ferritin is adventitious, surface bound and the metallic core can be reconstituted in the absence of phosphate groups with no change in the X-ray powder diffraction pattern(1), replacement of bridging phosphate by bridging carboxylate groups should not influence the three dimensional structure of the core. Calculations show that ~409 Fe_{11} units could fill the apoferritin inner cavity. Further details can be found in reference 17.

Intermediate and Higher Nuclearity Aggregates. Penta-, hepta-, nona- and decanuclear aggregates have not been synthesized yet, although $\{[Fe_5O(O_2CCH_3)_{12}](O_2CCH_3)\}$ has been detected by mass spectrometry.(3) Larger, polymeric aggregates have been reported but structural information is lacking.(3c,18)

It can be concluded that Fe(III) can generate a diversity of structural types despite the apparent low "plasticity" of its coordination sphere. In only one case, **6**, is the metal less than octahedrally coordinated. The presence of structural invariants like the kernel of tetranuclear aggregates suggest the possibility that some substructures are favored and therefore likely to occur in other, yet unknown, aggregates or even in the ferritin core. Both mono- and dinuclear μ-oxo units (perhaps the trinuclear ones too) can serve as monomers, leading to polynuclear compounds. This formal description is illustrated by some of the synthetic approaches described below.

Synthetic Methods

Type A trinuclear carboxylates are synthesized in general by basic hydrolysis of mononuclear salts.(3) The synthesis of **3B** marked the introduction of the stepwise approach described in Scheme 1. The $[Fe_2OCl_6]^{2-}$ anion was used as the dinuclear component but, since it has a half-life of 1.3 hours in acetonitrile(8b), it can also be a source for mononuclear Fe^{3+} building blocks. Aqueous hydrolysis of $Fe(OH)_3$ was used to synthesize **4E**, while **8** was prepared by hydrolysis of the mononuclear salt, $(C_6H_{15}N_3)FeCl_3$ in the presence of NaBr. The basic hydrolysis of the dinuclear $[Fe_2(OH)(MeHXTA)(H_2O)_2]$ gives **4D**, while the slow air evaporation of an aqueous solution of the mixed valence trinuclear aggregate $[Fe_3O(O_2CCF_3)_6(H_2O)_3]\cdot 3.5\ H_2O$ yields **4A**. In general, these reactions seem to be under thermodynamic rather than kinetic

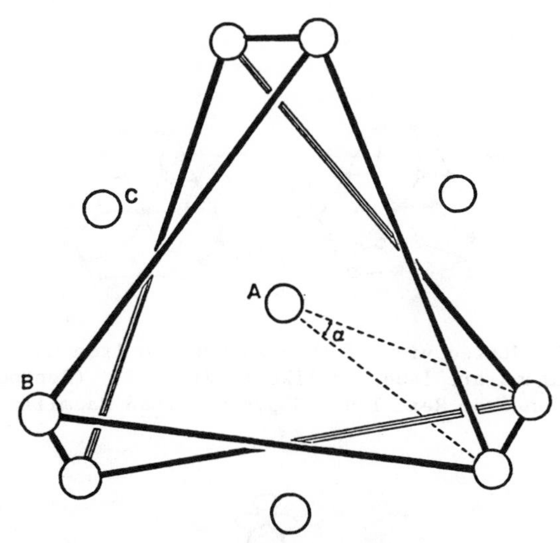

Fig. 11. Arrangement of iron atoms for **11B**, $[Fe_{11}O_6(OH)_6(O_2CPh)_{15}]$, as viewed down the C_3 symmetry axis. The top and bottom atoms are eclipsed. The atoms labelled A, B and C are crystallographically unique: A on the C_3 axis, B on general position and C on C_2 axes. The twist angle α is the projection of the angle between AB vectors onto the mean plane through the prism. (Reproduced from Ref. 17b. Copyright 1987 American Chemical Society.)

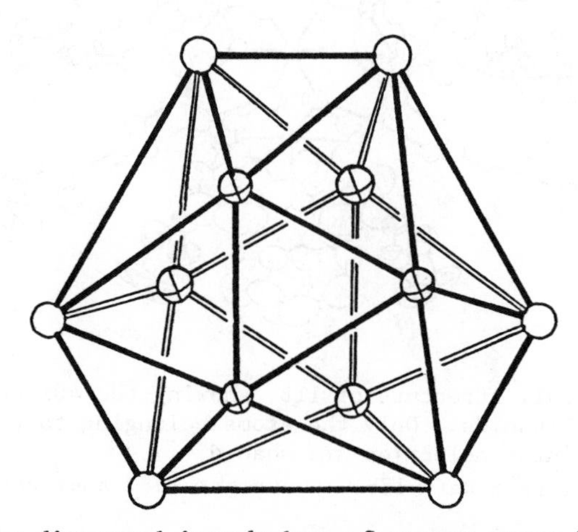

Fig. 12. The distorted icosahedron of oxygen atoms in **11B** viewed down the C_3 symmetry axis. Spherical crosses and open circles denote the μ_3-oxo and μ_3-hydroxo oxygen atoms, respectively. (Reproduced from Ref. 17b. Copyright 1987 American Chemical Society.)

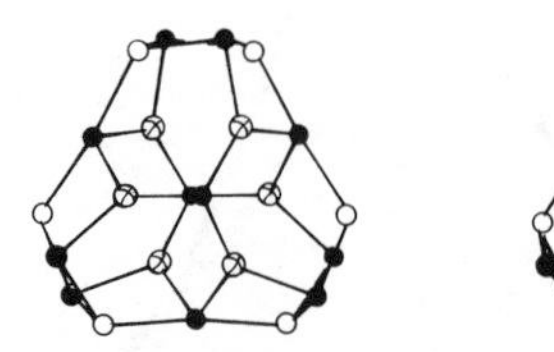

Fig. 13. Stereo view of the oxo-hydroxo iron core of **11b**. Oxygen atoms are labelled like in Fig. 12. (Reproduced with permission from Ref. 17a. Copyright 1986 Macmillan Journals, Ltd.)

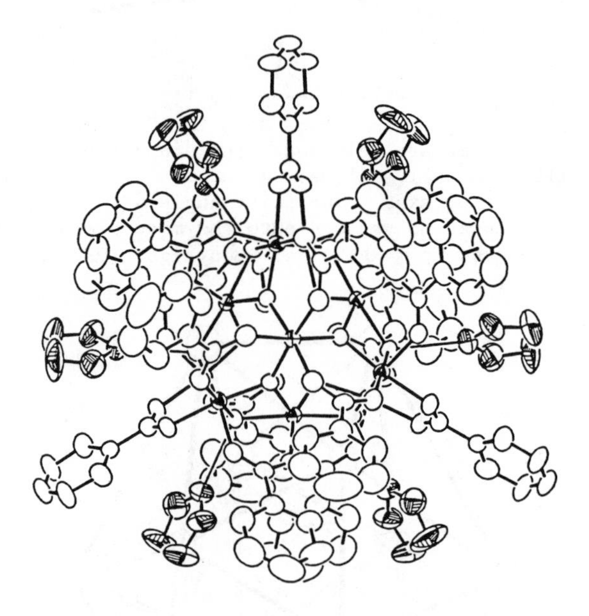

Fig. 14. Full structure of **11B**, showing the 40% probability thermal ellipsoids. Only the atoms belonging to the six tetrahydrofuran molecules are shaded.
(Reproduced from Ref. 17b. Copyright 1987 American Chemical Society.)

control, the nuclearity of the starting material being less important, (i.e., aggregates can easily interconvert) compared to the type of ligands used.

In the synthetic approaches described so far water was the reaction medium and, therefore the pH plays an important regulatory role. This is illustrated by the stepwise formation of the tetranuclear **4D** and by the assembly of the octanuclear aggregate **8** which takes place at pH = 9.

We have introduced a different approach by using aprotic solvents as the reaction medium.(5) In this case the water which may be needed for hydrolysis and an organic base are added in limited quantities in a controlled fashion. The synthesis of **4B** and **4C** were achieved, according to Scheme 2, in non-aqueous solvents. The same approach was used to synthesize **6** by boiling $\{Fe_3O(O_2CC(CH_3)_3](H_2O)_3\}[O_2CC(CH_3)_3]\cdot 2[(CH_3)_3CCOOH]$, **3A1**, a trinuclear aggregate, in tetradecane at 200-220°C. It is interesting to note that the hexanuclear aggregate was first detected in the mass spectrum of **3A1**.(3f) The undecairon aggregates are made by slowly hydrolyzing a yet unknown intermediate (possibly one with an even number of metal atoms(8b)) that results from the reaction of sodium benzoate with $(Et_4N)_2[Fe_2OCl_6]$ in aprotic solvents. The yields are strongly dependent upon the rate of hydrolysis which is controlled by the rate of water diffusion and the relative ratio water:organic solvent.

This approach emphasizes the role of water as a source of oxo ligands. It is important to note that crystalline tetranuclear **4B** and undecanuclear **11** aggregates form only if the amounts and concentrations of water vary within certain narrow limits. Perhaps the number of potentially bridging oxo and hydroxo groups that can form in solution determine the nuclearity of the aggregates.

Clearly, much work remains to be done in this area; ultimately one wishes to be able to predict the nuclearity of the oxo aggregate as a function of the metal source, ligands and reaction conditions. Nature already "knows" through evolution how to control the polymerization of the oxo-iron core of ferritin. It can be speculated that in the synthesis of the undecairon aggregates the hydrophobic solvent and the fifteen phenyl groups mimic the hydrophobic part of the ferritin protein sheath while the fifteen hydrophillic carboxylate groups, oriented toward the interior, bind the oxo-hydroxo iron core which is formed upon hydrolytic oligomerization of the starting mono- and/or dinuclear μ-oxo iron species. A very recent report describes the use of a phospholipid membrane for the same purposes.(19) The aqueous and organic reaction media seem to be somewhat complementary. While the "aqueous" approach allows a rigorous control of the pH and therefore of the protonation/deprotonation process involved in the interconversion of the oxo/hydroxo groups, the use of organic solvents allows the quantitative control of the above species. In principle, both methods provide means of regulating the polymerization of the various oxo/hydroxo iron complexes; the use of organic solvents, however, presents the advantage that a certain degree of kinetic control is achieved by the regulation of the rate of water diffusion. A judicious choice of ligands allows one to predict to a certain extent the course of the hydrolysis. At the two extremes, it is obvious that polydentate, encapsulating ligands will favor mononuclear complexes while the monodentate ones (like H_2O) or "innocent" counterions will not prevent the formation of large aggregates. The synthesis of the mixture of polymers known as the "Spiro-Saltman" ball(18) illustrates the latter point. Intermediate size aggregates form when the ligand denticity and the ligand/metal ratios have low values but so far no quantitative correlation has been established.

Physical Properties

The electronic spectroscopy of high-spin Fe(III) complexes is dominated by the presence of charge transfer bands. The position and intensities of these bands is a function of the organic ligand and μ-oxo or μ-hydroxo groups present, thus allowing the monitoring of the aggregate accretion process (*vide infra*). Charge transfer bands that extend well into the visible region and confer a brown color have been detected in the electronic spectra of Type A trinuclear aggregates(3c,20), in contrast to the yellow Type B complex **3B** which shows only an intense UV tail extending into the visible region.(5b)

The bands associated with tetranuclear oxo-iron aggregates listed in Table II are not found in the trinuclear aggregates and could serve as fingerprints.

Table II. Visible Spectroscopic Absorptions of Compounds Containing the $\{Fe_4O_2\}$ Oxo Core

Compound	λ_{max} (ϵ_{Fe}, M^{-1} cm^{-1})[a]	Ref.
$[Fe_4O_2(BICOH)_2(BICO)_2(O_2CPh)_4]Cl_2$	575 (sh), 476 (sh)	8
$(Et_4N)[Fe_4O_2(O_2CPh)_7(H_2Bpz_2)_2]$	565 (sh, 75), 467 (470)	9
amarantite, $\{Fe_4O_2\}$ core	500, 442, 434, 242	26
leucophosphite, $\{Fe_4(OH)_2\}$ core	538, 441.5, 428	26

[a] sh = shoulder.

Nothing has been reported about the electronic properties of **6** or **8** except that they are orange-red and brown respectively. Compounds **11A** and **11B** present time independent charge transfer bands below 490 nm in CH_2Cl_2 and a prominent shoulder at 492 nm which shifts to 480 nm in acetonitrile. The time-independence of these bands is contrasted with that of a 470 nm shoulder present in aqueous solutions of iron (III), which intensity increases in time as the polymers age and more μ-oxo groups are formed(18b). A near IR band at 881 nm(ϵ = 6 $M^{-1}cm^{-1}$) could be the $^6A_1$ - $^4T_1$ (^4G) enhanced d-d transition characteristic of octahedrally coordinated high-spin Fe(III) ions. A similar band is observed in the synthetic polymer $[Fe_4O_3(OH)_4(NO_3)_2]_n \cdot 1.5H_2O$ (approximate composition) and in ferritin at 900 nm(ϵ = 4.4 $M^{-1}cm^{-1}$) suggesting the predominance of octahedrally coordinated Fe(III).(21) Unfortunately, no quantitative relationship has been established so far between the size of the aggregates and UV-visible spectroscopy partly because of the lack of definitive assignments of the observed transitions.

Infrared and Raman spectroscopic techniques have not been widely used for oxo-iron aggregates. Only recently d^5-pyridine (Py) and ^{18}O substituted $[Fe_3O(O_2CCH_3)_6(Py)_3]^+$ were used to assign the 600 cm^{-1} and 300 cm^{-1} bands associated with the $\{Fe_3O\}$ core as ν_{as} δ_s respectively.(22) A 746 cm^{-1} resonance enhanced Raman band observed in the spectrum of **4C** has been tentatively assigned to the asymmetric Fe-O-Fe stretch.(9) A 425 cm^{-1} Raman band observed in the spectrum of **4D** has been assigned to the ν_S(Fe-O-Fe), in accord with theoretical calculations.(10a) Since there is little or no information available on the vibrational properties of the other oxo-iron aggregates there is a clear need for more work, especially using isotopes, in order to identify the "signatures" of the various cores.

Magnetic and Mössbauer properties of small aggregates are better understood due to recent progress in this area. In general, the values of the effective magnetic moments per iron are lower than 5.92 Bohr magnetons, the value expected for mononuclear high-spin Fe(III) complexes, indicating overall antiferromagnetic (AF) coupling. Since no direct metal-metal orbital overlap is possible, a superexchange mechanism based on the Heisenberg-Dirac-van Vleck formalism (HDVV) was invoked to explain the magnetic properties of the smallest, Type A, trinuclear aggregates.(23) Using a simple vectorial coupling model one can write:

$$\vec{S}_T = \sum_{i=1}^{n} \vec{S}_i \tag{4}$$

where $\vec{S}_i$ are the individual spin values of the n ions that form the aggregate and $\vec{S}_T$ is the total spin. Assuming that the AF exchange is isotropic, the spin Hamiltonian is:

$$H = -2 \sum_{i>j=1}^{n} J_{ij} \vec{S}_i \cdot \vec{S}_j \tag{5}$$

i.e., the aggregate is decomposed into a sum of interacting <u>pairs</u> of ions, J_{ij} being the exchange coupling constant between the ith and the jth metal ion. In principle, for trinuclear centers $J_{12} \neq J_{13} \neq J_{23}$ but for type A basic iron carboxylates, which have D_{3h} or C_{3v} point symmetry, $J_{12} = J_{13} = J_{23}$ in the first approximation. In this case, the ground state $|\vec{S}_T|$, which can range from 1/2 to 15/2, equals 1/2. The J values range from 20 to 30 cm^{-1}. For type B, the point symmetry is reduced to C_{2v} and $J_{13} = J_{23} \neq J_{12}$ (see Figure 2). The distortion of the equilateral triangle formed by the iron atoms toward an isosceles triangle results in a $|\vec{S}_T|$ = 5/2 ground state as determined by magnetization studies. Calculations have shown that for $|J_{12}|/|J_{13}| \geq 3.5$, $|\vec{S}_T|$ = 5/2 and $\vec{S}_1 + \vec{S}_2 = 0$, i.e., Fe(1) and Fe(2) are AF coupled in a diamagnetic ground state. Since the $|\vec{S}_T|$ = 5/2 ground state may also arise from a magnetically independent Fe(3), variable temperature magnetic susceptibility data were required. The results presented in Figure 15 show that the unique Fe(3) subsite is bound, albeit weakly, to the dinuclear Fe(1)-O-Fe(2) subsite. In this case J_{12} = $-55cm^{-1}$ and J_{13} = $-8cm^{-1}$. The 4.2 K magnetic Mössbauer spectra presented in Figure 16 also show the two subsites in ratio 2:1, as expected. The reverse in the ordering of the low lying spin state levels explains the dip in the observed effective moment that occurs at ~60 K (see Figure 15b). While for the type A the spin levels increase monotonically, S_T = 1/2, 3/2, 5/2,...15/2,for Class B the S_T decreases initially from 5/2 to 3/2, a state which is 54 cm^{-1} higher in energy. At higher temperature the levels are "normally" ordered and their spin state values increase monotonically.

Tetranuclear aggregates, comprising an even number of AF coupled metal atoms, should have a diamagnetic ground state. This is indeed the case for **4B** and **4C**.(8,9) Variable temperature magnetic data have been analyzed for **4E** using the molecular field approximation(11); as discussed previously, due to the presence of a center of symmetry, only two J's need to be considered, one describing the intradinuclear coupling and one describing the coupling between two dinuclear units.

Mössbauer spectroscopy confirms the diamagnetic ground state for **4B** and **4C**, but, in these cases, the two unique subsites are too similar to be resolved. This is not the case for **4A**, where two subsites can be resolved, in accord with the crystallographic model described previously. Interestingly, a Mössbauer analysis of the tetranuclear aggregate **4A**, using an approach suited for extended solid state structures, failed to yield sensible results, thus suggesting that four metals are not a good approximation for a solid.(24) In contrast, **11** shows a complex magnetic and Mössbauer behavior reminiscent of small (<100 Å) particles of iron oxides and also of the ferritin core(17b), thus suggesting that eleven iron atoms in an aggregate may be enough to model certain properties of a solid. Unfortunately, no Mössbauer data are available for the hexa- or octanuclear aggregates. We have recently discovered a relationship between the magnitude of the exchange coupling interactions and crystallographically determined structural parameters, the Fe-μ-O distances in particular.(25) This relationship is valid not only for binuclear but also for trinuclear and tetranuclear complexes. Thus, unlike other "sporting" spectroscopic techniques discussed so far, the magnetic and Mössbauer studies allow a direct correlation of an observable with both the nuclearity and geometry of an aggregate. This is but the first step in understanding the interactions in polynuclear iron aggregates.

Biological Relevance of Oxo-iron Aggregates and Conclusions

The main target of modelling studies in this area is the oxo-hydroxo iron core of iron storage proteins. Although the processes that govern the initiation and accretion of this core

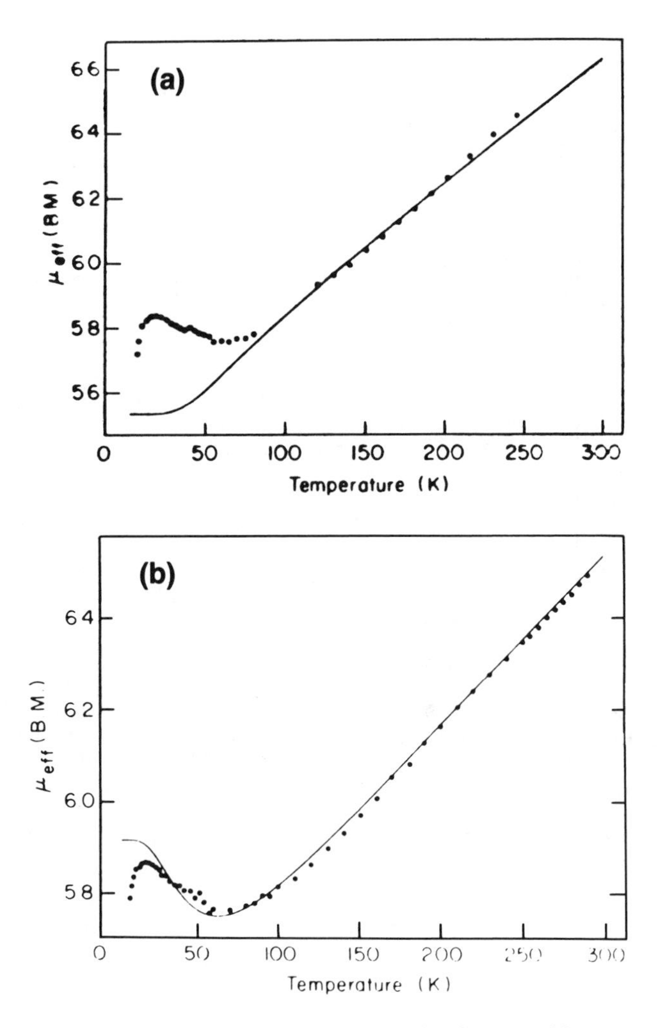

Fig. 15. Plots of experimental and calculated effective magnetic moments of **3B** as a function of temperature. See also Fig. 2a. (a) Fe(3) independent of Fe(1), Fe(2). (b) Fe(3) interacting with Fe(1), Fe(2). (Reproduced from Ref. 5b. Copyright 1987 American Chemical Society.)

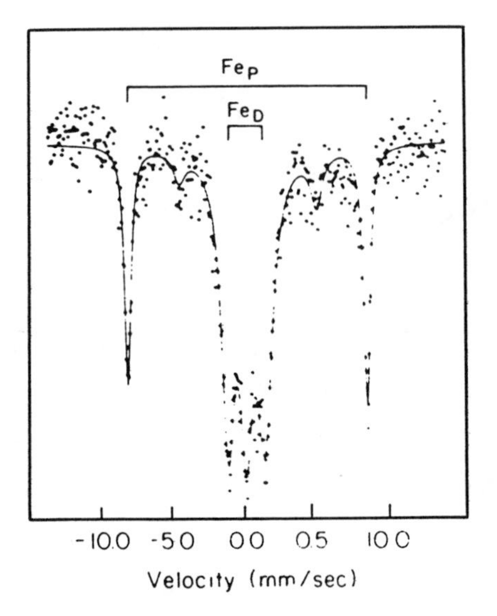

Fig. 16. Magnetic Mössbauer spectra of **3B** at 4.2 K and 60 kOe. The overbars designate splitting of the outermost absorption lines of the diamagnetic (Fe_D, Fe(1) and Fe(2)) and paramagnetic (Fe_P, Fe(3)) iron sites. The solid line represents a least-squares fit of the experimental data to Lorentzian lines. (Reproduced from Ref. 5b. Copyright 1987 American Chemical Society.)

remain largely unknown, the aggregates discussed here may represent intermediate stages. The judicious choice of solvents, ligands and reaction conditions allowed some control over the polymerization process. Much work remains to be done in this area as well as in the physical characterization of the aggregates. X-ray crystallography, Mössbauer and magnetic studies are the most valuable tools so far,but EXAFS is likely to play a major role in the future. Last but not least the chemistry of these interesting molecules is virtually unknown and kinetic studies on iron incorporation and release may provide novel insights into Fe uptake and release *in vivo*.

Acknowledgments

The author wishes to thank Drs. S. J. Lippard and L. Que, Jr. for preprints of their publications.

Literature Cited

1. (a) Theil, E. C. *Adv. Inorg. Biochem.* **1983**, *5*, 1 and reference cited therein; (b) Ford, G. C.; Harrison, P. M.; Rice, D. W.; Smith, J. M. A.; Treffry, A.; White, J. L.; Yariv, J. *Rev. Port. Quim.* **1985**, *27*, 119; (b) Ibidem, *J. Phil. Trans. R. Soc. London B* **1984**, *304*, 551.
2. (a) Towe, K. M.; Lowenstam, H. A.; Nesson, M. H. *Science* **1963**, *142*, 63; (b) Mann, S.; Frankel, R. B.; Blakemore, R. P. *Nature (London)* **1984**, *310*, 405.
3. (a) Lippard, S. J. *Chem. Brit.* **1986**, *222*. (b) Weinland, R. F.; Hohn, A. *Z. Anorg. Chem.* **1926**, *152*, 1; (c) Welo, L. A. *Philos. Mag. S7* **1928**, *6*, 481; (d) Catterick, J.; Thornton, P. *Adv. Inorg. Chem. Radiochem.* **1977**, *20*, 291 and references cited therein; (e) Holt, E. M.; Holt, S. L.; Alcock, N. W. *Cryst. Struct. Commun.* **1982**, *11*, 505; (f) Blake, A. B.; Fraser, L. R.; *J. Chem. Soc., Dalton Trans.* **1975**, 193.
4. Giacovazzo, C.; Scordari, F.; Menchetti, S. *Acta. Cryst. Sect B.* **1975**, *B31*, 2171.
5. (a) Gorun, S. M.; Lippard, S. J. *J. Amer. Chem. Soc.* **1985**, *107*, 4568; (b) Gorun, S. M.; Papaefthymiou, G. C.; Frankel, R. B.; Lippard, S. J. *J. Amer. Chem. Soc.* **1987**, *109*, 4244.
6. (a) Woehler, S. E.; Wittebort, R. J.; Oh, S. M.; Kambara, T.; Hendrickson, D. N.; Inniss, D.; Strouse, C. E. *J. Amer. Chem. Soc.* **1987**, *109*, 1063; (b) Oh, S. M.; Wilson, S. R.; Hendrickson, D. N.; Woehler, S. E.; Wittebort, R. J., Inniss, D.; Strouse, C. E. *J. Amer. Chem. Soc.* **1987**, *109*, 1073; (c) Meesuk, L.; Jayasooriya, U. A. ; Cannon, R. D. *J. Amer. Chem. Soc.* **1987**, *109*, 2009.
7. Ponomarev, V. I.; Atovmyan, L. O.; Bobkova, S. A.; Turté, K. I. *Dokl. Akad. Nauk SSSR* **1984**, *274*, 368.
8. (a) Gorun, S. M.; Lippard, S. J. *Inorg. Chem.* **1987**, in press; (b) Gorun, S. M. *Ph.D. Thesis*, Massachusetts Institute of Technology, Cambridge, **1986**.
9. Armstrong, W. H.; Roth, M. E.; Lippard, S. J. *J. Amer. Chem. Soc.* **1987**, *109*, 6318.

10. (a) Murch, B. P.; Bradley, F. C.; Boyle, P. D.; Papaefthymiou, V.; Que, Jr., L. *J. Amer. Chem. Soc.* **1988**, in press; (b) Murch, B. P.; Boyle, P. D.; Que, Jr., L. *J. Amer. Chem. Soc.* **1985**, *107*, 6728.
11. Jameson, D. L.; Xie, C.-L.; Hendrickson, D. N.; Potenza, J. A.; Schugar, H. J. *J. Amer. Chem. Soc.*, **1987**, 109, 740.
12. Süsse, P. *Zeit. für Kristall.* **1968**, *127*, 261.
13. Moore, P. B. *Amer. Mineral.* **1972**, *57*, 397.
14. Toftlund, H.; Murray, K. S.; Zwack, P. R.; Taylor, L. F.; Anderson, O. P. *J. Chem. Soc. Chem. Commun.* **1986**, 191.
15. Gerbeleu, N. V.; Batsanov, A. S.; Timko, G. A.; Struchkov, Y. T.; Indrichan, K. M.; Popovich, G. A. *Dokl. Akad. Nauk. SSSR*, **1987**, *293*, 364.
16. Wieghardt, K.; Pohl, K.; Jibril, I.; Huttner, G. *Angew. Chem. Int. Ed. Engl.* **1984**, *23*, 77.
17. (A) Gorun, S. M.; Lippard, S. J. *Nature* **1986**, *319*, 666; (b) Gorun, S. M.; Papaefthymiou, G. C.; Frankel, R. B.; Lippard, S. J. *J. Amer. Chem. Soc.* **1987**, *109*, 3337.
18. (a) Spiro, T. G.; Allerton, S. E.; Denner, J.; Terzis, A.; Bils, R.; Saltman, P. *J. Amer. Chem. Soc.* **1966**, *88*, 2721; (b) Sommer, B. A.; Margerum, D. W.; Reuner, J.; Saltman, P.; Spiro, T. G. *Bioinorg Chem.* **1973**, *2*, 295.
19. Mann, S.; Hannington, J. P.; Williams, R. J. P. *Nature*, **1987**, *324*, 565.
20. Long, G. J.; Robinson, W. T.; Tappmeyer, W. P.; Bridges, D. L. *J. Chem. Soc., Dalton Trans.* **1973**, 573.
21. Webb, J.; Gray, H. B. *Biochim Biophys. Acta* **1974**, *351*, 224.
22. Montri, L.; Canon, R. D.; *Spectrochimica Acta*, Part A **1985**, *41A*, 643.
23. Kambe, K. J. *Phys. Soc. Jpn.* **1950**, *5*, 48.
24. Stukan, R. A.; Ponomarev, V. I.; Nifontov, V. P.; Turté, K. I.; Atovmyan, L. O. *Zhur. Strukt. Khim.* **1985**, *26*, 62.
25. (a) Gorun, S. M.; Lippard, S. J. *Rec. Trav. Chim. Pays-Bas*, **1987**, *106*, 417; (b) Gorun, S. M.; Lippard, S. J., submitted for publication.
26. Rossman, G. R. *Amer. Miner.* **1976**, *61*, 933.

RECEIVED February 29, 1988

Chapter 11

Involvement of Manganese in Photosynthetic Water Oxidation

Gary W. Brudvig

Department of Chemistry, Yale University, New Haven, CT 06511

A multinuclear Mn complex functions in photosystem II to accumulate oxidizing equivalents and also to bind water and catalyze its four-electron oxidation. The Mn complex can exist in five oxidation states called S_i states (i = 0-4). Electron paramagnetic resonance (EPR), Mn K-edge X-ray absorption, and ultraviolet-absorption spectroscopies have been applied to study the structure and function of the Mn complex. The application of these methods to probe the Mn complex in photosystem II is briefly reviewed. Considering the results of both the X-ray absorption and EPR studies, a distortion of an oxo-bridged "cubane"-like Mn tetramer seems to best account for the arrangement of Mn ions in the S_2 state. Based on the known properties of the Mn complex in photosystem II and the coordination chemistry of Mn, structures were proposed for the five intermediate oxidation states of the Mn complex; these structures were incorporated into a molecular mechanism for the formation of an O-O bond and the displacement of O_2 from the S_4 state (Brudvig, G.W. and Crabtree, R.H. *Proc. Natl. Acad. Sci. USA* **1986**, *83*, 4586-88). This mechanism and the structure of the Mn complex are considered in light of recent studies of the Mn complex in photosystem II.

Although Mn occurs widely in biological systems, only a small number of enzymes have been identified that utilize oxidation states of Mn higher than +2. These redox-active Mn enzymes include the Mn superoxide dismutase, which is the only Mn enzyme in this group for which an X-ray crystal structure is available (1), the Mn-containing pseudocatalase (2), and photosystem II (for reviews see 3-6). Photosystem II is unique among this group of enzymes in that other transition metals have not been found to function in place of Mn, whereas alternate naturally-occurring forms of superoxide dismutase and catalase exist which contain Fe instead of Mn.

The function of photosystem II is to oxidize water and reduce

0097–6156/88/0372–0221$06.00/0

plastoquinone (PQ). In addition, photosystem II generates a pH gradient across the thylakoid membrane by producing and consuming protons on opposite sides of the membrane. The overall reaction, which requires four light-induced charge separations in the photosystem II reaction center, is:

$$2H_2O + 2PQ + 4H^+_{out} \rightarrow O_2 + 2PQH_2 + 4H^+_{in} \quad (1)$$

where in and out refer to the inside and outside of the thylakoid vesicle, respectively.

Mn was first shown to play an important role in photosynthetic O_2 evolution by nutritional studies of algae (7). The stoichiometry of Mn in photosystem II was determined by quantitating Mn released from thylakoid membranes by various treatments (8). These experiments established that Mn is specifically required for water oxidation and that four Mn ions per photosystem II are required for optimal rates of O_2 evolution (9). More recently, photosystem II preparations with high rates of O_2 evolution have been isolated from a variety of sources (for a review see 10). The isolation of an O_2-evolving photosystem II has proved to be a major step forward in both the biochemical and spectroscopic characterization of the O_2-evolving system. These preparations contain four Mn ions per photosystem II (11), thus confirming that four Mn ions are functionally associated with each O_2-evolving center.

Photooxidation of the Manganese Complex

To account for the periodicity of four in the yield of O_2 in a series of flashes (12-13), Kok and coworkers (14) proposed that photosystem II cycles through five states during flash illumination. These intermediate oxidation states are referred to as S_i states ($i = 0$-4) with the subscript denoting the number of oxidizing equivalents accumulated. The sequential advancement of the S states occurs via the light-induced charge separation in photosystem II.

A large body of evidence now supports the basic model put forward by Kok and coworkers (3-6,15). The S_4 state rapidly releases a molecule of O_2 and regenerates the S_0 state. The S_2 and S_3 states are unstable and are reduced in the dark to the S_1 state with half-times on the order of one minute at room temperature. Further, the S_0 state is oxidized in the dark to the S_1 state with a half-time on the order of ten minutes at room temperature (16-17). Hence, samples that are incubated in the dark for more than thirty minutes at room temperature contain only the S_1 state. In contrast, continuously illuminated samples contain equal fractions of states S_0 through S_3 which decay within a few minutes in the dark at room temperature to a mixture of S_0 and S_1 in a 1:3 ratio.

The question of the molecular basis for the S states has existed since the original proposal by Kok and coworkers. As first formulated, the S state designation referred to the oxidation state of the O_2-evolving center which could, in principle, include all of photosystem II and its associated components. Indeed, there are a number of redox-active components on the electron-donor side of photosystem II in addition to the Mn complex, such as the tyrosine radical that gives rise to EPR signal II_s, and cytochrome b_{559}.

However, a change in the oxidation state of these species does not alter either the period-four oscillation of O_2 yields in a series of flashes, provided that the flashes are sufficiently closely spaced (16), or the EPR spectral properties of the S_2 state (18). Moreover, EPR (17-31) and X-ray absorption (32-38) studies have shown that Mn is oxidized in the S_1 to S_2 transition. Hence, it appears that the S states should be interpreted in terms of distinct intermediate oxidation states of the Mn complex (see below).

EPR Studies of the Manganese Complex

In general, one expects to observe an EPR signal from a transition metal complex whenever the complex possesses an odd number of unpaired electrons. If the complex possesses an even number of unpaired electrons, then spin-spin interactions may give rise to large zero-field splittings of the spin levels which will prevent the observation of EPR signals with conventional EPR instrumentation. Because each S-state transition involves the removal of one electron from the O_2-evolving center, one predicts that alternate S states will have an odd number of unpaired electrons. These alternate S states should, in principle, be detectable by EPR spectroscopy. Indeed, the S_2 state exhibits a multiline EPR signal (Figure 1) from a multinuclear Mn complex (19). The S_2-state multiline EPR signal was one of the first direct probes of the Mn complex in photosystem II and much of the information on the structure and function of the Mn complex has come from analyses of the S_2-state EPR signals (17-31). However, a major limitation to the use of EPR spectroscopy to study the Mn complex is that measurements have been restricted to the S_2 state. Based on the argument that EPR signals are expected from alternate S states, one predicts that the S_0 state is a good candidate for detection by EPR spectroscopy. Nonetheless, an EPR signal has not yet been detected from the S_0 state.

Two distinct EPR signals have been observed from the S_2 state (Figure 1). The first is the multiline EPR signal centered at about $g = 2.0$. This EPR signal arises from an $S = 1/2$ state of a mixed-valence multinuclear Mn complex; the numerous hyperfine lines arise from the coupling of the unpaired electron to the nuclear spins of several Mn ions (each Mn ion has $I = 5/2$). The second EPR signal from the S_2 state exhibits a turning point at $g = 4.1$. The $g = 4.1$ EPR signal is generated by illumination of photosystem II membranes at 130 K, but is unstable and is converted into the multiline EPR signal upon warming to 200 K (18,22). However, the $g = 4.1$ EPR signal can be stabilized by the addition of various exogenous molecules including amines (29), fluoride (22), and sucrose (25).

Two interpretations of the identity of the species that gives rise to the S_2-state multiline and $g = 4.1$ EPR signals have been proposed (24-26). Both EPR signals are proposed to arise from Mn centers. The difference in interpretations concerns the number of Mn ions involved in each paramagnetic species.

One view is that three distinct Mn centers function in the water oxidation process: two mononuclear Mn centers and a binuclear Mn center (26). It was proposed (26) that the binuclear center gives rise to the multiline EPR signal, whereas, one of the

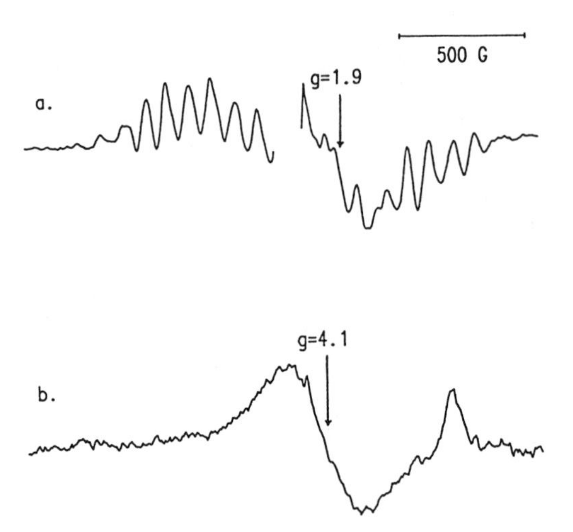

Figure 1: S_2 state EPR spectra: a) multiline EPR signal produced by illumination at 200 K; b) g = 4.1 EPR signal produced by illumination at 130 K. Experimental conditions are as in (24).

mononuclear Mn ions gives rise to the g = 4.1 EPR signal. In order to explain the stabilization of the g = 4.1 EPR signal by exogenous molecules, Hansson et al. (26) proposed that the g = 4.1 EPR signal arises from a mononuclear Mn(IV) in redox equilibrium with a binuclear Mn species. Hence, in the S_2 state, it was proposed that either the mononuclear Mn is oxidized to Mn(IV) and gives a g = 4.1 EPR signal or the binuclear Mn center is oxidized to Mn(III)-Mn(IV) and gives a multiline EPR signal. The reduction potentials of these two species must be comparable in the cases when the g = 4.1 EPR signal is stabilized. In this model, a further oxidation of the system to the S_3 state should leave both the binuclear and mononuclear centers oxidized. However, no EPR signal is observed from the Mn complex in the S_3 state which, in this model, requires that the g = 4.1 and multiline EPR signal species are magnetically coupled in the S_3 state. Hence, the redox equilibrium model requires that these two Mn species must be very close together.

The alternate proposal is that both the multiline and g = 4.1 EPR signals arise from the same tetranuclear Mn complex (18,24-25). The conversion of the g = 4.1 EPR signal into the multiline EPR signal upon incubation at 200 K in the dark can then be explained by a temperature-dependent structural change in the Mn site upon formation of the S_2 state, which alters the exchange couplings between the Mn ions (24). Generation of the S_2 state at 130 K may not allow such a rearrangement to occur rapidly and, hence, the g = 4.1 EPR signal could be viewed as an EPR signal arising from an S_1-state conformation that is in the S_2 state. The structural difference between the "g = 4.1" and "multiline" conformations need not be large; the conversion of the g = 4.1 EPR signal into the multiline EPR signal can be understood as a small rearrangement of the Mn complex into its preferred conformation in the higher oxidation state. Stabilization of the g = 4.1 EPR signal by exogenous molecules is explained in this model by stabilization of the "g = 4.1" conformation.

These two models may not be significantly different. The main difference between them is the magnitude of the magnetic coupling proposed to exist between the four Mn ions in the S_2 state. However, the different Mn centers must be very close together in order to account for the absence of an EPR signal from the S_3 state.

The S_2-state EPR signals have also been used to probe the coordination of exogenous ligands to the Mn complex (27-31). Ammonia, but not more bulky amines, dramatically alters the lineshape of the S_2-state multiline EPR signal, indicating that ammonia binds directly to Mn in the S_2 state (28). The binding of ammonia to the Mn complex has been proposed to be a nucleophilic addition reaction and to model the binding of substrate water (27-29). Also, ^{17}O-labeled water slightly broadens the S_2-state multiline EPR signal (30). The broadening of the S_2-state multiline EPR signal in the presence of ^{17}O-labeled water under the conditions of these experiments indicates that exchangeable water is a ligand to Mn in the S_1 state. One can envision that water coordinates to the Mn site either as the substrate or in the form of a structural oxo-bridge. (The mode of binding may be equivalent in these two cases; the distinction can be made on the basis of the species that gives rise to O_2). Several studies, however, indicate that the substrate water is not bound to the Mn site in the S_2 state (for a discussion see 27). By measuring the effect of ^{17}O-labeled water on the S_2-state multiline EPR signal following incubation with a specific S state, one may be able to make a distinction between substrate water, which appears to bind to the Mn complex in one of the higher S states, and structural oxo-bridges between the Mn ions, which may be exchangeable but remain bound throughout the S-state cycle. ^{17}O-labeled water also broadens the altered S_2-state multiline EPR signal formed in the presence of ammonia (31). One interpretation of this result is that the Mn complex in the ammonia-bound derivative of photosystem II contains exchangeable oxo-bridges between Mn ions that are not displaced by ammonia in the S_2 state.

X-ray Absorption Studies of the Manganese Complex

The oxidation states and ligation of Mn in photosystem II have been probed by X-ray absorption edge and extended X-ray absorption edge fine structure (EXAFS) measurements (32-38). EXAFS analyses have been done for Mn in the S_1 state present in dark-adapted thylakoid membranes (33,37) and in dark-adapted photosystem II membranes (35). These studies indicate that each Mn ion is probably six coordinate and ligated to oxygen and/or nitrogen ligands. Each Mn ion also sees 0.77 (35) or 2 - 3 (37) neighboring Mn ions at a distance of 2.7 Å. The EXAFS data for the S_2 state look essentially the same as for the S_1 state (36) and, therefore, it appears that the Mn site does not undergo a substantial structural reorganization during the S_1 to S_2 transition.

The EXAFS of Mn in photosystem II looks very much like that of a mixed-valence di-μ-oxo-bridged Mn dimer model compound (33,35). In particular, the Mn-Mn distance of 2.7 Å in photosystem II is characteristic of a di-μ-oxo-bridged structure. Recent EXAFS data have indicated that a second Mn-Mn distance of 3.3 Å may also be present (37-38). Although a clear picture of all of the Mn-Mn distances is not yet available, the EXAFS results are consistent with a structure in which two di-μ-oxo bridged Mn dimers are present in close proximity.

Of note is the apparent lack of chloride in the first coordination shell of Mn in either the S_1 or the S_2 state as revealed by EXAFS studies of Mn (35). This observation is of particular interest because chloride is required for optimal O_2 evolution rates (39) and has been proposed to act as a bridging ligand in a polynuclear Mn complex (40). Recent EPR studies, however, also suggest that chloride is not bound to Mn in the S_1 or S_2 state (41).

The energy of the X-ray absorption edge reflects the electron density about Mn which, in turn, reflects the oxidation state and ligation of Mn. The energy of the X-ray absorption edge of Mn in the S_1 state is in the range observed for Mn(III) model compounds (32,34). The X-ray absorption edge of Mn shifts to higher energy in the S_2 state and is in the range observed for Mn(III) and Mn(IV) model compounds (34). In light of the conclusion from the EXAFS studies that the coordination of Mn does not change substantially in the S_1 to S_2 transition (36), this result indicates that Mn is oxidized in the S_1 to S_2 transition. However, the shift in the Mn X-ray absorption edge is very small in the S_2 to S_3 transition, suggesting that Mn itself is not oxidized in this step (34,38).

These results seem to indicate that oxidation of Mn occurs in some, but not all, of the S-state transitions. This has led to proposals that redox-active centers other than Mn are involved in the storage of oxidizing equivalents during some of the S-state transitions. However, the energy of the Mn X-ray absorption edge reflects electron density about Mn and not directly the oxidation state. It is possible that the changes in the energy of the Mn X-ray absorption edge could be accounted for by a single oxidation of a Mn ion in both the S_1 to S_2 and S_2 to S_3 transitions if there is a change in ligation of the Mn ions in the S_2 to S_3 transition. Indeed, there is evidence that the substrate water coordinates to the Mn complex in the S_3 state (for a discussion see 27). The coordination of a Lewis base, such as water, to the Mn complex in the S_3 state would be expected to cause the shift of the Mn X-ray absorption edge in the S_2 to S_3 transition to be small even though the Mn complex is oxidized by one electron.

Ultraviolet Absorption Studies of the Manganese Complex

Flash-induced UV-difference spectral data indicate that a chromophore associated with the S states can be monitored in the 300-350 nm range (42-46). The difficulty with these measurements is that several other components in photosystem II also exhibit spectral changes in this region of the spectrum. After correcting for the absorbance changes due to redox changes of Q_A, Q_B (or an exogenous electron acceptor), and Z in photosystem II, Dekker et al. (42) found that the absorbance changes associated with the S_0 to S_1, S_1 to S_2, and S_2 to S_3 transitions were all equivalent and resembled the difference in absorbance between a Mn(III) and a Mn(IV)-gluconate model complex. Dekker et al. (42) concluded that one Mn(III) is oxidized to Mn(IV) in each of the first three S-state transitions. More recent work has considerably complicated this interpretation (43-46). It now seems probable that the absorbance changes are not all the same for the different S-state transitions. The most recent UV-spectral data are consistent with a single oxidation of Mn in each of the S-state transitions (45-46), although the assignment of the UV-difference spectra to the oxidation of Mn(III) to Mn(IV) is not clear because the spectral changes for the oxidation of Mn(II) to Mn(III) may be similar (47).

Structure of the Manganese Complex

The Mn-Mn distance of 2.7 Å determined by EXAFS is diagnostic of a di-μ-oxo-bridged Mn dimer in photosystem II (33,35). However, one could imagine a number of arrangements of the four Mn ions in photosystem II which contain a di-μ-oxo-bridged structure. Moreover, the observation of two different Mn-Mn distances by EXAFS, could only be accounted for by one of the following three arrangements of Mn: a Mn trimer plus one isolated (more than 3.3 Å away) Mn mononuclear center, two isolated inequivalent Mn dimers, or a Mn tetramer.

In order to distinguish between these possibilities, one must consider the EPR properties of the S_2 and S_3 states. Measurements have been made of the temperature dependence of the S_2-state multiline (24,48-49) and $g = 4.1$ (24) EPR signals in order to determine whether the EPR signals arise from the ground state or from a thermally-populated state. The S_2-state g = 4.1 EPR signal exhibits Curie-law behavior characteristic of an EPR transition from a ground state or from a system in the high-temperature limit (24). It seems probable that the g = 4.1 EPR signal arises from a ground S = 3/2 state of a Mn species. The S_2-state multiline EPR signal has been assigned to either a ground or low-lying S = 1/2 state of a multinuclear mixed-valence Mn species (24,48-49); and the S_3 state does not exhibit an EPR signal. These assignments severely restrict the possible arrangements of Mn in photosystem II and also provide important information on the structure of the Mn complex.

On the basis of a similarity of the ^{55}Mn nuclear hyperfine couplings in the S_2-state multiline EPR signal with those of EPR signals from Mn dimer model complexes, it has been suggested that the S_2-state multiline EPR signal arises from the S = 1/2 state of an antiferromagnetically exchange coupled mixed-valence Mn dimer (20,26). This assignment has led to proposals of a variety of models in which a Mn dimer is the catalytic site for water oxidation. Therefore, consider the magnetic properties of a binuclear mixed-valence Mn complex. A number of inorganic mixed-valence Mn dimers have been synthesized and their structures and magnetic properties have been determined (50-52). The unpaired electrons on each Mn ion are correlated through exchange interactions that arise from either direct orbital overlap or orbital overlap that is mediated by the bridging ligands (53). For a binuclear Mn complex, a single isotropic exchange coupling constant, J, usually is sufficient to account for the magnetic properties. Using the Heisenberg exchange Hamiltonian (Equation 2) one obtains the energy levels (Equation 3) for a pair of Mn ions with electron spins of S_A and S_B, respectively.

$$\hat{H}_{ex} = J\,\hat{S}_A \cdot \hat{S}_B = \tfrac{1}{2}J\,(\hat{S}^2 - \hat{S}_A{}^2 - \hat{S}_B{}^2) \qquad (2)$$

$$\text{Energy} = E = \tfrac{1}{2}J\,[\,S(S+1) - S_A(S_A+1) - S_B(S_B+1)\,] \qquad (3)$$

Figure 2 depicts the energies of the spin states that arise from either antiferromagnetic or ferromagnetic exchange coupling of a Mn(III)-Mn(IV) dimer. For di-μ-oxo-bridged Mn(III)-Mn(IV) model complexes, the exchange coupling is typically antiferromagnetic, in which case the ground state has S = 1/2. Observation of a Curie-law behavior for the temperature dependence of the S_2-state multiline EPR signal is also consistent with this EPR signal arising from the ground S = 1/2 state of an antiferromagnetically exchange-coupled Mn dimer (48).

The assignment of the S_2-state multiline EPR signal to a Mn dimer must be considered in light of the assignment of the S_2-state g = 4.1 EPR signal to a ground S = 3/2 state of a Mn species. From Figure 2, it is apparent that the S_2-state g = 4.1 EPR signal does not arise from a magnetically-isolated mixed-valence Mn dimer. Several assignments of the S_2-state g = 4.1 EPR signal have been proposed. The g = 4.1 EPR signal could arise from a mononuclear high-spin Mn(IV) which has S = 3/2 (26), from an S = 3/2 state of an exchange coupled trinuclear Mn complex (54), or from an S = 3/2 state of an exchange coupled tetranuclear Mn complex (24). All of

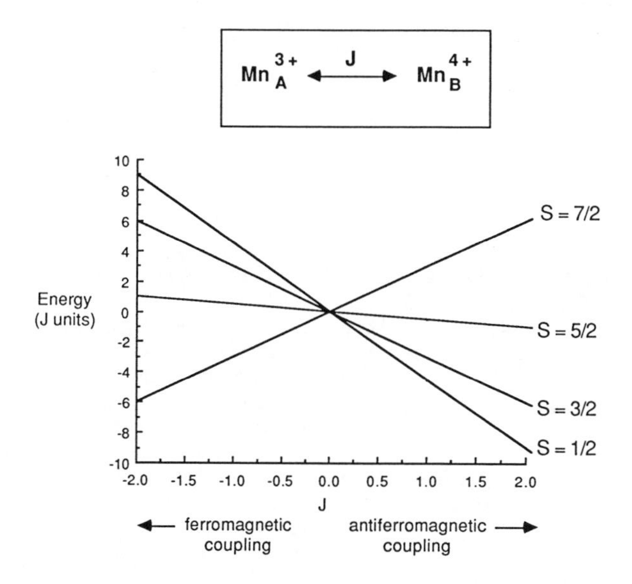

Figure 2: Energy levels given by Equation 3 for an exchange coupled Mn(III)-Mn(IV) dimer.

these possibilities are supported by observations of EPR signals near $g = 4$ from well-characterized Mn model complexes.

If we return to the possible arrangements of Mn in photosystem II that are consistent with the EXAFS data, one can rule out the possibility that the Mn ions are arranged as two isolated inequivalent Mn dimers because this arrangement cannot account for the S_2-state $g = 4.1$ EPR signal. This leaves two possible arrangements of the Mn ions: a Mn trimer plus a mononuclear Mn center, or a Mn tetramer. Consequently, one should look for an assignment for the S_2-state multiline EPR signal from either a mixed-valence Mn trimer or tetramer rather than from a mixed-valence Mn dimer.

The ^{55}Mn nuclear hyperfine couplings for the S_2-state multiline EPR signal strongly resemble those for mixed-valence Mn dimers. However, the ^{55}Mn nuclear hyperfine couplings for a Mn trimer or tetramer depend on the projection of each Mn ion's nuclear hyperfine coupling tensor on the total spin of the system. It has been shown that only two Mn ions will contribute significantly to the ^{55}Mn nuclear hyperfine couplings for a Mn tetramer if the tetramer is composed of two antiferromagnetically exchange coupled Mn dimers that are ferromagnetically exchange coupled (24). Consequently, it is not always possible to determine the number of Mn ions in a multinuclear Mn complex simply on the basis of the ^{55}Mn nuclear hyperfine couplings.

Consider next, the magnetic properties of a Mn tetramer. For a Mn tetramer, six isotropic exchange couplings are required to model the magnetic properties. In order to reduce the number of variables in a simulation of the magnetic properties, it has been assumed that some of the exchange couplings are equal (24). Consider a Mn tetramer composed of two Mn dimers in which the interdimer Mn exchange couplings are all equivalent (Figure 3). In this case, the number of exchange coupling constants is reduced to three and the energies can be solved analytically (55). Call one pair of Mn ions A and B and the second pair of Mn ions C and D. The exchange Hamiltonian is given in Equation 4.

$$\hat{H}_{ex} = J_{AB}\, \hat{S}_A \cdot \hat{S}_B + J_{CD}\, \hat{S}_C \cdot \hat{S}_D + J(\hat{S}_A \cdot \hat{S}_C + \hat{S}_A \cdot \hat{S}_D + \hat{S}_B \cdot \hat{S}_C + \hat{S}_B \cdot \hat{S}_D) \quad (4)$$

Making the substitution of $\hat{S}' = \hat{S}_A + \hat{S}_B$ and $\hat{S}^* = \hat{S}_C + \hat{S}_D$ gives:

$$\hat{H}_{ex} = \tfrac{1}{2}J_{AB}(\hat{S}'^2 - \hat{S}_A{}^2 - \hat{S}_B{}^2) + \tfrac{1}{2}J_{CD}(\hat{S}^{*2} - \hat{S}_C{}^2 - \hat{S}_D{}^2) + \tfrac{1}{2}J(\hat{S}^2 - \hat{S}'^2 - \hat{S}^{*2}) \quad (5)$$

The energies for this system are given in Equation 6.

$$\begin{aligned} E(S,S',S^*) = {} & \tfrac{1}{2}J_{AB}\,[\,S'(S'+1) - S_A(S_A+1) - S_B(S_B+1)\,] \\ & + \tfrac{1}{2}J_{CD}\,[\,S^*(S^*+1) - S_C(S_C+1) - S_D(S_D+1)\,] \\ & + \tfrac{1}{2}J\,[\,S(S+1) - S'(S'+1) - S^*(S^*+1)\,] \end{aligned} \quad (6)$$

Three possible combinations of Mn oxidation states are compatible with the EPR data from the S_2 state: Mn(II)-Mn(III)$_3$, Mn(III)$_3$-Mn(IV), and Mn(III)-Mn(IV)$_3$ (56-57). The most probable oxidation state is Mn(III)-Mn(IV)$_3$ (57). For a Mn(III)-Mn(IV)$_3$ complex the energy expression reduces to that shown in Equation 7.

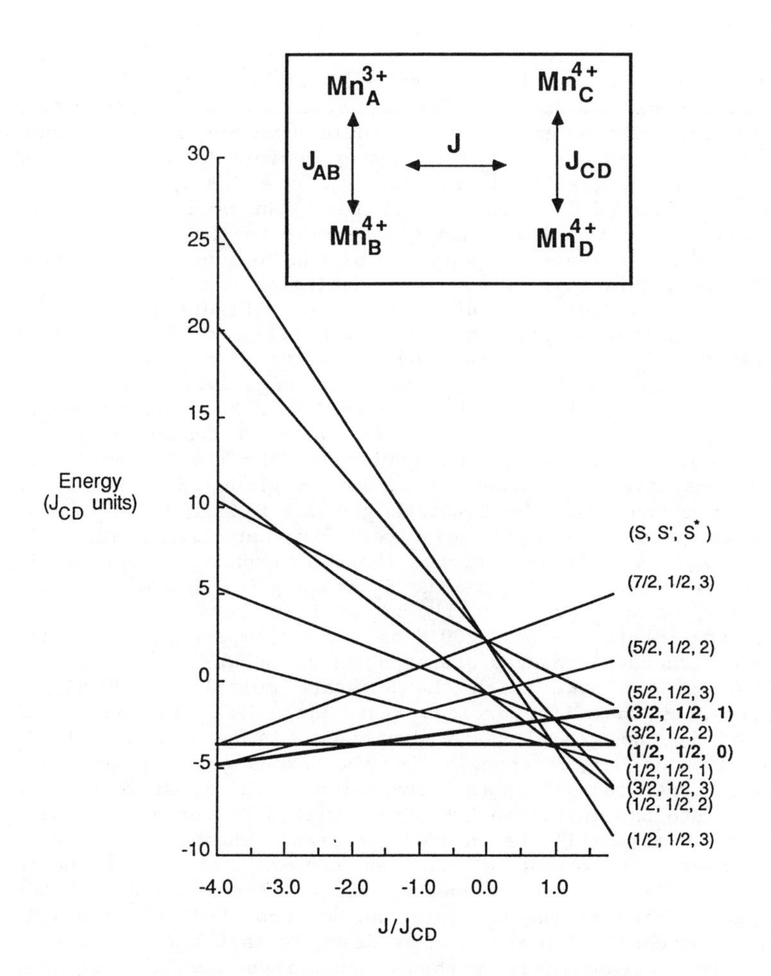

Figure 3: Lowest energy levels given by Equation 7 for an exchange coupled Mn(III)-$Mn(IV)_3$ tetramer in the limit where J_{AB} is large and positive. The (S,S',S^*) = (3/2,1/2,1) and (1/2,1/2,0) states, shown in bold, are proposed to give rise to the S_2-state g = 4.1 and multiline EPR signals, respectively.

$$E(S,S',S^*) = \tfrac{1}{2}J_{AB}\,[\,S'(S'+1) - 39/4\,] + \tfrac{1}{2}J_{CD}\,[\,S^*(S^*+1) - 15/2\,] + \tfrac{1}{2}J\,[\,S(S+1) - S'(S'+1) - S^*(S^*+1)\,] \quad (7)$$

Three criteria must be met for a Mn tetramer to account for the properties of the S_2-state EPR signals. For one conformation of the Mn complex, the ground state should have $S = 3/2$ to account for the $g = 4.1$ EPR signal; for another conformation, the ground state or a low-lying excited state should have $S = 1/2$ to account for the multiline EPR signal; and the ^{55}Mn nuclear hyperfine interaction in the $S = 1/2$ state giving rise to the multiline EPR signal should be dominated by only two of the Mn ions. It has been shown (24) that this last criterion is satisfied for the $(S,S',S^*) = (1/2,1/2,0)$ state of a Mn tetramer (Figure 3).

If the exchange coupling between the Mn(III)-Mn(IV) dimer in a Mn tetramer, J_{AB}, is large and positive (antiferromagnetic coupling), and if J_{CD} is positive and J is negative, then the EPR properties of the S_2 state can be explained by a Mn tetramer model. Figure 3 gives the energy levels for a Mn tetramer when the exchange couplings, J and J_{CD} are varied, in the limit when J_{AB} is large and positive. In order to produce a ground $S = 3/2$ state which could account for the S_2-state $g = 4.1$ EPR signal, J_{CD} must be positive, and J must be negative; further, J/J_{CD} must lie between -2 and -4. It is expected that the exchange couplings in the two conformations that give the S_2-state $g = 4.1$ and multiline EPR signals, respectively, will be similar because these two conformations interconvert at 200 K or below, temperatures at which only minor structural changes are expected to occur. Note that the (1/2,1/2,0) state, from which the S_2-state multiline EPR signal could arise, is the first excited state when J/J_{CD} is between -2 and -3 and is the ground state when J/J_{CD} is between 1 and -2. Consequently, no major changes in the exchange couplings are required in order to explain the conversion of the S_2-state $g = 4.1$ EPR signal into the multiline EPR signal with a Mn tetramer model.

Further work will be needed to test whether or not a trimer/monomer arrangement of Mn can account for the magnetic properties of the S_2 state. However, it is clear that all of the magnetic properties of the S_2 state can be accounted for with a Mn tetramer. One conclusion that can be drawn is that both antiferromagnetic and ferromagnetic exchange couplings must be present simultaneously in order for a tetramer model to account for the magnetic properties of the Mn complex in the S_2 state. This unusual combination of exchange couplings has been previously observed for "cubane"-like complexes (58-59), and this analogy suggests that the tetrameric Mn complex may have a "cubane"-like structure in the S_2 state (24). An oxo-bridged "cubane"-like Mn tetrameric complex was proposed for the structure of the Mn complex in the S_2 state (24,57). A distortion of an oxo-bridged "cubane"-like Mn tetramer seems to best account for the arrangement of Mn ions as seen by both EXAFS and EPR in the S_2 state.

Mechanism of Water Oxidation

Many proposals have been made for the role of Mn in photosynthetic water oxidation (for reviews see 3-6). It is clear that Mn functions both to accumulate the oxidizing equivalents produced during S-state advancement and also to bind water and catalyze its oxidation. The EXAFS and EPR data described in the previous sections are best accounted for by a tetrameric Mn complex with a distorted oxo-bridged "cubane"-like structure in the S_2 state. How might this site catalyze the water oxidation reaction?

The key step in the formation of an O_2 molecule is the activation of a bound water in order to form an O-O bond. This step is likely to be energetically the most demanding, and it is quite reasonable that the formation of the O-O bond does not occur until the most highly oxidized state of Mn is attained. An analogy can be drawn between photosynthetic water oxidation and the chemistry of cytochrome P450; in both cases a water molecule must be activated for reaction with a nucleophile. In cytochrome P450 model systems, oxidation of Fe or Mn to Fe(V) or Mn(V), respectively, is required to activate water to abstract a hydrogen atom from a hydrocarbon or to add to an olefin to form an epoxide (60). On this basis, a Mn(V) oxidation state may be needed to trigger the oxidation of water in photosystem II. The most probable oxidation state of Mn in the S_2 state is Mn(III)-Mn(IV)$_3$ (57). This oxidation state of the Mn complex in the S_2 state is also consistent with recent Mn X-ray absorption edge studies (38). Sequential one-electron oxidations of the Mn complex would then produce the S_4 state containing Mn(IV)$_3$-Mn(V).

A number of proposals for the mechanism of O_2 formation involve the generation of an O-O bond before the S_4 state is produced (reviewed in 6,61). These proposals seem less likely based on a consideration of the energetic requirements for water activation (62). There is also evidence from mass spectrometric studies of the isotopic composition of O_2 evolved from photosystem II that the water molecules that are oxidized to O_2 do not bind to the O_2-evolving center, or that they exchange readily with bulk water, in the S_0, S_1, S_2, and S_3 states (63). This result would be difficult to accommodate if the O-O bond is formed in one of the earlier S-state transitions.

A consideration of the evidence available on the natural system, as well as the coordination chemistry of Mn, led us to propose the model for water oxidation shown in Figure 4 (57). EPR spectral data obtained from the S_2 state are characteristic of a Mn_4O_4 "cubane"-like structure (24) and EXAFS studies of Mn indicate that the Mn site does not undergo a substantial structural change on going from S_1 to S_2 (36). Hence, the tetrameric Mn complex was proposed to exist in a Mn_4O_4 "cubane"-like structure in the lower oxidation states (S_0 to S_2). Studies of ammonia binding to the O_2-evolving center show that the Mn complex in the lower oxidation states (S_0 and S_1) is inert to nucleophilic addition, whereas the higher oxidation states (S_2 and S_3) are reactive to nucleophilic addition of ammonia (28,64). The activation of the Mn complex for nucleophilic addition in the higher S states is accounted for in this model by the sequential oxidation of Mn without accommodation of the ligand environment. This leads to a progressive increase in electron deficiency of the Mn complex with a corresponding increase in reactivity of the Mn complex to nucleophilic addition with increasing S state. Upon reaching the S_3 state it was suggested that the Mn complex coordinates two O^{2-} or OH^- ions from water

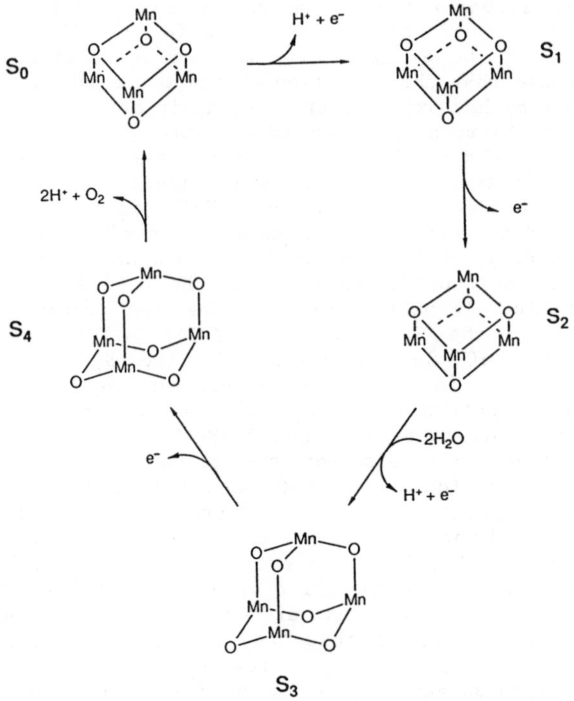

Figure 4: A proposed scheme for the structures of Mn associated with the S state transitions. In this scheme, O denotes either O^{2-} or OH^- ligands. Each Mn ion in the tetrameric complex is proposed to also be coordinated to the protein via three O or N ligands, as indicated by EXAFS studies of Mn in the S_1 state (33,35), although the protein-derived ligands are not shown.

molecules and undergoes a structural rearrangement to form a Mn_4O_6 "adamantane"-like structure. Further oxidation of the Mn complex generates the S_4 state which was proposed to be initially formed with a Mn_4O_6 structure. The electron withdrawal by high-valent Mn from the O ligands then triggers the formation of an O-O bond. A molecule of O_2 is released via reduction of the Mn complex and conversion of the μ_2-oxo ligands into μ_3-oxo ligands.

I have presented an overview of the current state-of-the-art in studies of the Mn complex in photosystem II. There are many unresolved questions and a clear picture of the structure and function of Mn in photosynthetic water oxidation is still not available. One useful approach to help determine the structure of the Mn complex in photosystem II involves the synthesis and characterization of Mn model complexes for comparison with the properties of the Mn complex in photosystem II. Recently, several tetrameric high-valent Mn-oxo complexes have been reported (see the chapter in this volume by G. Christou). Further characterization of existing and new high-valent tetrameric Mn-oxo model complexes, especially EPR and EXAFS measurements, will no doubt help clarify the present uncertain picture of the structure of the Mn complex in photosystem II.

Acknowledgments

I thank Warren Beck for help with the preparation of the figures and Warren Beck and Lynmarie Thompson for helpful comments on this manuscript. This work was supported by the National Institutes of Health (GM32715). G.W.B. is the recipient of a Camille and Henry Dreyfus Teacher/Scholar Award (1985-1990) and an Alfred P. Sloan Foundation Research Fellowship (1986-1988).

Literature Cited

1. Stallings, W.C.; Pattridge, K.A.; Strong, R.K.; Ludwig, M.L. *J. Biol. Chem.* **1984**, *259*, 10695-99.
2. Beyer, W.F., Jr.; Fridovich, I. *Biochemistry* **1985**, *24*, 6460-67.
3. Amesz, J. *Biochim. Biophys. Acta* **1983**, *726*, 1-12.
4. Dismukes, G.C. *Photochem. Photobiol.* **1986**, *43*, 99-115.
5. Babcock, G.T. In *New Comprehensive Biochemistry: Photosynthesis*; Amesz, J., Ed.; Elsevier: Amsterdam, 1987, pp. 125-58.
6. Brudvig, G.W. *J. Bioenerg. Biomemb.* **1987**, *19*, 91-104.
7. Pirson, A. *Z. Bot.* **1937**, *31*, 193-267.
8. Cheniae, G.M.; Martin, I.F. *Biochim. Biophys. Acta* **1970**, *197*, 219-39.
9. Cheniae, G.M. *Ann. Rev. Plant Physiol.* **1970**, *21*, 467-98.
10. Ghanotakis, D.F.; Yocum, C.F. *Photosyn. Res.* **1985**, *7*, 97-114.
11. Murata, N.; Miyao, M.; Omata, T.; Matsunami, H.; Kubawara, T. *Biochim. Biophys. Acta* **1984**, *765*, 363-69.
12. Joliot, P.; Barbieri, G.; Chabaud, R. *Photochem. Photobiol.* **1969**, *10*, 309-29.
13. Kok, B.; Forbush, B.; McGloin, M. *Photochem. Photobiol.* **1970**, *11*, 457-75.
14. Forbush, B.; Kok, B.; McGloin, M. *Photochem. Photobiol.* **1971**, *14*, 307-21.

15. Joliot, P.; Kok, B. In *Bioenergetics of Photosynthesis*; Govindjee, Ed.; Academic: New York, 1975, pp. 387-412.
16. Vermaas, W.F.J.; Renger, G.; Dohnt, G. *Biochim. Biophys. Acta* **1984**, *764*, 194-202.
17. Styring, S.; Rutherford, A.W. *Biochemistry* **1987**, *26*, 2401-05.
18. de Paula, J.C.; Innes, J.B.; Brudvig, G.W. *Biochemistry* **1985**, *24*, 8114-20.
19. Dismukes, G.C.; Siderer, Y. *FEBS Lett.* **1980**, *121*, 78-80.
20. Dismukes, G.C.; Siderer, Y. *Proc. Natl. Acad. Sci. USA* **1981**, *78*, 274-77.
21. Hansson, Ö.; Andréasson, L.-E. *Biochim. Biophys. Acta* **1982**, *679*, 261-68.
22. Casey, J.L.; Sauer, K. *Biochim. Biophys. Acta* **1984**, *767*, 21-28.
23. Zimmermann, J.-L.; Rutherford, A.W. *Biochim. Biophys. Acta* **1984**, *767*, 160-67.
24. de Paula, J.C.; Beck, W.F.; Brudvig, G.W. *J. Amer. Chem. Soc.* **1986**, *108*, 4002-09.
25. Zimmermann, J.-L.; Rutherford, A.W. *Biochemistry* **1986**, *25*, 4609-4615.
26. Hansson, Ö.; Aasa, R.; Vänngård, T. *Biophys. J.* **1987**, *51*, 825-32.
27. Beck, W.F. Ph.D. Thesis, Yale University, New Haven, 1988.
28. Beck, W.F.; de Paula, J.C.; Brudvig, G.W. *J. Amer. Chem. Soc.* **1986**, *108*, 4018-22.
29. Beck, W.F.; Brudvig, G.W. *Biochemistry* **1986**, *25*, 6479-86.
30. Hansson, Ö.; Andréasson, L.-E.; Vänngård, T. *FEBS Lett.* **1986**, *195*, 151-54.
31. Andréasson, L.-E.; Hansson, Ö. In *Progress in Photosynthesis Research* vol. 1; Biggins, J., Ed.; Nijhoff: Dordrecht, 1987, pp. 503-10.
32. Kirby, J.A.; Goodin, D.B.; Wydrzynski, T.; Robertson, A.S.; Klein, M.P. *J. Amer. Chem. Soc.* **1981**, *103*, 5537-42.
33. Kirby, J.A.; Robertson, A.S.; Smith, J.P.; Thompson, A.C.; Cooper, S.R.; Klein, M.P. *J. Amer. Chem. Soc.* **1981**, *103*, 5529-37.
34. Goodin, D.B.; Yachandra, V.K.; Britt, R.D.; Sauer, K.; Klein, M.P. *Biochim. Biophys. Acta* **1984**, *767*, 209-16.
35. Yachandra. V.K.; Guiles, R.D.; McDermott, A.E.; Britt, R.D.; Dexheimer, S.L.; Sauer, K.; Klein, M.P. *Biochim. Biophys. Acta* **1986**, *850*, 324-32.
36. Yachandra. V.K.; Guiles, R.D.; McDermott, A.E.; Cole, J.L.; Britt, R.D.; Dexheimer, S.L.; Sauer, K.; Klein, M. P. *Biochemistry* **1987**, *26*, 5974-5981.
37. George, G.N.; Prince, R.C.; Cramer, S.P. In Stanford Synchrotron Radiation Laboratory Users Meeting Report 1987.
38. Guiles, R.D.; Yachandra, V.K.; McDermott, A.E.; Britt, R.D.; Dexheimer, S.L.; Sauer, K.; Klein, M.P. In *Progress in Photosynthesis Research* vol. 1; Biggins, J., Ed.; Nijhoff: Dordrecht, 1987, pp. 561-64.
39. Kelley, P.; Izawa, S. *Biochim. Biophys. Acta* **1978**, *502*, 198-210.
40. Sandusky, P.O.; Yocum, C.F. *FEBS Lett.* **1983**, *162*, 339-43.
41. Yachandra. V.K.; Guiles, R.D.; Sauer, K.; Klein, M.P. *Biochim. Biophys. Acta* **1986**, *850*, 333-42.

42. Dekker, J.P.; van Gorkom, H.J.; Wensink, J.; Ouwehand, L. *Biochim. Biophys. Acta* **1984**, *767*, 1-9.
43. Lavergne, J. *Photochem. Photobiol.* **1986**, *43*, 311-17.
44. Renger, G.; Hanssum, B.; Weiss, W. In *Progress in Photosynthesis Research* vol. 1; Biggins, J., Ed.; Nijhoff: Dordrecht, 1987, pp. 541-44.
45. Saygin, Ö.; Witt, H.T. *Biochim. Biophys. Acta* **1987**, *893*, 452-69.
46. Lavergne, J. *Biochim. Biophys. Acta* **1987**, *894*, 91-107.
47. Vincent, J.B.; Christou, G. *FEBS Lett.* **1986**, *207*, 250-52.
48. Aasa, R.; Andréasson, L.-E.; Lagenfelt, G.; Vänngård, T. *FEBS Lett.* **1987**, *221*, 245-48.
49. de Paula, J.C.; Beck, W.F.; Miller, A.-F.; Wilson, R.B.; Brudvig, G.W. *J. Chem. Soc., Faraday Trans. 1* **1987**, *83*, 3635-51.
50. Cooper, S.R.; Dismukes, G.C.; Klein, M.P.; Calvin, M.C. *J. Amer. Chem. Soc.* **1978**, *100*, 7248-52.
51. Wieghardt, K.; Bossek, U.; Ventur, D.; Weiss, J. *J. Chem. Soc., Chem. Commun.* **1985**, 347-49.
52. Sheats, J.E.; Czernuszewicz, R.S.; Dismukes, G.C.; Rheingold, A.L.; Petroleas, V.; Stubbe, J.; Armstrong, W.H.; Beer, R.H.; Lippard, S.J. *J. Amer. Chem. Soc.* **1987**, *109*, 1435-44.
53. Cairns, C.J.; Busch, D.H. *Coord. Chem. Rev.* **1986**, *69*, 1-55.
54. Pecoraro, V.L.; Kessissoglou, D.P.; Li, X.; Butler, W.M. In *Progress in Photosynthesis Research* vol. 1; Biggins, J., Ed.; Nijhoff: Dordrecht, 1987, pp. 725-28.
55. Sinn, E. *Coord. Chem. Rev.* **1970**, *5*, 313-347.
56. Dismukes, G.C.; Ferris, K.; Watnick, P. *Photobiochem. Photobiophys.* **1982**, *3*, 243-256.
57. Brudvig, G.W.; Crabtree, R.H. *Proc. Natl. Acad. Sci. USA* **1986**, *83*, 4586-88.
58. Noodleman, L.; Norman, J.G., Jr.; Osborne, J.H.; Aizman, A.; Case, D.A. *J. Amer. Chem. Soc.* **1985**, *107*, 3418-26.
59. Haase, W.; Walz, L.; Nepveu, F. In *The Coordination Chemistry of Metalloenzymes*; Bertini, I., Drago, R.S., Luchinat, C., Eds.; D. Reidel Publishing Company: Holland, 1983, pp. 229-34.
60. Sheldon, R.A.; Kochi, J.K. *Metal Catalyzed Oxidations of Organic Compounds*; Academic: New York, 1981.
61. Renger, G.; Govindjee *Photosynth. Res.* **1985**, *6*, 33-55.
62. Brudvig, G.W.; de Paula, J.C. In *Progress in Photosynthesis Research* vol. 1; Biggins, J., Ed.; Nijhoff: Dordrecht, 1987, pp. 491-98.
63. Radmer, R.; Ollinger, O. *FEBS Lett.* **1986**, *195*, 285-89.
64. Velthuys, B.R. *Biochim. Biophys. Acta* **1975**, *396*, 392-401.

RECEIVED February 3, 1988

Chapter 12

Structural Types in Oxide-Bridged Manganese Chemistry

Toward a Model of the Photosynthetic Water Oxidation Center

George Christou and John B. Vincent

Department of Chemistry, Indiana University, Bloomington, IN 47405

A survey is presented of the structural types encountered during synthetic efforts in oxide-bridged manganese chemistry. These materials have resulted from efforts to model the Mn aggregate at the water oxidation center (WOC) of green plants. Complexes with Mn nuclearities in the range 2-12 are currently known, and their structural parameters are compared with those deduced by study of the WOC. The 'short' (_ca_ 2.7Å) Mn...Mn separation of the latter has been seen to date only in synthetic species containing _two_ oxide bridges between a Mn_2 pair, whereas, the 'long' (_ca_ 3.3Å) separation is compatible with at most only one oxide bridge. Particular emphasis is given to the dinuclear and tetranuclear complexes which most closely reproduce features observed or deduced in the WOC, and the potential correspondence of these complexes to the various S_n states of the latter is discussed.

It is now well established that manganese (Mn) is an essential component of the photosynthetic water oxidation center (WOC) (1). A variety of techniques have been applied to its study, and some pertinent conclusions can be briefly summarized: (i) the WOC contains between 2-4 Mn atoms (2,3); (ii) Mn...Mn separations of _ca_ 2.7Å and possibly _ca_ 3.3Å are present (3,4); (iii) the Mn atoms are bridged by O^{2-} (or OH^-) groups with Mn-O distances of _ca_ 1.75Å, and peripheral ligation is provided by O/N-based groups (from amino acid side chains) at distances of _ca_ 1.98Å (4); (iv) the Mn aggregate can adopt various oxidation levels (the S_n states) involving combinations of Mn oxidation states of II, III, and/or IV (1); and (v) the S_2 level is EPR-active and displays a spectrum containing a Mn-hyperfine-structured ("multiline") signal at g≈2 (2). Fuller details of the native manganese assembly can be found elsewhere in this volume.

0097-6156/88/0372-0238$06.00/0

One consequence of the intense effort being concentrated on the WOC has been a greater level of exploration by inorganic chemists of the chemistry of higher oxidation state Mn. The objective is the synthesis of Mn/O complexes which possess the appropriate structural features, oxidation levels, and spectroscopic and physical properties to allow them to be considered synthetic models of the WOC in its various S_n states. Their attainment should result in a greater level of understanding of the structural changes and general mechanism of action of the native site during the water oxidation cycle, including mode of substrate (H_2O) binding and conversion to O_2.

The purpose of this chapter is to review the current status of this model approach and to describe the structural types of discrete oxide-bridged Mn aggregates which have been synthesized to date. For completeness, all known nuclearities will be described, in terms of increasing nuclearity. Space restrictions preclude a complete survey and we shall concentrate mainly on the structural features of these materials; the reader is referred to the original literature for more detail, such as variable-temperature magnetic susceptibility, electrochemical and EPR data.

Dinuclear Complexes

The possibility of a dinuclear unit in the WOC has prompted much effort at this nuclearity and a variety of dinuclear oxide-bridged complexes are now known, containing either one or two bridging O^{2-} groups. They have been made employing a variety of procedures, including direct assembly from appropriate mononuclear reagents and appropriate modification of preformed dinuclear materials. Examples of the latter approach include chemical oxidation of $[Mn_2^{III}O(O_2CMe)_2(TACN)_2]^{2+}$ to the trication with $S_2O_8^{2-}$ (10), and its hydrolytically-induced oxidation to $[Mn_2O_2(O_2CMe)(TACN)_2]^{2+}$ (13). Some representative structures are reproduced in Figures 1 and 2. As can be seen from Table I, three types of bridging arrangements have been encountered; linear Mn-O-Mn, bent Mn-O-Mn and $Mn_2(\mu\text{-}O)_2$.

The complexes with linear oxide bridges contain no other bridging ligands and are characteristic of species with extensive Mn-O π-bonding. The linear arrangement necessitates a large Mn...Mn separation and observed distances are in the 3.42-3.54Å range. In contrast, complexes with a single bent oxide bridge invariably contain additional bridging ligands, and structurally-characterized materials possess the triply-bridged $[Mn(\mu\text{-}O)(\mu\text{-}O_2CMe)_2Mn]^{2+,3+}$ core. The TACN and Me_3-TACN species display an electrochemically reversible one-electron couple relating the $Mn_2^{III}/Mn^{III}Mn^{IV}$ levels, and examples of both have been isolated and structurally characterized. The $[Mn_2O(O_2CMe)_2]^{2+,3+}$ species have a characteristic Mn...Mn separation in the range 3.08-3.23Å. Interestingly, the $Mn^{III}Mn^{IV}$ complex has the longer rather than the shorter Mn...Mn separation that one might have expected.

The di-μ-oxide-bridged complexes have been observed in two types: (i) with no additional bridging ligands, and (ii) with an additional bridging $MeCO_2^-$ group. The former has been seen with both $Mn^{III}Mn^{IV}$ and Mn_2^{IV} oxidation states, while the latter only

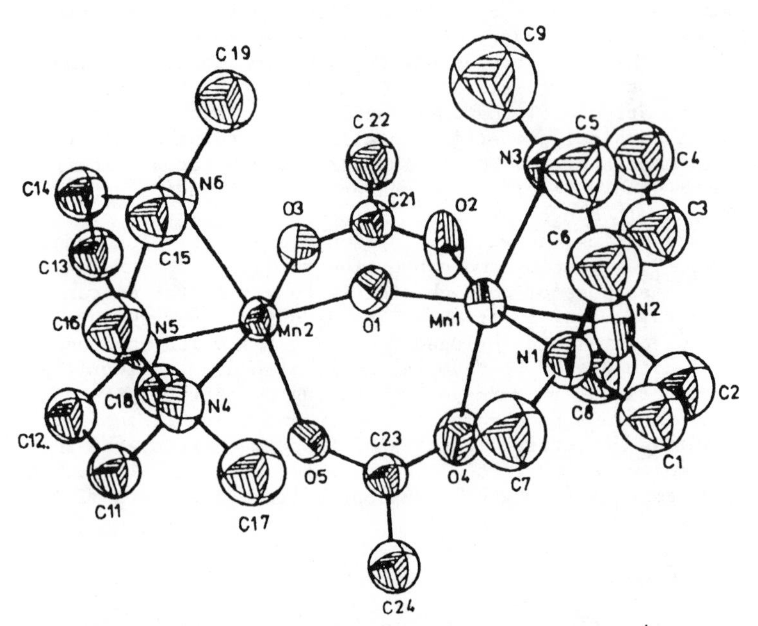

Figure 1. The structure of $[Mn_2O(O_2CMe)_2(Me_3\text{-}TACN)_2]^{3+}$; the other $[Mn_2O(O_2CMe)_2]$ complexes in Table I differ structurally only in the identity of the terminal ligands. (Reproduced with permission from ref. 10. Copyright 1986 VCH Verlagsgesellschaft)

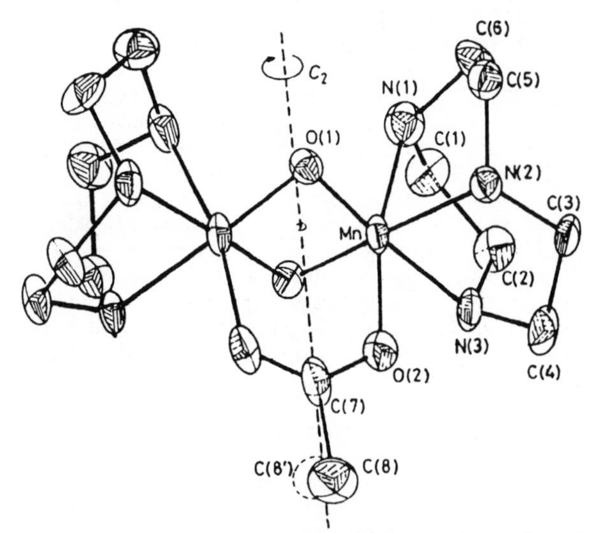

Figure 2. The structure of $[Mn_2O_2(O_2CMe)(TACN)_2]^{2+}$; the dashed line indicates the imposed two-fold axis. (Reproduced with permission from ref. 13. Copyright 1987 Royal Society of Chemistry)

for $Mn^{III}Mn^{IV}$. The latter also display slightly shorter Mn...Mn separations (2.59, 2.67Å) than the former (2.70-2.75Å). The $[Mn_2O_2(phen)_4]^{3+,4+}$ pair allows a useful structural comparison of the effect of oxidation level on exactly isostructural species, and again the higher oxidation state yields the slightly longer Mn...Mn separation (2.75 *vs.* 2.70Å).

TABLE I. Structural Parameters (Å) of Dinuclear Complexes

Complex	Mn...Mn	Mn-O	ref
Linear Oxide Bridges			
$[Mn_2^{III}O(CN)_{10}]^{6-}$	3.446(4)	1.723(4)	5
$[Mn_2^{III}O(phthal)_2]$	3.42(1)	1.71(1)	6
$[Mn_2^{IV}O(TPP)_2]^{2+}$	3.537(4)	1.743(4),1.794(4)	7
Bent Oxide Bridges			
$[Mn_2^{III}O(O_2CMe)_2(TPB)_2]$	3.159(1)	1.773(2),1.787(2)	8
$[Mn_2^{III}O(O_2CMe)_2(bipy)_2Cl_2]$	3.153(1)	1.777(12),1.788(11)	a
$[Mn_2^{III}O(O_2CMe)_2(TACN)_2]^{2+}$	3.084(3)	1.80(1)	9
$[Mn_2^{III,IV}O(O_2CMe)_2(Me_3\text{-}TACN)_2]^{3+}$	3.230(3)	1.826(6),1.814(6)	10
Di-μ-oxide Bridges			
$[Mn_2^{III,IV}O_2(bipy)_4]^{3+}$	2.716(2)	1.784-1.856	11
$[Mn_2^{III,IV}O_2(phen)_4]^{3+}$	2.700(1)	1.808-1.820	12
$[Mn_2^{III,IV}O_2(O_2CMe)(TACN)_2]^{2+}$	2.588(2)	1.817(5),1.808(4)	13
$[Mn_2^{III,IV}O_2(O_2CMe)(bipy)_2Cl_2]$	2.667(2)	1.793-1.843	a
$[Mn_2^{IV,IV}O_2(phen)_4]^{4+}$	2.748(2)	1.794-1.805	12
$[Mn_2^{IV,IV}O_2(pic)_4]$	2.747(2)	1.819(3)	a

a. unpublished

Should the WOC site prove to be dinuclear (or two separated dinuclear units), the question arises as to which known dinuclear complexes can be considered good models. As noted above, the Mn...Mn separations of Table I reflect the identity of the bridge and are not (seriously) dependent on the metal oxidation states. Thus, only the Mn_2O_2-containing complexes possess Mn...Mn separations in the ca 2.7Å range suggested by EXAFS data on the WOC. This can be taken as some evidence that a similar Mn_2O_2 unit might be present in the latter. In accord with this, a striking similarity has been noted between the EXAFS spectrum of $[Mn_2O_2(bipy)_4]^{3+}$ (and its phen analogue) and the EXAFS spectrum of the WOC (14). Strong correspondence was found between the number, type and distances of the ligand atoms in the first coordination sphere of the Mn from both sources. It would thus seem that it is the Mn_2O_2 complexes which represent structural models of the WOC. The $Mn^{III}Mn^{IV}$ complexes display a Mn-hyperfine-structured EPR signal in the g = 2 region for which 16 hyperfine lines are clearly resolvable (1, 13, 15). The latter bears similarity to the S_2 multiline signal suggesting that the $Mn^{III}Mn^{IV}$ complexes represent models of the S_2 level and, therefore, that the Mn_2^{IV} complexes might correspond to S_3; a reversible one-electron redox couple has been observed between these two oxidation levels in the synthetic complexes. Models for S_0 and S_1 would then presumably require $Mn^{II}Mn^{III}$ and Mn_2^{III} dinuclear complexes, but no di-μ-oxide bridged species at these oxidation states are currently known and must represent objectives of future work. A complex with a $Mn_2^{III}(OH)_2$ core has been reported (Mn...Mn = 2.72Å) (16), but there is some debate as to whether this might really be a $Mn_2^{IV}O_2$ species.

Should the longer (ca 3.3Å) Mn...Mn separation suggested by the EXAFS data on the WOC turn out to be real, this could either be interpreted as two of the above Mn_2O_2 units separated by ca 3.3Å (no example currently known) or, alternatively, a single Mn_2O_2 unit with one or two additional (and well separated) Mn atoms at ca 3.3Å from the central Mn_2O_2 unit (*vide infra*). The important point, however, is that two distinctly different Mn...Mn separations would not be possible for two isolated Mn_2O_2 units or a highly symmetrical trinuclear or tetranuclear species. Overall, the current status of dinuclear modelling studies suggests that Mn_2O_2 complexes represent a good minimal representation of the WOC.

Trinuclear Complexes

All reported trinuclear complexes possessing oxide bridges which have been characterized to date have been prepared from MnO_4^- in either aqueous or non-aqueous media, and possess the "basic carboxylate" structure *viz*. a μ_3-oxide-centered Mn_3 triangle. Both $Mn^{II}Mn_2^{III}$ and Mn_3^{III} species are currently known, and their pertinent structural parameters are listed in Table II.

TABLE II. Structural Parameters (Å) of Trinuclear Complexes

Complex	Mn...Mn	Mn-O	ref
$Mn_3O(O_2CMe)_6(py)_3$ [a]	3.363(1)	1.941(1)	17
$Mn_3O(O_2CMe)_6(py)_3$ [b]	3.351(1)	1.9364(18)	18
$Mn_3O(O_2CMe)_6(3\text{-}Cl\text{-}py)_3$	n.r.	1.863-2.034	19
$Mn_3O(O_2CPh)_6(py)_2(H_2O)$	3.214-3.418	1.798-2.154	18
$[Mn_3O(O_2CPh)_6(ImH)_3]^+$	3.259(1)-3.270(1)	1.888(4)-1.881(8)	c

a. room temperature b. -50°C c. unpublished n.r. = not reported

A representative structure is shown in Figure 3. In all cases, Mn...Mn separations are 'long' (ca 3.3Å), similar to values found for the mono-oxide-bridged Mn_2 complexes described above. Even in the trapped-valence complex $Mn_3O(O_2CPh)_6(py)_2(H_2O)$ (18) shown in Figure 3, where the Mn_3 unit is isosceles rather than equilateral, the Mn...Mn separations are not distinctly different, and no separation is found in the ca 2.7Å range. It could thus be concluded that an Mn_3O-type unit cannot represent the complete

Mn aggregate found in the WOC; this conclusion is admittedly based only on the synthetic trinuclear complexes currently known.

Tetranuclear Complexes

The possibility of a tetranuclear unit in the WOC has also prompted intense effort at this nuclearity. The first characterized tetranuclear Mn/O complex was $[Mn_4O_6(TACN)_4]^{4+}$, reported by Wieghardt and coworkers (20). This compound possesses a central Mn_4O_6 core with an adamantane-like structure (shown in Figure 4) and all Mn centers in the +4 oxidation level. The Mn...Mn separations (Table III) are all in the ca 3.2Å range and are thus again too long to model the ca 2.7Å separation seen in the WOC. Note that each Mn...Mn edge of the central Mn_4 tetrahedron is bridged by only one μ-O^{2-} group; the observed Mn...Mn separation is thus consistent with those observed in dinuclear and trinuclear complexes possessing only one μ-O^{2-} group between a Mn_2 pair.

In 1987, tetranuclear oxide-bridged Mn complexes became available which possess two distinct types of Mn...Mn separations. Vincent *et al.* showed that the trinuclear complexes containing $[Mn_3O]^{6+,7+}$ cores will react with 2,2′-bipyridine (bipy) to give complexes containing the $[Mn_4O_2]^{6+,7+,8+}$ core in high yield. With $[Mn_3{}^{III}O(O_2CMe)_6(py)_3]^+$, the product was $[Mn_4O_2(O_2CMe)_7(bipy)_2]^+$ shown in Figure 5 (21). All Mn are in the +3 oxidation level, and the Mn_4O_2 core can be considered as a

TABLE III. Structural Parameters (Å) of Tetranuclear Complexes

Complex	Mn...Mn (short)	Mn...Mn (long)	Mn-O	ref
$[Mn_4O_6(TACN)_4]^{4+}$	----	3.21(1)[a]	1.79(1)[a]	20
$[Mn_4O_2(O_2CMe)_7(bipy)_2]^+$	2.848(5)	3.299- 3.385	1.804- 1.930	21
$[Mn_4O_2(O_2CMe)_6(bipy)_2]$	2.7787(12)	3.288(1) 3.481(1)	1.8507(20)[b] 1.8560(21)[b] 2.1032(21)[c]	22
$[Mn_4O_3Cl_6(O_2CMe)_3(ImH)]^{2-}$	2.806- 2.818	3.246- 3.323	1.797- 2.002	23
$[Mn_4O_3Cl_4(O_2CMe)_3(py)_3]$	2.8145(15)	3.269(1)	1.865- 1.966	d

a. average distances, b. Mn(III)-O, c. Mn(II)-O, d. unpublished

"butterfly" with a μ_3-O^{2-} bridging each triangular face. An alternative way of viewing the structure emphasizes its Mn_3O parenthood; the product can be considered as two triangular units sharing an edge. Extension of the bipy reaction to the mixed-valence trinuclear complexes leads to products whose oxidation level depends on the identity of the carboxylate group. Thus, $Mn_3O(O_2CPh)_6(py)_2(H_2O)$ reacts with bipy to yield $Mn_4O_2(O_2CPh)_7$

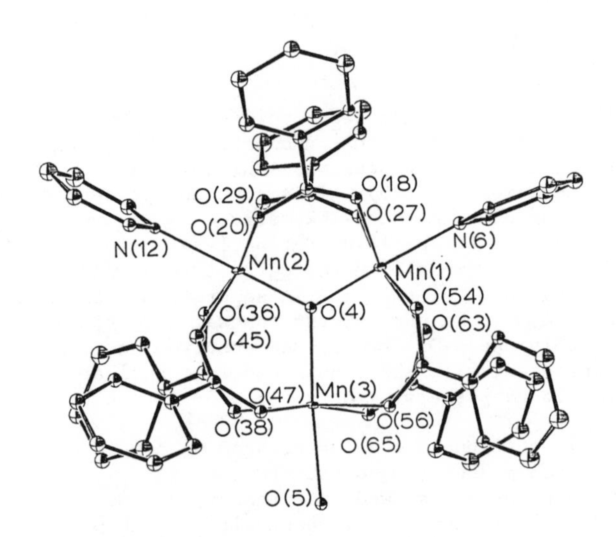

Figure 3. The structure of $Mn_3O(O_2CPh)_6(py)_2(H_2O)$ illustrating the unique terminal ligand asymmetry; Mn(3) is the Mn^{II} center possessing a terminal H_2O ligand. (Reproduced from ref. 8. Copyright 1987 American Chemical Society.)

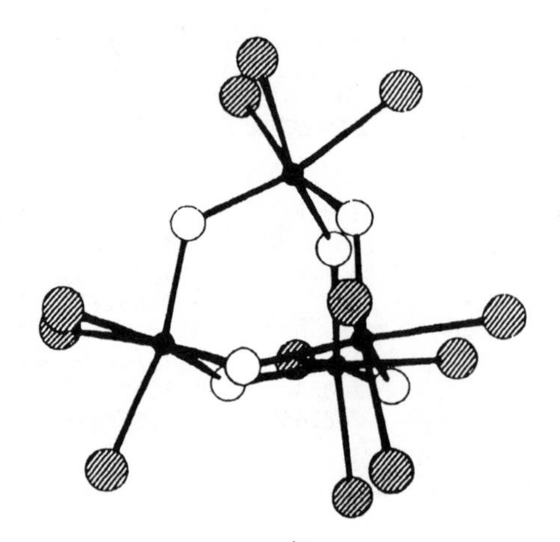

Figure 4. The structure of $[Mn_4O_6(TACN)_4]^{4+}$; only the nitrogen atoms of the TACN ligands are shown. (Reproduced with permission from ref. 20. Copyright 1983 VCH Verlagsgesellschaft.)

(bipy)$_2$ containing $Mn^{II}Mn_3^{III}$ (21). However, use of $Mn_2O(O_2CMe)_6$ (py)$_3$ leads instead to $Mn_4O_2(O_2CMe)_6(bipy)_2$ containing $Mn_2^{II}Mn_2^{III}$ (22). The structure of this latter complex, shown in Figure 6, shows it to be similar to $[Mn_4O_2(O_2CMe)_7(bipy)_2]^+$ except that the four Mn atoms are now planar and the unique carboxylate group bridging the central Mn_2 pair is absent.

The pertinent structural parameters of $[Mn_4O_2(O_2CMe)_7(bipy)_2]^+$ and $Mn_4O_2(O_2CMe)_6(bipy)_2$ are included in Table III. The two-hinge Mn atoms in the former complex are separated by 2.848(5)Å, whereas the remaining Mn...Mn separations are much longer, in the range 3.299(5)-3.385(5)Å. Note that the short Mn...Mn separation possesses two μ-oxide bridges, whereas the longer separations possess only one. $Mn_4O_2(O_2CMe)_6(bipy)_2$ has a similar arrangement with the corresponding separations being 2.779(1) and 3.288(1)-3.481(1)Å. The central Mn_2O_2 rhomb in these complexes is thus akin to those in the dinuclear systems containing two oxide bridges.

As for the dinuclear complexes, one might now ask how closely do the synthetic materials correspond overall to the native site should it contain an Mn_4 aggregate. There are indeed several features which again suggest that close approach to the native site may have been achieved: (i) four Mn atoms in two inequivalent pairs with two distinct Mn...Mn separations (24); (ii) bridging O^{2-} atoms between the metal centers; (iii) peripheral ligation by carboxylate and bipy corresponding to the metal-binding functions of aspartic/glutamic acids and histidine (the pyridine ring is a conservative replacement for imidazole); and (iv) metal oxidation levels corresponding to those in the S_n states.

Point (iv) warrants further comment. The EPR-active S_2 state is thought to contain $Mn_3^{III}Mn^{IV}$ (2). (The alternative possibility is $Mn^{III}Mn_3^{IV}$; the latter cannot be ruled out with certainty but the results of NMR solvent relaxation studies (25) and UV/vis difference spectroscopy (26,27) suggest the former to be the more likely.) Based on this and assuming the Mn aggregate in the WOC is indeed tetranuclear, the Mn oxidation states in the S_n levels can be defined: S_0, Mn^{II}, $3Mn^{III}$; S_1, $4Mn^{III}$; S_2, $3Mn^{III}$, Mn^{IV}; S_3, $2Mn^{III}$, $2Mn^{IV}$. Additionally, the putative S_{-1} level (28-30), accessible under certain conditions but not participating in the water oxidation cycle, would contain $2Mn^{II}$, $2Mn^{III}$. The synthetic Mn_4O_2 complexes would therefore have the following oxidation state correspondence to the S_n states:

$Mn_4O_2(O_2CMe)_6(bipy)_2$	$Mn_4O_2(O_2CPh)_7(bipy)_2$	$[Mn_4O_2(O_2CMe)_7(bipy)_2]^+$
S_{-1}	S_0	S_1

That the Mn_4O_2 is capable of adopting at least three isolable oxidation levels thus lends support, albeit circumstantial, to the possibility that it is to be found as the redox component in the WOC, if only at the lower S_n states. In support of this, the various oxidation levels of the synthetic materials have been found to be electrochemically interconvertible, and a $[Mn_4O_2]^{9+}$ level has been detected electrochemically (31).

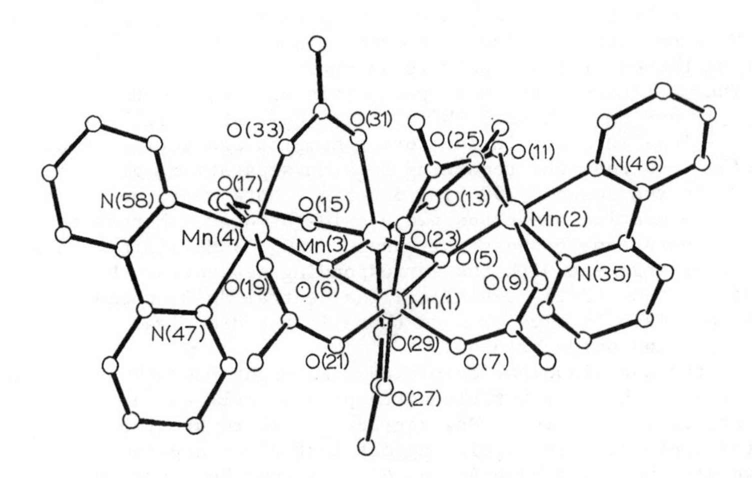

Figure 5. The structure of $[Mn_4O_2(O_2CMe)_7(bipy)_2]^+$. (Reproduced with permission from ref. 21. Copyright 1987 Royal Society of Chemistry.)

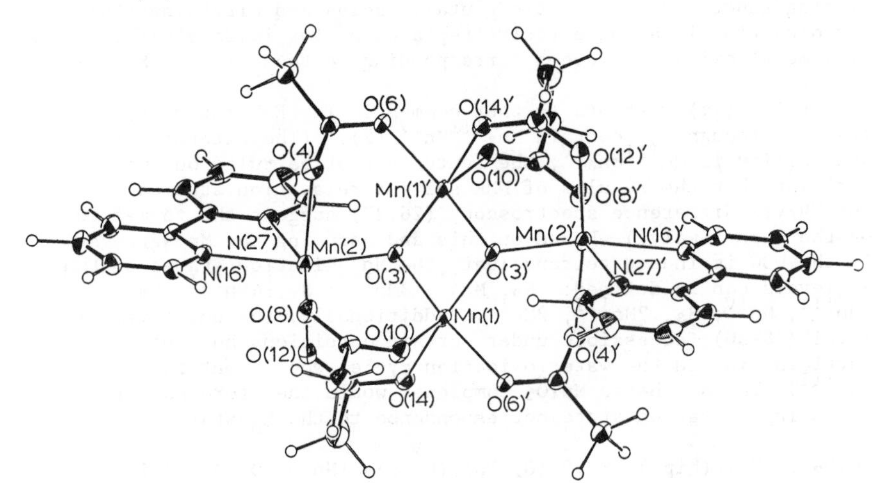

Figure 6. The structure of $Mn_4O_2(O_2CMe)_6(bipy)_2$; unlike the structure in Figure 5, the Mn_4 core is exactly planar. (Reproduced with permission from ref. 22. Copyright 1987 Royal Society of Chemistry).

More recent work has resulted in synthetic access to higher oxidation state tetranuclear complexes. Bashkin *et al.* have shown that treatment of "manganic acetate" with Me_3SiCl and imidazole (HIm) leads to formation of $[Mn_4O_3Cl_6(O_2CMe)_3(ImH)]^{2-}$ shown in Figure 7 (23). The structure is best described as a Mn_4 pyramid with a μ_3-Cl^- bridging the basal plane and three μ_3-O^{2-} atoms bridging each remaining Mn_3 plane. Each vertical Mn_2 pair is thus bridged by two O^{2-} atoms, and, not surprisingly from the discussion above, the Mn...Mn separation is 'short' (ca 2.8Å). In contrast, the basal Mn_2 pairs are bridged by only one O^{2-} atom and have longer Mn...Mn separations (ca 3.3Å); the Mn-μ_3-Cl^- bonds are quite long (ca 2.6Å) and do not serve to contract the Mn...Mn separation. The complex has a $Mn_3^{III}Mn^{IV}$ oxidation state and the apical Mn is assigned as the Mn^{IV} center by inspection of structural parameters. In addition, it displays an EPR spectrum showing a 16-line Mn-hyperfine-structured feature at $g \approx 2$; it has thus been proposed as a potential model of S_2.

The Mn_4O_2 and Mn_4O_3 complexes both have two distinct Mn...Mn separations and oxide bridges, and are also more structurally related than might be apparent at first glance. It can readily be seen that the Mn_4O_3 core can be obtained from an Mn_4O_2 butterfly unit by incorporation of a third μ_3-O^{2-} to bridge the two "wing-tip" Mn and one of the "hinge" Mn atoms of the Mn_4O_2 "butterfly". This leads to an interesting mechanistic proposal for water oxidation by the WOC (*vide infra*).

The Me_3SiCl reaction system has since been extended to other starting materials. Treatment of $[Mn_3O(O_2CMe)_6(py)_3]^+$ with Me_3SiCl leads to the isolation of neutral $Mn_4O_3Cl_4(O_2CMe)_3(py)_3$ (31). The Mn_4O_3Cl core of this complex is essentially identical to that of the previous complex with respect to its structural parameters and also contains $Mn_3^{III}Mn^{IV}$.

Higher Nuclearity Complexes

To complete our survey of structurally-characterized oxide-bridged Mn complexes, we describe species with nuclearities >4. Three distinct types have been reported; hexanuclear, nonanuclear and dodecanuclear. The hexanuclear complex $Mn_6O_2(piv)_{10}(pivH)_4$ results from the treatment of $MnCO_3$ with pivalic acid in refluxing toluene (32). Its structure is shown in Figure 8. The complex has an average Mn oxidation state of +2.33 ($4Mn^{II}$, $2Mn^{III}$), and the two central Mn atoms have been assigned as the Mn^{III} centers. The Mn...Mn separations were not reported. The complex can be structurally related to the Mn_4O_2 core of $Mn_4O_2(O_2CMe)_6(bipy)_2$ by noting that the two μ_3-O^{2-} atoms of the latter complex have become μ_4 by binding to two additional Mn^{II} centers to yield the Mn_6O_2 core of the hexanuclear species.

The reaction of $Mn_3O(O_2CPh)_6(py)_2(H_2O)$ with salicylic acid ($salH_2$) leads to formation of the nonanuclear species $Mn_9O_4(sal)_4(salH)_2(O_2CPh)_8(py)_4$, with an average Mn oxidation state of +2.89 ($8Mn^{III}$, Mn^{II}) (33). The structure is shown in Figure 9. It can be conveniently described as two Mn_4O_2 units held together by a central 'bridging' Mn^{II} atom; the latter is eight-coordinate by being bound to both oxygens of four salicylate carboxylate groups. The Mn_4O_2 units have the same "butterfly" arrangement as

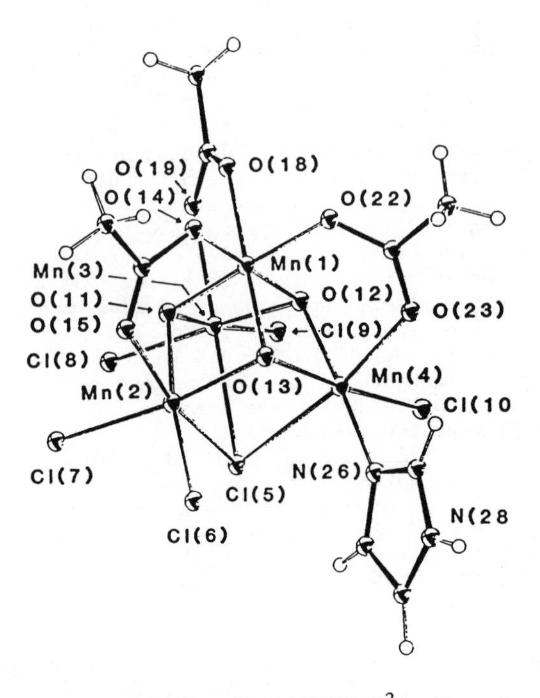

Figure 7. The structure of $[Mn_4O_3Cl_6(O_2CMe)_3(HIm)]^{2-}$. (Reproduced from ref. 23. Copyright 1987 American Chemical Society.)

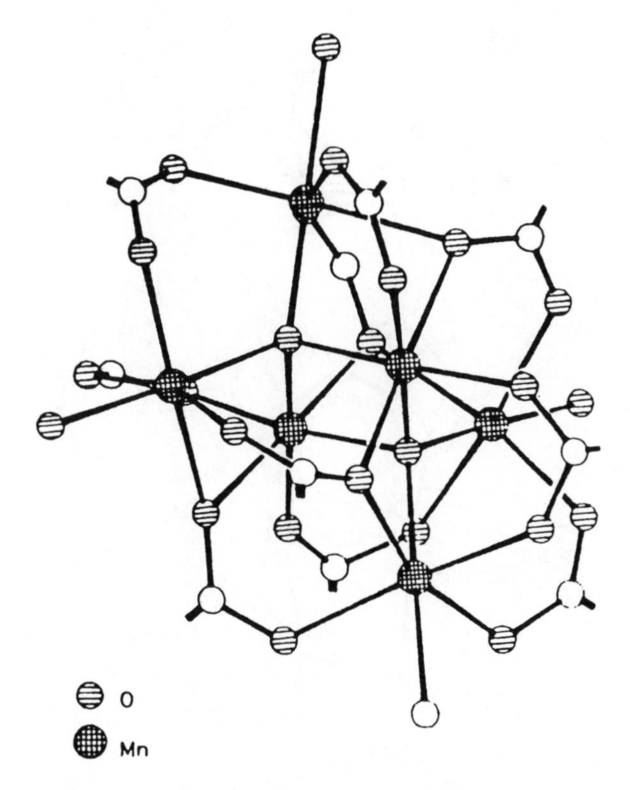

Figure 8. The structure of $Mn_6O_2(piv)_{10}(pivH)_4$; only portions of the peripheral ligands are shown. (Reproduced with permission from ref. 32. Copyright 1986 Royal Society of Chemistry).

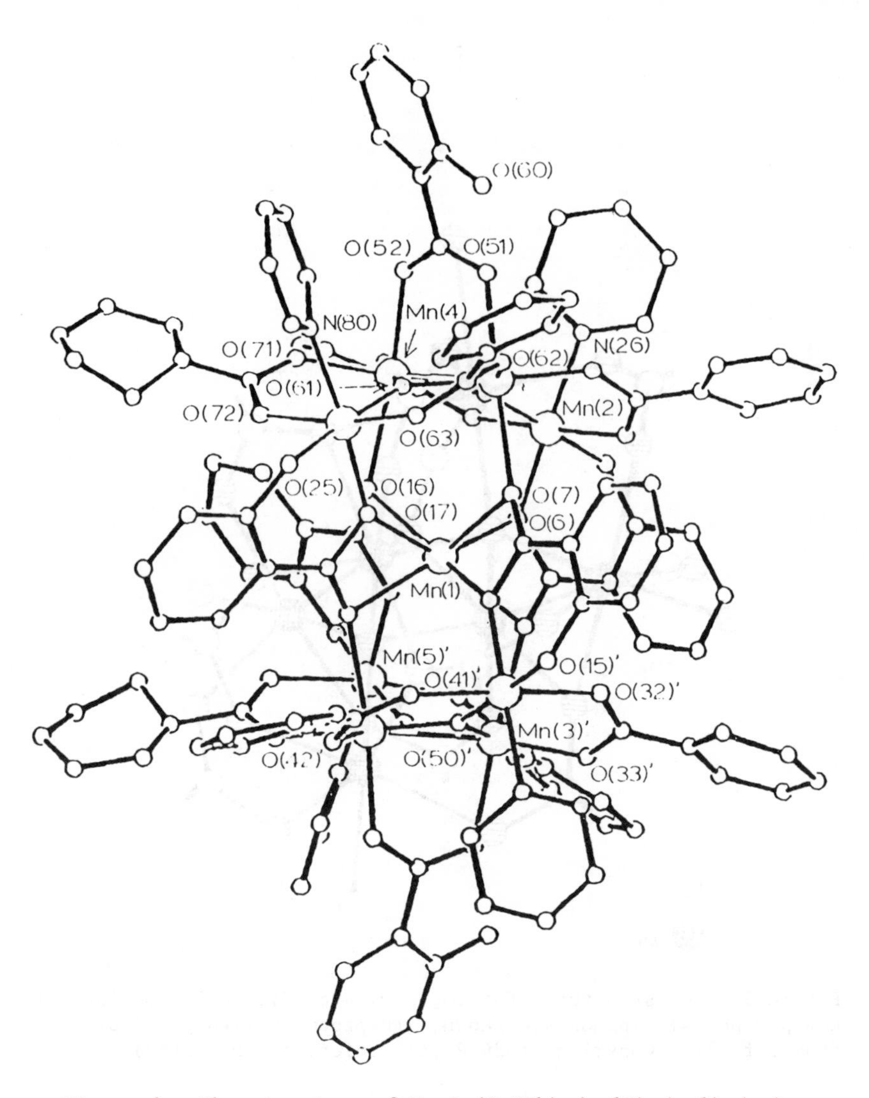

Figure 9. The structure of $Mn_9O_4(O_2CPh)_8(salH)_2(sal)_4(py)_4$. (Reproduced with permission from ref. 33. Copyright 1987 VCH Verlagsgesellschaft).

found in $[Mn_4O_2(O_2CMe)_7(bipy)_2]^+$, with 'short' and 'long' Mn...Mn separations of 2.817(6) and 3.406(6)-3.443(6)Å, respectively.

The largest oxide-bridged Mn aggregate known is $Mn_{12}O_{12}(O_2CR)_{16}(H_2O)_4$ (R = Me, Ph) (34). The structure of the benzoate complex is shown in Figure 10 (31). This remarkable complex contains a central Mn_4O_4 cubane enclosed within a non-planar ring of eight Mn centers *via* the intermediacy of additional μ_3-O^{2-} bridges. The complex is mixed-valence ($8Mn^{III}$, $4Mn^{IV}$), and structural parameters indicate the 'cubane' Mn atoms to be the Mn^{IV} centers. Again, Mn_2 pairs bridged by two oxide groups have Mn...Mn separations noticeably shorter (2.8-3.0Å) than those bridged by only one oxide group (3.3-3.5Å).

As for the nonanuclear complex, the $Mn_{12}O_{12}$ core can be structurally related to the Mn_4O_2 complexes by noting that the former can be considered as four Mn_4O_2 butterfly units fused together such that a Mn^{IV} center occupies a 'hinge' position of each butterfly. Four additional μ_3-O^{2-} bridges then complete the central cubane yielding the complete $Mn_{12}O_{12}$ core.

Summary of Structural Results

The current status of synthetic efforts in Mn/O^{2-} chemistry is summarized below, emphasizing some general observations and trends which have become apparent:

1. Synthetic procedures to Mn/O^{2-} complexes with a variety of nuclearities (2,3,4,6,9,12) are now at hand. The larger complexes can be considered aggregates of the smaller nuclearity species.

2. Structurally-characterized materials contain average Mn oxidation levels from +2.33 - +4, encompassing the complete range thought to be occurring in the WOC in its various S_n states.

3. In most cases, peripheral ligation is by O- and/or N-based groups as believed to be present in the WOC and provided by amino-acid side chain functions.

4. The 'short' (ca 2.7Å) Mn...Mn separation of the WOC deduced by EXAFS would appear to be associated with the presence of two bridging O^{2-} groups between a Mn_2 pair. Only when such an arrangement is present has a Mn...Mn separation in this range been seen to date.

5. The 'long' (ca 3.3Å) separation would appear to be consistent with at most only one bridging O^{2-} group between a Mn_2 pair; we say at most because bridging carboxylate or phenoxide groups (and no O^{2-}) can also result in a lengthened Mn...Mn separation.

6. Two types of tetranuclear complexes, containing Mn_4O_2 and Mn_4O_3 cores, are known which possess both 'short' and 'long' Mn...Mn separations; together these complexes span the oxidation range believed to be possessed by the WOC in its S_{-1}--S_2 states.

Mechanistic Proposals for the Water Oxidation Cycle

Two proposals for the water oxidation cycle involving Mn/O assemblies established in synthetic complexes have been published to date. These have attempted to describe the structural rearrangements of the Mn_4 core and concomitant substrate binding

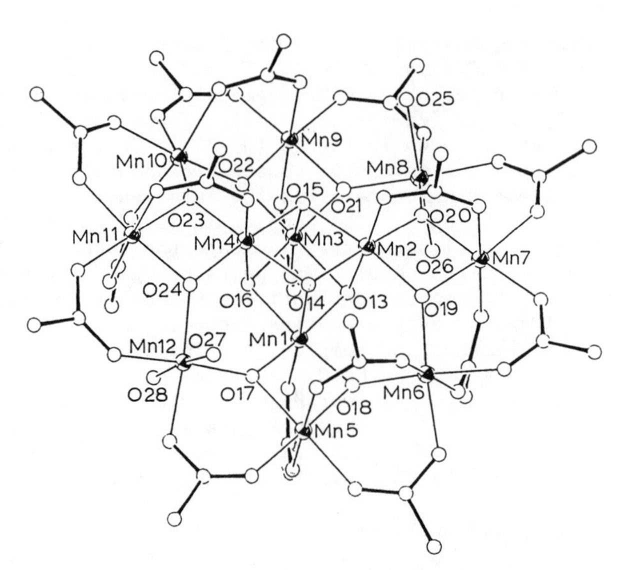

Figure 10. The structure of $Mn_{12}O_{12}(O_2CPh)_{16}(H_2O)_4$. The four H_2O molecules are bound to Mn atoms Mn(8) and Mn(12); only portions of the peripheral ligands are shown.

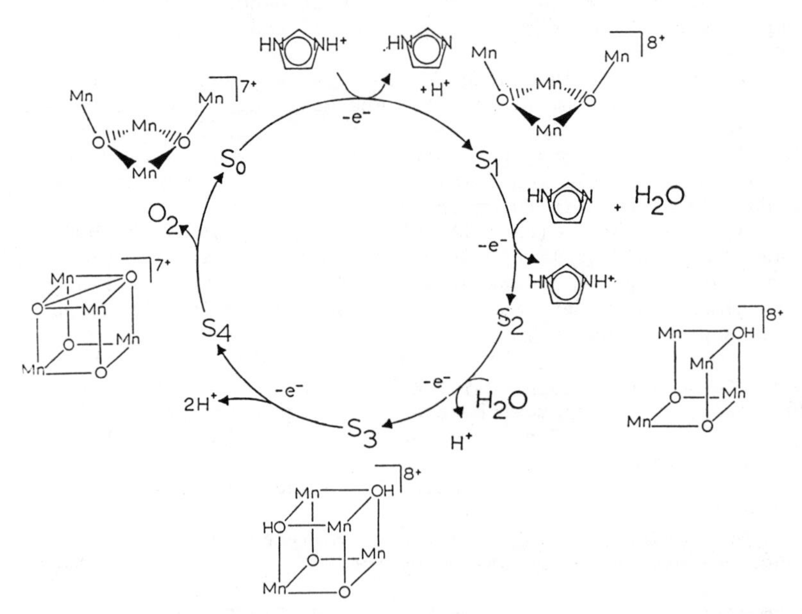

Figure 11. Proposed mechanistic scheme for water oxidation employing the Mn_4O_2 and Mn_4O_3 cores established in model complexes.

and transformation to O_2 which may be occurring during the catalytic cycle.

The first published proposal involved an Mn_4O_4 cubane at the lower S_n states converting to a Mn_4O_6 adamantane-like unit in the higher S_n states on incorporation of the two H_2O molecules (35). Subsequent formation and evolution of O_2 returns the unit to a Mn_4O_4 core ready for recycling. The second proposal employs the Mn_4O_2 butterfly unit at the lower S_n states converting, on incorporation of the two H_2O molecules, into a Mn_4O_4 cubane at the higher S_n states (36). Subsequent formation and evolution of O_2 converts the cubane back into a Mn_4O_2 unit ready for recycling.

Since the second proposal was formulated, the synthesis of Mn_4O_3- containing complexes was achieved. The structural relationship of this unit to the Mn_4O_2 core (*vide supra*) has led to an alternative scheme which is presented for the first time in Figure 11. It differs from the published version only in the identity of S_2. Essentially, incorporation of only one substrate molecule could convert the butterfly Mn_4O_2 unit at S_1 into a Mn_4O_3 unit (shown as $Mn_4O_2(OH)$) at S_2. Subsequent incorporation of the second H_2O molecule now completes the Mn_4O_4 ($Mn_4O_2(OH)_2$) core, followed by evolution of O_2 and return to a Mn_4O_2 unit at S_0. Whether the modified proposal is any more relevant to the actual mechanism of photosynthetic water oxidation is uncertain given our current knowledge of the latter, but it does have the attraction of incorporating more structurally-established structural units than the previous proposals. Indeed, only its proposed S_3 structure has yet to be found in a discrete tetranuclear model complex.

In conclusion, it is apparent that considerable progress has been made in the synthesis and characterization of higher oxidation state Mn/O complexes. A sufficiently large pool of complexes is now available to aid in the inter-disciplinary investigation of this important biological center, and structural and mechanistic proposals can now be made with more confidence and precedence. More work in the model area obviously remains to be performed, but the required synthetic procedures and knowledge of Mn/O chemistry is now at hand to allow and guide this extension.

Acknowledgments

Research performed at Indiana University was supported by the National Science Foundation.

Abbreviations

phthal = phthalocyanine
TPP = tetraphenylporphyrin
TBP = hydrido-*tris*(pyrazolyl)borate
TACN = 1,4,7-triazacyclononane
piv = pivalate
Me_3-TACN = 1,4,7-trimethyl-1,4,7-triazacyclononane
bipy = 2,2′-bipyridine
py = pyridine
ImH = imidazole

phen = 1,10-phenanthroline
pic = picolinate

Literature Cited

1. Amesz, J. *Biochim. Biophys. Acta* **1983**, *726*, 1. Dismukes, G.C. *Photochem. Photobiol.* **1986**, *43*, 99. Govindjee; Kambara, T.; Coleman, W. *Photochem. Photobiol.* **1985**, *42*, 187.
2. Dismukes, G.C.; Ferris, K.; Watnick, P. *Photochem. Photobiol.* **1982**, *3*, 243.
3. Guiles, R.D.; Yachandra, V.K.; McDermott, A.E.; Britt, R.D.; Dexheimer, S.L.; Sauer, K.; Klein, M.P. *Prog. Photosynth. Res., Proc. Int. Congr. Photosyn., 7th* **1987**, *1*, 561.
4. Yachandra, V.K.; Guiles, R.D.; McDermott, A.; Britt, R.D.; Dexheimer, S.L.; Sauer, K.; Klein, M.P. *Biochim. Biophys. Acta* **1986**, *850*, 324.
5. Ziolo, R.F.; Stanford, R.H.; Rossman, G.R.; Gray, H.B. *J. Am. Chem. Soc.* **1974**, *96*, 7910.
6. Vogt, L.H., Jr.; Zalkin, A.; Templeton, D.H. *Inorg. Chem.* **1967**, *6*, 1725.
7. Schardt, B.C.; Hollander, F.J.; Hill, C.L. *J. Am. Chem. Soc.* **1982**, *104*, 3964.
8. Sheats, J.E.; Czernuszewicz, R.S.; Dismukes, G.C.; Rheingold, A.L.; Petrouleas, V.; Stubbe, J.; Armstrong, W.H.; Beer, R.H.; Lippard, S.J. *J. Am. Chem. Soc.* **1987**, *109*, 1435.
9. Wieghardt, K.; Bossek, U.; Ventur, D.; Weiss, J. *J. Chem. Soc., Chem. Commun.* **1985**, 347.
10. Wieghardt, K.; Bossek, U.; Bonvoisin, J.; Beauvillain, P.; Girerd, J.-J.; Nuber, B.; Weiss, J.; Heinze, J. *Angew. Chem. Int. Ed. Engl.* **1986**, *25*, 1030.
11. Plaskin, P.M.; Stoufer, R.C.; Mathew, M.; Palenik, G.J. *J. Am. Chem. Soc.* **1972**, *94*, 2121.
12. Stebler, M.; Ludi, A.; Burgi, H.-B. *Inorg. Chem.* **1986**, *25*, 4743.
13. Wieghardt, K.; Bossek, U.; Zsolnai, L.; Huttner, G.; Blondin, G.; Girerd, J.-J.; Babonneau, F. *J. Chem. Soc., Chem. Commun.* **1987**, 651.
14. Kirby, J.A.; Robertson, A.S.; Smith, J.P.; Thompson, A.C.; Cooper, S.R.; Klein, M.P. *J. Am. Chem. Soc.* **1981**, *103*, 5529.
15. Dismukes, G.C.; Siderer, Y. *Proc. Natl. Acad. Sci. USA* **1981**, *78*, 274.
16. Maslen, H.S.; Waters, T.N. *J. Chem. Soc., Chem. Commun.* **1973**, 760.
17. Baikie, A.R.E.; Hursthouse, M.B.; New, D.B.; Thornton, P. *J. Chem. Soc., Chem. Commun.* **1978**, 62.
18. Vincent, J.B.; Chang, H.-R.; Folting, K.; Huffman, J.C.; Christou, G.; Hendrickson, D.N. *J. Am. Chem. Soc.* **1987**, *109*, 5703.
19. Baikie, A.R.E.; Hursthouse, M.B.; New, L.; Thornton, P.; White, R.G. *J. Chem. Soc., Chem. Commun.* **1980**, 684.

20. Wieghardt, K.; Bossek, U.; Gebert, W. *Angew. Chem. Int. Ed. Engl.* **1983**, *22*, 328.
21. Vincent, J.B.; Christmas, C.; Huffman, J.C.; Christou, G.; Chang, H.-R.; Hendrickson, D.N. *J. Chem. Soc., Chem. Commun.* **1987**, 236.
22. Christmas, C.; Vincent, J.B.; Huffman, J.C.; Christou, G.; Chang, H.-R.; Hendrickson, D.N. *J. Chem. Soc., Chem. Commun.* **1987**, 1303.
23. Bashkin, J.S.; Chang. H.-R.; Streib, W.E.; Huffman, J.C.; Hendrickson, D.N.; Christou, G. *J. Am. Chem. Soc.* **1987**, *109*, 6502.
24. Murata, N.; Miyao, M. *Trends Biochem. Sci.* **1985**, *10*, 122.
25. Srinivasan, A.N.; Sharp, R.R. *Biochim. Biophys. Acta* **1986**, *851*, 369.
26. Vincent, J.B.; Christou, G. *FEBS Lett.* **1986**, *207*, 250.
27. Witt, H.T.; Schlodder, E.; Brettel, K.; Saygin, O. *Photosyn. Res.* **1986**, *9*, 453.
28. Velthuys, B.; Kok, B. *Biochim. Biophys. Acta* **1978**, *502*, 211.
29. Pistorius, E.K.; Schmid, G.H. *Biochim. Biophys. Acta* **1987**, *890*, 352.
30. Bader, K.P.; Thibault, P.; Schmid, G.H. *Z. Naturforsch. Teil C.* **1983**, *38*, 778.
31. Vincent, J.B.; Christou, G., unpublished results.
32. Baikie, A.R.E.; Howes, A.J.; Hursthouse, M.B.; Quick, A.B.; Thornton, P. *J. Chem. Soc., Chem. Commun.* **1986**, 1587.
33. Christmas, C.; Vincent, J.B.; Huffman, J.C.; Christou, G.; Chang, H.-R.; Hendrickson, D.N. *Angew. Chem. Int. Ed. Engl.* **1987**, *26*, 915.
34. Lis, T. *Acta Cryst.* **1980**, *B36*, 2042.
35. Brudvig, G.W.; Crabtree, R.H. *Proc. Natl. Acad. Sci. USA* **1986**, *83*, 4586.
36. Vincent, J.B.; Christou, G. *Inorg. Chim. Acta* **1987**, *136*, L41.

RECEIVED January 20, 1988

METAL–SULFUR CLUSTERS

Chapter 13

Synthetic Strategies for Modeling Metal–Sulfur Sites in Proteins

B. A. Averill

Department of Chemistry, University of Virginia, Charlottesville, VA 22901

Four distinct types of Fe-S center have now been found in proteins, ranging from mono- to tetranuclear; in addition, a novel Mo-Fe-S cluster is present in the enzyme nitrogenase. Synthetic analogs of most of these have been prepared and used to provide insight into the intrinsic properties of the metal-sulfur centers in the absence of protein-imposed constraints. The strategies used to prepare both Fe-S and Mo-Fe-S clusters are described; they range from spontaneous self-assembly to the designed synthesis of clusters with specific structural features.

Proteins containing iron-sulfur clusters are ubiquitous in nature, due primarily to their involvement in biological electron transfer reactions. In addition to functioning as simple reagents for electron transfer, protein-bound iron-sulfur clusters also function in catalysis of numerous redox reactions (e.g., H_2 oxidation, N_2 reduction) and, in some cases, of reactions that involve the addition or elimination of water to or from specific substrates (e.g., aconitase in the tricarboxylic acid cycle) (1).

At present, four distinct types of Fe-S center have been established in such proteins by X-ray crystallographic methods; schematic views of these are given in Figure 1. Although these units vary from mono- to tetranuclear, all exhibit a common theme of essentially tetrahedral iron atoms ligated by sulfur donors, either terminal cysteine thiolates or bridging sulfides. In view of the diversity of structures and biological functions, two of the major goals of contemporary metallobiochemistry are to understand why particular proteins utilize a specific type of Fe-S cluster, and how proteins containing apparently identical Fe-S units can modulate their properties to achieve a particular type of reactivity. Such questions are, however, difficult to answer even by extensive studies of the proteins themselves. Fortunately, it has proven possible in recent years to prepare synthetic

0097–6156/88/0372–0258$09.50/0

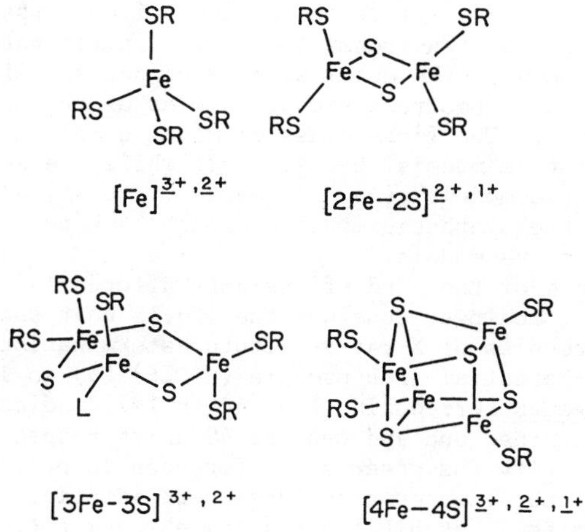

$[Fe]^{\underline{3+},\underline{2+}}$ $[2Fe{-}2S]^{\underline{2+},1+}$

$[3Fe{-}3S]^{3+,2+}$ $[4Fe{-}4S]^{\underline{3+},\underline{2+},\underline{1+}}$

Figure 1. Schematic views of the structurally characterized Fe-S centers found in proteins to date. Core oxidation states are indicated by superscripts; underlines indicate those structures and oxidation states for which structurally characterized synthetic analogs are available.

analogs (2) of the mono-, bi-, and tetranuclear centers shown in Figure 1; these have in some cases provided at least tentative answers to questions such as those posed above. The synthetic strategies used to prepare such model complexes constitute the subject of this chapter.

I. RATIONALE

Before plunging into a discussion of how such complexes are prepared, it is perhaps worthwhile to consider explicitly the rationale for such activity. The synthesis and characterization of accurate model complexes for a given metal site in a protein or other macromolecule allows one to: (i) determine the intrinsic properties of the metal site in the absence of perturbations provided by the protein environment; or (ii) in favorable cases, deduce the structure of the metal site by comparison of corresponding physical and spectroscopic properties of the model and metalloprotein (3). The first class of model complexes has been termed "corroborative models" by Hill (4), while the second are termed "speculative models" (4). To date, virtually all the major achievements of the synthetic model approach have been in development of corroborative models.

As an example of the kind of insights affordable by examination of synthetic analogs, consider the 4Fe-4S unit shown in Figure 1. The results of X-ray structure determinations of *Chromatium* high-potential iron protein (HiPIP) (5) in 1971 and *Peptococus aerogenes* ferredoxin (Fd) (6) in 1972 indicated that the proteins contained one and two 4Fe-4S units respectively, but did little to clarify the observed differences in protein physical properties since the cluster structures were identical within experimental limits. Yet HiPIP's possess a redox potential of ≈+0.35V and exhibit EPR signals at $g = 2.1$ only in the oxidized form (7), while the Fd's possess a redox potential of ≈-0.42V and are EPR-active, with $g_{av} = 1.94$, only in the *reduced* form (8). It was initially difficult to reconcile these facts. The synthesis of the $[Fe_4S_4(SR)_4]^{2-}$ clusters in 1972, the observed equivalence in spectroscopic properties and inferred isoelectronic nature of $[Fe_4S_4(SR)_4]^{2-}$, oxidized Fd, and reduced HiPIP (9, 10), and the demonstration of the existence of a four-membered electron series ($[Fe_4S_4(SR)_4]^{n-}$, n = 1-4) (9) led immediately to a straightforward explanation, as shown in Scheme 1 below. The $[Fe_4S_4(SR)_4]^{2-}$ cluster, containing an $[Fe_4S_4]^{2+}$ core with formally $2Fe^{III}$ and $2Fe^{II}$, is capable of both accepting (Fd) and donating (HiPIP) an electron; this provided a physical basis for the three-state hypothesis advanced independently by Carter, et al. (12).

$$[Fe_4S_4(SR)_4]^{4-} \underset{}{\overset{\pm e^-}{\rightleftarrows}} [Fe_4S_4(SR)_4]^{3-} \overset{\pm e^-}{\rightleftarrows} [Fe_4S_4(SR)_4]^{2-} \overset{\pm e}{\rightleftarrows} [Fe_4S_4(SR)_4]^{1-}$$

$$Fd_{red} \rightleftarrows Fd_{ox}$$

$$HP_{superred} \rightleftarrows HP_{red} \rightleftarrows HP_{ox}$$

Scheme 1

II. SYNTHETIC STRATEGIES FOR Fe-S CLUSTERS

The basic strategy involved in synthesis of Fe-S clusters from simple inorganic reagents is that of "spontaneous self-assembly" (3). That is, the nature of the products (nuclearity, oxidation state) is controlled by thermodynamic considerations, which dictate that the observed product is generally that which is most stable under the particular conditions utilized. As we shall see subsequently, only in one or two cases does one obtain a kinetically controlled product that spontaneously rearranges to give a different cluster. In practice, these considerations mean that the particular product obtained in a given reaction is controlled by the nature and ratios of the reagents, by the solvent, and in some cases by the temperature or even by nature of the quaternary cation present in the reaction mixture. This generally necessitates a substantial amount of exploratory research to define the optimal conditions for producing the desired cluster product.

In many cases, however, a cluster in a particular oxidation state is not accessible by direct synthetic methods using simple ligands or by straightforward oxidation or reduction of the cluster product. Solutions that have been developed for this problem include the use of sterically hindered ligands or ligands with varied donor atoms.

Similarly, clusters with atypical ligands may be desireable to model specific details of a particular system. Such clusters may not be accessible via a direct synthesis for any of a variety of reasons or they may be prohibitively expensive (e.g., synthetic peptides) for use in a reaction whose yield is significantly less than quantitative. Methods for substitution of terminal ligands on Fe-S clusters have now been developed to the point that virtually any desired ligands or combination of ligands can be incorporated via ligand exchange reactions (2).

Spontaneous Self-assembly Reactions

Iron-sulfide-thiolate systems. Reaction of iron salts with sulfide or sulfur and appropriate terminal ligands is known to result in products ranging in nuclearity from 1 to 8. Although many of the reactions used to produce these clusters are unique to a given system, some general statements regarding certain systems are possible. This is in large part due to work from Holm's laboratory that has been devoted to a systematic examination of the products arising from reaction of iron halides, sulfide or sulfur, and thiolates in non-aqueous solution. For systems containing $FeCl_2$ and various ratios of thiolate with a 1:1 ratio of sulfur:iron, the major products obtained in acetonitrile solution are indicated in Scheme 2 (13). Structures of products **4** and **5** are shown in Figure 2, while schematic views of the protein clusters corresponding to analogs **1**, **2**, and **6** were provided in Figure 1. Thus, structurally characterized synthetic models for the Fe^{n+} (n = 3,2), $[Fe_4S_4]^{n+}$ (n = 1,2,3), and $[Fe_2S_2]^{n+}$ (n = 2) centers are now available, with only the $[Fe_2S_2]^{+}$ unit remaining as a synthetic objective. Discussion of the trinuclear clusters **3** and **7** is deferred to a subsequent section devoted to models for 3Fe-nS sites in proteins.

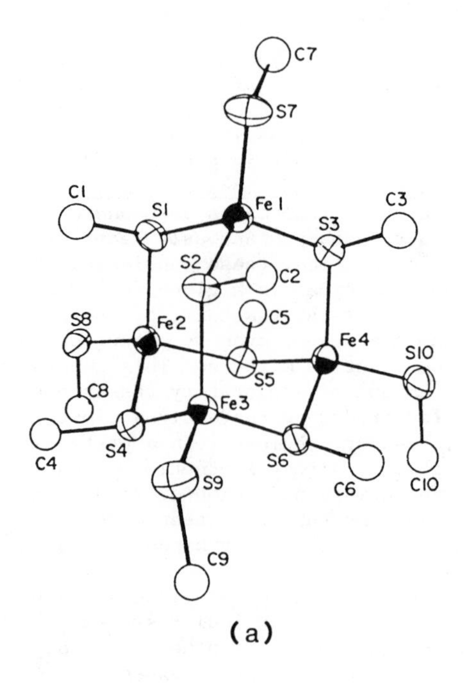

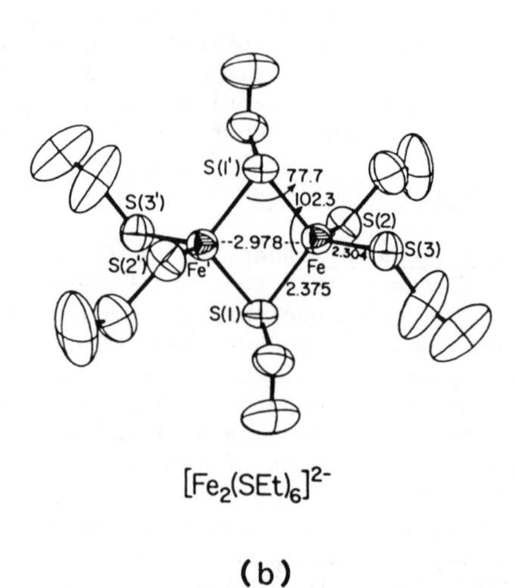

Figure 2. Structures of the known oligonuclear binary Fe^{2+}-thiolate complexes. (a) $[Fe_4(SPh)_{10}]^{2-}$. Only α-carbons of phenyl rings are shown. (Reproduced from ref. 84. Copyright 1980 the authors.) (b) $[Fe_2(SEt)_6]^{2-}$. (Reproduced from ref. 13. Copyright 1982 American Chemical Society.)

It is immediately apparent from Scheme 2 that the nature of the product resulting from reaction of $FeCl_2$ with simple thiolates depends crucially on the $RS^-/FeCl_2$ ratio, and that these iron thiolates also react with sulfur to give different Fe-S-thiolate cluster products (13). Thus, high $RS^-/FeCl_2$ ratios ($\geq$ 4:1) favor the tetrahedral $[Fe(SR)_4]^{2-}$, corresponding to reduced rubredoxin, and as the ratio decreases the degree of aggregation increases, until the adamantane-like $[Fe_4(SR)_{10}]^{2-}$ cage is obtained at a 2.5:1 ratio. Decreasing the ratio still further does not result in formation of even larger aggregates, probably due to unfavorable steric interactions between thiolates in any such structures, but instead produces the cyclic trimer **7**. Reaction of the mono- or binuclear thiolates with one equivalent of sulfur per iron results in formation of the binuclear sulfide-bridged species **2**, although the former also produces significant amounts of the linear trinuclear cluster **3**. In contrast, reaction of the adamantane cage with sulfur produces the tetranuclear cluster **6** in an all-or-none fashion under similar reaction conditions (14). Conversion of **2** to **6** occurs smoothly in protic solvents (15), while **3** is converted to **6** by $FeCl_2$ in MeCN (16).

As might be expected, the reaction between a given iron-thiolate complex and sulfur is also sensitive to stoichiometry. Shown in Scheme 3 are the products obtained from reaction of $[Fe(SEt)_4]^{2-}$ with varying amounts of sulfur under various conditions (16). Reaction of the mononuclear tetrathiolate with one equivalent of sulfur in MeCN produces mainly the binuclear complex **2** (R = Et). Increasing the S:Fe ratio to 1.4 and use of acetone optimizes formation of the linear trinuclear species **3** (R = Et). Clusters **2** and **3** are actually metastable species that undergo further oligomerization upon heating in aprotic solvents to produce the tetranuclear cluster **6** and a new hexanuclear cluster $[Fe_6S_9(SR)_2]^{4-}$ (**8**), respectively. Complex **8** can also be obtained by direct reaction at elevated temperatures. Two views of **8** are provided in Figure 3; it contains a single quadruply-bridging sulfide, two triply-bridging sulfides, and four doubly-bridging sulfides that virtually saturate the coordination sites of the tetrahedrally-ligated iron atoms, leaving only two sites available for terminal thiolates.

As a further example of the influence of reaction mixture stoichiometry and of metastable clusters, consider the reaction shown in the top portion of Scheme 4. Reaction of $FeCl_2$ with 1.5 equivalents of PhS^- has been shown previously to give a mixed-ligand adamantane structure, $[Fe_4(SPh)_6Cl_4]^{2-}$, analogous to **5** (17). Reaction of this species in situ with one equivalent of sulfur per iron atom produces different clusters depending upon the nature of the quaternary cation Q^+ (18). For $Q^+ = Ph_4P^+$ or Bu_4N^+, the product is a tetranuclear cluster with terminal chloride ligands, $[Fe_4S_4Cl_4]^{2-}$, previously prepared via a ligand exchange reaction of $[Fe_4S_4(SR)_4]^{2-}$ with acyl chlorides (19). For $Q^+ = Et_4N^+$, however, a novel hexanuclear product is obtained with anion stoichiometry $[Fe_6S_6Cl_6]^{3-}$ (18). A view of the structure of the bromo analog of this hexagonal prismane cluster (**9**) is given in Figure 4. It seems clear that the major influence of the Et_4N^+ ion is likely to be in crystal packing energetics that fortui-

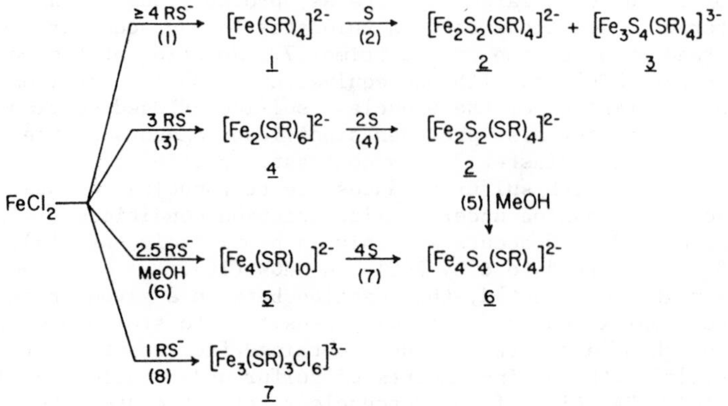

$FeCl_2 \xrightarrow[(1)]{\geq 4\ RS^-} [Fe(SR)_4]^{2-}$ (**1**) $\xrightarrow[(2)]{S} [Fe_2S_2(SR)_4]^{2-}$ (**2**) $+ [Fe_3S_4(SR)_4]^{3-}$ (**3**)

$FeCl_2 \xrightarrow[(3)]{3\ RS^-} [Fe_2(SR)_6]^{2-}$ (**4**) $\xrightarrow[(4)]{2S} [Fe_2S_2(SR)_4]^{2-}$ (**2**)

$[Fe_2S_2(SR)_4]^{2-}$ (**2**) $\xrightarrow[(5)]{MeOH} [Fe_4S_4(SR)_4]^{2-}$ (**6**)

$FeCl_2 \xrightarrow[MeOH\ (6)]{2.5\ RS^-} [Fe_4(SR)_{10}]^{2-}$ (**5**) $\xrightarrow[(7)]{4S} [Fe_4S_4(SR)_4]^{2-}$ (**6**)

$FeCl_2 \xrightarrow[(8)]{1\ RS^-} [Fe_3(SR)_3Cl_6]^{3-}$ (**7**)

Scheme 2

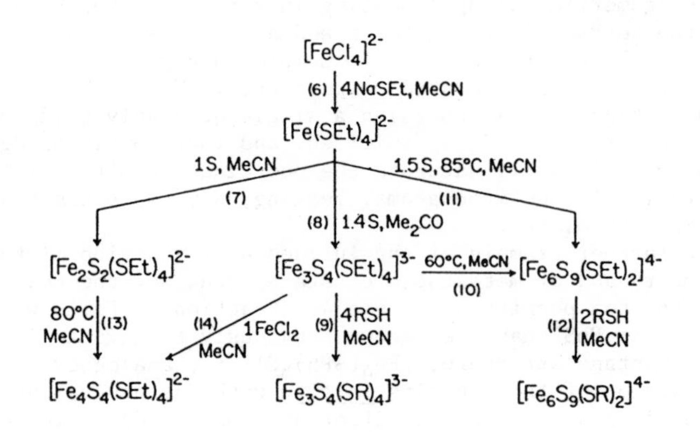

Scheme 3

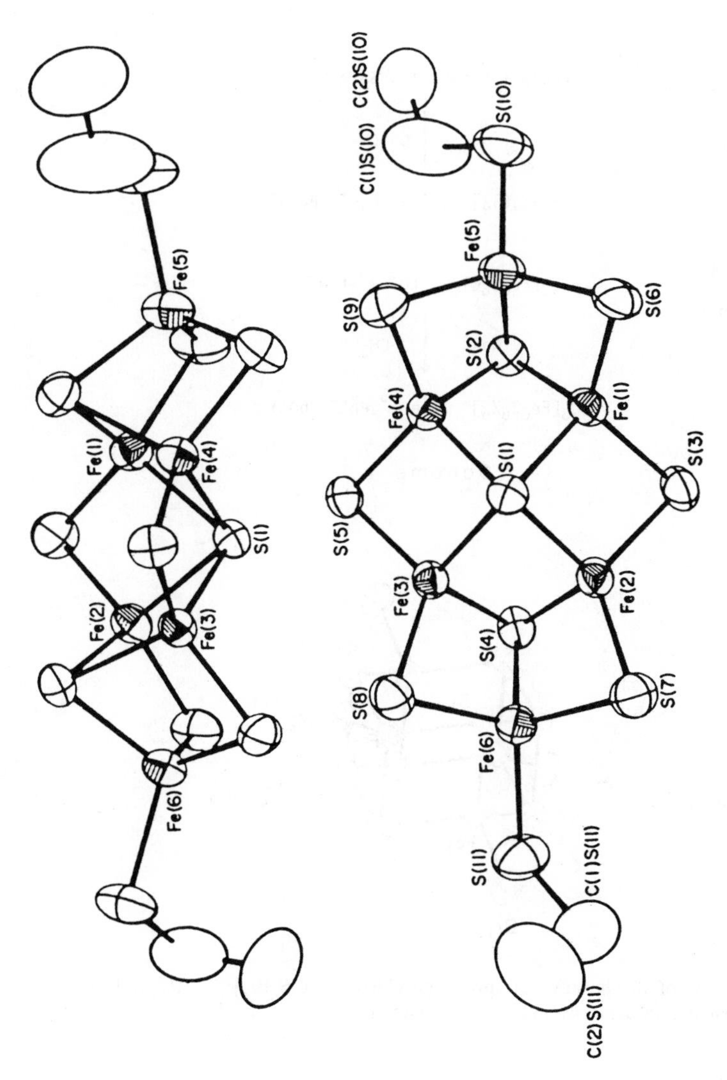

Figure 3. Two views of the structure of the hexanuclear cluster, $[Fe_6S_9(SEt)_2]^{4-}$. (Reproduced from ref. 16. Copyright 1983 American Chemical Society.)

$$4\ FeCl_2 + 6NaSPh + \tfrac{1}{2}\ S_8 + 2Q^+Cl^-$$

$$\downarrow\ CH_3CN$$

$$Q_2[Fe_4S_4Cl_4]\quad (Q^+ = Ph_4P^+,\ Bu_4N^+)$$

or

$$Q_3[Fe_6S_6Cl_6]\quad (Q^+ = Et_4N^+)$$

$$NaX\ \downarrow\ CH_3CN$$

$$Q_3[Fe_6S_6X_6]\quad (X = I^-, PhS^-, PhO^-)$$

Scheme 4

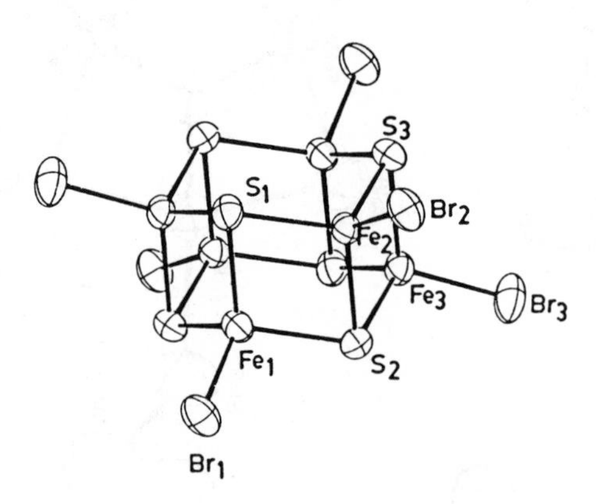

Figure 4. A view of the hexanuclear prismane cluster, $[Fe_6S_6Br_6]^{3-}$. (Reproduced from ref. 20. Copyright 1986 American Chemical Society.)

tously stabilize the hexanuclear cluster, but its presence in solution during the reaction with sulfur is also crucial. If Et_4NCl is added after S_8, only the Et_4N^+ salt of $[Fe_4S_4Cl_4]^{2-}$ is obtained. The chloride hexamer is metastable, converting to the tetramer slowly upon standing in solution or rapidly upon being heated (18). It is, however, sufficiently stable to allow exchange of terminal ligands (18, 20), as shown in the bottom portion of Scheme 4. The thiolate-ligated hexamers are even less stable than the halide derivatives, but phenoxide ligands appear to afford substantial stabilization of the prismane core. This is also illustrated by the direct formation of the phenoxide analogs of **9** from $[Fe_4S_4(SR)_4]^{2-}$ salts and excess phenols, in a ligand-induced rearrangement to the hexameric species (21).

Other Fe-S Clusters. Spontaneous self-assembly reactions using phosphines as terminal ligands have been reported to yield a variety of new Co-S (22) and Ni-S clusters (23). Extension to Fe-S clusters was initially effected by the following reaction in heterogeneous solution (24).

$$[Fe(H_2O)_6](BF_4)_2 + 3\ PEt_3 \xrightarrow[CH_2Cl_2/H_2O]{H_2S} \underset{\mathbf{10}}{[Fe_6S_8(PEt_3)_6]^{2+}}$$

The structure of **10**, isolated as its BPh_4^- salt, is shown in Figure 5a; it consists of a regular octahedron of Fe atoms with each triangular face capped by a triply-bridging sulfide (24). A similar reaction using $(Me_3Si)_2S$ rather than H_2S as a sulfide source has recently been reported (25) to produce a related cluster, the $Fe_7S_6(PEt_3)_4Cl_3$ ion (**11**), whose structure is shown in Figure 5b. It is derived from a distorted Fe_8 cube with quadruply-bridging sulfides on each face by simple removal of an Fe atom from one vertex. The parent cubic Fe_8 cluster has also been prepared as the $[Fe_8S_6I_8]^{3-}$ ion (**12**), but not from a self-assembly reaction. Instead, it was reportedly obtained by capping the open faces of the $[Fe_6S_6I_6]^{2-}$ ion, which has a structure very similar to that shown for **9**, using Fe metal, I^-, and I_2 (26); its structure is shown in Figure 5c.

The biological relevance of such clusters remains unclear. The stability of the closed Fe_6 and Fe_8 clusters and the Fe_7 fragment suggest that it would not be surprising to encounter one or more of these in a biological milieu, but to date there is no evidence for the occurrence of simple Fe-S clusters of nuclearity higher than 4 in any biological system. Their utility may well prove to be as conceptual models and possible starting materials for more complex heterometallic systems such as the FeMo-cofactor of nitrogenase.

Three-iron clusters. Despite the availability of an X-ray crystallographic structure determination of *Azotobacter vinelandii* ferredoxin I (27), which contains both a normal 4Fe-4S cluster and a 3Fe cluster, the fundamental properties and even the structures

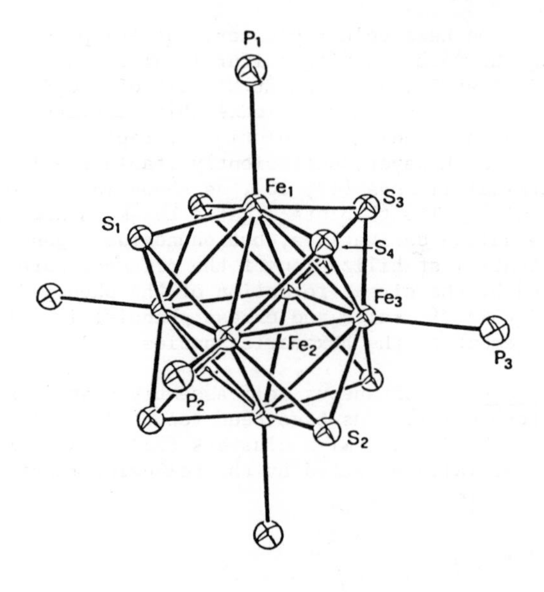

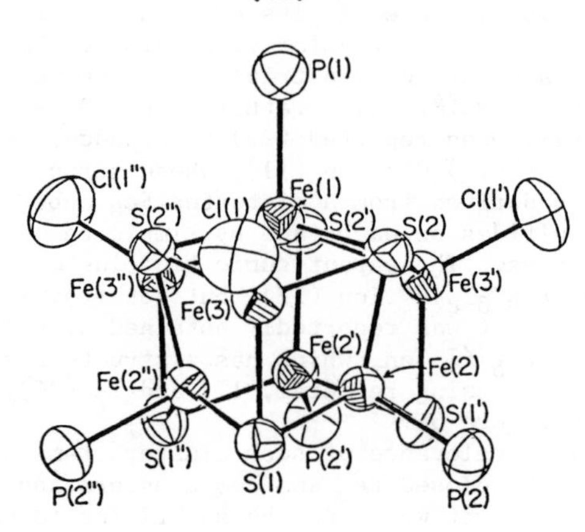

Figure 5. (a) Structure of $[Fe_6S_8(PEt_3)_6]^{2+}$. (Reproduced from ref. 24. Copyright 1985 American Chemical Society.) (b) Structure of $Fe_7S_6(PEt_3)_4Cl_3$. (Reproduced from ref. 25. Copyright 1986 American Chemical Society.)

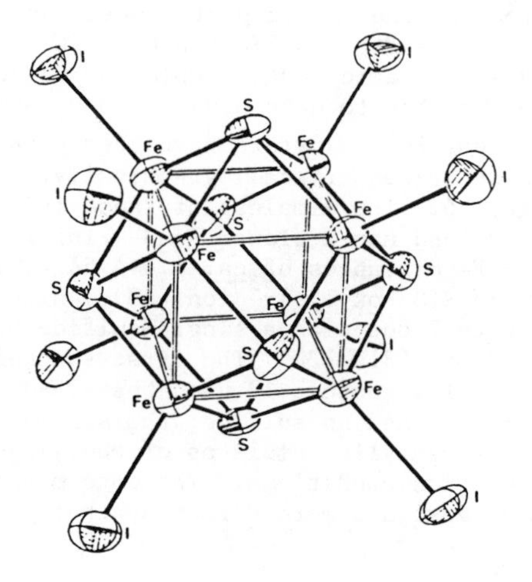

$Fe_8S_6I_8^{3-}$

(c)

Figure 5. *Continued.* (c) Structure of $Fe_8S_6I_8^{3-}$. (Reproduced from ref. 26. Copyright 1984 the authors.)

of such units in other proteins remain controversial. This is due to the fact that spectroscopic measurements indicate a very high degree of congruence between 3Fe centers in all proteins examined while structural studies indicate substantial differences. Thus, the X-ray structure shows a cyclohexane chair-like conformation with Fe--Fe distances in excess of 4Å (27), while EXAFS studies on Desulfovibrio gigas ferredoxin (29) and aconitase (30) indicate short Fe--Fe distances of ca. 2.7Å, consistent with a structure closely related to that of an Fe_4S_4 cluster. In all cases, however, spectral and magnetic data are consistent with an all-ferric description for the oxidized 3Fe center and Fe^{3+}-Fe^{3+}-Fe^{2+} oxidation states in the reduced 3Fe center. It would obviously be valuable to have a synthetic model complex of related structure to serve as a benchmark, but to date none has been prepared.

Three trinuclear Fe-S clusters have been prepared by self-assembly and characterized crystallographically; their structures are shown in Figure 6. The simplest of these is cluster **7**, which contains no sulfide and has a planar Fe_3S_3 ring with bridging thiolates and Fe--Fe distances of ca. 4.4-4.5Å (13). The second of these is the $[Fe_3S(S_2\text{-}\underline{o}xyl)_3]^{2-}$ ion (**13**), which has a somewhat similar structure to **7** but with a single sulfide bridging the triangle of iron atoms (31, 32). The first synthesis reported (31) for a cluster of structure **13** used the 4,5-dimethyl analog of o-xylyldithiol, $FeCl_2$, and no sulfide; the sulfide presumably originated from the benzylic thiolates of the ligand under the strongly basic reaction conditions. The determination of the structure of **13** suggested a more direct synthetic route, shown below.

$$3\ FeCl_2 + 3\ Na_2(S_2\text{-}\underline{o}\text{-xyl}) + Li_2S \xrightarrow[\text{MeCN/EtOH}]{} [Fe_3S(S_2\text{-}\underline{o}\text{-xyl})_3]^{2-}$$

A third type of trinuclear cluster duplicates the Fe_3S_4 stoichiometry that is most accepted for protein-bound 3Fe clusters and in fact appears to be a structural isomer of the 3Fe cluster present in aconitase at least. The $[Fe_3S_4(SR)_4]^{3-}$ ion (**3**) is obtained by reaction of $[Fe(SR)_4]^{2-}$ with 1.4 equivalents of sulfur in MeCN, as indicated above (13, 16). It contains a central Fe(III) ligated tetrahedrally by four sulfides that connect it to two terminal $Fe(III)(SR)_2$ units. The spectroscopic and magnetic properties of **3** are virtually identical with those of the 3Fe form of aconitase at high pH, but substantially different from those of aconitase under normal physiological conditions (23). Possible structures for the latter that are consistent with available data are shown in Figure 7.

Since a common route to 3Fe derivatives of proteins containing 4Fe-4S clusters is via oxidation (34), this implies that it might be possible to prepare a synthetic 3Fe-4S cluster by simple oxidation of an $[Fe_4S_4L_4]^{2-}$ cluster. A recent attempt to implement this strategy examined oxidation of $[Fe_4S_4(S\text{-}\underline{t}\text{-Bu})_4]^{2-}$ by ferricyanide in DMF/H_2O solution at $T < -40°C$ (35). Although a species with EPR and Mössbauer parameters similar to those of protein-bound 3Fe centers was generated (in $\leq$ 30% yield), it was

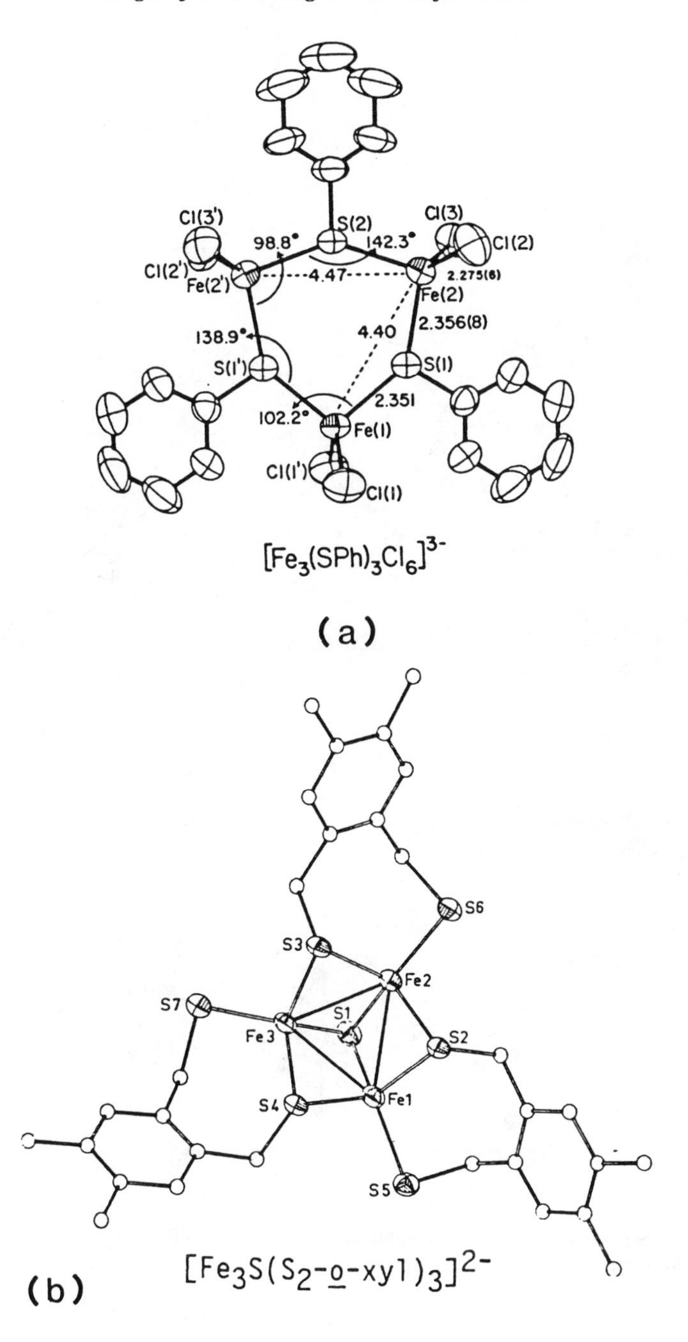

Figure 6. Views of known trinuclear Fe-S clusters. (a) $[Fe_3(SPh)_3Cl_6]^{3-}$. (Reproduced from ref. 13. Copyright 1982 American Chemical Society.) (b) $[Fe_3S(S_2\text{-}4,5\text{-}Me_2\text{-}o\text{-}xyl)_3]^{2-}$ (Reproduced from ref. 31. Copyright 1981 the authors.) *Continued on next page.*

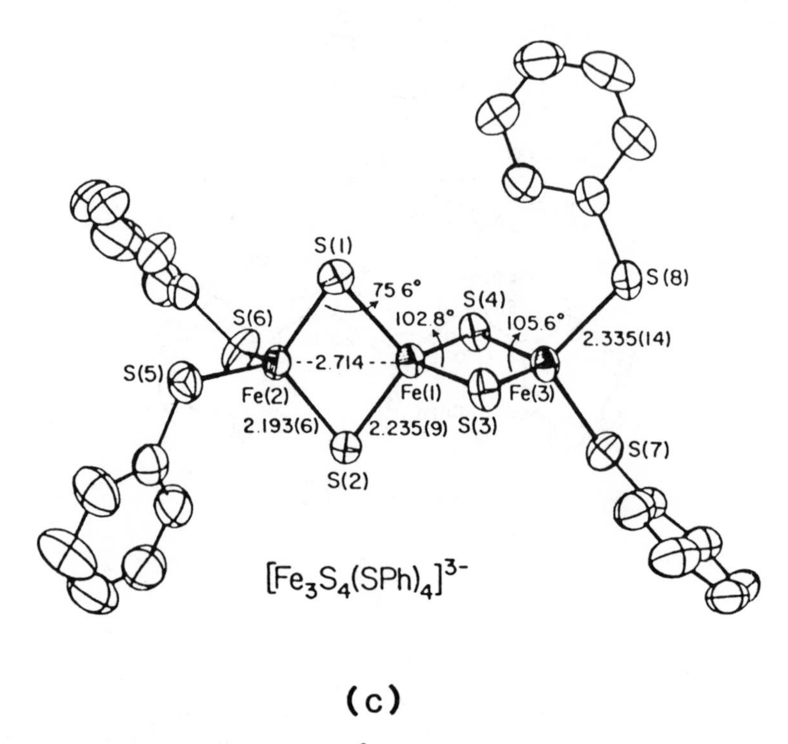

Figure 6. *Continued.* (c) $[Fe_3S_4(SPh)_4]^{3-}$. (Reproduced from ref. 13. Copyright 1982 American Chemical Society.)

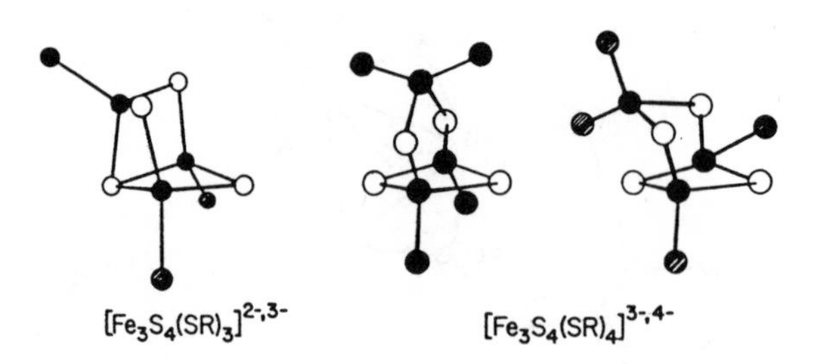

Figure 7. Schematic drawings of possible structures for 3Fe-4S clusters in proteins. (Data taken from ref. 32.)

unstable above -40°C and could not be purified or isolated (35). These results and the lack of success in direct syntheses suggest that a 3Fe-4S cluster with a structure similar to those shown in Figure 7 may be a relatively high energy structure that is stabilized by a particular arrangement of ligating groups within the protein, which may or may not be possible to simulate in a more complex organic ligand.

Use of Bulky Ligands to Stabilize Synthetic Analogs in High Oxidation States

Despite electrochemical results indicating that oxidation of the $[Fe_4S_4(SR)_4]^{2-}$ to the monoanion (isoelectronic with oxidized HiPIP) is quasireversible with certain thiolates (e.g., t-BuS^-) (11), efforts to obtain the monoanion in crystalline form were unsuccessful until recently. Similarly, the $[Fe(S_2\text{-}\underline{o}\text{-xyl})_2]^{1-,2-}$ ions were crystallized in 1976 (36), but attempts to prepare unconstrained analogs of the oxidized rubredoxin site, $[Fe(SR)_4]^{1-}$, were also unsuccessful until recently. One reason for the lack of success with the tetranuclear centers is clearly the tendency of the oxidized cluster to lose iron, producing an (unstable) 3Fe cluster (35). This lack of stability of the higher oxidation states of analogue clusters is a reflection of a general lack of stability of thiolate complexes of relatively oxidizing metals. One way of overcoming this problem is by use of sterically hindered thiolates, which apparently stabilize higher oxidation state metal complexes via both kinetic and thermodynamic reasons (e.g., destabilization of the disulfide oxidation product). This strategy has been applied successfully to the synthesis of both oxidized HiPIP and oxidized rubredoxin models by Millar, Koch, and coworkers.

Thus, oxidation of the readily prepared $[Fe_4S_4(SR)_4]^{2-}$ complexes with the sterically hindered thiolates 2,3,5,6-tetramethylthiophenolate and 2,4,6-triisopropylthiolphenolate by ferricinium ion proceeds with smooth formation of the corresponding monoanion (37), isolated as the crystalline Bu_4N^+ salt for SR^- = 2,4,6-i-$Pr_3C_6H_2S^-$; its structure is shown in Figure 8a.

Analogs of oxidized rubredoxin, $[Fe(SR)_4]^-$, can be prepared via either of two strategies that employ steric hindrance. The first is by simple reaction of $FeCl_3$ with 4 equiv. of bulky thiolate such as 2,3,4 6-tetramethylthiophenolate (38). This produces stable crystalline materials in high yields; the structure of one such anion is shown in Figure 8b. An alternative strategy suggests that the difficulties in direct synthesis of $[Fe(SR)_4]^-$ ions may be kinetic rather than thermodynamic in nature. Reactions of sterically hindered phenoxide complexes, $[Fe(OR)_4]^-$, with simple alkyl and aryl thiols produces the corresponding $[Fe(SR)_4]^-$ complexes in high yield; these appear to be quite stable even for R groups as simple as R = Et (39).

Syntheses Via Ligand Exchange Reactions

Terminal ligand exchange reactions are by now well established for the synthetic analogue clusters **2** and **6** (1). The key characteristics of such reactions are the following. (i) Reaction of a cluster containing a given ligand L^- with a second ligand HL' is driven by protonation of the most basic ligand present, and results in coordination of the conjugate base of the most acidic ligand to the Fe-S core. (ii) Ligand substitution can be affected with a variety of electrophiles EL, where E = H, acyl or sulfonyl and L = halide, carboxylate, sulfonate, thiolate, or phenolate. (iii) A synthetically useful variant is to utilize the chloride-ligated species, e. g., $[Fe_2S_2Cl_4]^{2-}$ or $[Fe_4S_4Cl_4]^{2-}$, and the sodium salt of a given anionic ligand, with precipitation of NaCl driving the reaction to completion. (iv) Multiple substitution of ligands on a given Fe-S core occurs in a stepwise fashion, with essentially statistical values of the equilibrium constants for formation of the mixed-ligand species in all systems examined to date. Application of these concepts to two specific problems is described below.

The Rieske Protein. The Rieske Fe-S protein isolated is an essential component of the electron transport chain of mammalian mitochondria and organisms such as *Thermus thermophilus*; it contains two 2Fe-2S centers with atypical spectroscopic and redox properties (40,41) (e.g., E_o' of +150mV vs. -400mV for typical plant ferredoxins). A variety of studies have shown that each 2Fe-2S cluster contains only two cysteinyl thiolate ligands, with the two remaining terminal ligand positions occupied by nitrogen, possibly from histidine imidazoles (42) to form a structure similar to that shown schematically in Figure 9.

Efforts toward developing synthetic models for the Rieske Fe-S centers focussed initially on preparing Fe_2S_2 cores with non-thiolate ligands, and have centered on nitrogenous ligands since the realization of their probable occurrence in the Rieske protein. In addition to the $[Fe_2S_2(OAr)_4]^{2-}$ ions (Ar = aryl) (43), a variety of binuclear Fe-S clusters with non-thiolate ligation have been prepared and are shown in Figure 10. All were prepared by reaction of the disodium salts of the ligands LL' with $[Fe_2S_2Cl_4]^{2-}$. Ligand donor atom sets include O_4 (43,44), N_4 (44,45), S_2O_2 (46), S_2N_2 (45), and O_2N_2 (45). To date, however, none have achieved the C_{2v} structure shown in Figure 9, with the two nitrogenous ligands on the same Fe as indicated by a recent resonance Raman study (47). In view of the tendency to obtain statistical mixtures of all possible mixed ligand Fe-S clusters, the challenge posed by an accurate synthetic model of the Rieske center appears formidable indeed.

The P clusters of nitrogenase. The enzyme nitrogenase consists of two proteins: the Fe protein (m.w. ≈ 55,000), which contains a single 4Fe-4S center, and the more complex MoFe protein (m.w. 220,000) (48,49). The minimum functional unit of the latter appears to be the half molecule, an asymmetric dimer containing 1 Mo, 14-16 Fe, and 16-18 sulfides. Application of a vast array of spectroscopic methods to the MoFe protein in a variety of oxidation states has led to the conclusion that it contains two types of metal-sulfur cluster in a 2:1 ratio: unusual Fe_4S_4 units termed P clusters, and the protein-bound form of the FeMo-cofactor (50).

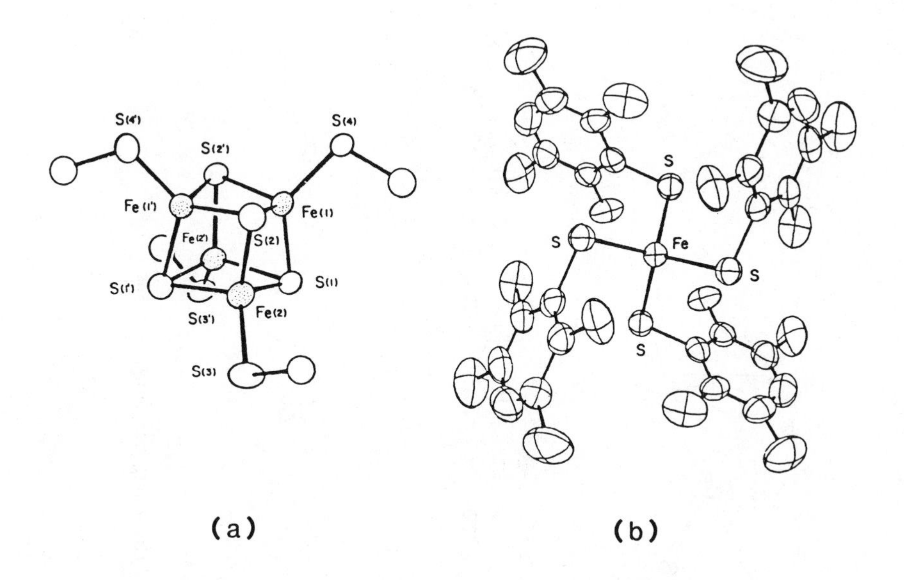

Figure 8. (a) Structure of $[Fe_4S_4(S\text{-}2,4,6\text{-}i\text{-}Pr_3C_6H_2)_4]^-$. (Reproduced from ref. 37. Copyright 1985 American Chemical Society.) (b) Structure of $[Fe(S\text{-}2,3,5,6\text{-}Me_4C_6H)_4]^-$. (Reproduced from ref. 38. Copyright 1982 American Chemical Society.)

(Cys)S S Im(His)
Fe Fe
(Cys)S S Im(His)

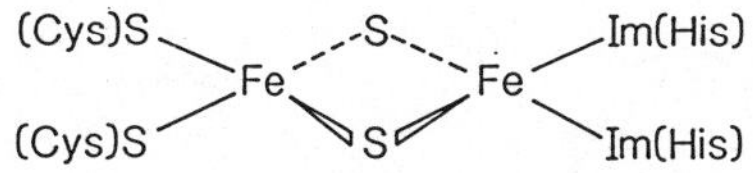

Figure 9. A schematic view of the probable structure of the Rieske Fe-S center (based on the data in ref. 47.)

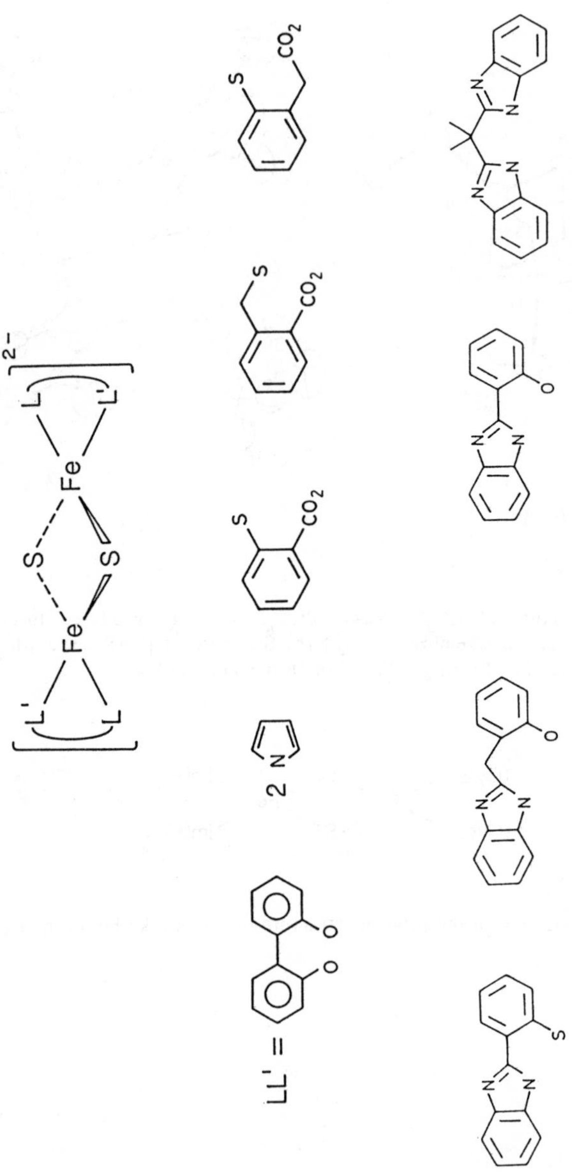

Figure 10. Schematic drawings of 2Fe-2S clusters containing nonthiolate ligands. (Data taken from refs. 44–46.)

Discussion of the properties of the latter and synthetic models thereof is deferred until the next section.

The P clusters are defined by their unique Mössbauer properties. In the resting state, they each contain a unique site with spectroscopic properties typical of high-spin Fe^{2+} (hence designated site "Fe^{2+}") in a tetrahedral sulfur environment ($\delta = 0.69$ mm/s, $\Delta E_Q = 3.02$ mm/s). Each "Fe^{2+}" iron is spin coupled to a set of three similar Fe atoms with average parameters $\delta = 0.64$ mm/s, $\Delta E_Q = 0.81$ mm/s (designated sites D) to give a net $S = 0$ ground state. Upon oxidation of the MoFe-protein with artificial electron acceptors, the spectra of both the D and Fe^{2+} sites are converted in parallel to a complex magnetic spectrum typical of an overall $S \geq 5/2$ ground state (48,49); the high ground state spin is supported by MCD magnetization data (49).

The generally accepted, but not unique, interpretation of the available data is that two very similar P clusters exist per half molecule of the MoFe protein, and that each consists of an unusual variant of an Fe_4S_4 cluster in which three Fe atoms are somehow distinguished from the fourth. Possible explanations of the data have focussed on the presence of non-thiolate ligands in place of or in addition to cysteine thiolates (56). This hypothesis is supported by the relative paucity of conserved cysteinyl residues in the polypeptides (eight per asymmetric dimer) (52); there simply are not enough cysteines to ligate all of the iron atoms in the FeMo-cofactor and the P clusters. A major unresolved question concerns the oxidation state of the Fe_4S_4 core in the P clusters. The two possibilities that lead to an $S = 0$ ground state are the $[4Fe\text{-}4S]^{2+}$ state familiar from oxidized ferredoxins and the tetranuclear synthetic analogues, and the $[4Fe\text{-}4S]^{0}$ state, which has only been detected as a transient in electrochemistry of the synthetic analogs and for which there is no protein-bound precedent. Arguments in favor of the latter include the high average isomer shift of the iron atoms, typical of ferrous iron, and the observation of transient $g = 1.94$ signals indicative of the $[4Fe\text{-}4S]^{+}$ state during oxidation of the MoFe protein (53,54). Arguments in favor of the former include the very low, assuredly non-physiological potential required for reduction of the synthetic $[Fe_4S_4(SR)_4]^{3-}$ species to the tetranion (ca. 700 mV more negative than for the $[4Fe\text{-}4S]^{2+/1+}$ process! (11)) and the fact that increased coordination number at a vertex of $[4Fe\text{-}4S]^{2+}$ core leads to a significant increase in isomer shift (55,56), suggesting a potential explanation for the observed properties of the P clusters. It is clear that the problem is eminently suitable for attack via the use of "speculative models".

One of the hypotheses initially advanced for the P clusters (51) was the simple replacement of three thiolate ligands by more electronegative O- or N-donors. Although the lability of ligands on the $[4Fe\text{-}4S]^{2+}$ core makes it impossible to prepare pure samples of such mixed-ligand clusters for study in solution, replacement of thiolates by carboxylate (57) and phenolate (58) ligands has been examined, and the structures of both fully and doubly substituted phenoxide-ligated $[4Fe\text{-}4S]^{2+}$ clusters determined (58,59). It is clear from the observed properties of these materials that simple ligand substitution of this sort is not adequate to account for the spectroscopic and redox properties of the P clusters.

As alluded to above, an alternative hypothesis has been the presence of additional, non-thiolate ligands to three of the Fe atoms of a $[4Fe\text{-}4S]^{2+}$ core. Again, no such models are available due to the lability of the Fe_4S_4 core. Two examples of clusters with 5-coordinate Fe sites are, however, known in the solid state: $[Fe_4S_4(SPh)_2(Et_2dtc)_2]^{2-}$ (55) and $[Fe_4S_4(SC_6H_4\text{-}2\text{-}OH)_4]^{2-}$(**14**) (56). The structure of the latter is shown in Figure 11, which demonstrates that the <u>ortho</u> -OH of one ligand has coordinated via a long Fe-O bond to afford a 5-coordinate site. (In solution, NMR studies show that the -OH is not coordinated, and all four ligands are equivalent (56).) The Mössbauer parameters of the unique Fe site of **14** show a ca. 0.2 mm/s increase in isomer shift relative to the other three sites.

Detailed exploration of either of the above hypotheses via synthetic model complexes has been precluded by the lability of ligands on the Fe_4S_4 core. Although in a few isolated instances (17,55,59), it has proven possible to obtain mixed ligand clusters, these are due to selective crystallization of a particular complex from a dynamic mixture of all possible mixed ligand species and are difficult to predict in advance. Indeed, such an approach has not yet yielded a cluster with a single thiolate ligand. Consequently, the synthesis of large tridentate ligands has been explored in order to stabilize a 3:1 mixed ligand cluster $[Fe_4S_4L_3L']^{2-}$. Entropic considerations suggest that use of a rigid or semi-rigid tripodal ligand that juxtaposes three donor groups in an appropriate fashion to coordinate to three vertices of an Fe_4S_4 core will shift the solution equilibria to favor the desired mixed ligand species. This concept has been explored via two different organic frameworks.

Our group has prepared the trisubstituted triptycene tris(phenol) ligand **15**, shown in Figure 12a (60). Molecular models indicate that the oxygen donor atoms are within 0.5Å of the optimal position for coordination to three vertices of an Fe_4S_4 core. We have found (61) that reaction of **15** with $[Fe_4S_4(SEt)_4]^{2-}$ proceeds smoothly with replacement of three thiolates. Unfortunately, however, **15** induces rearrangement of the $[4Fe\text{-}4S]^{2+}$ core to the isoelectronic $[6Fe\text{-}6S]^{3+}$ core, as shown by the appearance of new contact-shifted ^1H NMR signals in the range typical of the phenolate-ligated $[6Fe\text{-}6S]^{3+}$ core; the latter also has three-fold symmetry and is most stabilized by phenoxide ligands. Preparation of the tris(thiophenol) analogue of **15** shown in Figure 12b, which should eliminate this problem, is in progress.

A similar approach in Holm's laboratory has utilized the tendency of certain hexasubstituted benzenes to form cavities comprised of substituents disposed in an alternating up and down array (62). Reaction of the symmetrically substituted trithiol benzene derivative **16** ($L\cdot(SH)_3$) (Figure 12c) with $[Fe_4S_4(SEt)_4]^{2-}$, followed by selective displacement of the remaining EtS^- with Cl^- using Me_3COCl, affords the selectively substituted cluster $[Fe_4S_4(L\cdot S_3)Cl]^{2-}$ (**17**), the structure of which is shown in Figure 13 (62). The three thiol-containing arms of the tripod have folded inward, forming a cavity in which the cluster is ligated.

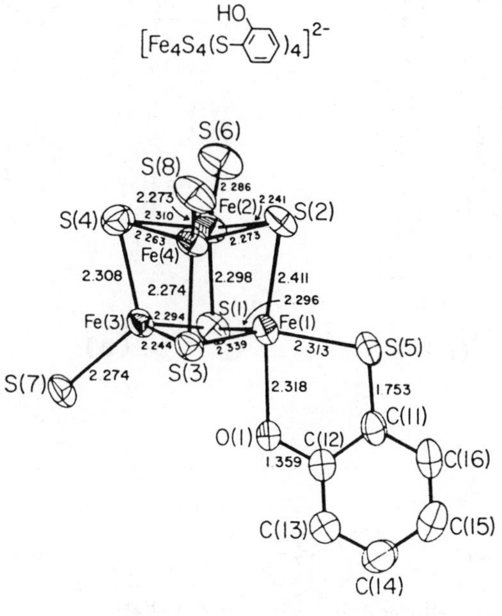

$[Fe_4S_4(S\text{-}C_6H_4\text{-}o\text{-}OH)_4]^{2-}$

Figure 11. Structure of a $[4Fe\text{-}4S]^{2+}$ cluster with one 5-coordinate Fe site. (Reproduced from ref. 56. Copyright 1983 American Chemical Society.)

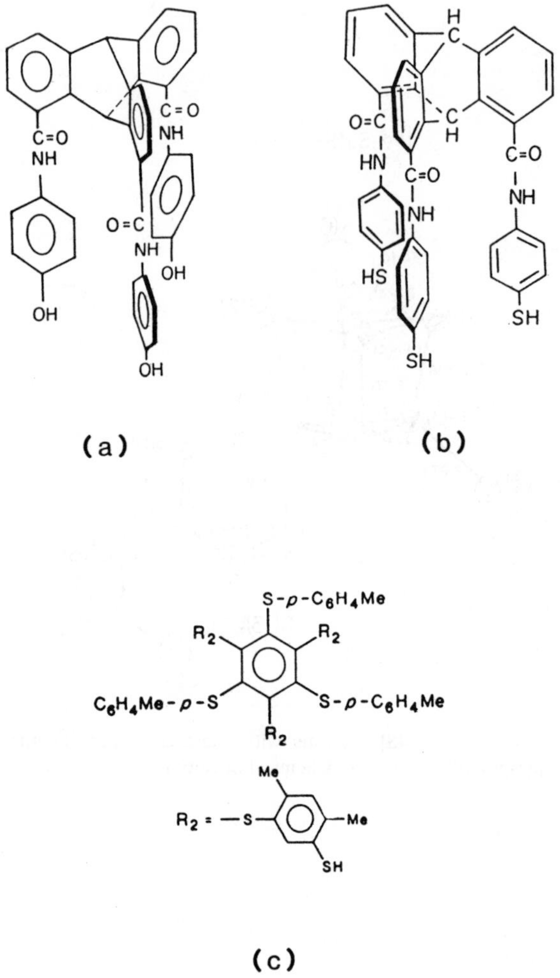

(a)

(b)

(c)

Figure 12. Schematic drawings of macrotridentate ligands capable of coordinating to three vertices of a 4Fe-4S core. [(a) reproduced from ref. 60. Copyright 1986 American Chemical Society; (c) reproduced from ref. 62. Copyright 1987 American Chemical Society.]

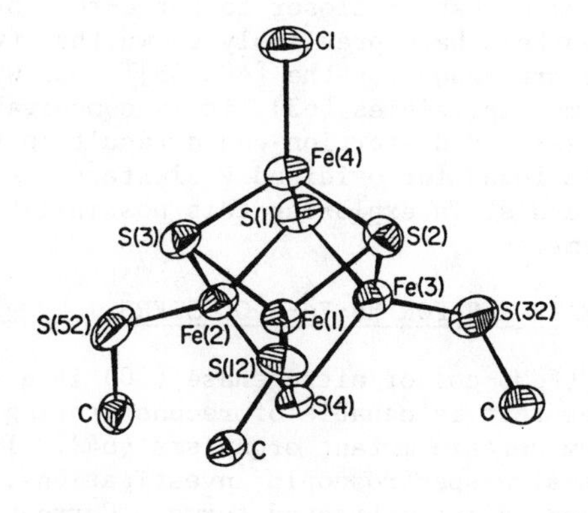

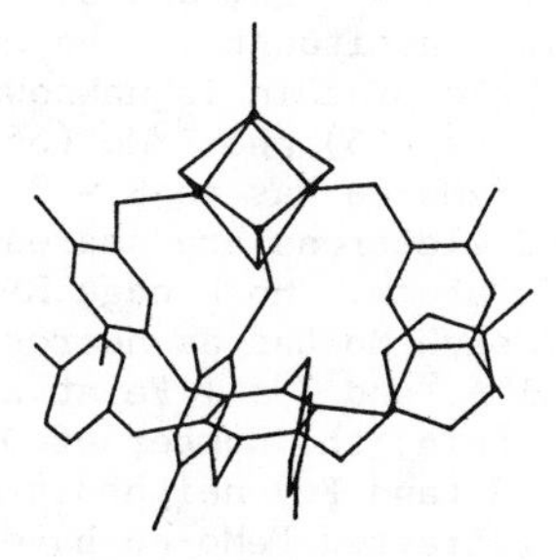

Figure 13. Structure of a 4Fe-4S cluster in which selective substitution of one Fe site has been achieved by use of a macrotridentate ligand. (Reproduced from ref. 62. Copyright 1987 American Chemical Society.)

Selective substitution of ligands ($Cl^- \rightleftarrows RS^-$) at the unique Fe has also been demonstrated (62).

Although **17** is not per se a reasonable model for the P clusters, it does demonstrate for the first time that it is possible via rational means to prepare clusters in which three of the Fe sites are distinguished from the fourth. Elaboration of **16** (and possibly **15**) may well allow one to explore systematically the hypotheses put forward above. In addition, it is found that ligand **16** abolishes the normal D_{2d} distortion of a $[4Fe\text{-}4S]^{2+}$ core (62), producing a core that is closer to three-fold symmetry. Since Holm and coworkers have previously shown that two different types of structure are found for the $[4Fe\text{-}4S]^+$ core with different ground state spin multiplicities (63), it is conceivable that an imposed pseudo-three-fold distortion could result in the S = 5/2 or 7/2 ground state found for oxidized P clusters. Ligands such as **15** and **16** will assist in exploring this possibility in synthetic model systems.

III. SYNTHETIC STRATEGIES FOR Mo-Fe-S CLUSTERS

The FeMo-cofactor (FeMo-co) of nitrogenase (50) is a small extractable cluster that is capable of reconstituting inactive enzyme derived from certain mutant organisms (64). It has been the object of extensive spectroscopic investigations, both within the MoFe-protein and in its extracted forms. Current wisdom is that FeMo-co contains 1 Mo, 6-7 Fe, and 8-9 S (48,50); there appear to be no organic constituents. The means by which FeMo-co is coordinated to the MoFe protein is unknown. EPR (48,50), ^{57}Fe Mössbauer (48,50) and ^{57}Fe (65) and ^{95}Mo (66) ENDOR spectroscopy have demonstrated that FeMo-co has an S = 3/2 ground state and that the three unpaired electrons are shared by the Mo and at least 5 (probably 6) Fe atoms. Mo K-edge EXAFS studies of FeMo-co have demonstrated that each Mo has as nearest neighbors 4-5 S at 2.35 Å, 1-2 O(N) at 2.1 Å, and 3 ± 1 Fe at 2.7Å (67,68); upon extraction from the protein, the number of O(N) ligands increases to 3 and the number of S (and Fe) neighbors decreases (68). Fe K-edge EXAFS studies of extracted FeMo-co have shown that each Fe has on the average 3-4 S at 2.25Å, 2-3 Fe at 2.7Å, 0.4 Mo at 2.76Å, and 1 O(N) at 1.8Å as nearest neighbors (69,70). In addition, evidence has been presented for a second set of Fe neighbors at 3.75Å (70). Taken together, these results indicate that FeMo-co is a novel Mo-Fe-S cluster, with 2-3 iron atoms linked to Mo via bridging sulfur atoms and the rest of the iron atoms being more distant. Ligation by oxygen or nitrogen donors to both Fe and Mo is likely.

Spontaneous Self-assembly Reactions

Synthetic efforts at preparing model complexes for the FeMo-cofactor have largely focussed on two types of Mo-Fe-S cluster, both of which are prepared via self-assembly reactions using tetrathiomolybdate as starting material. The first of these is the "linear" type of cluster, containing the MoS_2Fe unit formed by coordination of discrete $MoS_4{}^{2-}$ units to Fe. The second is the "cubane" type of cluster, containing the $MoFe_3S_4$ unit and formed from MoS_4^{2-} by an induced redox reaction. Both are prepared from iron salts or complexes and MoS_4^{2-}, with the presence of excess thiolate in the reaction mixture favoring the cubane products. The latter are generally obtained as dimeric "double cubanes", which must be cleaved to produce the single cubane clusters. The preparation and properties of both linear (51,71,72) and cubane (51,71) Mo-Fe-S clusters have been reviewed in detail, and will not be discussed further here.

Although both types of cluster exhibit interesting chemistry in their own right, neither provides an accurate model for FeMo-co. A brief analysis of their merits and shortcomings in this regard is worthwhile. First and most obviously, neither possesses the correct Fe/Mo ratio. Although it is possible in principle to design clusters of these types containing approximately the correct ratio of components, in fact this has not been achieved synthetically. Second, while the cubane clusters contain Mo in a reasonable oxidation state (ca. Mo^{3+}-Mo^{4+}) and hence have Mo-S distances comparable to those observed by EXAFS in FeMo-co, the linear compounds contain Mo in a relatively high oxidation state (Mo^{6+}-Mo^{5+}), which is reflected in Mo-S distances that are ca. 0.1Å shorter. Third, the linear clusters exhibit no reaction chemistry with small molecules at Mo, but such chemistry is well-established for the single cubane $MoFe_3S_4$ clusters (71). It is clear that new strategies to produce synthetic Mo-Fe-S clusters with proper stoichiometry, Mo oxidation state, and reactivity are necessary; two such approaches are described below.

Designed Syntheses

Despite a great deal of work (mostly unpublished) aimed at generating more accurate FeMo-co models based on the linear or cubane structures, none have been forthcoming. This has prompted us to analyze these systems to determine what factors are limiting the scope of the synthetic efforts. Our conclusion is that the use of tetrathiomolybdate is largely responsible. Despite many attempts at reducing the linear clusters to generate new species in which the Mo site is more reduced and hence expected to have an increased affinity for small molecules, all such efforts produce either known $MoFe_3S_4$ cubane derivatives or $[Fe(S_2MoS_2)_2]^{3-}$ (73). Consequently, we and others have begun to explore new synthetic routes utilizing starting materials other than MoS_4^{2-}. The strategy is shown in Scheme 5. Instead of forming the Mo-S-Fe unit via nucleophilic attack of a soluble $Mo\text{-}S^{=}$ species (of which there are few examples) on an iron halide or related species, we and others have chosen to explore the reverse strategy, in which the Mo-S-Fe unit is formed by nucleophilic attack of a soluble $Fe\text{-}S^{=}$ species on molybdenum. As we shall see, this approach allows one to build Mo-Fe-S clusters containing specific structural features.

One example of this strategy is the synthesis of the $Mo(CO)_3$-capped $[Fe_6S_6L_6]^{n-}$ clusters described in detail by Coucouvanis elsewhere in this volume (74). In this approach, the triply-bridging sulfides of an $[Fe_6S_6L_6]^{3-}$ prismane act as a tridentate ligand to $Mo(CO)_3$ fragments, forming $[Fe_6S_6L_6(Mo(CO)_3)_2]^{n-}$ (n = 3,4) clusters. To date, it has not proven possible to isolate a cluster in which only one of the trigonal faces is capped by an $Mo(CO)_3$ fragment, although such species are in equilibrium with the dicapped cluster in solution (74). This system represents one of the very few Mo-Fe-S clusters in which the detailed structure of the product was anticipated prior to its synthesis. If the problems associated with forming a monocapped species and subsequent oxidative decarbonylation at Mo can be overcome, these complexes will indeed be the closest synthetic representations to FeMo-co yet achieved.

Our group is exploring an alternative synthetic strategy based on the chemistry of $Fe_2S_2(CO)_6$ (**18**). Seyferth has demonstrated that the bridging S_2^{2-} unit in **18** can be cleaved reductively to form the $[Fe_2S_2(CO)_6]^{2-}$ dianion (**19**) (75), which exhibits reactivity typical of an organic dithiolate. In particular, reaction of **19** with MCl_4 (M = Ge, Sn) produces the corresponding $M(S_2Fe_2(CO)_6)_2$ clusters (76), suggesting that similar reaction with Mo halides should produce clusters containing the novel triangular MoS_2Fe_2 unit. This expectation has been borne out by the synthesis of the $[MoOFe_5S_6(CO)_{12}]^{2-}$ ion (**20**) from $MoCl_5$ or $MoOCl_3{\cdot}2THF$ and **19** (77); the structure of **20** is shown in Figure 14. In addition to the expected MoS_2Fe_2 unit, cluster **20** also contains two unexpected structural features: an Mo=O unit arising (in the case of $MoCl_5$) from oxygen atom abstraction from the THF solvent, and an Fe_3S_2 cluster containing two triply bridging sulfides and linked via one Fe atom to the Mo by two doubly bridging sulfides. The latter structural feature has as its only precedent the $[MoFe_3S_6(CO)_6]^{2-}$ ion, itself prepared by reaction of **19** with the linear $[S_2MoS_2FeCl_2]^{2-}$ anion (78). Thus, although the desired MoS_2Fe_2 structural feature has been achieved in **20**, unexpected fragmentation and rearrangement reactions of the Fe_2S_2 core of **19** are also occurring. This is reflected in the isolation of several new Fe-S-CO clusters, such as the $[Fe_6S_6(CO)_{12}]^{2-}$, $[Fe_5S_4(CO)_{12}]^{2-}$, and $[Fe_4S_4(CO)_{12}]^{2-}$ ions, from these and related reaction mixtures (77,79). Cluster **20** may thus be viewed as the product of a combination of the directed synthesis and spontaneous self assembly approaches.

The synthesis of compounds such as **20** in effect uses CO as a protecting group to solubilize the $Fe_2S_2^{2-}$ unit. Because CO is notably absent from the list of constituents of FeMo-co and because it stabilizes the bulk of the iron in **20** in the Fe^{I} oxidation state, which is of dubious biological relevance, a successful strategy for synthesis of realistic FeMo-co models must include provision for its removal. (This is analogous to but more difficult than the need to decarbonylate the Mo atoms of the capped hexamers discussed above.) We have now developed methodology for controlled oxidative decarbonylation of Mo-Fe-S-CO and Fe-S-CO clusters, based on the observation (80) that $Fe_2S_2(CO)_6$ itself is converted to the tetranuclear cluster **6** by reaction with thiolate/disulfide mixtures. Use of appropriate nucleophile/oxidant pairs has resulted in the synthesis from $[MoOFe_5S_6(CO)_{12}]^{2-}$ of two new Mo-Fe-S clusters that contain no CO. Structural and physical characterization is in progress. Similarly, the Mo=O unit would appear to preclude reaction of small molecules at Mo.

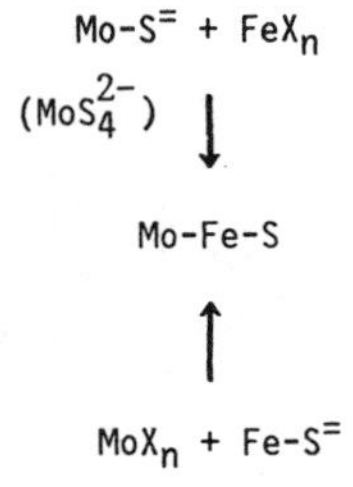

$Mo\text{-}S^{=} + FeX_n$

(MoS_4^{2-}) ↓

Mo-Fe-S

↑

$MoX_n + Fe\text{-}S^{=}$

Scheme 5

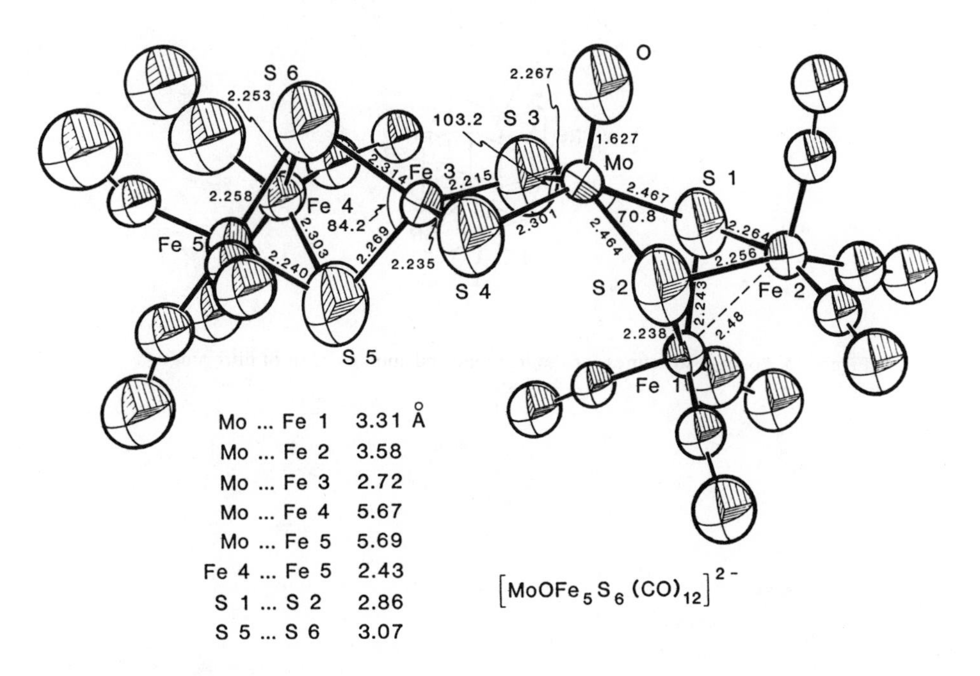

Figure 14. Structure of the $[MoOFe_5S_6(CO)_{12}]^{2-}$ ion. (Reproduced from ref. 77. Copyright 1986 the authors.

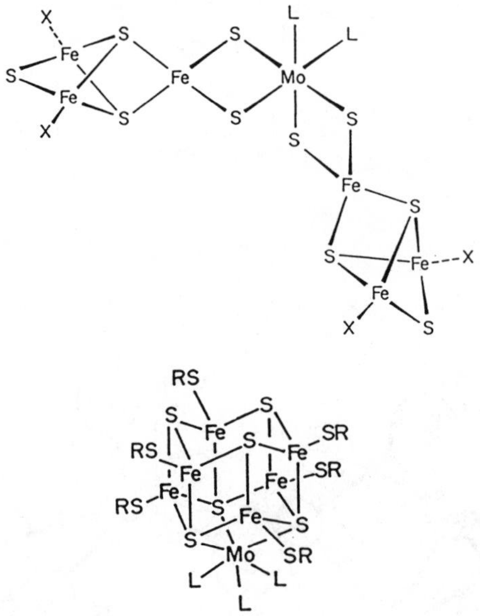

Figure 15. Possible structures for the iron–molybdenum cofactor of nitrogenase.

We are, however, focussing on using such a unit as a means of introducing possible analogs of intermediates in N_2 fixation such as hydrazines, via modification of routes proven for such transformations in simple mononuclear Mo=O complexes (81). In addition, synthetic routes to clusters containing more readily replaceable ligands on Mo (such as CO) are currently being explored.

Conclusions

Despite extensive efforts in many laboratories, an accurate synthetic analogue of the FeMo-cofactor has not yet been prepared. Indeed, the structure of the FeMo-cofactor itself is unknown and the subject of lively debate and investigation. Based on currently available results on both the biological system and known synthetic M-S clusters, it would appear that the two leading candidates for the structure of the FeMo-cofactor are those shown in Figure 15. The top structure (**21**) is a "bent" variant of a linear type of structure that is based on our recent synthetic results (77). It contains two Fe_3 clusters linked to Mo via bridging sulfides. The bottom structure (**22**), a variant of that originally proposed by Holm (82), is based on the structures of the mineral pentlandite, $[Fe_8S_6I_8]^{3-}$ (26), $[Co_8S_6(SPh)_8]^{4-,5-}$ (22) and the $Mo(CO)_3$-capped $[Fe_6S_6L_6]^{n-}$ clusters (74).

Both structures accommodate most of the available data on the FeMo-cofactor in a chemically reasonable fashion. Both are in reasonable agreement with the observed stoichiometry, and both have six Fe atoms that separate naturally into two groups of three as suggested by the ^{57}Fe Mössbauer and ENDOR results. In addition, both structures agree reasonably well with reported Mo and Fe EXAFS data, although the latest work indicating the presence of 3-4 Fe's as nearest neighbors to Mo (68) favors structure **22**. Finally, both structures have available cis coordination sites for reaction with acetylene, which presumably binds in a side-on fashion, and for formation of H_2 from a cis-dihydride upon coordination of N_2, as is suggested from pre-steady state kinetics studies (83). Substantial efforts directed towards the synthesis of clusters related to structures **21** and **22** are in progress in several laboratories. It will be interesting to see whether the synthetic work using speculative models or crystallographic studies of the biological materials will be the first to elucidate the structure of the FeMo-cofactor.

ACKNOWLEDGMENTS

Research in the author's laboratory in this area has been generously supported by the National Science Foundation, the National Institutes of Health, and the U. S. Department of Agriculture's Competitive Research Grants Program over the past ten years.

REFERENCES

1. "Iron-Sulfur Proteins", Spiro, T. G., Ed.; Wiley-Interscience: New York, **1982**.

2. Berg, J. M.; Holm, R. H. In Ref. 1; Ch. 1.
3. Ibers, J. A.; Holm, R. H. Science **1980**, 209, 233.
4. Hill, H. A. O. Chem. Brit. **1976**, 12, 119.
5. Carter, C. W., Jr.; Freer, S. T.; Xuong, Ng. H.; Alden, R. A.; Kraut, J. Cold Spring Harbor Symp. Quant. Biol. **1972**, 36, 381.
6. Sieker, L. C.; Adman, E.; Jensen, L. H. Nature **1972**, 235, 40.
7. Dus, K.; DeKlerk, H.; Sletten, D.; Bartsch, R. G. Biochim. Biophys. Acta **1967**, 140, 291.
8. Stombaugh, N. A.; Sundquist, J. E.; Burris, R. H.; Orme-Johnson, W. H. Biochemistry **1976**, 15, 2633.
9. Herskovitz, T.; Averill, B. A.; Holm, R. H.; Ibers, J. A.; Phillips, W. D.; Weiher, J. F. Proc. Natl. Acad. Sci. U.S.A. **1972**, 69, 2437.
10. Averill, B. A.; Herskovitz, T.; Holm, R. H.; Ibers, J. A. J. Am. Chem. Soc. **1973**, 95, 3523.
11. DePamphilis, B. V.; Averill, B. A.; Herskovitz, T.; Que, L., Jr.; Holm, R. H. J. Am. Chem. Soc. **1974**, 96, 4159.
12. Carter, C. W.; Kraut, J.; Freer, S. T.; Alden, R. A.; Sieker, L. C.; Adman, E.; Jensen, L. H. Proc. Natl. Acad. Sci. U.S.A. **1972**, 68, 3526.
13. Hagen, K. S.; Holm, R. H. J. Am. Chem. Soc. **1982**, 104, 5496.
14. Hagen, K. S.; Reynolds, J. G.; Holm, R. H. J. Am. Chem. Soc. **1981**, 103, 4054.
15. Cambray, J.; Lane, R. W.; Wedd, A. G.; Johnson, R. W.; Holm, R. H. Inorg. Chem. **1977**, 16, 2565.
16. Hagen, K. S.; Watson, A. D.; Holm, R. H. J. Am. Chem. Soc. **1983**, 105, 3905.
17. Coucouvanis, D.; Kanatzidis, M.; Simhon, E.; Baenziger, D. C. J. Am. Chem. Soc. **1982**, 104, 1874.
18. Kanatzidis, M. G.; Hagen, W. R.; Dunham, W. R.; Lester, R. K.; Coucouvanis, D. J. Am. Chem. Soc. **1985**, 107, 953.
19. Wong, G. B.; Bobrik, M. A.; Holm, R. H. Inorg. Chem. **1978**, 17, 578.
20. Kanatzidis, M. G.; Salifoglou, A.; Coucouvanis, D. Inorg. Chem. **1986**, 25, 2460.
21. Cleland, W. E., Jr.; Averill, B. A. Inorg. Chim. Acta **1985**, 106, L17.
22. (a) Christou, G.; Hagen, K. S.; Holm, R. H. J. Am. Chem. Soc. **1982**, 104, 1744. (b) Christou, G.; Hagen, K. S.; Bashkin, J. K.; Holm, R. H., Inorg. Chem. **1985**, 24, 1010.
23. Fenske, D.; Hachgenie, J.; Ohmer, J. Angew. Chem. Int. Ed. Engl. **1985**, 24, 706.
24. Agresti, A.; Bacci, M.; Cecconi, F.; Ghilardi, C. A.; Midollini, S. Inorg. Chem. **1985**, 24, 689.
25. Noda, I.; Snyder, B. S.; Holm, R. H. Inorg. Chem. **1986**, 25, 3853.
26. Pohl, S.; Saak, W. Angew. Chem. Int. Ed. Engl. **1984**, 23, 907.
27. Ghosh, D.; Furey, W. F.; O'Donnell, S.; Stout, C. D. J. Biol. Chem. **1981**, 256, 4185.
28. Johnson, M. K.; Czernuszewicz, R. S.; Spiro, T. G.; Fee, J. A.; Sweeney, W. V. J. Am. Chem. Soc. **1983**, 105, 6671.
29. Antonio, M. R.; Averill, B. A.; Moura, I.; Moura, J. J. G.; Orme-Johnson, W. H.; Teo, B.-K.; Xavier, A. V. J. Biol. Chem. **1982**, 257, 6646.

30. Beinert, H.; Emptage, M. H.; Dreyer, J.-L.; Scott, R. A.; Hahn, J. E.; Hodgson, K. O.; Thomson, A. J. Proc. Natl. Acad. Sci. U.S.A. **1983**, 80, 393.
31. Henkel, G.; Tremel, W.; Krebs, B. Angew. Chem. Int. Ed. Engl. **1981**, 20, 1033.
32. Hagen, K. S.; Christou, G.; Holm, R. H. Inorg. Chem. **1983**, 22, 309.
33. Kennedy, M. C.; Kent, T. A.; Emptage, M.; Merkle, H.; Beinert, H.; Münck, E. J. Biol. Chem. **1984**, 259, 14463.
34. (a) Thomson, A. J.; Robinson, R. E.; Johnson, M. K.; Cammack, R.; Rao, K. K.; Hall, D. O. Biochim. Biophys. Acta **1981**, 637, 423. (b) Johnson, M. K.; Spiro, T. G.; Mortenson, L. E. J. Biol. Chem. **1982**, 257, 2447.
35. Weterings, J. P.; Kent, T. A.; Prins, R. Inorg. Chem. **1987**, 26, 324.
36. Lane, R. W.; Ibers, J. A.; Frankel, R. B.; Papaefthymiou, G. C.; Holm, R. H. J. Am. Chem. Soc. **1977**, 99, 84.
37. O'Sullivan, T.; Millar, M. M. J. Am. Chem. Soc. **1985**, 107, 4096.
38. Millar, M.; Lee, J. F.; Koch, S. A.; Fikar, R. Inorg. Chem. **1982**, 21, 4105.
39. Koch, S. A.; Maelia, L. E.; Millar, M. J. Am. Chem. Soc. **1983**, 105, 5944.
40. Rieske, J. S.; MacLennan, D. H.; Coleman, R. Biochem. Biophys. Res. Commun. **1964**, 15, 338.
41. Fee, J. A.; Findling, K. L.; Yoshida, T.; Hille, R.; Tarr, G. E.; Hearshen, D. O.; Dunham, W. R.; Day, E. P.; Kent, T. A.; Münck, E. J. Biol. Chem. **1984**, 259, 124.
42. Cline, J. A.; Hoffman, B. M.; Mims, W. B.; LaHaie, E.; Ballou, D. P.; Fee, J. A. J. Biol. Chem. **1985**, 260, 3251.
43. Cleland, W. E., Jr.; Averill, B. A. Inorg. Chem. **1984**, 23, 4192.
44. Coucouvanis, D.; Salifoglou, A.; Kanatzidis, M. G.; Simopoulous, A.; Papaefthymiou, V. J. Am. Chem. Soc. **1984**, 106, 6081.
45. Beardwood, P.; Gibson, J. F. J. Chem. Soc. Chem. Comm. **1985**, 1345.
46. Beardwood, P.; Gibson, J. F. J. Chem. Soc. Chem. Comm. **1986**, 490.
47. Kuila, D.; Fee, J. A.; Schoonover, J. R.; Woodruff, W. H.; Batie, C. J.; Ballou, D. P. J. Am. Chem. Soc. **1987**, 109, 1559.
48. Nelson, M. J.; Lindahl, P. A.; Orme-Johnson, W. H. Adv. Inorg. Biochem. **1982**, 4, 1.
49. Stephens, P. J. In "Molybdenum Enzymes", Spiro, T. G., Ed; Wiley-Interscience: New York, **1985**, Ch. 3.
50. Stiefel, E. I.; Cramer, S. P. In "Molybdenum Enzymes", Spiro, T. G., Ed; Wiley-Interscience: New York, **1985**, Ch. 2.
51. Averill, B. A. Struct. Bonding (Berlin) **1983**, 53, 59.
52. Hase, T.; Wakabayashi, S.; Nakano, T.; Zumft, W. G.; Matsubara, H. FEBS Lett. **1984**, 166, 39.
53. Smith, B. E.; Lowe, D. J.; Chen, G. X.; O'Donnell, M. J.; Hawkes, T. R. Biochem. J. **1983**, 209, 207.

54. Orme-Johnson, W. H.; Lindahl, P.; Meade, J.; Worren, W.; Nelson, M.; Groh, S.; Orme-Johnson, N. R.; Münck, E.; Huynh, B. H.; Emptage, M.; Rawlings, J.; Smith, J.; Roberts, J.; Hoffman, B.; Mims, W. B. In "Current Perspectives in Nitrogen Fixation", Gibson, A. H.; Newton, W. E., Eds.; Austral. Acad. Science: Canberra, **1981**, pp. 79-83.
55. Kanatzidis, M. G.; Ryan, M.; Coucouvanis, D.; Simopoulos, A.; Kostikas, A. Inorg. Chem. **1983**, 22, 179.
56. Johnson, R. E.; Papaefthymiou, G. C.; Frankel, R. B.; Holm, R. H. J. Am. Chem. Soc. **1983**, 105, 7280.
57. Johnson, R. W.; Holm, R. H. J. Am. Chem. Soc. **1987**, 100, 5338.
58. Cleland, W. E.; Holtman, D. A.; Sabat, M.; Ibers, J. A.; DeFotis, G. C.; Averill, B. A. J. Am. Chem. Soc. **1983**, 105, 6021.
59. Kanatzidis, M. G.; Baenziger, N. C.; Coucouvanis, D.; Simopoulos, A.; Kostikas, A. J. Am. Chem. Soc. **1984**, 106, 4500.
60. Rogers, M. E.; Averill, B. A. J. Org. Chem. **1986**, 51, 3308.
61. Rogers, M. E.; Averill, B. A., unpublished results.
62. Stack, T. D. P.; Holm, R. H. J. Am. Chem. Soc. **1987**, 109, 2546.
63. (a) Carney, M. J.; Holm, R. H.; Papaefthymiou, G. C.; Frankel, R. B. J. Am. Chem. Soc. **1986**, 108, 3519. (b) Laskowski, E. J.; Reynolds, J. G.; Frankel, R. B.; Foner, S.; Papaefthymiou, G. C.; Holm, R. H. J. Am. Chem. Soc. **1979**, 101, 6562.
64. Shah, V. K.; Brill, W. S. Proc. Natl. Acad. Sci. U.S.A. **1977**, 74, 3249.
65. Hoffman, B. M.; Venters, R. A.; Roberts, J. E.; Nelson, M. J.; Orme-Johnson, W. H. J. Am. Chem. Soc. **1982**, 104, 4711.
66. Hoffman, B. M.; Roberts, J. E.; Orme-Johnson, W. H. J. Am. Chem. Soc. **1982**, 104, 860.
67. Cramer, S. P.; Gillum, W. O.; Hodgson, K. O.; Mortenson, L. E.; Stiefel, E. I.; Chisnell, J. R.; Brill, W. S.; Shah, V. K. J. Am. Chem. Soc. **1978**, 100, 3814.
68. Newton, W. E.; Burgess, B. KI.; Cummings, S. C.; Lough, S.; McDonald, J. W.; Rubinson, J. F.; Conradson, S. D.; Hodgson, K. O. In "Advances in Nitrogen Fixation", Veeger, C.; Newton, W. E., Eds.; Nijhoff/Junk: Pudoc, **1984**, pp. 103-114.
69. Antonio, M. R.; Teo, B. K.; Orme-Johnson, W. H.; Nelson, M. J.; Groh, S. E.; Lindahl, P. A.; Kauzlarich, S. M.; Averill, B. A. J. Am. Chem. Soc. **1982**, 104, 4703.
70. Arber, J. M.; Flood, A. C.; Garner, C. D.; Hasnain, S. S.; Smith, B. E. J. de Physique **1986**, 47, C8-1159.
71. Holm, R. H.; Simhon, E. D. In "Molybdenum Enzymes", Spiro, T. G., Ed.; Wiley-Interscience; New York, **1985**, Ch. 1.
72. Coucouvanis, D. Acc. Chem. Res. **1981**, 14, 201.
73. Coucouvanis, D.; Simhon, E. D.; Baenziger, N. C. J. Am. Chem. Soc. **1980**, 102, 6644.
74. Coucouvanis, D. Chapter 19 in this volume.

75. Seyferth, D.; Henderson, R. S.; Song, L.-C. Organometallics **1982**, 1, 125.
76. Nametkin, N. S.; Tyurin, V. D.; Aleksandrov, G. C.; Kuzmin, O. V.; Nekhaev, A. I.; Andrianov, V. G.; Mavlonov, M.; Struchkov, Yu. T. Izv. Akad. Nauk SSSR, Ser. Khim. **1979**, 1353.
77. Bose, K. S.; Lamberty, P. E.; Kovacs, J. A.; Sinn, E.; Averill, B. A. Polyhedron **1986**, 5, 393.
78. Kovacs, J. A.; Bashkin, J. K.; Holm, R. H. J. Am. Chem. Soc. **1985**, 107, 1784.
79. (a) Bose, K. S.; Sinn, E.; Averill, B. A. Organometallics **1984**, 3, 1126. (b) Lilley, G. L.; Sinn, E.; Averill, B. A. Inorg. Chem. **1986**, 25, 1073.
80. Mayerle, J. J.; Denmark, S. E.; DePamphilis, B. V.; Ibers, J. A.; Holm, R. H. J. Am. Chem. Soc. **1975**, 97, 1032.
81. Dahlstrom, P. L.; Dilworth, J. R.; Shulman, P.; Zubieta, J. Inorg. Chem. **1982**, 21, 933.
82. Holm, R. H. Chem. Soc. Rev. **1981**, 10, 455.
83. Thorneley, R. N. F.; Lowe, D. J. In "Molybdenum Enzymes", Spiro, T. G., Ed.; Wiley-Interscience: New York, **1985**, Ch. 5.
84. Hagen, K.S.; Berg, J.M.; Holm, R.H. Inorg. Chim. Acta 1980, 45, L17.

RECEIVED April 5, 1988

Chapter 14

Importance of Peptide Sequence in Electron-Transfer Reactions of Iron–Sulfur Clusters

Akira Nakamura and Norikazu Ueyama

Department of Macromolecular Science, Faculty of Science, Osaka University, Toyonaka, Osaka 560, Japan

Some characteristic properties of iron-sulfur clusters in proteins result from interactions with invariant amino acid fragments around the clusters. In particular, clusters in peptide environments exhibit positive shifts in redox potentials relative to those in non-peptide environments. Such shifts are observed for a variety of oligopeptide model complexes of 1Fe, 2Fe-2S, and 4Fe-4S proteins.

A large data base is now available for the peptide sequences of iron-sulfur proteins. A close examination of these sequences reveals various invariant sequences, the extent being dependent on the similarity of the proteins. In general, the sequences in the neighborhood of cysteine residues which directly bind to iron(II/III) ions are more or less strongly conserved. For example, simple sequences, e.g. Cys-Pro, have been frequently found for many ferredoxins. In some cases, sequences have been recognized where amino acids of similar alkyl side chains are found near the coordinating cysteines, e.g.Cys-Pro-Val and Cys-Pro-Leu (1). We have been interested in the chemical implications of these invariant sequences and examined the spectroscopic as well as the electrochemical properties of ferredoxin model complexes prepared from the oligopeptides consisting of these invariant sequences. We have already reported a) Fe complexes of chelating oligopeptide as models of rubredoxin which has the tetrapeptide sequence, Cys-Pro-Leu-Cys (2-5), b) a variety of peptide model complexes of plant-type 2Fe ferredoxins ranging from tetra- to twenty-peptides (6-12), and c) a series of peptide complexes which model 4-Fe bacterial and high potential iron-proteins (HiPIP) (13-16). Since these models are air-sensitive and some of them are thermally unstable, the preparative reaction generally involves ligand exchange reactions using $FeCl_2$, $[Fe_2S_2(S\text{-}t\text{-}Bu)_4]^{2-}$, or $[Fe_4S_4(S\text{-}t\text{-}Bu)_4]^{2-}$ as starting material in carefully deaerated DMF under Ar. Figure 1 shows the amino acid sequences of various rubredoxins, where the Cys-X-Y-Cys part is invariant (17). Figures 2 and 3 show the amino acid sequences of plant-type ferredoxins containing an invariant sequence of Cys-A-B-C-D-Cys-X-Y-Cys (18) and the

0097–6156/88/0372–0292$06.00/0

Clostridium pasteurianum[3] F-Met-Lys-Lys-Tyr-Thr-[Cys]-Thr-Val-[Cys]-Gly-Tyr-Ile-Tyr-Asp-Pro-
Peptostreptococcus elsdenii[1] Met-Asp-Lys-Tyr-Glu-[Cys]-Ser-Ile-[Cys]-Gly-Tyr-Ile-Tyr-Asp-Glu-
Micrococcus aerogenes[1] Met-Gln-Lys-Phe-Glu-[Cys]-Thr-Leu-[Cys]-Gly-Tyr-Ile-Tyr-Asp-Pro-
Desulfovibrio vulgaris[29] Met-Lys-Lys-Tyr-Val-[Cys]-Thr-Val-[Cys]-Gly-Tyr-Glu-Tyr-Asp-Pro-

(a) Glu(16)-Asp-Gly-Asp-Pro(20)-Asp-Asp-Gly-Val-Asn-Pro-Gly-Thr-Asp-Phe(30)-Lys-Asp-Ile-Pro-Asp-
(b) Ala-Glu-Gly-Asp——Asp-Gly-Asn-Val-Ala-Ala-Gly-Thr-Lys-Phe-Ala-Asp-Leu-Pro-Ala-
(c) Ala-Leu-Val-Gly-Pro-Asp-Thr-Pro-Asn-Gln-Asn-Gly——Ala-Phe-Glu-Asp-Val-Ser-Glu-
(d) Ala-Glu-Gly-Asp-Pro-Thr-Asn-Gly-Val-Lys-Pro-Gly-Thr-Ser-Phe-Asp-Asp-Leu-Pro-Ala-

(a) Asp(36)-Trp-Val-[Cys]-Pro(40)-Leu-[Cys]-Gly-Val-Gly-Lys-Asp-Glu-Phe-Glu(50)-Glu-Val-Glu-Glu(54)
(b) Asp-Trp-Val-[Cys]-Pro-Thr-[Cys]-Gly-Ala-Asp-Lys-Asp-Ala-Phe-Val-Lys-Met-Asp
(c) Asp-Trp-Val-[Cys]-Pro-Leu-[Cys]-Gly-Ala-Gly-Lys-Glu-Asp-Phe-Glu-Val-Tyr-Glu-Asp
(d) Asp-Trp-Val-[Cys]-Pro-Val-[Cys]-Gly-Ala-Pro-Lys-Ser-Glu-Phe-Glu-Ala-Ala

Figure 1. Amino acid sequences of various rubredoxins.

Ala(1)-Thr-Tyr-Lys-Val-Thr-Leu-Ile-Ser-Glu(10)-Ala-Glu-Gly-Ile-Asn-Glu-Thr-Ile-Asp-Cys(20)-
Asp(21)-Asp-Asp-Thr-Tyr-Ile-Leu-Asp-Ala-Ala(30)-Glu-Glu-Ala-Gly-Leu-Asp-Leu-Pro-Tyr-Ser(40)-
[Cys](41)-Arg-Ala-Gly-Ala-[Cys]-Ser-Thr-[Cys]-Ala(50)-Gly-Lys-Ile-Thr-Ser-Gly-Ser-Ile-Asp-Gln(60)-
Ser(61)-Asp-Gln-Ser-Phe-Leu-Asp-Asp-Asp-Gln(70)-Ile-Glu-Ala-Gly-Tyr-Val-Leu-Thr-[Cys]-Val(80)-
Ala(81)-Tyr-Pro-Thr-Ser-Asp-Cys-Thr-Ile-Gln(90)-Thr-His-Gln-Glu-Glu-Gly-Leu-Tyr(98)

Figure 2. Amino acid sequence of *Spirulina maxima* ferredoxin.

Ala(1)-Tyr-Val-Ile-Asn-Asp-Ser-[Cys]-Ile-Ala(10)-[Cys]-Gly-Ala-[Cys]-Lys-Pro-Glu-[Cys]-Pro-Val(20)-
Asn(21)-Ile-Gln-Gln-Gly-Ser-Ile-Tyr-Ala-Ile(30)-Asp-Ala-Asp-Ser-[Cys]-Ile-Asp-[Cys]-Gly-Ser(40)-
[Cys](41)-Ala-Ser-Val-[Cys]-Pro-Val-Gly-Ala-Pro(50)-Asn-Pro-Glu-Asp(54)

Figure 3. Amino acid sequence of *Peptococcus aerogenes* ferredoxin.

amino acid sequences of bacterial ferredoxins which contain Cys-X-Y-Cys-Gly-Ala-Cys-A-B-C-Cys, respectively (19).

Rubredoxin Models

Rubredoxin is an electron-transfer protein with an Fe(III)/Fe(II) redox couple at -0.31 V (SCE) in water (20). Our peptide model, [Fe(Z-**Cys**-Pro-Leu-**Cys**-OMe)$_2$]$^{2-}$ (Z = benzyloxycarbonyl) (21) exhibits its Fe(III)/Fe(II) redox couple at -0.50 V (SCE) in Me_2SO (9). This is similar to the value observed for the native protein when the difference of the solvent is taken into account. When the model complex is solubilized in water by formation of micelles with addition of the non-ionic detergent, Triton X-100, we also observed a quasi-reversible redox couple at -0.37 V (SCE) (5). The Fe(III) complexes of Cys-X-Y-Cys peptides also exhibit a characteristic MCD band at 350 nm due to ligand-to-metal charge transfer which has also been found in oxidized rubredoxin (4).

The other simple peptide complex e.g. [Fe(Z-**Cys**-Ala-OMe)$_4$]$^{2-}$ did not exhibit such a reversible redox couple under similar conditions. The Fe(III) complexes of simple peptide thiolates or cysteine alkyl esters are found to be thermally quite unstable and decompose by oxidation at the thiolate ligand by intramolecular electron transfer. Thus the macro-ring chelation of the Cys-Pro-Leu-Cys ligand appears to stabilize the Fe(III) state. The stability of the Fe(III) form as indicated by the cyclic voltammogram measurements and by the visible spectra of the Fe(III) peptide complexes suggests that the peptide prevents thermal and hydrolytic decomposition of the Fe-S bond because of the hydrophobicity and steric bulk of the Pro and Leu residues (3,4).

The chelation of the Cys(1)-X-Y-Cys(2) fragment induces a distortion from D_{2d} symmetry. We examined the variation of the overlap population of Fe-S bond with various Fe-S(C) torsion angles by extended Hückel MO calculations (22). Restriction of Fe-S torsion angle in Cys(2) due to the preferable conformation of the chelating peptide leads to a C_2 distortion of the Fe(III) active site. This distortion was found to increase the overlap population of the corresponding Fe-S bond and thus to cause the positive shift of the redox potential.

2-Fe Ferredoxin Models

Complete sequences of ca. 50 different plant-type ferredoxins(Fd) are known. The invariant sequences nearest to the 2-Fe core are confirmed to be Pro-Tyr-Ser-Cys-Arg-Ala-Gly-Ala-Cys-Ser-Thr-Cys-Ala-Gly and Leu-Thr-Cys-Val. 2Fe-2S complexes of oligopeptides with the Cys-X-Y-Cys sequence have been synthesized by ligand exchange reactions (7,23). We have examined the redox potentials of these model complexes, and the results are shown in Table I. The reversibility improved remarkably and the potential approached the value of the native proteins as the sequence more closely simulated that of the proteins. It is conjectured that hydrogen bonds from the peptide N-H's to thiolate and/or sulfide groups increase the stability of the reduced cluster. It is likely that peptide sequences like those found in the proteins favor the formation of such hydrogen bonds.

The Fe_2S_2 complex derived from the 20-peptide exhibits a visible spectrum typical of plant-type 2-Fe proteins (10). (*See* Figure 4.) But at least two isomeric 2-Fe species are detected by the cyclic voltammetry (CV) and also by the differential-pulse polarography in DMF. One isomer has an $E_{1/2}$ at -0.64 V (SCE), which is very similar to that of the native protein, and this complex presumably adopts a structure similar to that found in the protein, with a peptide conformation that favors some hydrogen bonding between coordinated sulfurs and peptide NH's. The other isomer has an $E_{1/2}$ at -0.96 V (SCE) and may have the peptide in a less restricted conformation with less NH---S hydrogen bonding. The formation of isomeric complexes under the ligand exchange reaction conditions probably reflects kinetic control of the Cys-thiolate coordination to the metal as opposed to the conformational preference of the peptide.

The nature of solvent was found to be important for the electron transfer behavior. The Fe_2S_2 complex with two Cys-Gly-Ala-Gly-Ala-Cys ligands shows $E_{1/2}$ values of -0.95 V in DMF and -0.76 V in CH_3CN. The positive shift of the

Table I. Redox Potentials of the [2Fe-2S]Peptide Model Complexes

Complex	$E_{1/2}$ V vs SCE (i_{pa}/i_{pc})[a]
$[Fe_2S_2$(S-*t*-Bu)$_4]^{2-}$	-1.46(0.80)
$[Fe_2S_2$(Z-Ala-**Cys**-OMe)$_4]^{2-}$	-1.04(0.30)
	-1.06(0.67)-PC[b]
$[Fe_2S_2$(Z-Val-**Cys**-Val-OMe)$_4]^{2-}$	-1.21 (0.80)
$[Fe_2S_2$(Z-**Cys**-Ala-Ala-**Cys**-OMe)$_2]^{2-}$	-1.06(0.80)
	-1.04(0.80)-PC[b]
	-1.06(0.62)-CH_3CN
$[Fe_2S_2$(Z-**Cys**-Thr-Val-**Cys**-OMe)$_2]^{2-}$	-1.09(0.90)
$[Fe_2S_2$(Z-**Cys**-Pro-Leu-**Cys**-OMe)$_2]^{2-}$	-1.18(0.70)
$[Fe_2S_2$(Z-**Cys**-Val-Val-**Cys**-OMe)$_2]^{2-}$	-1.41(0.80)
$[Fe_2S_2$(Z-**Cys**-Gly-Ala-Gly-Ala-**Cys**-OMe)$_2]^{2-}$	-0.95(0,73)
	-0.85(0.86)-PC[b]
	-0.76(0.65)-CH_3CN
$[Fe_2S_2$(Z-**Cys**-Ala-Ala-Gly-Ala-**Cys**-OMe)$_2]^{2-}$	-0.94(0.92)
	-1.10(0.5)-PC[b]
$[Fe_2S_2$(Z-**Cys**-Ala-Tyr-Ala-Gly-**Cys**-OMe)$_2]^{2-}$	0.94(0.62)
$[Fe_2S_2$(Ac-**Cys**-Ala-Tyr-Ala-Gly-**Cys**-OMe)$_2]^{2-}$	-0.98(0.67)
$[Fe_2S_2$(Z-**Cys**-Gly-Ala-Gly-Ala-**Cys**- -Ala-Ala-**Cys**-OMe)(S-*t*-Bu$]^{2-}$	-0.92(0.73)
$[Fe_2S_2$(Z-**Cys**-Gly-Ala-Gly-Ala-**Cys**- -Ala-Ala-**Cys**-OMe)(Z-Val-**Cys**-Val-OMe)$]^{2-}$	-0.94(0.33)
$[Fe_2S_2$(20-peptide)$]^{2-}$	-0.64(0.95)
	-0.96(0.94)

[a]Measured in DMF unless otherwise noted
[b]PC: propylene carbonate

potential in CH_3CN suggests that the less polar solvent promotes the formation of more hydrogen bonds; this shift is similar to that observed for the (2-)/(3-) redox potential of $[Fe_4S_4$(Z-**Cys**-Gly-Ala-OMe$)_4]^{2-}$ in CH_2Cl_2 (13). Further evidence is derived from CD spectroscopy. The intensity of CD extrema increased remarkably in the case of $[Fe_2S_2$(Z-cys-X-Y-cys-OMe$)_2]^{2-}$ in CH_3CN in comparison to the CD spectra in DMF. This indicates that the less polar solvent restricts the possible conformations adopted by the oligopeptide and is consistent with the hydrogen bonding model. The X-ray structure of *Spirulina platensis* Fd indicates a hydrophobic environment near the Fe_2S_2 core and therefore our observation of the positively shifted potential in weakly polar solvent provides a reasonable explanation of the redox behavior of native Fds (24).

4-Fe Ferredoxin Models

The invariant sequence for *P. aerogenes* Fd near its [4Fe-4S] cluster core involves a heptapeptide, Cys-Ile-Ala-Cys-Gly-Ala-Cys, with three cysteine thiolato coordinating sites. The $[Fe_4S_4]^{2+}$ complexes of this and related sequences have been prepared by ligand exchange reactions with $[Fe_4S_4(S\text{-}t\text{-}Bu)_4]^{2-}$ in DMF, a strategy similar to that employed by Que et al in 1974 (25). In the earlier study, complexes of Gly-Cys-Gly were prepared and found to exhibit a positive shift in the (2-)/(3-) redox potential as four tripeptides were replaced with a 12-peptide. We have examined the dependence of the redox couple upon the peptide structure, in a DMF or a CH_2Cl_2 solution using freshly prepared cluster complexes under Ar atmosphere at room temperature. The results are shown in Table II. In general, the (2-)/(3-) potentials of the complexes showed a dependence of the number of cysteines on a particular oligopeptide, shifting to more positive values as the number of cysteines increased.

Some of the complexes were found to exhibit redox potentials that shifted with temperature (230-300 K). The complexes with Cys-Gly-Ala and Cys-Gly-Ala-Cys sequences were found to be particularly susceptible to temperature variations in CH_2Cl_2 (13,14) with positive shifts at (ca. 0.10 V). These shifts were rationalized by the formation of NH---S hydrogen-bonded conformers (see Fig. 5), which are expected to stabilize the reduced forms of the complexes Such a hydrogen-bonded structure has been deduced from the X-ray analysis of several bacterial Fds (26).

The presence of a Gly residue just after a coordinated Cys residue is crucial for the observed turn of the peptide at the Gly portion which corresponds to a preferable ß-turn conformation for achieving hairpin turn structure. When an additional cysteine residue is attached to this sequence as in $[Fe_4S_4$(Z-**Cys**-Gly-Ala-**Cys**-OMe$)_2]^{2-}$, two chelate rings are formed with more highly oriented NH---S hydrogen bonds within the ring. The redox potential was found to exhibit positive shift even at room temperature from that of $[Fe_4S_4$(Z-**Cys**-Gly-Ala-OMe$)_4]^{2-}$. In this case, two hydrogen bonds are formed at most since there are two chelate rings around the cluster. In $[Fe_4S_4$(Z-**Cys**-Gly-Ala-OMe$)_4]^{2-}$, four hydrogen bonds are possible when all the tripeptide ligands assume the folded conformation as shown in Fig. 5. Actually at low temperature (220 K), the redox potentials of all these complexes were almost the same. For further positive shift, sequences such as Cys-Gly-Ala-Cys-Gly-Ala may be suitable, but this has not yet been found in the known invariant sequences of bacterial ferredoxin.

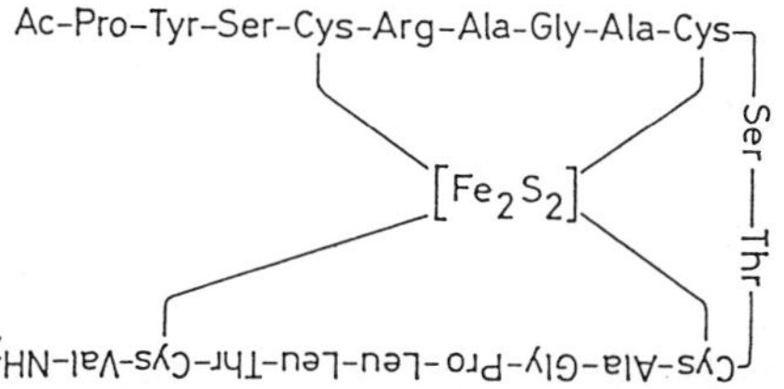

Figure 4. Proposed structure of 20-peptide model complex.

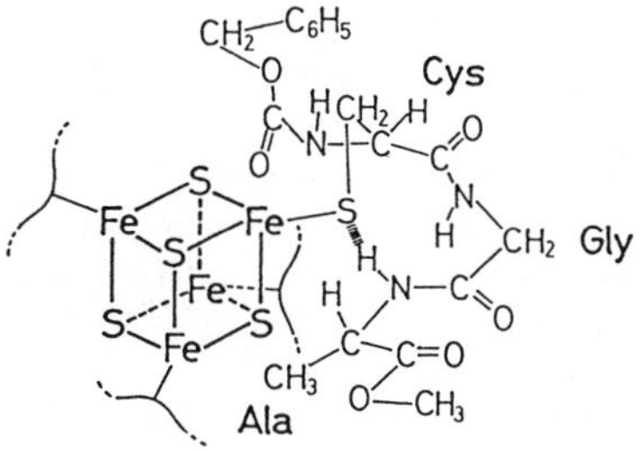

Figure 5. Proposed structure of $[Fe_4S_4(\text{Z-Cys-Gly-Ala-OMe})_4]^{2-}$ in dichloromethane.

Table II. Redox potentials (3-/2-) for [4Fe-4S] peptide model complexes in dichloromethane at room temperature

Complex	Redox potential (V vs SCE)
$[Fe_4S_4$(Z-**Cys**-Ile-Ala-**Cys**-Gly-Ala-**Cys** OMe)(S-*t*-Bu)$]^{2-}$	-0.80 (-0.92)[a] (-1.10)[a]
$[Fe_4S_4$(Z-**Cys**-Ile-Ala-**Cys**-Gly-Ala-**Cys** OMe)(Z-**Cys**-Pro-Leu-OMe)$]^{2-}$	-0.83
$[Fe_4S_4$(Z-**Cys**-Gly-Ala-**Cys**-OMe$)_2]^{2-}$	-0.91
$[Fe_4S_4$(Z-**Cys**-Gly-Ala-**Cys**-OMe$)_2]^{2-}$	-0.91
$[Fe_4S_4$(Z-**Cys**-Ile-Ala-**Cys**-OMe$)_2]^{2-}$	-0.91
$[Fe_4S_4$(Z-**Cys**-Ile-Ala-**Cys**-Gly-OMe$)_2]^{2-}$	-0.91
$[Fe_4S_4$(Z-**Cys**-Gly-OMe$)_4]^{2-}$	-1.01
$[Fe_4S_4$(Z-**Cys**-Gly-Ala-OMe$)_4]^{2-}$	-1.00
$[Fe_4S_4$(Z-**Cys**-Pro-Gly-OMe$)_4]^{2-}$	-1.07
$[Fe_4S_4$(Z-**Cys**-Pro-Leu-OMe$)_4]^{2-}$	-1.19

[a]The redox potential values in parenthesis refer to those of isomers (see Text).

The best peptide model complex involving all the amino acid residues within 5 Å of the cluster was prepared using a heptapeptide, Z-Cys-Ile-Ala-Cys-Gly-Ala-Cys-OMe, and a tripeptide, Z-Cys-Pro-Val-OMe. The redox potential of this complex in CH_2Cl_2 was -0.83 V (SCE) which is very close to that of the native protein when the difference of solvent is considered (cf. Table II). A similar heptapeptide complex with a S-*t*-Bu ligand was found to give three CV peaks, which indicates the presence of isomers.

The first half of the sequence, i.e. Cys-Ile-Ala, was also prepared and the redox reaction of the cluster with four Z-Cys-Ile-Ala-OMe ligands was examined. This sequence was found to enhance the stability of the superoxidized state of 4Fe-cluster probably by its hydrophobic peptide side chains. Thus, the CV results indicated a quasi-reversible redox couple at +0.12 V (SCE) in DMF (*see* Table III). Thus the complex may be regarded as a model of HiPIP. Simple model complexes, $[Fe_4S_4$(S-*t*-Bu$)_4]^{2-}$ (27), $[Fe_4S_4$(2,4,6-trimethylbenzenethiolato$)_4]^{2-}$ (28), and $[Fe_4S_4$(2,4,6-triisopropylbenzenethiolato$)_4]^{2-}$ (16), have been synthesized, and the latter has been crystalligraphically characterized (29).

The protein environment of HiPIP consists of many aromatic rings together with some aliphatic groups, with a smaller extent of NH---S hydrogen bonding found. Since we have noticed the presence of a sequence, Trp-Cys, at the coordination site near the 4Fe-cluster, $[Fe_4S_4]^{2+}$ complexes with tripeptides, such as Z-Trp-Cys-Ala-OMe and Z-Trp-Cys-Val-OMe, were prepared. The cyclic voltammetry of $[Fe_4S_4$(Z-Trp-**Cys**-Val-OMe$)_4]^{2-}$ in Me_2SO showed a positively

Table III. Redox Potentials of Peptide Model Complexes and Bulky Thiolate Complexes for 2-/1- and 3-/2- in DMF

Complexes	Redox potentials (V vs. SCE) 2-/1-	3-/2-
$[Fe_4S_4(\text{Z-Cys-Ile-Ala-OMe})_4]^{2-}$	+0.12	-1.00
$[Fe_4S_4(\text{Z-Cys-Pro-Leu-OMe})_4]^{2-}$	-0.02a)	-1.19
$[Fe_4S_4(\text{Z-Cys-Pro-Val-OMe})_4]^{2-}$	irreversible[a]	-1.18
$[Fe_4S_4(\text{Z-Cys-Pro-Gly-OMe})_4]^{2-}$	n.o.[a]	-1.07
$[Fe_4S_4(\text{S-2,4,6-trimethylbenzenethiolato})_4]^{2-}$	+0.02	-1.27
$[Fe_4S_4(\text{S-2,4,6-triisopropylbenzenethiolato})_4]^{2-}$	-0.03	-1.38

[a]In Me_2SO

shifted redox couple at +0.06 V (SCE) compared with those of simple model peptide complexes. The proximity of Trp indole rings to the Fe_4S_4 cluster was also evidenced by the effective quenching of the Trp fluorescence upon excitation at 295 nm. The quenching is probably due to the energy transfer from the indole ring to the $[Fe_4S_4]^{2+}$ core. The fluorescence intensity at 340 nm indicated that the most effective quenching occurs in the case of the Z-Trp-Cys-Ala-OMe complex.

The stability of $[Fe_4S_4]^{3+}$ core was utilized for electron-transfer oxidation catalysis. With 1,4-benzoquinone as oxidant, benzoin was catalytically oxidized by

Ph-CH(OH)-CO-Ph + 1,4-benzoquinone $\xrightarrow{[Fe_4S_4(SR)_4]^{2-}}$ Ph-CO-CO-Ph + hydroquinone

the peptide/Fe_4S_4 complexes under ambient conditions (3). $[Fe_2S_2(SR)_4]^{2-}$ also has a fairly high catalytic oxidation activity for benzoin. In this case, however, formation of $[Fe_4S_4(SR)_4]^{2-}$ from the reduced species of $[Fe_2S_2(SR)_4]^{2-}$ is probably involved in the process of slow conversion from $[Fe_2S_2(SPh)_4]^{2-}$ to $[Fe_4S_4(SPh)_4]^{2-}$ in protic media as has been reported (31). No catalytic activity except for the stoichiometric reaction was observed in the presence of $[Fe(S_2\text{-}o\text{-}xyl)_2]^{2-}$. Pertinent data are shown in Table IV. The effectiveness of catalysis was found to be parallel with the thermal stability of the oxidized $[Fe_4S_4]^{2+}$ core, which can be estimated by the CV data. For example, a complex with Z-Cys-Ile-Ala-OMe was the best among the complexes examined so far. Non-peptide thiolate/$[Fe_4S_4]^{2+}$ complexes also work as efficient catalysts (see Table 4). Here the complexes with bulky aromatic thiolate were found to be the most promising. The hydrophobicity around the iron cluster core is thus the most important for the effective catalytic cycle. The aromatic rings seem to promote the reversiblity of $[Fe_4S_4]^{2+/3+}$ redox cycle.

Table IV. Catalytic oxidation of benzoin by 1,4-benzoquinone in the presence of 4Fe-, 2Fe-, 1Fe-complexes

Complexes	Yields of benzil (%)
none	trace[b]
$[Fe_4S_4$(Z-**Cys**-Ile-Ala-OMe$)_4]^{2-}$	16,800[b]
$[Fe_4S_4$(Z-**Cys**-Pro-Gly-OMe$)_4]^{2-}$	10,400[b]
$[Fe_4S_4$(Z-**Cys**-Pro-Gly-OMe$)_4]^{2-}$	2,400[b]
$[Fe_4S_4$(S-*t*-Bu$)_4]^{2-}$	8,100[b]
$[Fe_4S_4$(2,4,6-triisopropylbenzenethiolato$)_4]^{2-}$	17,000[b]
$[Fe_4S_4$(2,4,6-trimethylbenzenethiolato$)_4]^{2-}$	8,000[b]
$[Fe(S_2$-*o*-xyl$)_2]^{2-}$	1,750[b]

[a]Reaction conditions: [benzoin], 2.5 x 10^{-1}M; [1,4-benzoquinone], 5 x 10^{-1}M; [complex], 1 x 10^{-3}M; 2 ml of THF/DMF (1:0.1); 3 h, 25°C.
[b]Reaction conditions: [benzoin], 2.5 x 10^{-1}M; [1,4-benzoquinone], 2.5 x 10^{-1} M; [(Fe-complex$)_4$], 1 x 10^{-3}M; 5 ml of THF/DMF (1:1); 10 h, 25°C.

We have emphasized the chemical importance of invariant peptide fragments around the active site of iron-sulfur proteins. Significant interactions between the invariant peptide fragment and Fe^{3+}, $Fe_2S_2^{2+}$, or $Fe_4S_4^{2+}$ core have been observed. Interactions such as NH---S hydrogen bonding, the chelation effect of Cys-X-Y-Cys, and steric effects around the Cys thiolate ligand are found. These interactions result in a positive shift of the redox potential and the stabilization of the higher oxidation state. Further studies on synthetic oligopeptide model complexes having a specific conformation in nonpolar solvents will shed light on the pathways of electron transfer in native proteins. The invariant peptide fragments have the potential for lowering the activation energy of the electron transfer.

Literature Cited

1. M. O. Dayhoff, "Atlas of Protein Sequence and Structure" National Biomedical Foundation, Washington, D. C. Vol. 5 (1972); Vol. 5, Suppl. 2 (1976).
2. Ueyama, N.; Nakata, M.; Nakamura, A. *Bull. Chem. Soc. Jpn.*, **1981**, *54*, 1727.
3. Nakata, M.; Ueyama, N.; Nakamura, A. *Bull. Chem. Soc. Jpn.*, **1983**, *56*, 3641.
4. Ueyama, N.; Nakata, M.; Fuji, M.; Terakawa, T.; Nakamura, A. *Inorg. Chem.*, **1985**, *24*, 2190.
5. Nakata, M.; Ueyama, N.; Fuji, M.; Nakamura, A.; Wada, K.; Matsubara, H. *Biochim. Biophys. Acta*, **1984**, *788*, 306.
6. Ueyama, N.; Nakata, M.; Nakamura, *Polymer J.*, **1985**, *17*, 721.
7. Ueyama, N.; Ueno, S.; Nakata, M.; Nakamura, A. *Bull. Chem. Soc. Jpn.*, **1984**, *57*, 984.

8. Ueno, S.; Ueyama, N.; Nakamura, A.; Tsukihara, T. *Inorg. Chem.*, **1986**, *25*, 1000.
9. Ueyama, N.; Ueno, S.; Nakamura, A. *Bull. Chem. Soc. Jpn.*, **1987**, *60*, 283.
10. Ueno, S.; Ueyama, N.; Nakamura, A.; Wada, K.; Matsubara, H.; Kumagai, S.; Sakakibara, S.; Tsukihara, T. *Pept. Chem.*, **1984**, 133.
11. Ueno, S.; Ueyama, N.; Nakamura, A. *Pept. Chem.*, **1985**, 269.
12. Ueyama, N.; Ueno, S.; Nakamura, A. *Bull. Chem. Soc. Jpn.*, **1987**, *60*, 283.
13. Ueyama, N.; Terakawa, T.; Nakata, M.; Nakamura, A. *J. Am. Chem. Soc.*, **1983**, *105*, 7098.
14. Ueyama, N.; Kajiwara, A.; Terakawa, T.; Ueno, S.; Nakamura, A. *Inorg. Chem.*, **1985**, *24*, 4700.
15. Ueyama, N.; Fuji, M.; Sugawara, T.; Nakamura, A. *Pept. Chem.*, **1985**, 301.
16. Ueyama, N.; Terakawa, T.; Sugawara, T.; Fuji, M.; Nakamura, A. *Chem. Lett.*, **1984**, 1284.
17. Orme-Johnson, W. H. *Annu. Rev. Biochem.*, **1973**, *42*, 159.
18. Hase, T.; Ohmiya, N.; Matsubara, H. Mullinger, R. N.; Rao, K. K.; Hall, D. O. *Biochem. J.*, **1976**, *159*, 55.
19. Matsubara, H.; Sasaki, R. M.; Tsuchiya, D. K.; Evans, M. C. W. *J. Biol. Chem.*, **1970**, *245*, 2121.
20. Lovenberg, W.; Sobel, B. E. *Proc. Natl. Acad. Sci. U.S.A.*, **1965**, *54*, 193.
21. Coordinating cysteine residue is represented in bold letters, i.e. **Cys**.
22. Ueyama, N.; Sugawara, T.; Tatsumi, K.; Nakamura, A. *Inorg. Chem.*, **1987**, *26*, 1978.
23. Balasubramaniam, A.; Coucouvanis, D. *Inorg. Chem.*, **1983**, *78*, L35.
24. Tsukihara, T.; Fukuyama, K.; Tahara, H.; Katsube, Y.; Matsuura, Y.; Tanaka, N.; Kakudo, M.; Wada, K.; Matsubara, H. *J. Biochem.*, **1978**, *84*, 1645.
25. Que, L., Jr.; Anglin, J. R.; Bobrik, M. A., Davison, A.; Holm, R. H. *J. Am. Chem. Soc.*, **1974**, *96*, 6042.
26. Adman, E. T.; Sieker, L. C.; Jensen, L. H., *J. Biol. Chem.*, **1977**, *251*, 3801.
27. (a) DePamphilis, B. V.; Averill, B. A.; Herskovitz, T.; Que, L., Jr.; Holm, R. H. *J. Am. Chem. Soc.*, **1974**, *96*, 4159. (b) Mascharak, P. K. L.; Hagen, K. S.; Spence, J. T.; Holm, R. H. *Inorg. Chim. Acta*, **1983**, *80*, 157.
28. Ueyama, N.; Sugawara, T.; Fuji, M.; Nakamura, A.; Yasuoka, N. *Chem. Lett.*, **1985**, 175.
29. O'Sullivan, T.; Millar, M. M. *J. Am. Chem. Soc.*, **1985**, *107*, 4095.
30. Ueyama, N.; Sugawara, T.; Kajiwara, A.; Nakamura, A. *J. Chem. Soc. Chem. Commun.*, **1986**, 434.
31. Que, L., Jr.; Holm, R. H.; Mortenson, L. E. *J. Am. Chem. Soc.*, **1975**, *97*, 463.

RECEIVED April 5, 1988

Chapter 15

Double Exchange in Reduced Fe_3S_4 Clusters and Novel Clusters with MFe_3S_4 Cores

E. Münck[1], V. Papaefthymiou[1], K. K. Surerus[1], and J.-J. Girerd[2]

[1]Gray Freshwater Biological Institute, University of Minnesota, Navarre, MN 55392
[2]Laboratoire de Spectrochimie des Elements de Transition, Equipé de Recherche Associee au Centre National de la Recherche Scientifique, No. 420, Université de Paris-Sud, 91405 Orsay, France

Mössbauer studies of the electron transport protein ferredoxin II from *Desulfovibrio gigas* have shown that the reduced Fe_3S_4 cluster contains one trapped valence Fe^{3+} site and one delocalized Fe^{2+}-Fe^{3+} pair. The two Fe of the delocalized pair are indistinguishable, and the pair has a dimer spin $S_{12} = 9/2$, suggesting ferromagnetic coupling. The observation of electron delocalization together with ferromagnetic spin alignment suggests that the spin coupling of the mixed-valence dimer is dominated by double exchange, a mechanism introduced more than 30 years ago by Anderson and Hasegawa. This article presents the salient experimental results and introduces a spin Hamiltonian which includes a term describing double exchange. This Hamiltonian provides an excellent description of the data. We discuss also the spectral properties of the novel clusters [MFe_3S_4] obtained by incubating ferredoxin II in the presence of excess M = Co^{2+}, Zn^{2+}, and Cd^{2+}.

Iron-sulfur clusters serve a variety of functions in biological systems, predominately in electron transport. Most clusters participate in one-electron redox reactions and they can thus exist in states of mixed-valence, that is in states containing Fe^{3+} as well as Fe^{2+} sites. Such mixed-valence states occur for instance in $[Fe_2S_2]^{1+}$ clusters, reduced Fe_3S_4 clusters and in structures with the cubane $[Fe_4S_4]^{1+,2+,3+}$ cores. Mössbauer spectroscopy is particularly suited for the study of these systems, because this technique provides a wealth of information from which details about the electronic structure of the cluster core can be deduced. Mössbauer studies have shown that iron-sulfur clusters contain sites to which a distinct valence can be assigned (localized or trapped valences). For instance, the two iron sites of reduced Fe_2S_2 clusters exhibit Mössbauer parameters which are typical of ferric and ferrous FeS_4 anions, respectively. Fe_4S_4 cubanes, on the other hand, exhibit strong electron delocalization; thus, the parameters for the quadrupole splitting, ΔE_Q, and the isomer shift, δ, are observed to be roughly the average of those observed for $Fe^{3+}S_4$ and $Fe^{2+}S_4$ anions.

0097–6156/88/0372–0302$07.00/0

The magnetic properties of iron-sulfur clusters are determined by strong exchange interactions which couple the spins of the individual ions to a resultant system spin S. Not suprisingly, the magnetic properties are intimately tied to the extent of electron delocalization. Thus, for Fe_2S_2 clusters the spin coupling between the Fe^{3+} site ($S_1 = 5/2$) and the Fe^{2+} site ($S_2 = 2$) is well described by the Heisenberg Hamiltonian $\mathbf{H} = J\mathbf{S}_1 \cdot \mathbf{S}_2$, as testified by the successful model put forward by Gibson and coworkers (1-4). The spin-coupling problem of Fe_4S_4 clusters, on the other hand, has thus far not been solved. Mössbauer studies of $[Fe_4S_4]^{3+}$ (5,6) and $[Fe_4S_4]^{1+}$ (7,8) have shown that the four iron sites occur in pairs of identical irons. One pair in $[Fe_4S_4]^{3+}$, for instance, contains two identical irons at, roughly, the $Fe^{2.5+}$ level. Thus, the pair contains an itinerant electron delocalized evenly between a ferric and a ferrous site. We will discuss in this article that the delocalized electron provides a coupling mechanism (double exchange) which is different from the familiar exchange mechanism described by $\mathbf{H} = J\mathbf{S}_1 \cdot \mathbf{S}_2$.

The decisive experimental data requiring the inclusion of double exchange were obtained from an analysis of Mössbauer studies of reduced Fe_3S_4 clusters. In the following section we will discuss the observation which forced us to amend the familiar Heisenberg Hamiltonian, namely the observation of valence delocalization in conjunction with an apparent ferromagnetic coupling in a subdimer of the Fe_3S_4 cluster. The concept of double exchange was proposed in 1951 by Zener (9) and the fundamental features of the mechanism were analyzed in 1955 by Anderson and Hasegawa (10). In the third section of this article we discuss the basic features of double exchange. The subsequent section treats briefly the application of a new spin Hamiltonian (11) containing Heisenberg as well as double exchange terms for the analysis of the Fe_3S_4 cluster data. In the final section we describe successful attempts to produce novel MFe_3S_4 clusters (M = Co, Zn, Cd) from protein-bound Fe_3S_4 precursors. The new clusters are not only interesting from a bioinorganic perspective, they also provide us with opportunities to explore the three spin coupling problem (using the diamagnetic Zn^{2+} or Cd^{2+}) further. Since we have only just begun to (11) incorporate double exchange into the analyses, this article should be viewed in the spirit of a progress report.

Experimental Evidence for Double Exchange in Ferredoxin II

Ferredoxin II (Fd II) isolated from *Desulfovibrio gigas* is a small tetramer consisting of four identical subunits, each of molecular mass 6,000 daltons. Each subunit contains *one* Fe_3S_4 cluster. A variety of studies have shown that the four clusters are identical and that they do not interact with each other. Thus, for our purpose, we can consider Fd II as a monomer with one Fe_3S_4 cluster.

The Fe_3S_4 cluster of Fd II has been studied with a variety of physical techniques which include EPR (12,13), EXAFS (14), MCD (15) and Mössbauer spectroscopy (11,12,16). In the following we briefly summarize those results which pertain to problems addressed in this article.

Mössbauer studies of Fd II have established that the protein contains a 3-Fe cluster, and that the three iron sites reside in a predominantly tetrahedral environment of sulfur ligands. The cluster has a Fe_3S_4 stoichiometry (see Beinert and Thomson (17)), and EXAFS data (14) suggest a compact core with an Fe-Fe distance of 2.7 Å, just as observed for Fe_4S_4 cubanes. The structure is perhaps close to that shown on the left of Figure 1. Thus, we picture the Fe_3S_4 cluster as a cubane Fe_4S_4 cluster from which one Fe has been removed. This view is supported by the observation that the cluster is easily converted into a structure with an Fe_4S_4 core when the protein is incubated with excess Fe^{2+} (16).

Fd II can be studied in two oxidation states. In the oxidized state the cluster has electronic spin S = 1/2. This spin results from antiferromagnetic coupling of three high-spin ferric ($S_1 = S_2 = S_3 = 5/2$) iron sites. The magnetic hyperfine parameters obtained from an analysis of the low temperature Mössbauer spectra have been analyzed (18) in the framework of the Heisenberg Hamiltonian.

$$\mathbf{H} = J_{12}\mathbf{S}_1 \cdot \mathbf{S}_2 + J_{23}\mathbf{S}_2 \cdot \mathbf{S}_3 + J_{13}\mathbf{S}_1 \cdot \mathbf{S}_3 \qquad (1)$$

The analysis shows that all three coupling constants are positive (antiferromagnetic coupling) and about equal in magnitude (18). Gayda and coworkers (19) have studied the cluster with EPR and have deduced that the J-values are approximately 40 cm^{-1}. Figure 2A shows a Mössbauer spectrum of the oxidized cluster recorded at 100 K (under these conditons the electronic spin S = 1/2 relaxes fast and magnetic hyperfine interactions are averaged out). The spectrum consists of one doublet with quadrupole splitting, $\Delta E_Q = 0.53$ mm/s and isomer shift, $\delta = 0.28$ mm/s. These parameters are typical for $Fe^{3+}S_4$ anions. $Fe^{2+}S_4$ anions, on the other hand, typically exhibit $\Delta E_Q \approx 3$ mm/s and $\delta \approx 0.70$ mm/s (see ref (20)).

At this juncture we digress for a moment and turn our attention to the Mössbauer spectra of clusters with Fe_2S_2 cores. These clusters exhibit at high temperatures in the reduced state (S = 1/2; g = 1.94 EPR signal) Mössbauer spectra which clearly show the presence of an Fe^{3+} ($\Delta E_Q \approx 0.5$ mm/s, $\delta \approx 0.30$ mm/s) and an Fe^{2+} ($\Delta E_Q \approx 3.0$ mm/s, $\delta \approx 0.70$ mm/s) site. An example is shown in Figure 3A (21). Analyses of the low temperature EPR and Mössbauer spectra (Figure 3B) yield g-values and the two magnetic hyperfine tensors. The experimentally observed quantities have been analyzed successfully (see Sands and Dunham (22)) in the framework of the Heisenberg Hamiltonian

$$\mathbf{H} = J\mathbf{S}_1 \cdot \mathbf{S}_2 \qquad (2)$$

where $S_1 = 5/2$ is the spin of the ferric site and $S_2 = 2$ that of the ferrous site. Since the ground state of system has total spin S = 1/2, the coupling is antiferromagnetic. Note that the Fe_2S_2 cluster, in the mixed valence state, has localized Fe^{2+} and Fe^{3+} valence states with ΔE_Q and δ values typical of ferrous and ferric FeS_4 anions. Note also that the coupling is antiferromagnetic. These properties have been observed for all Fe_2S_2 systems studied thus far.

Upon reduction by one electron, the Fe_3S_4 cluster of Fd II exhibits a Mössbauer spectrum consisting of two doublets with a 2:1 intensity ratio (Figure 2B). The single iron of doublet II has $\Delta E_Q = 0.52$ mm/s and $\delta = 0.32$ mm/s; the site thus has parameters typical of $Fe^{3+}S_4$ anions. The two remaining irons, on the other hand, both exhibit $\Delta E_Q = 1.47$ mm/s and $\delta = 0.46$ mm/s. These parameters are just the average of typical Fe^{3+} and Fe^{2+} values. Thus, both iron sites are roughly at the $Fe^{2.5+}$ oxidation level, suggesting that the electron which has entered the complex is shared equally between the two irons of doublet I. We will, therefore, refer to these two irons as the Fe^{2+}-Fe^{3+} delocalized pair. Such a delocalized pair has been observed for every reduced Fe_3S_4 cluster studied thus far. Moreover, Fe^{2+}-Fe^{3+} pairs are observed for Fe_4S_4 cubanes in all three accessible oxidation states. In order to learn more about the electronic structure of this pair we turn our attention to Mössbauer studies in strong applied fields.

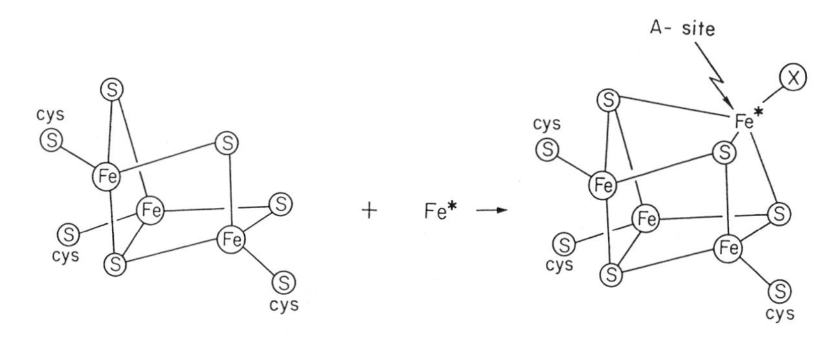

Figure 1. Putative structure of Fe_3S_4 cluster (left) and illustration of conversion into MFe_3S_4 with M = Fe, Co, Zn and Cd.

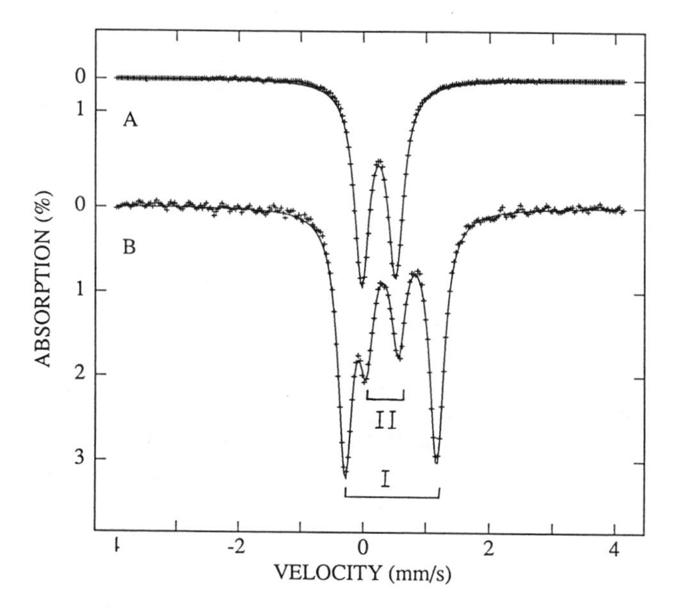

Figure 2. Mössbauer spectra of the Fe_3S_4 cluster of *D. gigas* Fd II. (A) Spectrum of the oxidized cluster (S = 1/2) taken at 100 K. (B) 4.2 K spectrum of the reduced cluster (S = 2) recorded in zero field; see also Figure 4.

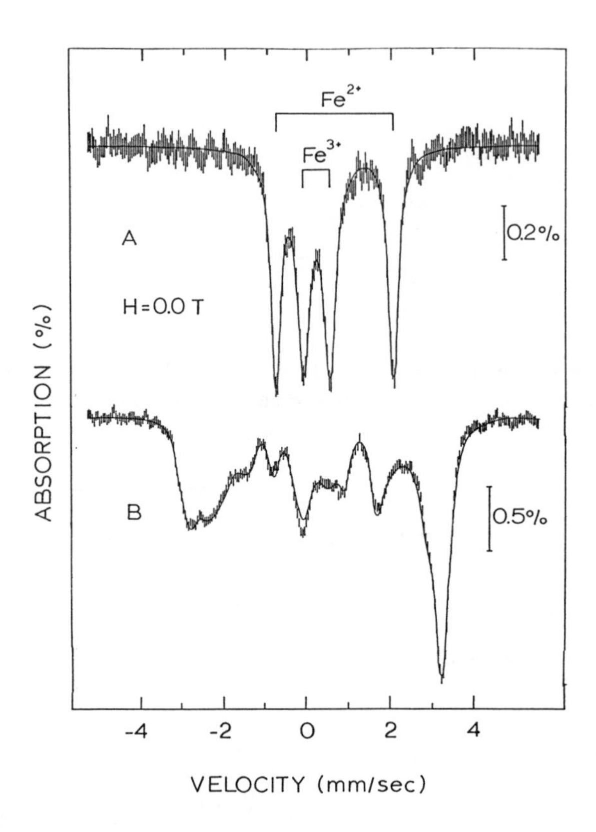

Figure 3. Mössbauer spectra of the reduced Rieske protein *Thermus Thermophilus*. (A) Spectrum taken at 230 K. The brackets indicate the doublets of the trapped-valence Fe^{2+} and Fe^{3+} sites. (B) 4.2 K spectrum of the same sample. The solid line is a spectral simulation based on an S = 1/2 spin Hamiltonian. S = 1/2 is the system spin resulting from coupling $\mathbf{S} = \mathbf{S}_A + \mathbf{S}_B$ according to $\mathbf{H} = J\mathbf{S}_A \cdot \mathbf{S}_B$ for $J > 0$; $S_A = 5/2$ and $S_B = 2$.

The reduced Fe_3S_4 cluster of Fd II has cluster spin S = 2 (15). This suggests that one can obtain information about the zero-field splitting and magnetic hyperfine interactions by studying the material in strong applied fields. We have studied samples in fields up to 6.0 Tesla over a wide range of temperatures. A typical spectrum, taken in a field of 1.0 Tesla at T = 1.3 K, is shown in Figure 4. We have described the details of our data analysis elsewhere (11) and have discussed how 16 fine structure and hyperfine parameters were determined. Here we focus solely on the magnetic hyperfine tensors $\tilde{A}$ of the delocalized pair and the ferric site. The two irons of the Fe^{2+}-Fe^{3+} pair are indistinguishable within the excellent resolution of the spectra; for instance, the corresponding components of the magnetic hyperfine tensors cannot differ by more than 5%. Secondly, the A-tensor of the Fe^{3+} site has positive components. This observation is highly significant. Since a tetrahedral sulfur environment always yields a high-spin configuration, the A-tensor of a ferric FeS_4 site is dominated by the Fermi contact term, which is isotropic and negative. This implies that the local spin of the Fe^{3+} site ($S_3 = 5/2$) is aligned to the system spin S = 2 in such a way that its projection onto the direction of **S** is negative. Figure 5 shows a vector coupling diagram constructed from **S**, $\mathbf{S_3}$ and the spin $\mathbf{S_{12}}$ of the delocalized pair. In a Mössbauer experiment one measures $A_3\langle\mathbf{S}\rangle$ where $\langle\mathbf{S}\rangle$ is the expectation value of the system spin **S** and A_3 the observed magnetic hyperfine coupling constant (which we take for simplicity as the average of A_x, A_y and A_z; see Table I).

Table I Hyperfine Parameters of Reduced Ferredoxin II

Site	A_x(MHz)	A_y(MHz)	A_z(MHz)	A_{av}(MHz)	A_{theory}(MHz)
I	-20.5	-20.5	-16.4	-19.1	$A_A = A_B = -19.2$
II	+13.7	+15.8	+17.3	+15.6	$A_C = +16.7$

From the relation $A_3\langle\mathbf{S}\rangle = a_3\langle\mathbf{S_3}\rangle$ we can relate the measured value for A_3 to the "uncoupled" value a_3, which is $a_3 \approx -20$ MHz for a ferric FeS_4 site (11). From the vector coupling diagram we obtain

$$\langle S_3\rangle = |S_3| \cos\alpha \cos\beta \ \mathbf{z} = \frac{S(S+1) + S_3(S_3+1) - S_{12}(S_{12}+1)}{2S(S+1)} \langle S\rangle$$

and therefore $\langle S_3\rangle = -(5/6)\langle S\rangle$ for $S_{12} = 9/2$ and $\langle S_3\rangle = -(1/12)\langle S\rangle$ for $S_{12} = 7/2$. For $S_{12} = 9/2$ we obtain $A_3 = +16.7$ MHz which compares well with the experimental value $A_3(av) = +15.6$ MHz. For $S_{12} = 7/2$ we obtain $A_3 = +1.6$ MHz which is clearly in conflict with the data. All other allowed values for S_{12} yield the wrong sign for A_3. Thus, we arrive at the result that the delocalized dimer has $S_{12} = 9/2$. When this analysis was suggested by Münck and Kent (23) it was not clear whether the vector coupling diagram was applicable for the present situation. We have shown subsequently, however, that S_{12} remains a good quantum number in a model containing double exchange (11). Note that the very presence of the localized Fe^{3+} site allows us to determine the spin S_{12} of the delocalized dimer.

Formally, $S_{12} = 9/2$ results from ferromagnetic coupling of a ferric ($S_1 = 5/2$) to a ferrous ($S_2 = 2$) ion. One thus could be tempted to describe the dimer with Equation

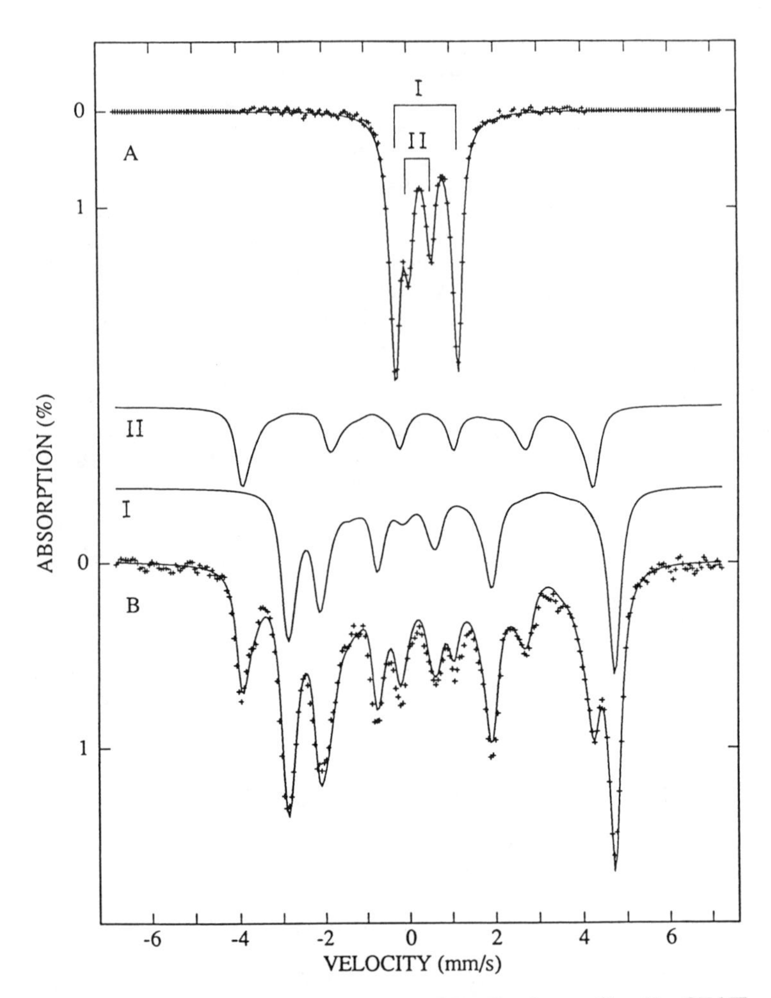

Figure 4. Mössbauer spectra of the reduced Fe_3S_4 cluster (S = 2) of Fd II recorded at 4.2 K in zero field (A) and at 1.3 K in parallel applied field of 1.0 Tesla (B). The solid lines in (B) are theoretical curves for the delocalized pair (I) and the Fe^{3+} site (II). The sum is drawn through the data. Parameters, fitting procedures and additional spectra are given in ref (11).

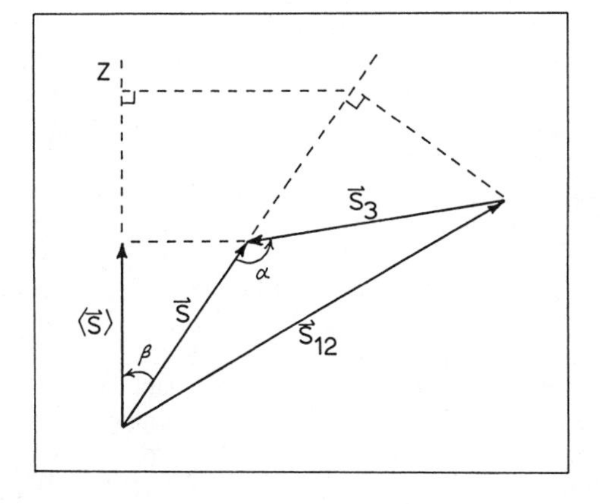

Figure 5. Vector coupling diagram for the three sites of the Fe_3S_4 cluster of Fd II. S = 2 is the system spin and S_3 = 5/2 is the spin of the Fe^{3+} site. S_{12} is the spin of the delocalized dimer. **z** is the direction for which <S> is evaluated.

(2) choosing $J < 0$. It is, however, not clear at all which iron would have $S_1 = 5/2$ and which iron would have $S_2 = 2$. Thus it is clear that some modifications are in order. Since Fe_2S_2 cluster have localized Fe^{2+} and Fe^{3+} sites which are antiferromagnetically coupled, and since the delocalized Fe^{2+}-Fe^{3+} pair exhibits ferromagnetic coupling, one suspects that valence delocalization and magnetism are linked in mixed valence clusters. Anderson and Hasegawa (10) have shown that a delocalized electron provides a coupling mechanism (called double exchange) which causes a ferromagnetic alignment of the spins.

Heisenberg Exchange and Double Exchange

In this section we outline, in a very simplified way, the main differences in spin coupling mechanisms between homovalent and mixed valence compounds. Consider a binuclear complex with metals A and B and assume that each metal contains one unpaired electron allocated to orbitals a and b, respectively (see Figure 6). Let us assume further that the states $|a\rangle$ and $|b\rangle$ contain the contributions of the ligands (including the bridging ligands) and that we have orthogonalized the two functions, $\langle a|b\rangle = 0$. If only those two "active" electrons are taken into account the electrostatic part of the Hamiltonian is

$$\mathbf{H} = \mathbf{h}(1) + \mathbf{h}(2) + r_{12}^{-1} \tag{3}$$

where $\mathbf{h}(i)$ is the one-electron Hamiltonian containing the kinetic energy of electron i and the potential energy of i due to the nuclei A and B and all "non-active" electrons; r_{12}^{-1} describes the repulsion between the two active electrons. By allocating one electron each to A and B, we can form the triplet $^3\Gamma_u$ [$|ab|$; $|\bar{a}\bar{b}|$; $(|a\bar{b}| + |\bar{a}b|)/\sqrt{2}$] and the singlet $|^1\Gamma_g\rangle = (|a\bar{b}| - |\bar{a}b|)/\sqrt{2}$. Here $|a\bar{b}|$ is the Slater determinant containing one "spin-up" electron in a and one "spin-down" electron in b. By accommodating two electrons either on A or on B we obtain the ionic configurations $|^1\Gamma_g\rangle^{ex} = (|a\bar{a}| + |b\bar{b}|)/\sqrt{2}$ and $|^1\Gamma_u\rangle^{ex} = (|a\bar{a}| - |b\bar{b}|)/\sqrt{2}$. These will be highly excited configurations (see Figure 6) at energy $\approx 2\alpha + U$, where $\alpha = \langle a(1)|\mathbf{h}(1)|a(1)\rangle = \langle b(1)|\mathbf{h}(1)|b(1)\rangle$ and where $U = \langle a\bar{a}|1/r_{12}|a\bar{a}\rangle = \langle b\bar{b}|1/r_{12}|b\bar{b}\rangle$ is the monocentric Coulomb repulsion integral. The Hamiltonian Equation (3) mixes the $^1\Gamma_g$ singlets, yielding the matrix

$\|^1\Gamma_g\rangle$	$\|^1\Gamma_g\rangle^{ex}$
$2\alpha + k + j$	2β
2β	$2\alpha + U + j$

where $\beta = \langle a(1)|\mathbf{h}(1)|b(1)\rangle$ is a transfer integral. (For a more complete treatment of the off-diagonal term see ref (44)). For completeness we have included the inter-atomic Coulomb (k) and exchange (j) integrals. Since $U >> |\beta|$ and since $k << U$ the energy of the ground singlet shifts by an amount $E_T - E_S = J = -2j + 4\beta^2/(U - k)$ relative to the triplet. The quantity j is positive and gives a ferromagnetic contribution whereas the second term lowers the energy of the singlet and gives an antiferromagnetic contribution. The energy difference between the ground singlet and triplet can be described formally by the Heisenberg Hamiltonian

$$H = JS_A \cdot S_B \quad (4)$$

where S_A and S_B are the spins of A and B (here $S_A = S_B = 1/2$). The energies of Equation (4) are given by

$$E(S) = (J/2)S(S+1) \quad (5)$$

where S = 0,1 are the system spins (we have dropped constant terms).

The theory of inter-atomic exchange interactions (super-exchange) has been reviewed in many excellent articles; see for instance Anderson (24), Martin (25), and Ginsberg (26) and the book edited by Willet, Gatteschi, and Kahn (27). Hay, Thibeault and Hoffmann (28) and Kahn and Briat (29) have developed theories which view the exchange interactions from a molecular orbital perspective.

Next, we consider the mixed valence case by adding one more electron to the system. We will call it the "extra" electron and assume that it is mobile, i.e. able to delocalize between A and B. By allocating two electrons on A into the orbitals a_1 and a_2 and keeping one electron on B in b_1 we can form a quartet and two doublets. In Figure 7 (left) we have indicated the distribution which gives the state with S=3/2 and M_S=+3/2. Let us designate this quartet state by $|A\rangle = |a_1b_1a_2|$ to indicate that the extra electron is localized on A. In this state the system has energy $\langle A|\mathbf{H}|A\rangle = \varepsilon_A(S = 3/2)$. If the excess electron occupies orbital b_2 we obtain the configuration $|B\rangle = |a_1b_1b_2|$ with energy $\langle B|\mathbf{H}|B\rangle = \varepsilon_B(S = 3/2)$. Amending Equation (3) for the present case the Hamiltonian matrix involving $|A\rangle$ and $|B\rangle$ can be written as

	$|A\rangle$	$|B\rangle$
$|A\rangle$	$\varepsilon_A(3/2)$	β_2
$|B\rangle$	β_2	$\varepsilon_B(3/2)$

(6)

where $\beta_2 = \langle a_2(1)|\mathbf{h}(1)|b_2(1)\rangle$ is a one-electron transfer or hopping integral (we have neglected contributions to β_2 from the r_{12}^{-1} term and we have also assumed that the matrix element $\langle a_1|\mathbf{h}|b_2\rangle = \langle a_2|\mathbf{h}|b_1\rangle = 0$). Since β_2 mixes the two configurations, it describes the tendency of the mixed valence system to delocalize. Two extreme cases are of interest. For $\varepsilon_B - \varepsilon_A >> |\beta_2|$ the lowest quartet state is $|A\rangle$ and the electron is thus localized on A. (ε_A and ε_B may differ because the sites are inequivalent or because of vibronic effects (31)). For $\varepsilon_A = \varepsilon_B$, on the other hand, the <u>two</u> quartets $|A\rangle$ and $|B\rangle$ mix to form the two <u>delocalized</u> states

$$\Phi_{1,2} = 1/(\sqrt{2})(|A\rangle \pm |B\rangle) \quad (7)$$

at energies $E_{1,2} = \varepsilon_A(3/2) \pm \beta_2$. Thus, if the two site energies are equal the two states may resonantly interact, and delocalization occurs readily for $\beta_2 \neq 0$. This is in strong contrast to the homovalent case where transfer of an electron requires expenditure of the large repulsion energy U ($\approx 10^5$ cm^{-1}). Since the transfer process is caused by electrostatic interactions, the extra electron does not change its spin upon hopping. Since intra-atomic exchange between the delocalized electron and the electrons in a_1 and in b_1 is ferromagnetic (Hund's rule), this mechanism favors a parallel alignment of the spins of the a_1 and b_1 electrons. More generally, the mobile electron mediates ferromagnetic coupling between the site spins S_0 of sites A and B.

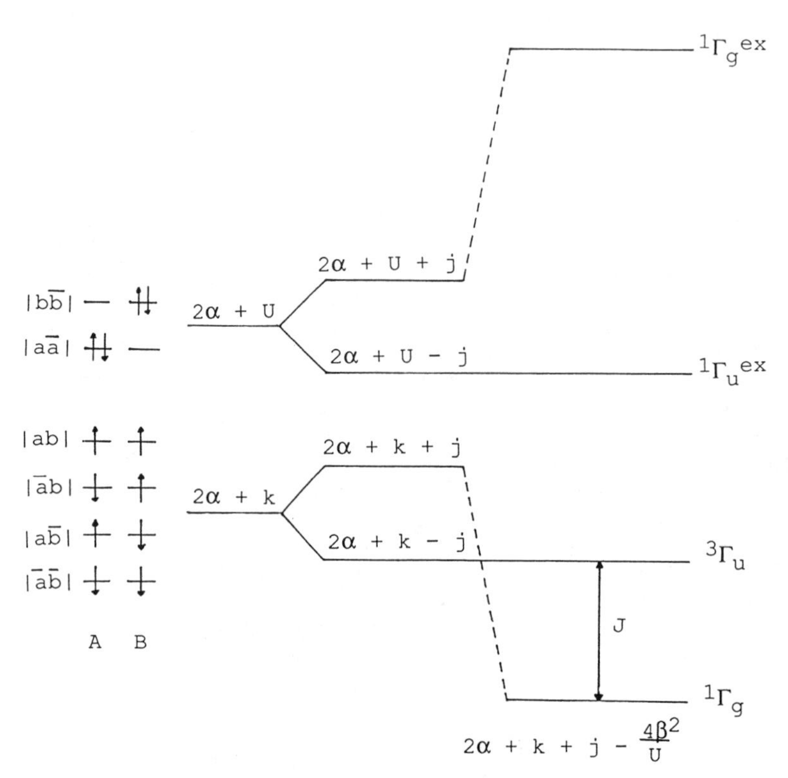

Figure 6. Illustration of exchange interactions in homovalent system consisting of two metal sites A and B. The system contains two electrons. The six distinct microstates are indicated on the left. The antiferromagnetic contribution results from mixing an excited state ionic configuration with the ground state singlet.

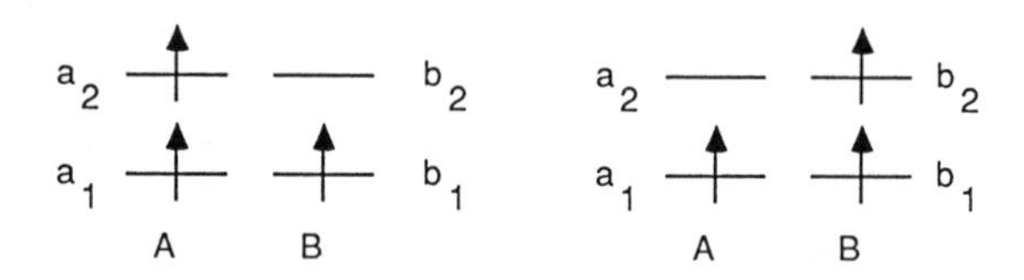

Figure 7. The two configurations resulting from placing the extra electron on A (left) and B (right).

Since only electrostatic interactions are considered, configurations with different system spin do not mix. One can therefore set up a separate matrix for each state with system spin S. If intra-atomic exchange of the mobile electron is large compared to the transfer integral β (this case is called double exchange) one obtains a 2x2 matrix like that of Matrix (6). Anderson and Hasegawa (10) have shown that the off-diagonal element in Matrix (6) can then be written as $\beta(S + 1/2)/(2S_0 + 1)$. This yields for $\varepsilon_A(S) = \varepsilon_B(S)$ a resonance splitting of $\pm B(S + 1/2)$ where $B = \beta/(2S_0 + 1)$ and $\beta = \langle a|\mathbf{h}|b\rangle$. (The factor $(S + 1/2)/(2S_0 + 1)$ is a Racah coefficient; for more details see below.) This dependence on $(S + 1/2)$ rather than $S(S + 1)$ as in the usual Heisenberg mechanism is the distinctive feature of double exchange [If a bridging ligand links A with B one possible transfer process, originally invoked by Zener (9), consists of two simultaneous movements of electrons (hence the name double exchange): transfer of an electron from A to the ligand coupled with a transfer from the ligand to B.].

Borshch et al. (30) have analyzed an Anderson-Hasegawa molecule containing three electrons, and they found that the Heisenberg term for inter-atomic exchange can be incorporated into the matrix; see also (31). They also have suggested that this scheme would be a good point of departure for a system with arbitrary site spins $S_A = S_B$ to which the mobile electron with spin s can couple. Thus Borshch et al. propose a matrix

$$\begin{bmatrix} J'\mathbf{S_A}\cdot\mathbf{s} + J(\mathbf{S_A} + \mathbf{s})\cdot\mathbf{S_B} & \beta\cdot\mathbf{1} \\ \beta\cdot\mathbf{1} & J'\mathbf{S_B}\cdot\mathbf{s} + J(\mathbf{S_B} + \mathbf{s})\cdot\mathbf{S_A} \end{bmatrix} \tag{8}$$

where J' describes the intra-atomic exchange of the mobile electron and where $\mathbf{1}$ is the unit operator (see also Appendix II of ref (10)). We will assume that J' is so large that non-Hund states can be ignored. The mobile electron couples then only in one way, rather than two, to $\mathbf{S_A}$ and $\mathbf{S_B}$. Writing $\mathbf{S_A} + \mathbf{s} = {}^A\mathbf{S_A}$, $\mathbf{S_B} + \mathbf{s} = {}^B\mathbf{S_B}$ and ${}^A\mathbf{S_A} + {}^A\mathbf{S_B} = {}^B\mathbf{S_A} + {}^B\mathbf{S_B} = \mathbf{S}$ the scheme can be written for a fixed S as

$$\begin{bmatrix} E_A + J^A\mathbf{S_A}\cdot{}^A\mathbf{S_B} & B(S + 1/2) \\ B(S + 1/2) & E_B + J^B\mathbf{S_A}\cdot{}^B\mathbf{S_B} \end{bmatrix} \tag{9}$$

where the superscript A (or B) indicates that the quantity is to be taken for the configuration where the extra electron is on A (or B). Thus, if the extra electron of a Fe^{2+}-Fe^{3+} dimer resides on A we have ${}^AS_A = 2$ and ${}^AS_B = 5/2$; if it resides on B we use ${}^BS_A = 5/2$ and ${}^BS_B = 2$.

In scheme (9) we could have substituted on the diagonal $S(S + 1)/2$ for ${}^A\mathbf{S_A}\cdot{}^A\mathbf{S_B}$ and ${}^B\mathbf{S_A}\cdot{}^B\mathbf{S_B}$. As it stands, the matrix is a hybrid because it contains matrix elements on the off-diagonal and operators on the diagonal. The present form of (9) is, however, despite its awkward form, a more useful starting point for setting up a spin Hamiltonian for the description of trinuclear and tetranuclear clusters containing delocalized dimers.

For ${}^AJ = {}^BJ = J$ and $E_A = E_B$ the Matrix (9) yields the eigenvalue spectrum (constant terms suppressed)

$$E(S) = (J/2)S(S + 1) \pm B(S + 1/2) \tag{10}$$

Figure 8 shows the energy levels for an Fe^{2+}-Fe^{3+} dimer according to Equation (10). For $B = 0$ and $J > 0$ (antiferromagnetic Heisenberg exchange) the familiar spin ladder results, although each multiplet occurs twice (because $E_A = E_B$). For $B \neq 0$ each multiplet of total spin S is split by double exchange into odd and even states. In our simple model this splitting is symmetric. Noodleman and Baerends (35) have pointed out that the double exchange splitting may be asymmetric, although the total splitting is still $2B(S + 1/2)$. For $B/J \geq 4.5$ the double exchange term dominates and the "ferromagnetic" state with $S = 9/2$ will be the ground state. Note that the double exchange splitting does not depend on the sign of B. This is generally true for dimers. For trimers and tetramers, however, the splittings may depend also on the sign of B; see, for instance, Belinskii (32) who has given correlation diagrams for trimers with D_{3h} symmetry.

An interesting situation occurs when $E_A - E_B$ is large compared to the double exchange term in Matrix (9), $E_A - E_B >> |B|(S + 1/2)$. Solving the secular equation and expanding the root shows that the double exchange term lowers the energy of the ground state by $-B^2(S + 1/2)^2/(E_A - E_B)$. Dropping the spin independent contribution, we can see that the double exchange term has a contribution proportional to $S(S+1)$ which we can combine with the Heisenberg term to obtain an effective coupling constant $J_{eff} = J - 2B^2/(E_A - E_B)$. Thus, the double exchange term gives a ferromagnetic contribution to the coupling constant. A double exchange contribution to J is difficult to recognize experimentally unless, in the case of iron compounds, one has information about delocalization from isomer shift and quadrupole splitting data.

The literature on double exchange is not very extensive. The main papers dealing with the problem discussed here are the original work by Anderson and Hasegawa (10), and papers by Karpenko (33), Borshch et al. (30), Belinskii et al. (34) and Girerd (31). Particularly noteworthy is the analysis by Noodleman and Baerends (35) in their theoretical study of the electronic structure of ferredoxins with Fe_2S_2 cores. Quite recently, Belinskii (32) has published an extensive theoretical study for mixed-valence trimers with D_{3h} symmetry.

The parameter B has not yet been determined for any iron compound. However, Noodleman and Baerends (35) have calculated this parameter for reduced Fe_2S_2 clusters and found $B \approx 500\ cm^{-1}$; experimentally determined J-values are $J \approx +200 cm^{-1}$.

Coupling Model for the Fe_3S_4 Cluster of Ferredoxin II

Mössbauer, EPR, ENDOR, low temperature MCD and susceptibility data are generally analyzed in the framework of a spin Hamiltonian. This Hamiltonian contains the spin operator **S** which for a cluster refers to the total (system) spin rather than to the local spins $\mathbf{S}_i$, of the individual metals constituting the cluster. Thus, the spin Hamiltonian parameters describing zero-field splittings (D, E), Zeeman interactions ($\tilde{g}$), and magnetic hyperfine interactions ($\tilde{A}$) refer to the system state. The purpose of a coupling model is to explain the experimental data by determining coupling constants J (and B) and to express the measured spin Hamiltonian parameters by parameters describing the individual sites. Thus, for reduced Fe_2S_2 clusters the measured magnetic hyperfine coupling constants $A(Fe^{3+})$ and $A(Fe^{2+})$ are related (22) to those of the "uncoupled" ions by $A(Fe^{3+}) = (7/3)a(Fe^{3+})$ and $A(Fe^{2+}) = -(4/3)a(Fe^{2+})$. For a tetrahedral sulfur coordination, which prevails in iron-sulfur clusters, one may use a-values which have been obtained for ferric and

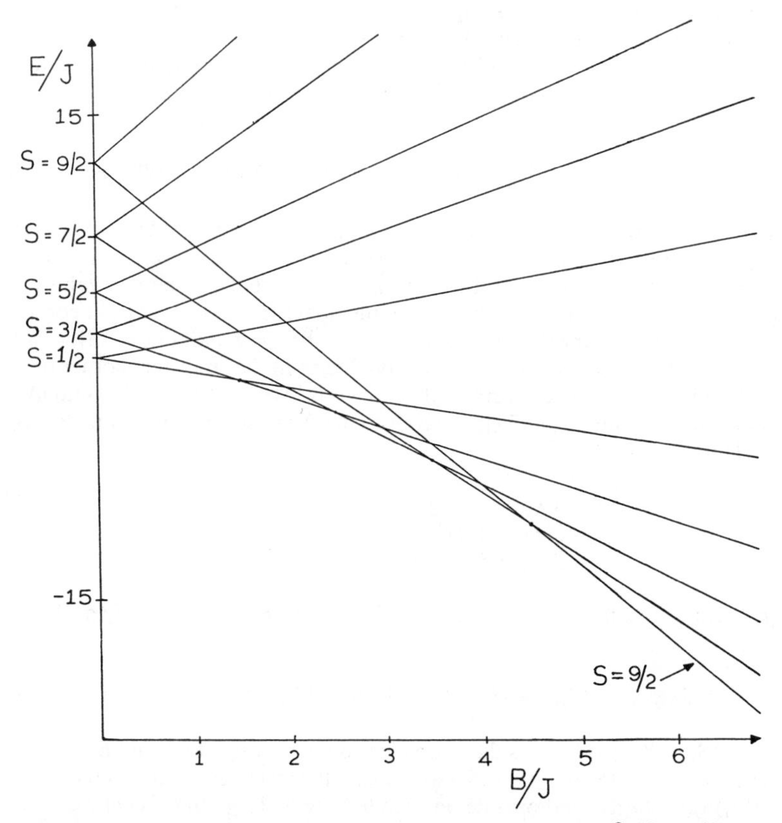

Figure 8. Energy level diagram for exchange coupled $Fe^{2+}(S_A = 2)$-$Fe^{3+}(S_B = 5/2)$ dimer plotted versus the delocalization (double exchange) parameter B. The diagram is symmetric around B = 0.

ferrous FeS_4 complexes. It is argued in ref (11) that $a(Fe^{3+}) \approx -20$ MHz and $a(Fe^{2+}) \approx -22.2$ MHz are reasonable choices.

Let us assume that we have an A-B dimer exhibiting double exchange in addition to the usual Heisenberg exchange. How can we incorporate a double exchange term into the spin Hamiltonian? The considerations of the preceding section suggest that we need a Hamiltonian that mixes the configurations $|A>$ and $|B>$. Let us write $|A> \equiv |{}^{A}S_A{}^{A}S_B;S_{AB}>$ and $|B> \equiv |{}^{B}S_A{}^{B}S_B;S_{AB}>$ for the configuration with the extra electron on A and B, respectively; ${}^{B}S_A$ is the spin of site A when the excess electron is on B and ${}^{A}S_A + {}^{A}S_B = {}^{B}S_A + {}^{B}S_B = \mathbf{S}_{AB}$. It is our goal to reproduce the Matrix of (9). The following effective Hamiltonian evaluated in the basis $|A>$ and $|B>$ will give the desired result.

$$\mathbf{H} = [{}^{A}J_{AB}{}^{A}S_A \cdot {}^{A}S_B + E_A]\mathbf{O}_A + [{}^{B}J_{AB}{}^{B}S_A \cdot {}^{B}S_B + E_B]\mathbf{O}_B + B\mathbf{T}_{AB} \qquad (11)$$

$\mathbf{O}_A$ and $\mathbf{O}_B$ are occupation operators which have the effect $\mathbf{O}_A|A> = |A>$, $\mathbf{O}_A|B> = 0$, $\mathbf{O}_B|B> = |B>$ and $\mathbf{O}_B|A> = 0$. With these operators we can control diagonal elements in Matrix (9). We need these operators because the numerical values for the site spins S_A and S_B are interchanged in the upper left and lower right corner of (9); for an Fe^{2+}-Fe^{3+} dimer we would have ${}^{A}S_A = {}^{B}S_B = 2$ and ${}^{B}S_A = {}^{A}S_B = 5/2$.

The term $B\mathbf{T}_{AB}$ reproduces the off-diagonal term in the "transfer" corner of Matrix (9). $\mathbf{T}_{AB}$ is a transfer operator that converts $|A>$ into $|B>$ (or $|B>$ into $|A>$) and "looks up" the dimer spin S_{AB} to produce the Anderson-Hasegawa factor (S_{AB} + 1/2). Thus

$$\begin{aligned} \mathbf{T}_{AB}|A> &= \mathbf{T}_{AB}|{}^{A}S_A{}^{A}S_B;S_{AB}> \\ &= (S_{AB} + 1/2)|{}^{B}S_A{}^{B}S_B;S_{AB}> \\ &= (S_{AB} + 1/2)|B>. \end{aligned} \qquad (12)$$

If $\mathbf{T}_{AB}$ operates in a three-spin problem on the state $|A> = |({}^{A}S_A{}^{A}S_B)S_{AB},{}^{A}S_C;S>$ we obtain

$$\mathbf{T}_{AB}|A> = (S_{AB} + 1/2)|({}^{B}S_A{}^{B}S_B)S_{AB},{}^{B}S_C;S> \qquad (13).$$

Note that ${}^{A}S_C = {}^{B}S_C$ as long as the extra electron is confined to A and B. If $\mathbf{T}_{AB}$ has to operate on $|{}^{A}S_A,({}^{A}S_B{}^{A}S_C)S_{BC};S>$ we expand the state vector first in the basis $|({}^{A}S_A{}^{A}S_B)S_{AB},{}^{A}S_C;S>$ and operate on each term with $\mathbf{T}_{AB}$. In ref (11) we have split the transfer operator into a pure transfer operator and an operator $\mathbf{V}_{AB}$ that produces the (S_{AB} + 1/2) factor; in hindsight this seems to be of little benefit.

It may be useful here to comment briefly on the origin of the (S + 1/2) factor. Designating the core spins of A and B as $\mathbf{S}_A{}^0$ and $\mathbf{S}_B{}^0$ we write the kets $|A>$ and $|B>$ as follows: $|A> = |(S_A{}^0 s_A)\ S_A{}^0 \pm 1/2,\ S_B{}^0;\ S>$ and $|B> = |S_A{}^0,\ (s_B S_B{}^0)\ S_B{}^0 \pm 1/2;\ S>$ where the subscript of the spin s = 1/2 of the extra electron designates its location. In taking $<A|\mathbf{H}|B>$, where $\mathbf{H} = \Sigma_i \mathbf{h}(i) + \Sigma_{i,j} r_{ij}{}^{-1}$ (see Equation (3)), the Hamiltonian transfers the extra electron from B to A, giving $<A|\mathbf{H}|B> = \beta<A|A'>$ with $|A'> = |S_A{}^0,\ (s_A S_B{}^0)\ S_B{}^0 \pm 1/2;\ S>$. The transfer process has transformed the problem of coupling the four spins $\mathbf{S}_A{}^0$, $\mathbf{S}_B{}^0$, $\mathbf{s}_A$ and $\mathbf{s}_B$ into one involving the three spins $\mathbf{S}_A{}^0$, $\mathbf{S}_B{}^0$, and $\mathbf{s}_A$. For a mixed-valence dimer we have $S_A{}^0 = S_B{}^0 = S_0$ and obtain

$$\begin{aligned} <A|A'> &= (-1)^{-2S_0-S+1/2}(2S'+1)W(S_0\ S'\ S'\ S_0;\ 1/2\ S) \\ &= (-1)^{1-2S}(S+1/2)/(2S_0+1) \end{aligned} \qquad (14)$$

where S' is either $S' = S_0 + 1/2$ or $S' = S_0 - 1/2$, and where W(a b c d ; e f) is the Racah recoupling coefficient. Since S is half-integral for a mixed valence dimer, the phase can be dropped in Equation (14). Heisenberg exchange, $\mathbf{H} = JS_A{\cdot}S_B = JS_AS_B\cos\theta$, is proportional to $\cos\theta$, where θ is the angle between $\mathbf{S_A}$ and S_B. Anderson and Hasegawa (10) have shown with semiclassical $[(S+1/2)/(2S_0+1) \approx S/2S_0]$ arguments that $S/2S_0 \cong \cos\theta/2$ where θ is the angle between the site spins $S_A{}^0$ and $S_B{}^0$.

For a symmetric dimer ($E_A = E_B$, ${}^AJ_{AB} = {}^BJ_{AB} = J$) Equation (11) gives the solution (10) with $S = S_{AB}$.

We have generalized Equation (11) for the description of trinuclear and tetranuclear clusters and have applied the new Hamiltonian to the Fe_3S_4 cluster of Fd II. Details have been given elsewhere (11). Here we simply state the assumptions and quote the results.

For Fd II we have assumed, in accord with the experimental evidence, that the Fe^{2+}-Fe^{3+} dimer (A-B) is the only delocalized pair, i.e. $B_{AB} = B$ and $B_{AC} = B_{BC} = 0$. We have further assumed that all coupling constants J are equal; this assumption can be somewhat relaxed without effecting the results in an essential way (11). With these assumptions the energy levels are given by

$$E(S) = (J/2)S(S + 1) \pm B(S_{AB} + 1/2) \tag{15}$$

where S is the system spin and where S_{AB} is the spin of the delocalized pair. The lowest energy levels are plotted versus B/J for J > 0 in Figure 9. For B/J > 2 we obtain the desired solution, namely a state with S = 2 and $S_{AB} = 9/2$. For this solution the state vector takes the form $|S_{AB};S\rangle_\pm = (1/\sqrt{2})[|A\rangle \pm |B\rangle]$ where $|A\rangle = |({}^AS_A{}^AS_B)S_{AB},S_C;S\rangle$ and $|B\rangle = |({}^BS_A{}^BS_B)S_{AB},S_C;S\rangle$. Our data do not allow us to discriminate between the symmetric and antisymmetric delocalized state, i.e. the sign of B is undetermined. By using an expression given in ref (11) we have computed with the above quoted values for a(Fe^{3+}) and a(Fe^{2+}) the magnetic hyperfine coupling constants. The results are listed in the last column of Table I. Considering the simplicity of the model, signs and magnitudes of the theoretical A-values are in excellent agreement with the experimental A-values.

At this juncture we may ask why Fe_2S_2 clusters exhibit localized valences whereas the subdimer of the Fe_3S_4 cluster is delocalized. At present we cannot answer this question fully; a contributing factor, however, seems to be the phenomenon of spin frustration. Consider a trinuclear cluster and the Hamiltonian of Equation (1) with all J's positive and assume that site A is Fe^{2+} ($S_A = 2$) and that sites B and C are Fe^{3+} ($S_B = S_C = 5/2$). If we couple the spins of A and B antiparallel according to $J_{AB}S_A{\cdot}S_B$, and then couple the spin of C antiparallel to that of B to satisfy $J_{BC}S_B{\cdot}S_C$, the spins of A and C come out parallel rather that antiparallel as desired by $J_{AC}S_A{\cdot}S_C$. In order to minimize the total energy each pair has to compromise. Assuming equal J's for simplicity the lowest energy of the system is attained for $S_{AB} = 5/2$ (rather than $S_{AB} = 1/2$) and total S = 0; as can be seen from Figure 9 the ground state is $|S_{AB} = 5/2, S = 0\rangle$ for B = 0. Since frustration has increased the spin of the A-B dimer already towards the ferromagnetic alignment, the dimer spin $S_{AB} = 9/2$ is attained for B/J > 2 for the trinuclear cluster whereas the isolated dimer of Figure 8 requires B/J > 4.5 to achieve the ferromagnetic ground state. Because of the (S_{AB} + 1/2) factor the trend towards delocalization is much stronger when the subdimer has $S_{AB} = 9/2$.

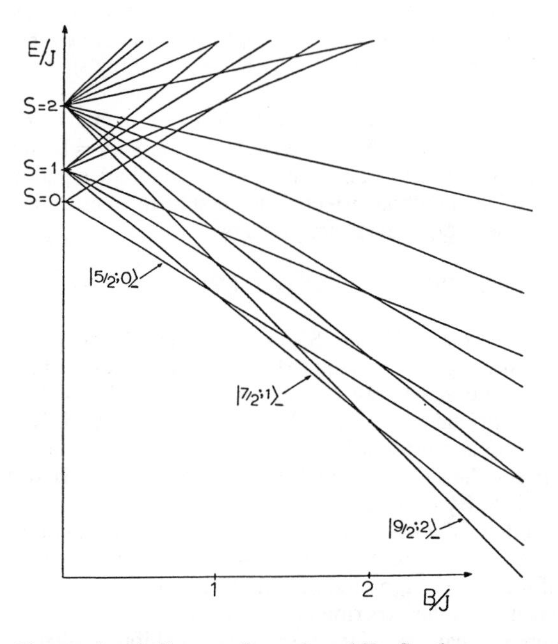

Figure 9. Energy level diagram for reduced Fe_3S_4 cluster according to Equation (12) plotted for antiferromagnetic ($J > 0$) Heisenberg exchange. For $B/J > 2$ the system ground state have $S = 2$ and $S_{AB} = 9/2$. The extra electron was allowed to delocalize between sites A and B.

Cubane Clusters with $[MFe_3S_4]$ Core ($M = Co^{2+}, Zn^{2+}, Cd^{2+}$)

In hindsight it seems obvious to describe the magnetic properties of the delocalized pair of Fd II with a theory which includes double exchange. In fact, for almost two years after the initial data were obtained we groped with the problem. During this time we developed, for lack of theoretical insight, an experimental strategy to elucidate further the electronic structure of cubanes. Since the Fe_3S_4 and Fe_4S_4 forms of the cluster could be readily interconverted (16), it seemed attractive to put a metal other than Fe into the open site of the Fe_3S_4 cluster. In particular, we anticipated that the diamagnetic Cu^{1+}, Zn^{2+}, and Ga^{3+}, with their full d-shell, would give interesting results. If the Fe_3S_4 cluster had indeed a cubane structure, the incorporation of Zn^{2+} would give minimal structural changes, perhaps perturbing the iron sites in a minor way. From a broader perspective one would look forward to studying novel clusters of the MFe_3S_4 type.

It turned out that Co^{2+} worked very nicely, yielding a cluster with a $CoFe_3S_4$ stoichiometry (36). Although the $CoFe_3S_4$ cluster yielded beautiful EPR spectra and well resolved Mössbauer spectra (36), the paramagnetism of the Co^{2+} ($S = 3/2$) left the problem of describing the magnetism of the cluster at the same complexity as that of Fe_4S_4.

The procedure of converting Fe_3S_4 into $ZnFe_3S_4$ is remarkably simple. First, the Fe_3S_4 cluster needs to be reduced, for instance by adding an excess of dithionite. Next, one adds an excess, typically 10 fold, of $Zn(NO_3)_2$ and incubates the mixture under anaerobic conditions for typically 2 hours. One can then study the product directly, i.e. without purification, or one can remove most of the excess Zn^{2+} with an anaerobic Sephadex column (our lowest Zn/Fe ratio had 1.3 Zn/3 Fe). Column purified and unpurified samples give identical Mössbauer and EPR spectra. For the quoted conditions the conversion is essentially complete, i.e. no discernible amount of unconverted Fe_3S_4 is left.

The incubation product is EPR-active, yielding reproducible spectra as shown in Figure 10. The observed resonances at $g = 9.8, 9.3, 4.8$, and ≈ 3.8 belong to a system with $S = 5/2$ (37). The lower panel in Figure 10 shows the association of the resonances (for $E/D \approx 0.25$) to the three Kramers doublets of the $S = 5/2$ system. Variable temperature EPR studies of the low field resonances at $g = 9.8$ and 9.3 (for an expanded view see Figure 1 of ref (37)) show that the doublet with $g_z = 9.8$ is the ground state and that the zero-field splitting parameter is $D \approx -2.7$ cm^{-1}. The sharp, derivative-type resonance at $g = 4.3$ belongs to a second $S = 5/2$ species; we have estimated from the EPR and Mössbauer spectra that it accounts for less than 10% of the total spin concentration.

Typical Mössbauer spectra of $ZnFe_3S_4$ cluster are shown in Figure 11A and B. We have recorded spectra over a wide range of temperature in magnetic fields up to 6.0 Tesla. The details of the analysis of the spectra have not been completed at this time; the following facts, however, have emerged. The 4.2 K spectrum (at this temperature only the electronic ground doublet is populated) consists of two spectral components appearing in a 2:1 intensity ratio. Such an intensity pattern is also observed at 50 K, under conditions where magnetic hyperfine interactions are averaged out because of fast electronic spin relaxation. Doublet 1 has $\Delta E_Q = 2.7$ mm/s and $\delta = 0.62$ mm/s, suggesting a ferrous FeS_4 site with a slightly smaller d-electron population than the ferrous ion of FeS_4 complexes. This site appears thus to be a trapped valence Fe^{2+} site. The two irons of site 2 have $\Delta E_Q \cong 1.6$ mm/s and $\delta =$

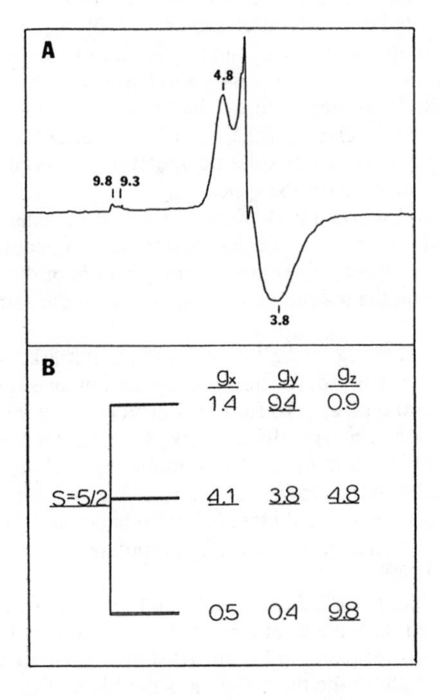

Figure 10. EPR spectrum of $[ZnFe_3S_4]^{1+}$ recorded at 8 K. The lower panel shows energy levels and effective g-values for an S = 5/2 system with D < 0 and E/D = 0.25, where D and E are the zero field splitting parameters.

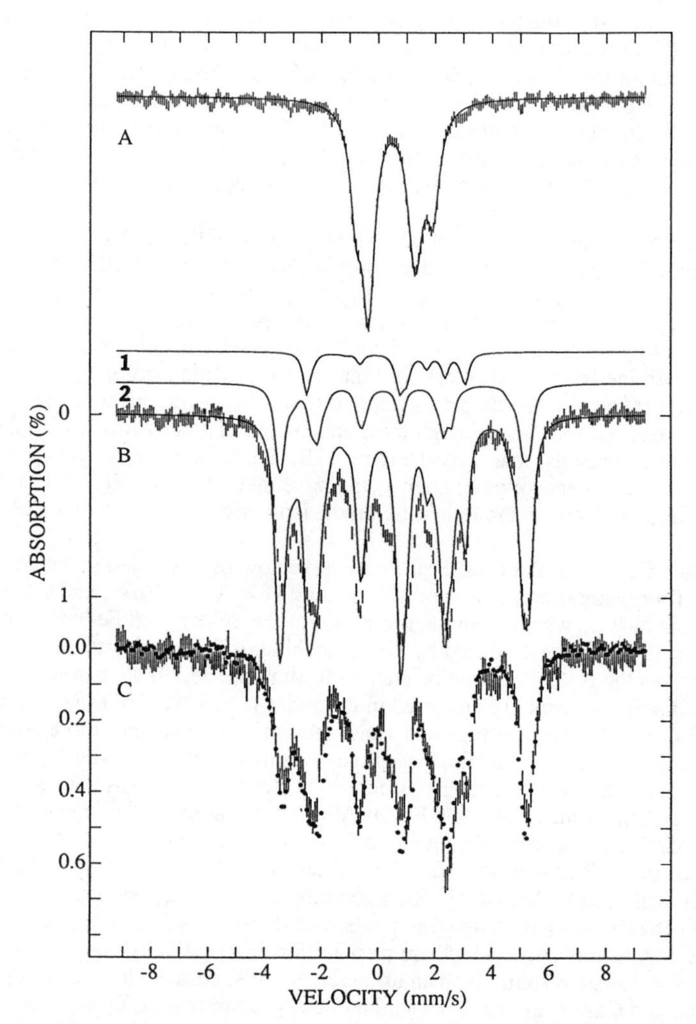

Figure 11. Mössbauer spectra of $[ZnFe_3S_4]^{1+}$ and $[CdFe_3S_4]^{1+}$. (A) and (B) show spectra of $[ZnFe_3S_4]^{1+}$ recorded at 50 K (A) and at 4.2 K (B). The spectra were recorded in a parallel field of 60 mTesla. Solid line in (A) is the result of least-squares fitting the spectrum to two doublets. Solid lines in (B) are theoretical curves for the delocalized pair (2) and the Fe^{2+} site (1). The sum is drawn through the data. For details see ref (37). In (C) the spectrum of $[ZnFe_3S_4]^{1+}$ (circles, same spectrum as in (B)) is compared with that (hatch marks) of $[CdFe_3S_4]^{1+}$. 25% of the Fe_3S_4 clusters in the Cd-treated sample remained unconverted, and we have subtracted their contribution from the raw data.

0.52 mm/s. These parameters suggest again an Fe^{2+}-Fe^{3+} pair as observed in the reduced Fe_3S_4 cluster, the pair having aquired a slightly increased ferrous character as witnessed by the increase of δ. With this information we can assign the core oxidation state of the cluster as $[ZnFe_3S_4]^{1+}$: we have formally two Fe^{2+} and one Fe^{3+} ion; since the cluster has S = 5/2, the system is in a Kramers state and, therefore, the Zn must be divalent (as expected). This suggests that the conversion process consists of incorporation of Zn^{2+} into the core of the reduced Fe_3S_4 cluster, followed by a one-electron reduction, the electron being supplied by the excess dithionite.

A closer inspection of the data show that the two Fe of the Fe^{2+}-Fe^{3+} pair are inequivalent. The current status of our analysis shows that ΔE_Q is different for both Fe of the pair; $\Delta E_Q(2a)$ = 1.7 mm/s and $\Delta E_Q(2b)$ = 1.5 mm/s. Similarily, the magnetic hyperfine interactions show a small, but discernible difference (37).

Since the lowest Kramers doublet of the S = 5/2 system is uniaxial, $g_z >> g_x, g_y$, one can determine from the spectrum in Figure 11B only the z-components of the magnetic hyperfine tensors. In strong applied fields, however, the spectra become also sensitive to A_x and A_y, although these spectra are distinctly less resolved (for well understood reasons) than that of Figure 11B. Since we have not yet completed the analysis of the A-tensor components, and have therefore not fully analysed the spin coupling, we focus in the following on the isomeric shifts and quadrupole splittings.

The three Fe sites of the Fd II Fe_3S_4 core have now been studied under conditions where the fourth site is empty and or occupied by Fe^{2+}, Co^{2+}, Zn^{2+}, and Cd^{2+}. Thus, we may ask to what extent the properties of the delocalized Fe^{2+}-Fe^{3+} pair and the single (third) site are modified by the occupation of the fourth site. Using the isomer shift as the primary oxidation state indicator (see 38,39) we can see that the delocalized pair is essentially unaltered in the series $[Fe_3S_4]^0$, $[CoFe_3S_4]^{2+}$ and $[Fe_4S_4]^{2+}$ (see Table II). On the other hand, the ferric site becomes more ferrous through the series. In fact, in $[Fe_4S_4]^{2+}$ it belongs to a second Fe^{2+}-Fe^{3+} pair, the two sites of which, however, can be discerned by their distinct ΔE_Q values. [Interestingly, the fourth and labile Fe has ΔE_Q = 1.32 mm/s. In aconitase (40), on the other hand, the iron with the smaller ΔE_Q (≈ 0.80 mm/s) is the one that can be removed from the Fe_4S_4 cluster. In other diamagnetic $[Fe_4S_4]^{2+}$ clusters the four sites are indistinguishable (8,41)]. Extrapolating from $[Fe_3S_4]^0$ suggests that the core of $[Fe_4S_4]^{2+}$ consists of two fragments with dimer spin $S_{AB} = S_{CD} = 9/2$ (with dominant double exchange) which are then antiferromagnetically coupled to give the observed S = 0 system spin. Such an arrangement of subdimers has been suggested by Aizman and Case (42) who have studied Fe_4S_4 cubanes with Xα valence bond theory. This suggests that the J-values published for cubanes (43) may need to be revised after a reanalysis of the data with a theory containing double exchange.

In $[Fe_4S_4]^{1+}$, $[CoFe_3S_4]^{1+}$ and $[ZnFe_3S_4]^{1+}$ the third site is formally Fe^{2+}. In $[Fe_4S_4]^{1+}$ the two Fe of the Fe^{2+}-Fe^{2+} pair exhibit less Fe^{2+} characteristics, as witnessed by the decrease in ΔE_Q relative to ferrous FeS_4 complexes. Some delocalization from the Fe^{2+}-Fe^{2+} pair into the Fe^{2+}-Fe^{3+} pair seems to have occurred. This is supported by an increase of δ for the Fe^{2+}-Fe^{3+} pair. This suggests that a spin coupling model for $[Fe_4S_4]^{1+}$ should include, in additon to the superexchange terms, double exchange terms linking the Fe^{2+}-Fe^{3+} pair with the two remaining ferrous sites. This, of course, becomes a formidable problem with many unknowns, and the reader will appreciate why the $[ZnFe_3S_4]^{1+}$ cluster is a very attractive "model" system.

Table II Isomer Shifts and Quadrupole Splittings of the Three Fe Sites of Fd II in Various States of the Cluster

Cluster	Formal Valences	δ mm/s	ΔE_Q mm/s	# of Fe	Cluster Spin	Ref
$[Fe_3S_4]^0$	Fe^{2+}~Fe^{3+}	0.46	1.47	2	S = 2	11
	Fe^{3+}	0.32	0.52	1		
$[Fe_4S_4]^{2+}$	Fe^{2+}~Fe^{3+}	0.45	1.32	2	S = 0	16
	Fe^{3+}--Fe^{2+}	0.45	1.32	1		
		0.41	0.55	1		
$[CoFe_3S_4]^{2+}$	Fe^{2+}~Fe^{3+}	0.44	1.35	2	S = 1/2	36
	Fe^{3+}--Co^{2+}	0.35	1.1	1		
$[Fe_4S_4]^{1+}$	Fe^{2+}~Fe^{3+}	0.51	1.07	2	S = 1/2	16
	Fe^{2+}--Fe^{2+}	0.60	1.67	2		
$[ZnFe_3S_4]^{1+}$	Fe^{2+}~Fe^{3+}	0.52	1.5,	1	S = 5/2	37
			1.7	1		
	Fe^{2+}--Zn^{2+}	0.62	2.7	1		
$[CoFe_3S_4]^{1+}$	Fe^{2+}~Fe^{3+}	0.53	1.28	2	S > 0	36
	Fe^{2+}--Co^{2+}	0.53	1.28	1		

We have started a program to incorporate other diamagnetic metals into the Fe_3S_4 core. In Figure 11C the Mössbauer spectrum of $[CdFe_3S_4]^{1+}$ (K. K. Surerus, I. Moura, J. J. G. Moura, J. LeGall and E. Münck, unpublished results) is compared to $[ZnFe_3S_4]^{1+}$. Both clusters have S = 5/2 and similar, but distinguishable EPR signals. Considering that the Cd^{2+} ion is much larger than Zn^{2+} we were surprised by the similarity of the Mössbauer spectra. In order to illuminate the effects of an additional positive charge on the electronic structure of the cluster, we have also attempted to incorporate a trivalent diamagnetic metal into the core. Incubation of reduced Fd II in the presence of Ga^{3+} yielded, in two preliminary experiments, EPR signals quite similar to those of $[ZnFe_3S_4]^{1+}$. It will be interesting to compare the Mössbauer spectra of $[GaFe_3S_4]^{2+}$ with those of the Zn-containing analog.

In summary, by combining the information obtained from the new mixed-metal clusters with the progress made in describing the spin coupling of delocalized mixed valence dimers with double exchange, we can anticipate new insight into the electronic structure of many clusters of interest in bio-inorganic chemistry.

Acknowledgments

We wish to acknowledge the contributions of Drs. Isabel Moura, José J. G. Moura and Jean LeGall who have worked with us for the past eight years on all projects involving ferredoxin II. This work was supported by a grant from the National Science Foundation and by a NATO Collaboration Research grant.

Literature Cited

1. Gibson, J. F.; Hall, D. O.; Thornley, J. H. M.; Whatley, F. R. *Proc. Natl. Acad. Sci.* **1966**, *56*, 987-990.
2. Dunham, W. R.; Bearden, A. J.; Salmeen, I. T.; Palmer, G.; Sands, R. H.; Orme-Johnson, W. H.; Beinert, H. *Biochem. Biophys. Acta* **1971**, *253*, 134.
3. Münck, E.; Debrunner, P. G.; Tsibris, J. C. M.; Gunsalas, I. C. *Biochemistry* **1972**, *11*, 855-863.
4. Bertrand, P.; Gayda, J. P. *Biochem. Biophys. Acta* **1979**, *579*, 107-121.
5. Middleton, P.; Dickson, D. P. E.; Johnson, C. E.; Rush, J. D. *Eur. J. Biochem.* **1980**, *104*, 289.
6. Papaefthymiou, V.; Millar, M. M.; Münck, E. *Inorg. Chem.* **1986**, *25*, 3010-3014.
7. Middleton, P.; Dickson, D. P. E.; Johnson, C. E.; Rush, J. D. *Eur. J. Biochem.* **1978**, *88*, 135-141.
8. Christner, J.A.; Janick, P. A.; Siegel, L. M.; Münck, E. *J. Biol. Chem.* **1983**, *258*, 11157-11164.
9. Zener, C. *Phys. Rev.* **1951**, *82*, 403-405.
10. Anderson, P. W.; Hasegawa, H. *Phys. Rev.* **1955**, *100*, 675-681.
11. Papaefthymiou, V.; Girerd, J.-J.; Moura, I.; Moura, J. J. G.; Münck, E. *J. Am. Chem. Soc.* **1987**, *109*, 4703-4710.
12. Huynh, B. H.; Moura, J. J. G.; Moura, I.; Kent, T. A.; LeGall, J.; Xavier, A.; Münck, E. *J. Biol. Chem.* **1980**, *255*, 3242-3244.
13. Gayda, J.-P.; Bertrand, P.; Theodule, F.-X.; Moura, J. J. G. *J. Chem. Phys.* **1982**, *77*, 3387-3391.
14. Antonio, M R.; Averill, B. A.; Moura, I.; Moura, J. J. G.; Orme-Johnson, W. H.; Teo, B.-K.; Xavier, A. V. *J. Biol. Chem.* **1982**, *257*, 6646-6649.
15. Thomson, A. J.; Robinson, A. E.; Johnson, M. K.; Moura, J. J. G.; Moura, I.; Xavier, A. V.; LeGall, J. *Biochim. Biophys. Acta* **1981**, *670*, 93-100.

16. Moura, J. J. G.; Moura, I.; Kent, T. A.; Lipscomb, J. D.; Huynh, B. H.; LeGall, J.; Xavier, A. V.; Münck, E. *J. Biol. Chem.* **1982**, *257*, 6259-6267.
17. Beinert, H.; Thomson, A. J. *Arch. Biochem. Biophys.* **1983**, *222*, 333-361.
18. Kent, T. A.; Huynh, B. H.; Münck, E. *Proc. Natl. Acad. Sci.* **1980**, *77*, 6574-6576.
19. Guigliarelli, B.; More, C.; Bertrand, P.; Gayda, J. P. *J. Chem. Phys.* **1986**, *85*, 2774-2778.
20. Winkler, H.; Kostikas, A.; Petrouleas, V.; Simopoulos, A.; Trautwein, X. A. *Hyp. Int.* **1986**, *29*, 1347-1350.
21. Fee, J. A.; Findling, K. L.; Yoshida, T.; Hille, R,; Tarr, G. E.; Hearshen, D. O.; Dunham, W. R.; Day, E. P.; Kent, T. A.; Münck, E. *J. Biol. Chem.* **1984**, *259*, 124-133.
22. Sands, R. H.; Dunham, W. R. *Quart. Rev. Biophys.* **1975**, *4*, 443-504.
23. Münck, E.; Kent, T. A. *Hyp. Int.* **1986**, *27*, 161-172.
24. Anderson, P. W.; *Magnetism*; Rado, G. T.; Suhl, H., Eds.; Academic Press, *Vol 1*, **1963**; Ch. 2.
25. Martin, R. L. *New Pathways in Inorganic Chemistry*; Ebsworth, E. A. V.; Maddock, A. G.; Sharpe, A. G. Eds.; Cambridge University Press: New York, **1968**; Ch. 9, pp 175-231.
26. Ginsberg, A. P. *Inorg. Chim. Acta Rev.* **1971**, *5*, 45.
27. *Magneto-Structural Correlations in Exchange Coupled Systems*; Willet, R. D.; Gatteschi, D.; Kahn, O., Eds.; NATO ASI Series C No. 140; D. Reidel Publishing Company: Boston, **1985**.
28. Hay, P. J.; Thibeault, G. C.; Hoffman, R. *J. Am. Chem. Soc.* **1975**, *97*, 4884-4899.
29. Kahn, O.; Briat, B. *J. Chem. Soc., Faraday II* **1976**, *72*, 268-281.
30. Borshch, S. A.; Kotov. I. N.; Bersuker, I. B. *Sov. J. Chem. Phys.* **1985**, *3*, 1009-1016.
31. Girerd, J.-J. *J. Chem. Phys.* **1983**, *79*, 1766-1775.
32. Belinskii, M. I. *Mol. Phys.* **1987**, *60*, 793-819.
33. Karpenko, B. V. *J. Magnetism and Magnetic Materials* **1976**, *3*, 267-274.
34. Belinskii, M. I.; Tsukerblat, B. S.; Gerbeleu, N. V. *Sov. Phys. Solid State* **1983**, *25*, 497-498.
35. Noodleman, L.; Baerends, E. J. *J. Am. Chem. Soc.* **1984**, *106*, 2316-2327.
36. Moura, I.; Moura, J. J. G.; Münck, E.; Papaefthymiou, V.; LeGall, J. *J. Am. Chem. Soc.* **1986**, *108*, 349-351.
37. Surerus, K. K.; Münck, E.; Moura, I.; Moura, J. J. G.; LeGall, J. *J Am. Chem. Soc.* **1987**, *109*, 3805-3807.
38. Wolff, T. E.; Berg, J. M.; Hodgson, K. O.; Frankel, R. B.; Holm, R. H. *J. Am. Chem. Soc.* **1979**, *101*, 4140-4150.
39. Noodleman, L.; Norman, Jr., J. G.; Osborne, J. H.; Aizman, A.; Case, D. A. *J. Am. Chem. Soc.* **1985**, *107*, 3418-3426.
40. Kent, T. A.; Dreyer, J.-L.; Kennedy, M. C.; Huynh, B. H.; Emptage, M. H.; Beinert, H.; Münck, E. *Proc. Natl. Acad. Sci.* **1982**, *79*, 1096-1100.
41. Averill, B. A.; Dwivedi, A.; Debrunner, P.; Vollmer, S. J.; Wong, J. Y.; Switzer, R. L. *J. Biol. Chem.* **1980**, *255*, 6007-6010.+
42. Aizman, A.; Case D. A. *J. Am. Chem. Soc.* **1982**, *104*, 3269-3279.
43. Papaefthymiou, G. C.; Laskowski, E. J.; Frota-Pessoa, S.; Frankel, R. B.; Holm, R. H. *Inorg. Chem.* **1982**, *21*, 1723-1728.
44. Girerd, J.-J.; Journeau, Y.; Kahn, O. *Chem. Phys. Lett.* **1981**, *82*, 534-538.

RECEIVED January 20, 1988

Chapter 16

Variable-Temperature Magnetic Circular Dichroism Studies of Metalloproteins

Michael K. Johnson

Department of Chemistry, University of Georgia, Athens, GA 30602

Variable temperature magnetic circular dichroism spectroscopy affords a means of investigating the electronic and magnetic properties of metal centers in metalloproteins. This pedagogical account discusses the experimental and theoretical aspects of the technique that are applicable to the study of metal centers and metal clusters in proteins. The type of information that is available includes resolving and assigning electronic transitions and determining ground state parameters such as spin state, g-values, zero field splitting parameters. These aspects are illustrated using Ni(II)-substituted rubredoxin, the [3Fe-4S] cluster in a bacterial ferredoxin, and the MoFe and VFe nitrogenase proteins as examples.

Magnetic circular dichroism (MCD) spectroscopy is being increasingly used to investigate the electronic and magnetic properties of metal centers in metalloproteins (for recent reviews see Refs. 1-3). The technique facilitates resolution and assignment of electronic transitions and is applicable to any chromophoric metal center. Moreover, since paramagnetic metal centers invariably afford temperature-dependent MCD bands, variable temperature studies enable selective identification and deconvolution of the optical transitions from individual chromophores, even in complex multicomponent metalloenzymes such as cytochrome *c* oxidase (4,5), succinate dehydrogenase (6), and nitrogenase (7,8). When paramagnetic metal centers are present, careful analysis of the temperature and/or magnetic field dependence of discrete transitions furnishes electronic ground state information such as spin state, g-values, zero field splitting parameters, and magnetic coupling constants. While similar information is often accessible via EPR, Mössbauer, or magnetic susceptibility studies, MCD offers certain advantages over each of these techniques in the study of metalloproteins. For example, unlike Mössbauer, it is not limited to Fe and does not require isotopic enrichment. Weak magnetic interactions or zero field splittings that can prevent the observation of EPR resonances, do not prevent characterization of paramagnetic center(s) by MCD spectroscopy. Magnetic susceptibility studies can be difficult to interpret for multicentered metalloproteins particularly when paramagnetic impurities are present, whereas with MCD spectroscopy the magnetic properties of individual centers can be investigated independently. However, it should be emphasized that MCD studies

0097–6156/88/0372–0326$06.00/0

are usually most informative when used in combination with one or more of the above techniques.

This pedagogical account is intended to provide a brief introduction for the non-specialist, to the theoretical and experimental aspects of variable temperature MCD spectroscopy that are applicable in the study of metalloproteins. This is followed by some individual examples of MCD studies of metalloproteins that have been chosen to illustrate the utility of the technique and the type of information that is available.

Theoretical Aspects

Electronic transitions between Zeeman components of the electronic ground and excited states are circularly polarized. MCD exploits this phenomena, measuring the differential absorption of left and right circularly polarized light, ΔA, as function of wavelength in the presence of a magnetic field applied parallel to the direction of light propagation. This difference technique, as opposed to the conventional Zeeman experiment, is necessary because the absorption band widths observed for metalloproteins are far too large to permit resolution of Zeeman splittings. Theoretical treatments (see for example ref. 9) lead to a general expression for MCD intensity for a transition A $\rightarrow$ J:

$$\Delta A_{|A\rightarrow J|} = \gamma\left[A_1\left(-\frac{df}{dE}\right) + \left(B_0 + \frac{C_0}{kT}\right) f\right]\beta Bbl \qquad (1)$$

$$\int \left(f(E_{JA},E)/E\right) dE = 1$$

where γ is a collection of physical constants, b is the molar concentration, l is the pathlength (cm), B is the magnetic flux, β is the Bohr magneton, $f(E_{JA},E)$ is the line shape which is a function of both the transition energy, $E_{JA} = E_J - E_A$, and the incident photon energy, E, and A_1, B_0, and C_0 are parameters that depend on the electric dipole selection rules for the absorption of circularly polarized light in a magnetic field. From this expression, it is apparent that MCD intensity is the sum of three terms, $-A_1(df/dE)$, B_0f, and C_0f/kT, which are called A, B, and C-terms, respectively.

In most cases, A, B, and C-terms may be distinguished by their dispersion and temperature dependence. A-terms arise when there is degeneracy in either the excited state only, or both the ground and excited states. They are temperature independent and give rise to derivative-shaped dispersion with a cross-over point at the energy of the zero field absorption maximum. B-terms originate from field-induced mixing of the ground and excited state with all other possible excited states. While non-overlapping B-terms will exhibit absorption-shaped dispersion, derivative-shaped B-terms (pseudo-A-terms) can result from overlapping transitions with oppositely signed B_0 parameters. B-terms will be temperature independent, except in the rare occurrence of a mixing state that becomes thermally populated over the temperature range of the experiment. C-terms require a degenerate or near-degenerate ground state. In common with B-terms, non-overlapping C-terms will exhibit absorption-shaped dispersion. However, when the excited state is split by an energy smaller than the bandwidth and the C-terms to each component are of opposite sign, derivative-shaped bands (pseudo-A-terms) can appear. This situation is frequently encountered for excited states that are split by spin-orbit coupling. Fortunately C-terms are readily distinguished from A- and B-terms by their inverse temperature dependence. If present, they usually dominate the MCD spectrum at low temperatures, since they can be enhanced up to 70-fold on going from room temperature to liquid helium temperatures. For biological chromophores, which generally possess only low symmetry, ground state degeneracy can usually be equated with spin degeneracy. Consequently low temperature MCD provides a selective optical probe for paramagnetic chromophores such as transition metal centers.

Equation 1 only applies to paramagnetic chromophores in the Curie law-limit, i.e. kT much greater than the ground state Zeeman splitting, $g\beta B$. However, on lowering the temperature or increasing the field so that $g\beta B$ becomes comparable or greater than kT, ΔA will become non-linear as a function of B/T and eventually independent of B/T. At this point only the lowest Zeeman component is populated and the system is completely magnetized. Plots of ΔA as a function of B/T to magnetic saturation are termed MCD magnetization plots. Analysis of such plots, after correction for temperature independent contributions, can lead to estimates of ground state g-values and hence spin state. The theory for analyzing MCD magnetization plots is well developed for ground states consisting of an 'isolated' Kramers doublet (i.e. no low lying states becoming thermally populated over the temperature range of the experiment) (10-12). While a detailed discussion of the theoretical expressions used in fitting magnetization data is beyond the scope of this article, it is important to realize that, unlike magnetization curves derived from magnetic susceptibility data, the steepness of MCD magnetization plots is dependent on the polarization of the electronic transition as well as the ground state g-values. Indeed the dependence on the polarization becomes particularly acute for ground states with large g-value anisotropy (12,13).

So far the discussion has been confined to 'isolated' Kramers doublet ground states and therefore is strictly applicable only to non-interacting centers with $S = 1/2$ ground states. Paramagnetic transition metal centers with $S > 1/2$ ground states are readily recognizable by their MCD magnetization characteristics. First, the curves deviate substantially from theoretical data for $g_{av} = 2$. Second, data points measured at different temperatures do not lie on a smooth curve, due to the population of low lying zero field components. Third, plots of MCD intensity as a function of 1/T only become linear at high temperatures when the spread of zero field components is very much less than kT. Complete analysis of magnetization data from $S > 1/2$ chromophores presents a complex theoretical problem, requiring the inclusion of field induced mixing of the zero field components, as well as zero field splitting parameters, g-values, and the polarization of the transitions from each doublet. Preliminary analysis can be effected by fitting the data at a temperature such that only the lowest doublet state is populated, using the expressions derived for an 'isolated' doublet ground state. While such a procedure neglects field induced mixing of zero field components, it does permit estimates of the effective g-values for the lowest doublet and hence the ground state spin.

Many transition metal centers exhibit extremely low lying energy states as a result of either zero field splitting or magnetic interaction. Provided the energies of these low lying states are in the range 1 - 100 cm^{-1}, they can be assessed by measuring the MCD intensity at small applied fields as a function of 1/T and fitting to a Boltzmann population distribution (14). In this way MCD studies afford quantitative assessment of zero field splitting parameters and/or magnetic coupling constants.

Experimental Aspects

Instrumentation for MCD measurements consists of a spectropolarimeter mated to either an electromagnet or superconducting magnet. Commercial spectropolarimeters are available for the wavelength range 180 - 2000 nm and recent technological advances have enabled home-built instruments to go out to 5000 nm. To obtain electronic ground state information via magnetization curves, it is necessary to use magnetic fields of 5 Tesla or more, with sample temperatures as low as 1.5 K. Superconducting magnets that facilitate such experiments are available commercially from Oxford Instruments. With such high magnetic fields, it is necessary to separate the magnet spatially from the spectrometer and to shield certain components such as the photomultiplier tube and the photoelastic modulator. Accurate control and absolute measurement of both the temperature and magnetic field are essential and are accomplished by multiple resistance thermometers (such as carbon glass resistors which exhibit negligible magnetic field effects) mounted both above and below the sample, and by a transverse Hall probe positioned in place of the sample.

The MCD experiment involves the detection of transmitted light, and consequently frozen samples must be in the form of a glass. With protein solutions this is achieved by addition of at least 40% (v/v) ethylene glycol, glycerol, or sucrose. It is essential to check that no spurious effects result from the medium, by comparing the spectroscopic properties (i.e. UV-visible absorption, room temperature MCD and CD, and EPR) and, wherever possible, the activity of the metalloprotein, both with and without the addition of the glassing agent. While in general the presence of the glassing agent does not appear to perturb the properties of the metal chromophore in metalloproteins, there are some notable exceptions. For example, the $[4Fe\text{-}4S]^{1+}$ center in nitrogenase Fe-protein consists of a mixture of species with S = 1/2 and 3/2 ground states in aqueous buffer solutions, whereas the addition of 50% ethylene glycol results in conversion of almost all of the clusters to the S = 1/2 form (15). It is not possible to generalize concerning the quantity of sample that is required for variable temperature MCD studies, since the intensity of the signals from different types of biological chromophores can vary by as much as three orders of magnitude. However, typically 200 - 300 μl of sample exhibiting a maximum OD in the wavelength region of interest of approximately 0.5, are usually required to produce optimal data. Sample cells for low temperature experiments consist of 0.1 cm or 0.2 cm quartz cuvettes that can be filled under anaerobic or aerobic conditions.

The measured spectra are a superposition of the natural and magnetically-induced CD. Unless otherwise indicated, published MCD spectra are shown after computer subtraction of the natural CD, with the intensity expressed as $\Delta\epsilon$ in units of $M^{-1}cm^{-1}$, which corresponds to the difference in the molar extinction coefficients for left and right circularly polarized light. In the Curie law limit the $\Delta\epsilon$ scale is often normalized per unit field, i.e. $\Delta\epsilon/B$. However, this procedure is not meaningful at high fields and low temperatures where saturation effects may be occurring, and in such cases the applied magnetic field is stated in the text or figure legend. Quantitation of the MCD intensity of frozen samples is accurate to ±10%. The major sources of error reside in changes in sample pathlength on freezing and depolarization of the light beam as a result of inhomogeneities in the frozen sample. The latter can be approximately assessed and corrected for measuring the CD of an optically active sample with and without the sample in position.

Examples

The results of variable temperature MCD studies of Ni(II)-substituted rubredoxin, the 7Fe ferredoxin from *Thermus thermophilus* and the MoFe and VFe nitrogenase proteins from *Azotobacter vinelandii* are discussed below. These specific examples have been chosen to illustrate how the technique can be used to, (a) resolve and assign overlapping electronic transitions, (b) determine ground state g-values and spin state, (c) provide quantitative information concerning the energies of low lying states, and (d) deconvolute the optical transitions from individual chromophores in multicomponent metalloproteins.

Ni(II)-substituted rubredoxin. In view of the propensity of Ni(II) to form square planar complexes with chelating ligands possessing sulfhydryl groups, it was of interest to determine whether or not the constraints imposed by folding of the rubredoxin polypeptide are sufficient to afford tetrahedral coordination geometry. Since square planar and tetrahedral coordination geometry for Ni(II) give rise to diamagnetic (S = 0) and paramagnetic (S = 1) ground states, respectively, this question can readily be addressed by variable temperature MCD spectroscopy. Figure 1 shows the room temperature absorption spectrum and MCD spectra recorded at temperatures between 4.2 K and 137 K with a magnetic field of 4.5 T for Ni(II)-substituted rubredoxin from *Desulfovibrio gigas* in the wavelength region 300 - 800 nm. The room temperature absorption spectrum is composed of two intense S → Ni(II) charge transfer bands centered at 360 nm and 450 nm and two weaker Ni(II) d-d bands at 670 nm and 720 nm. The corresponding MCD spectra exhibit bands in analogous regions that display unusual temperature dependence. Below 15 K, the MCD spectrum is independent of temperature, with all bands showing linear dependence on the magnetic field strength. Such behavior shows that a non-degenerate state lies lowest in energy. Based on their dispersion and intensity, the MCD bands originating from this non-degenerate level are assigned to B-terms. However, pronounced changes in the MCD spectrum occur at temperatures above 15 K. The negative bands at 437 nm and 718 nm and the positive bands at 338 nm and 363 nm all decrease with increasing temperature, concomitant with the appearance of a positive band at 437 nm. As shown in Figure 2, the negative band at 664 nm shows an initial increase and subsequent decrease with increasing temperature, with the maximum occurring around 50 K. This complex temperature dependence can readily be interpreted in terms of population of a degenerate low lying state that gives rise to C-terms exhibiting inverse temperature dependence.

The energy separation between the non-degenerate and degenerate states can be assessed by fitting the MCD intensity as a function of 1/T at a discrete wavelength to a Boltzmann population distribution for a two level system. Figure 2 shows the temperature dependence of the MCD intensity at 664 nm, with the solid line corresponding to the best fit. The best fit data are for an energy separation of 55 cm^{-1}, and assume a two-fold degenerate excited state. Since only two variable parameters are involved (i.e. the energy separation and the magnitude of the C-term from the degenerate level) and the data at this wavelength display temperature dependence with a maximum around 50 K due to opposite signs for the C and B-terms, a unique fit is readily obtained. Similar values for the energy separation can be obtained by fitting 1/T dependence data at other wavelengths, where the C and B-terms from the two levels are of the same sign and have comparable intensity. Using this energy separation, the temperature-independent and temperature-dependent MCD spectra, originating from the non-degenerate and low-lying degenerate states, respectively, can readily be deconvoluted (data not shown).

The presence of a low-lying degenerate state strongly suggests an S = 1 ground state. Accordingly, the coordination geometry is tetrahedral with a $^3T_1$ ground state to a first approximation. Removal of the ground state degeneracy in zero field to yield a non-degenerate ground state could result from spin orbit coupling alone or a combination of spin orbit coupling and low symmetry distortions. While it is difficult to decide between these alternatives based on the MCD data for Ni(II)-rubredoxin alone, parallel studies of the analog complex $[Ni(SPh)_4]^{2-}$ (data not shown) strongly suggest the latter alternative. Except for the S → Ni(II) charge transfer region being shifted to lower energy and a somewhat smaller energy separation (44 cm^{-1}) between the non-degenerate

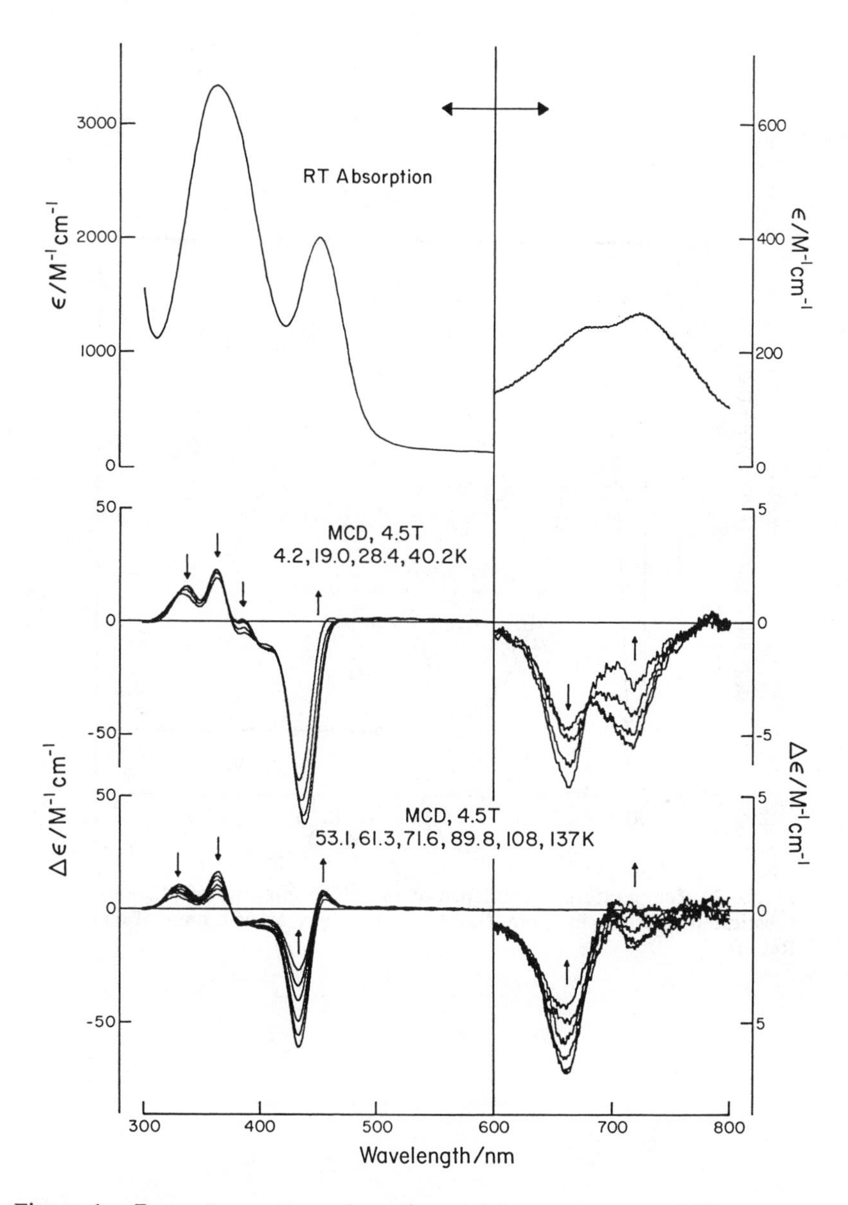

Figure 1. Room temperature absorption and low temperature MCD spectra of Ni(II)-substituted rubredoxin from *Desulfovibrio gigas*. Arrows indicate the direction of change of MCD intensity with increasing temperature. Taken from Ref. 16.

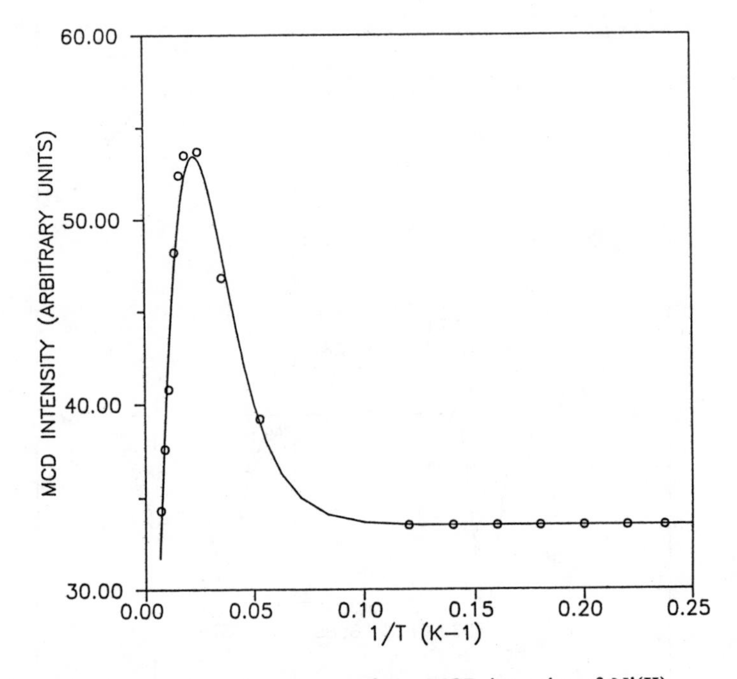

Figure 2. Temperature dependence of the MCD intensity of Ni(II)-substituted rubredoxin from *Desulfovibrio gigas* at 664 nm. Taken from Ref. 16.

and degenerate levels, the analog complex displays very similar MCD characteristics. Since X-ray crystallographic data for $[Ni(SPh)_4]^{2-}$ show a D_{2d} distorted NiS_4 core (17), a similar distortion may be inferred for Ni(II)-substituted rubredoxin. The ground state is therefore split into $^3A_2$ and ^3E components, with the former lying lower in energy and exhibiting large, positive axial zero field splitting. In terms of the spin Hamiltonian the energy separation between the $M_s = 0$ and $M_s = \pm 1$ levels corresponds to the axial zero field splitting parameter D. The d-d region of the absorption spectrum is therefore assigned, under D_{2d} symmetry to the $^3A_2 \rightarrow {}^3A_2$ and $^3A_2 \rightarrow {}^3E$ components of the parent tetrahedral transition $^3T_1(F) \rightarrow {}^3T_1(P)$ transition.

This example provides a good illustration of how variable temperature MCD measurements can be used to obtain zero field splitting parameters for non-Kramers ground states with $D > 0$. For ground states where a doublet state lies lowest in energy (i.e. Kramers systems or axial non-Kramers systems with $D < 0$), the general procedure involves fitting plots of MCD intensity versus $1/T$ at small applied fields (i.e. in the Curie law limit) to a Boltzmann population distribution over the ground state manifold, with the axial zero field splitting and the C-terms from each doublet as the variable parameters (14). The accuracy of this procedure depends on the number of parameters and the extent of deviation from linearity of the plots of MCD intensity versus $1/T$.

7Fe ferredoxin from *Thermus thermophilus*. Iron-sulfur proteins provide good examples of the utility of variable temperature MCD spectroscopy for resolving electronic transitions and determining ground state properties (1). This is well illustrated by MCD studies of the oxidized and reduced [3Fe-4S] cluster in the 7Fe ferredoxin from *T. thermophilus* (13). Figures 3(a) and 3(b) show room temperature absorption spectra and variable temperature MCD spectra for the oxidized ($[3Fe\text{-}4S]^{1+}$, $[4Fe\text{-}4S]^{2+}$) and partially-reduced ($[3Fe\text{-}4S]^{0}$, $[4Fe\text{-}4S]^{2+}$) ferredoxin, respectively. The shoulder at 310 nm in the absorption spectrum of the partially reduced sample is due to excess of the reductant, dithionite. In both redox states the tetranuclear cluster has a $S = 0$ ground state, as a result of antiferromagnetic coupling, with the lowest paramagnetic level > 100 cm^{-1} higher in energy. Consequently, while both clusters contribute to the broad and featureless UV-visible absorption spectra, the temperature dependent C-terms that are observed can be attributed exclusively to the trinuclear center, which therefore must be paramagnetic in both the oxidized and one electron reduced states.

The complexity of the low temperature MCD spectra of the oxidized and reduced trinuclear cluster shows the multiplicity of the predominantly $S \rightarrow Fe$ charge transfer transitions that contribute to the absorption envelope. While MCD spectroscopy provides a method of resolving the electronic transitions, assignment cannot be attempted without detailed knowledge of the electronic structure. However, the complexity of the low temperature MCD spectra is useful in that it furnishes a discriminating method for determining the type and redox state of protein bound iron-sulfur clusters. Each well characterized type of iron-sulfur cluster, i.e. [2Fe-2S], [3Fe-4S], and [4Fe-4S], has been shown to have a characteristic low temperature MCD spectrum in each paramagnetic redox state (1)

Inspection of the temperature dependence of the MCD bands of the oxidized and reduced [3Fe-4S] center in *T. thermophilus* ferredoxin, see Figure 3, reveals that the reduced cluster magnetizes more rapidly with decreasing temperature. This is shown more clearly in the magnetization data, which are shown as plots of MCD intensity, in the form of a percentage of that estimated

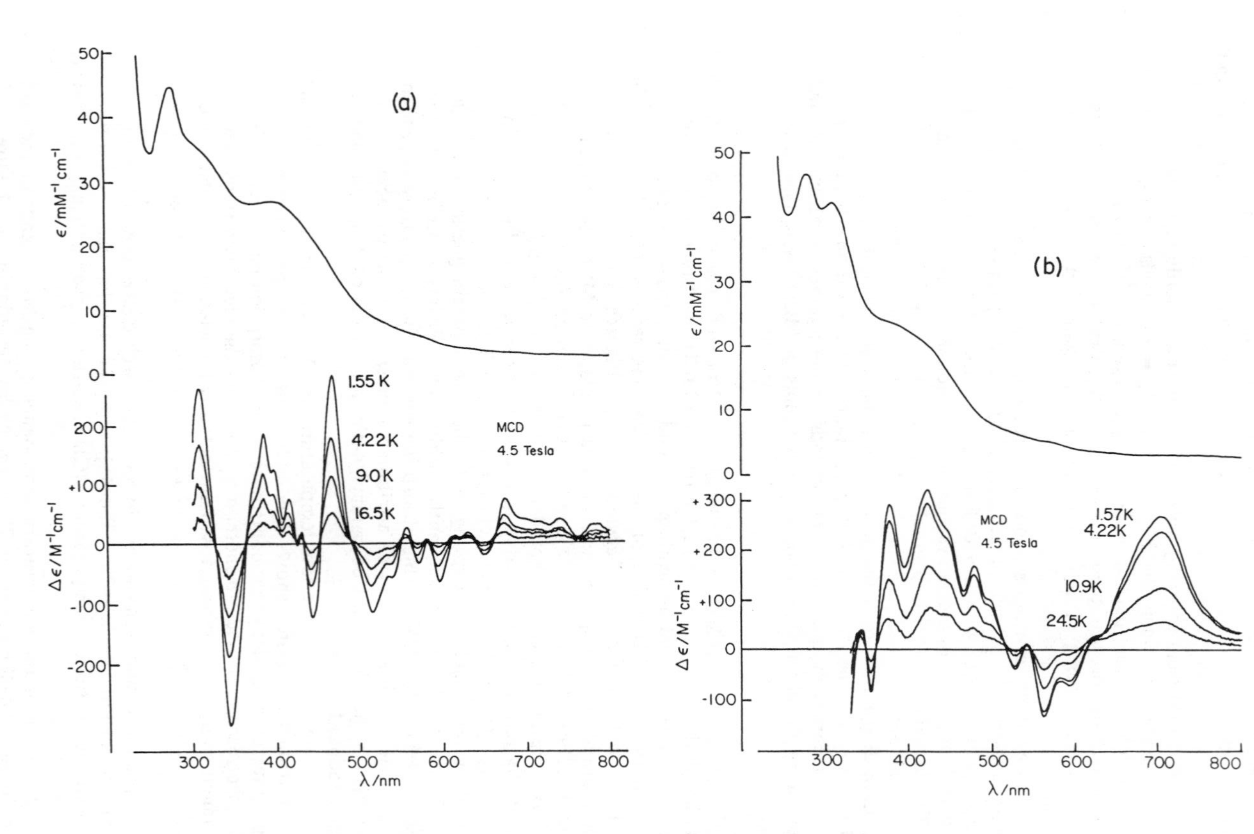

Figure 3. UV-visible absorption and variable temperature MCD spectra for (a) oxidized and (b) partially reduced *Thermus thermophilus* ferredoxin.

at magnetic saturation, as a function of $\beta B/2kT$. Analyses of the magnetization plots afford insight into the nature of electronic ground states of the oxidized and reduced cluster.

For the oxidized [3Fe-4S] cluster, the data points at all temperatures lie on a smooth curve that is well fit by theoretical data for an isotropic g-value of 2 (solid line in Figure 4(a)). Identical magnetization plots are found for each of the major bands in the spectrum, indicating that all of the observed electronic transitions originate from an isolated $S = 1/2$ ground state. Such a ground state would be expected to exhibit an EPR resonance, and a broad resonance centered around $g = 2$ accounting for 1 spin/molecule is indeed observed for oxidized *T. thermophilus* ferredoxin (13). The MCD magnetization data for the reduced [3Fe-4S] cluster is more complex, see Figure 4(b). First, data points obtained at different temperatures do not lie on a smooth curve and second, the lowest temperature data magnetize more steeply than that expected for a simple $S = 1/2$ ground state. Such behavior is indicative of a zero field split $S > 1/2$ ground state. Moreover the lowest temperature data are well fit by theoretical curves calculated for effective g-values of $g_{\parallel} = 8.0$ and $g_{\perp} = 0.0$, solid line in Figure 4(b). Since these g-values can only arise from the $M_s = \pm 2$ doublet of an $S = 2$ spin system, the ground state can be assigned to $S = 2$ with negative, axial zero field splitting, $D < 0$. In general, ground state doublets with large g-value anisotropy exhibit MCD magnetization data that are strongly dependent on the polarization of the electronic transition (12). However, in this instance, all temperature dependent MCD transitions must be xy-polarized since $g = 0$. In accord with this, identical magnetization data are obtained for each of the major temperature dependent MCD bands (13). Having established the nature of the ground state, the axial zero field splitting parameter can then be estimated by analyzing the temperature dependence of the MCD intensity at small applied fields using the procedure outlined above. In this case D was found to be -2.2 ± 1.0 cm^{-1} giving a separation between the $M_s = \pm 1$ and ± 2 doublets of 6.6 ± 3.0 cm^{-1} (13)

MoFe and VFe nitrogenase proteins from *Azotobacter vinelandii*. Finally we turn our attention to one of the most complex of all enzymes, to illustrate how variable temperature MCD spectroscopy can selectively investigate the electronic and magnetic properties of individual centers in a multicomponent metalloenzyme. The nature of the metal clusters in the MoFe nitrogenase protein were originally characterized by the combination of EPR and Mössbauer spectroscopies (18-20). Two main types of clusters were identified: four Fe-S centers called P clusters that are diamagnetic ($S = 0$ ground states) in the enzyme as isolated in the presence of dithionite, and paramagnetic with a half integer spin ($S > 1/2$ ground state) on oxidation with thionine; two Mo-Fe-S clusters or M centers that are paramagnetic ($S = 3/2$ ground state) as isolated and diamagnetic ($S = 0$ ground state) on thionine oxidation. Since variable temperature MCD is specifically a probe for paramagnetic chromophores, the electronic and magnetic properties of P and M centers can be investigated selectively, by studying the enzyme in each redox state.

Recently, a second or alternative nitrogenase has been isolated from *Azotobacter vinelandii* (21) and *Azotobacter chroococcum* (22) that contains vanadium as opposed to molybdenum. The MoFe and VFe nitrogenase proteins from *A. vinelandii* (called *Av*1 and *Av*1', respectively) are known to have different polypeptide structures and it obviously of interest to know to what extent the cluster composition is conserved. Variable temperature MCD studies of the as isolated and thionine oxidized proteins provided a convenient means of addressing this question.

Figures 5 and 6 compare the variable temperature MCD spectra of dithio-

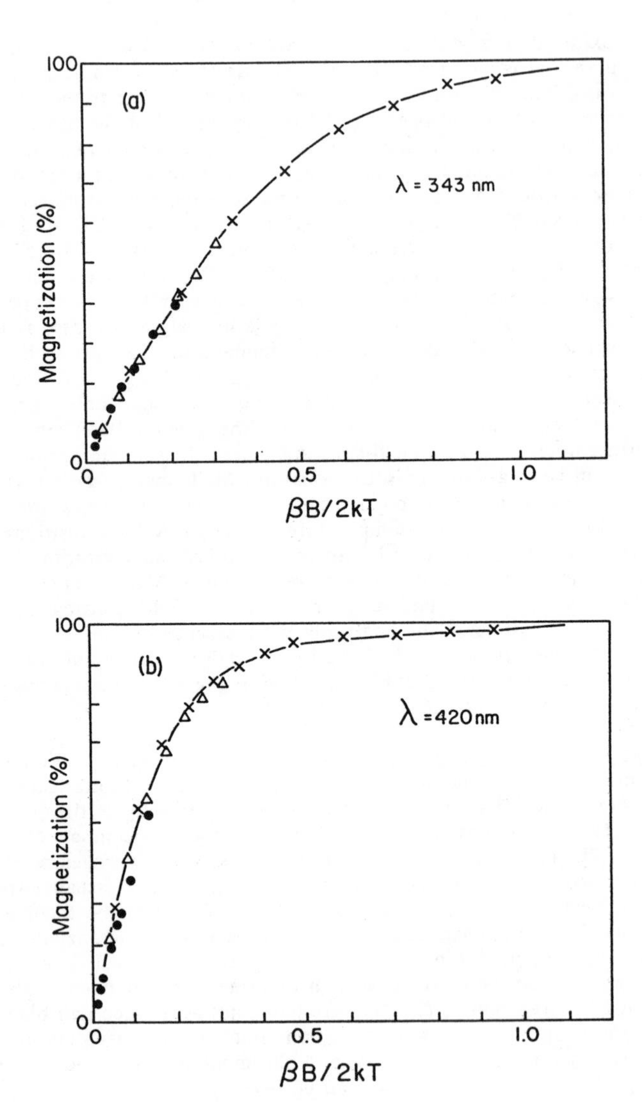

Figure 4. MCD magnetization plots for (a) oxidized and (b) partially reduced *Thermus thermophilus* ferredoxin. Data collected at 1.55 K (x), 4.22 K (Δ), and 9-100 K (•), with magnetic fields between 0 and 4.5 T.

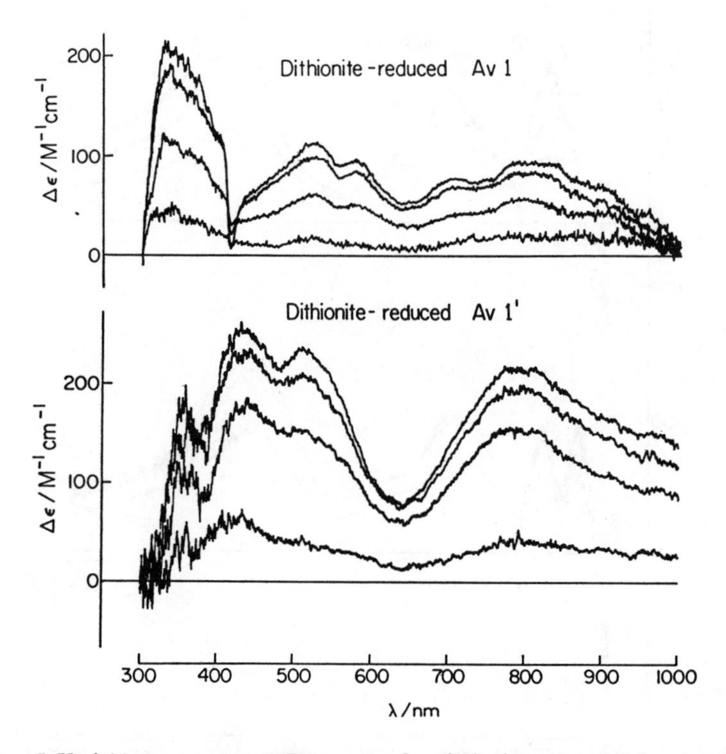

Figure 5. Variable-temperature MCD spectra for dithionite-reduced *Av*l and *Av*l' at 4.5 T. Temperatures: 1.64 K, 4.22 K, 15.0 K, and 120 K for *Av*l; 1.64 K, 4.22 K, 9.5 K, and 60 K for *Av*l'. MCD intensity increasing with decreasing temperature. (Reproduced from ref. 8. Copyright 1987 American Chemical Society.)

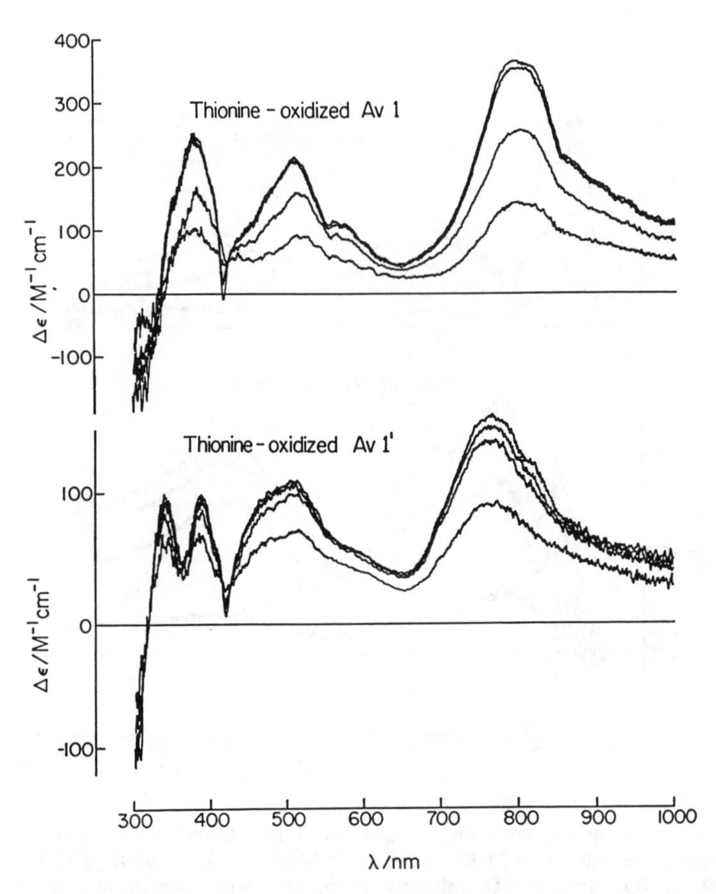

Figure 6. Variable-temperature MCD spectra for thionine-oxidized *Av*1 and *Av*1' at 4.5 T. Temperatures: 1.57 K, 4.22 K, 18.0 K, and 30.0 K for *Av*1; 1.58 K, 4.22 K, 7.6 K, and 20.0 K for *Av*1'. MCD intensity increasing with decreasing temperature. (Reproduced from ref. 8. Copyright 1987 American Chemical Society.)

nite-reduced *Av*1 and *Av*1', and thionine-oxidized *Av*1 and *Av*1', respectively. The temperature-dependent MCD bands of dithionite-reduced *Av*1 can be assigned exclusively to the paramagnetic M centers and provide an optical fingerprint of this novel Mo-Fe-S cluster. Almost identical MCD spectra are observed for the dithionite-reduced MoFe nitrogenase protein of *Klebsiella pneumoniae* (7), showing that this center is highly conserved in these two distinct nitrogen-fixing organisms. In contrast there are marked differences in the band positions and intensity of the low temperature MCD spectra of dithionite-reduced *Av*1 and *Av*1'. EPR studies indicate the presence of S = 3/2 paramagnets in both proteins, albeit with very different zero field splitting parameters: $D \simeq +5\ cm^{-1}$, $E/D \simeq 0.05$ for S = 3/2, M centers in *Av*1 (19); $D \simeq -0.7\ cm^{-1}$, $E/D \simeq 0.26$ for the S = 3/2 center in dithionite-reduced *Av*1' (B. J. Hales, unpublished results). MCD magnetization data for dithionite-reduced *Av*1 and *Av*1' (8) indicate that the EPR-detectable S = 3/2 paramagnets are also responsible for the temperature-dependent MCD transitions. Therefore, by analogy with the MoFe protein it seems probable that the S = 3/2 paramagnetic chromophore in dithionite-reduced *Av*1 is a V-Fe-S cluster. However, the EPR and MCD studies indicate that this cluster has magnetic and electronic properties distinct from that of the Mo-Fe-S clusters in conventional Mo nitrogenases.

No EPR signals were observed for the thionine-oxidized *Av*1 or *Av*1' used for MCD investigations. In contrast, both samples exhibit intense temperature-dependent MCD spectra in the wavelength range 300 to 1000 nm, see Figure 6. Since Mössbauer studies of thionine-oxidized *Av*1 (19,20) indicate that oxidized P-clusters are the only paramagnetic centers, the temperature-dependent MCD transitions may be similarly assigned. The MCD spectrum of thionine-oxidized *Av*1 is identical in both form and intensity to that observed under comparable conditions for thionine-oxidized *Kp*1 (7), showing the homogeneity of P-clusters in Mo-containing, conventional nitrogenases from diverse nitrogen-fixing organisms. In contrast, thionine-oxidized *Av*1' exhibits a low temperature MCD spectrum that is similar in form, but not identical to that of *Av*1 or *Kp*1.

A unique feature of the MCD characteristics of oxidized P- clusters that distinguishes them from all other known types of biological Fe-S clusters is the steepness of their magnetization curves. Figure 7 shows magnetization plots for the dominant bands in the spectra of thionine-oxidized *Av*1 and *Av*1'. In each case data points were obtained for fields between 0 and 4.5 T at four different temperatures in the range 1.57 to 30 K. The nesting of the data obtained at different temperatures (denoted by different symbols) and the steepness of magnetization plots attests to a ground state with $S > 1/2$. To facilitate comparison of the data, theoretical curves for an isolated ground state doublet with effective g-values of 10, 0, 0, (broken line) and 14, 0, 0, (solid line) are shown in both sets of data. These are the g-values anticipated for the $M_s = \pm 5/2$ or $\pm 7/2$ doublet ground state of an S = 5/2 or 7/2 spin system, respectively, with purely axial zero field splitting ($D < 0$). These correspond to the best descriptions of the ground state in light of the available MCD (7), Mössbauer (18-20), and magnetic susceptibility (23) data. The close correspondence in the magnetization data for thionine-oxidized *Av*1 and *Av*1', coupled with the absence of observed EPR signals indicates similar ground state properties for the paramagnetic chromophores in these two proteins. In both cases, the experimental data at the lowest temperature (only the lowest doublet significantly populated) lies between that expected for an S = 5/2 and 7/2 spin system. It is not possible at present to distinguish between these two alternatives for the ground state spin.

Based on the form of the low temperature MCD spectra and magnetization behavior, it can be concluded that clusters with similar electronic and

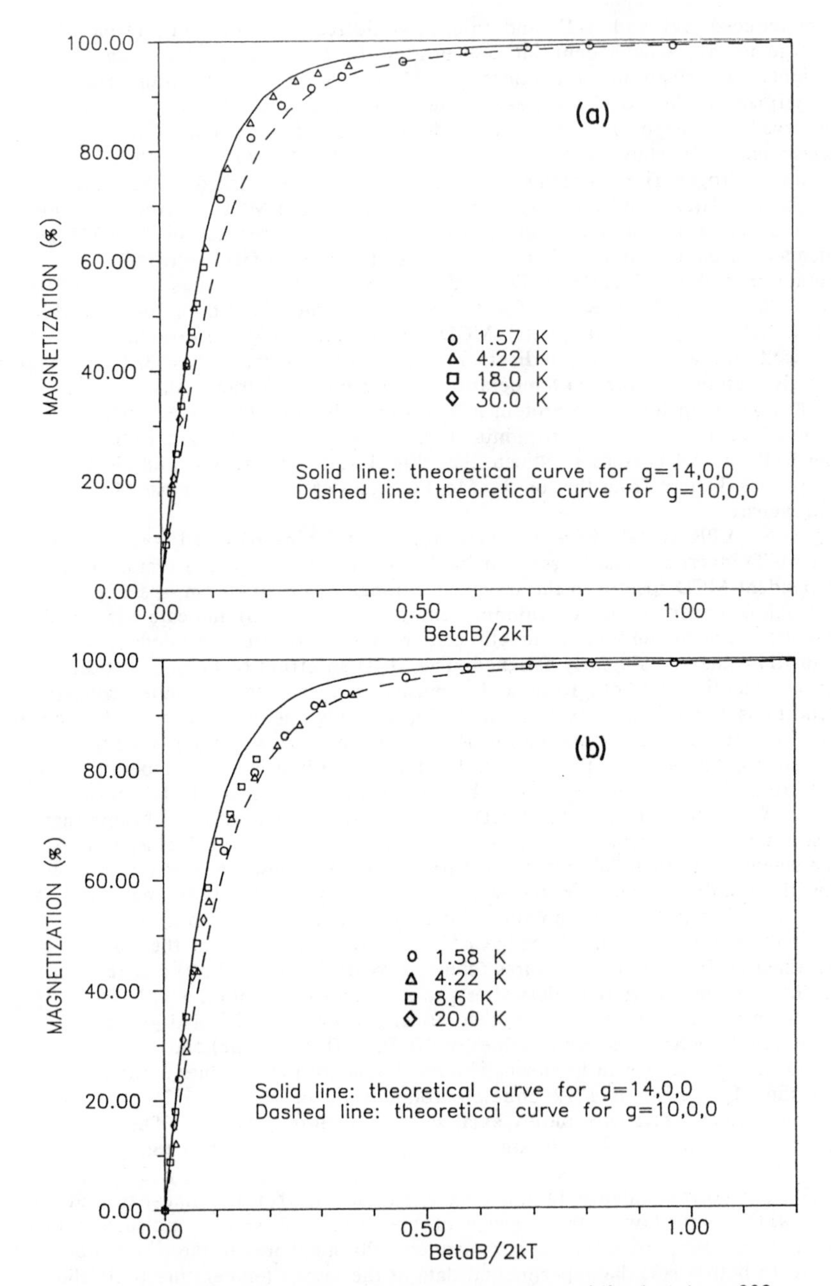

Figure 7. MCD magnetization plots for thionine-oxidized *Av*1 at 800 nm (a) and thionine oxidized *Av*1' at 760 nm (b). Magnetic fields between 0 and 4.5 T. Modified from Ref. 8.

magnetic properties to P-clusters (i.e. S = 5/2 or 7/2 ground state with D < 0) are present in the alternative enzyme. We have called these centers P'-clusters so as to distinguish them from conventional P-clusters (8). The intensity of the MCD spectrum for thionine-oxidized *Av*1' compared to *Av*1 or *Kp*1 suggests the alternative enzyme contains approximately two P'-clusters, as opposed to four P-clusters in the conventional enzyme (19,20). This observation is consistent with the lower Fe content *Av*1' compared to that of *Av*1 (21). In fact both the MCD and analytical data are consistent with *Av*1' containing one V-Fe-S and two P' clusters. An alternative explanation is that the cluster composition is analogous to that of the conventional enzyme with two V-Fe-S and four P' clusters. However, for this explanation to be correct, the samples used for spectroscopic investigations would need to have approximately 50% of the protein in the form of apoprotein. Investigations are in progress to discriminate between these two possibilities.

Acknowledgments

The work described above was greatly aided by the provision of samples from colleagues: *D. gigas* Ni(II)-substituted rubredoxin, I. Moura and J. J. G. Moura; *T. thermophilus* ferredoxin, J. A. Fee; VFe and MoFe nitrogenase proteins from *A. vinelandii*, B. J. Hales. Research in the author's laboratory is supported by grants from NIH (GM33806) and NSF (DMB8796212) and an Alfred P. Sloan Research Fellowship.

Literature Cited

1. Johnson, M. K.; Robinson, A. E.; Thomson, A. T. In *Iron-Sulfur Proteins*; Spiro, T. G., ed.; John Wiley and Sons: New York, **1982**, pp 367-406
2. Dooley, D. M.; Dawson, J. H. *Coord. Chem. Rev.* **1984**, *60*, 1-66.
3. Dawson, J. H.; Dooley, D. M. In *Iron Porphyrins, Part III*; Lever, A. B. P.; Gray, H. B., Eds.; Benjamin-Cummings, **1985**.
4. Thomson, A. J.; Johnson, M. K.; Greenwood, C.; Gooding, P. E. *Biochem. J.* **1981**, *193*, 687-697.
5. Greenwood, C.; Hill, B. C.; Barber, D.; Eglinton, D. G.; Thomson, A. J. *Biochem. J.* **1983**, *215*, 303-316.
6. Johnson, M. K.; Morningstar, J. E.; Bennett, D. E.; Ackrell, B. A. C.; Kearney, E. B. *J. Biol. Chem.* **1985**, *260*, 7368-7378.
7. Johnson, M. K.; Thomson, A. J.; Robinson, A. E.; Smith, B. E. *Biochim. Biophys. Acta* **1981**, *671*, 61-70.
8. Morningstar, J. E.; Johnson, M. K.; Case, E. E.; Hales, B. J. *Biochemistry* **1987**, *26*, 1795-1800.
9. Stephens, P. J. *Adv. Chem. Phys.* **1976**, *35*, 197-264.
10. Schatz, P. N.; Mowery, R. L.; Krausz, E. R. *Mol. Phys.* **1978**, *35*, 1537-1557.
11. Thomson, A. J.; Johnson, M. K. *Biochem. J.* **1980**, *191*, 411-420.
12. Bennett, D. E.; Johnson, M. K. *Biochim. Biophys. Acta* **1987**, *911*, 71-80.
13. Johnson, M. K.; Bennett, D. E.; Fee, J. A.; Sweeney, W. V. *Biochim. Biophys. Acta* **1987**, *911*, 81-94.
14. Browett, W. R.; Fucaloro, A. F.; Morgan, T. V.; Stephens, P. J. *J. Am. Chem. Soc.* **1983**, *105*, 1868-1872.
15. Lindahl, P. A.; Day, E. P.; Kent, T. A.; Orme-Johnson. W. H.; Münck, E. *J. Biol. Chem.* **1985**, *260*, 11160-11173.
16. Kowal, A. T.; Zambrano, I. C.; Moura, I.; Moura, J. J. G.; LeGall, J.; Johnson, M. K. *Inorg. Chem.*, in press.
17. Swenson, D.; Baenziger, N. C.; Coucouvanis, D. *J. Am. Chem. Soc.* **1978**, *100*, 1932-1935.

18. Smith, B. E.; Lang, G. *Biochem. J.* **1974**, *137*, 169-180.
19. Münck, E.; Rhodes, H.; Orme-Johnson, W. H.; Davis, L. C.; Brill, W. J.; Shah, V. K. *Biochim. Biophys. Acta* **1975**, *400*, 32-53.
20. Zimmermann, R.; Münck, E.; Brill, W. J.; Shah, M. T.; Henzl, M. T.; Rawlings, J.; Orme-Johnson, W. H. *Biochim. Biophys. Acta* **1978**, *537*, 185-207.
21. Hales, B. J.; Case, E. E.; Morningstar, J. E.; Dzeda, M. F.; Mauterer, L. A. *Biochemistry* **1986**, *25*, 7251-7255.
22. Robson, R. L.; Eady, R. R.; Richardson, T. H.; Miller, R. W.; Hawkins, M.; Postgate, J. R. *Nature* **1986**, *32*, 388-390.
23. Smith, J. P.; Emptage, M. H.; Orme-Johnson, W. H. *J. Biol. Chem.* **1982**, *257*, 2310-2313.

RECEIVED December 18, 1987

Chapter 17

Aconitase

Evolution of the Active-Site Picture

Mark H. Emptage

Experimental Station, Central Research and Development Department, E. I. du Pont de Nemours and Company, Wilmington, DE 19898

Aconitase, the second enzyme of the citric acid cycle, catalyzes the interconversion of citrate and isocitrate via a dehydration-rehydration reaction. Only within the past few years has it been recognized that aconitase is an Fe-S protein. This result not only called for a reassessment of the "ferrous wheel" model of aconitase, which featured a monomer iron active site, but also raised questions concerning the role of a redox active Fe-S cluster in an enzyme which catalyzes no apparent redox chemistry. Insight into the role of the Fe-S cluster has been gained from the information obtained from several techniques; most notably, EPR, Mössbauer, ENDOR and EXAFS spectroscopy. The results from these studies have led to the discovery of new Fe-S cluster structures and facile cluster interconversions which control enzyme activity. Moreover, aconitase has provided the first example of the direct participation of an Fe-S cluster in catalysis through the coordination of substrate to a single iron of a [4Fe-4S] cluster. Specific isotopic labeling of both the cluster and substrate has allowed their interaction to be studied in detail. Finally, recent studies suggest that several other enzymes, which also catalyze similar hydration and/or dehydration reactions, may utilize a Fe-S prosthetic group.

Aconitase [citrate(isocitrate) hydro-lyase, EC 4.2.1.3] is the second enzyme of the citric acid cycle, which plays a central role in metabolism for all aerobic organisms. This enzyme catalyzes a dehydration-rehydration reaction interconverting citrate and 2R,3S-isocitrate via the allylic intermediate *cis*-aconitate.

$$\begin{array}{c} H_2C-COO^- \\ | \\ HO-C-COO^- \\ | \\ H_2C-COO^- \end{array} \underset{+H_2O}{\overset{-H_2O}{\rightleftharpoons}} \begin{array}{c} HC-COO^- \\ \| \\ C-COO^- \\ | \\ H_2C-COO^- \end{array} \underset{-H_2O}{\overset{+H_2O}{\rightleftharpoons}} \begin{array}{c} H \\ HO-C-COO^- \\ | \\ C-COO^- \\ | \\ H_2C-COO^- \end{array} \quad (1)$$

0097–6156/88/0372–0343$08.25/0

Since this reaction requires no oxidation-reduction chemistry, the finding 15 years ago that aconitase is an Fe-S protein was quite unexpected. Until recently a paradigm for Fe-S proteins has been that they function primarily in reactions requiring electron transfer. Accordingly, the central question in the study of aconitase for the past several years has been: What is the function of its Fe-S cluster?

The rapid progress in the understanding of the active site of aconitase in the 1980's has primarily originated from the work of H. Beinert and his collaborators. Three essential factors contributed to the success of this work: 1) a ready and consistent source of enzyme (gram quantities), 2) a solid chemical and biochemical understanding of aconitase, and 3) close interactions with outstanding collaborators (most notably E. Münck's group for Mössbauer spectroscopy and B. M. Hoffman's group for electron nuclear double resonance (ENDOR) spectroscopy). This paper will review how the application of spectroscopy in conjunction with chemical and biochemical data has resulted in an increasingly more sophisticated, and one assumes increasingly accurate, picture of the active site of aconitase.

Historical Background

Aconitase was first described 50 years ago by Martius (1,2) and soon there after named by Breusch (3). The enzyme demonstrated the then surprizing ability to distinguish between the chemically identical acetyl arms of citrate (4). The stereospecificity of enzyme catalyzed reactions was not fully understood until the late 1940's when Ogston pointed out that as long as a substrate attaches to an asymmetric enzyme at three points, the enzyme can differentiate between two identical arms of a symmetrical molecule (5).

Early attempts to purify the enzyme brought the quick realization that aconitase is easily inactivated (6,7). In the early 1950's Dickman and Cloutier (8,9) found that inactivated aconitase could be reactivated by incubation with iron and a reductant. From kinetic analyses of the iron and reductant effects on enzyme activity, Morrison argued that both formed Michaelis-Menten complexes with the enzyme (10). This refuted the earlier idea that the sole role of the reductant was to maintain iron in a reduced state (9). Of several metal cations tried, only ferrous ion was found to be capable of this reactivation process (8,11). Because of the absolute requirement for iron in activation, the known chelation properties of citrate, and Ogston's 3-point attachment proposal, Speyer and Dickman proposed that the active site iron provides three coordination sites for substrate binding – one for hydroxyl and two for carboxyl groups (12).

Gawron *et al.* (13,14) determined the stereochemistry of natural isocitric acid by chemical means. The results require the *trans*-addition of water across the *cis*-aconitate intermediate double bond to produce either citrate or 2R,3S-isocitrate. Mass and NMR analyses of isotopically labeled citrate and isocitrate in the early 1960's (15-17), defined the stereospecificities of the dehydration steps. These results led Gawron to propose the binding of *cis*-aconitate to the active site in two orientations differing by a 180° rotation about the double bond, as shown in Equation 2. This allows for the protonation by a base (–BH) and hydroxylation of the double bond to occur on aconitase at single, separate loci for the formation of either citrate or isocitrate.

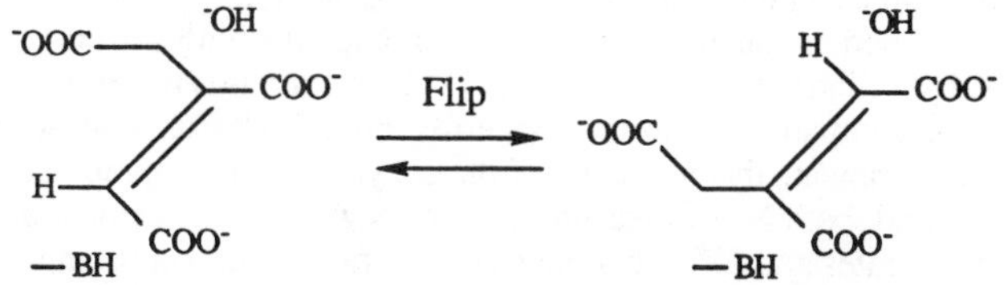

$$\text{(2)}$$

Classic isotopic tracer experiments by Rose and O'Connell (18) demonstrated that aconitase converted 3-[^{3}H]-isocitrate to 2-[^{3}H]-citrate without loss of tritium at early times. However, when an alternate substrate 3-[^{3}H]-2-methylisocitrate was combined with high concentrations of *cis*-aconitate, tritium was found in isocitrate. These results demonstrated both intra and intermolecular proton transfer between substrates. This implies that, indeed, there is a single base on the enzyme involved both in the proton removal from either citrate or isocitrate and protonation of *cis*-aconitate. Because of the demonstration of intramolecular proton transfer the proton exchange rate of the catalytic base with solvent must be surprisingly slower than the turnover number of aconitase (~15 s^{-1}). On the other hand, the hydroxyl group of citrate showed no transfer of label to isocitrate, indicating rapid exchange with solvent during the course of reaction.

The "Ferrous Wheel" Mechanism

In the late 1960's Glusker expanded upon the Speyer and Dickman model by bringing together the new stereochemical information and the crystallographically determined structures of citrate, *cis*-aconitate, and isocitrate-metal complexes (19, 20). The model that emerged became known as the "ferrous wheel" mechanism. As shown in Equation 3, the essence of this mechanism is that water and the two carboxyl groups of *cis*-aconitate can coordinate to a ferrous ion binding site in two orientations differing by a 90° rotation about the double bond. This results in the movement of the middle carboxyl group between two adjacent coordination sites. Attack by the iron-activated water leads to citrate in one orientation and isocitrate in the other. An amino acid residue is proposed to be positioned to protonate the adjacent carbon on the opposing face of the planar double-bond system.

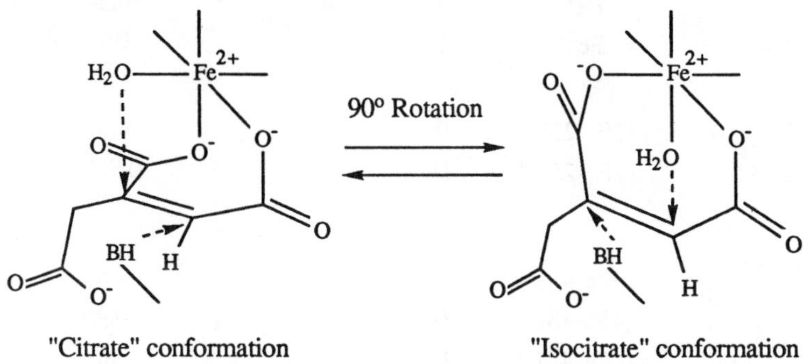

$$\text{(3)}$$

"Citrate" conformation "Isocitrate" conformation

The first spectroscopic studies of the enzyme itself began in the early 1970's after aconitase was purified to apparent homogeneity in large quantities from porcine heart tissue by Villafranca and Mildvan (21). However, this material was flawed in two ways as compared to current standards. First, both iron analyses and absorption spectra now indicate that about 2/3 of the enzyme from this preparation was apoenzyme. Second, both NMR measurements of enzyme enhancement of water proton relaxation rates and EPR now indicate the presence of a substantial amount of adventiously bound ferric ion. These early studies suggested that two binding sites on aconitase could accommodate either Fe(II) or Mn(II); however, the Mn(II) bound enzyme was inactive. It was assumed that these sites were the active sites of aconitase (11). Distances between the metal ion and substrate (and inhibitors) were estimated from the enhancement of relaxation times of substrate protons using both Fe(II)-enzyme and Mn(II)-enzyme (22). The estimated distances closely matched the ferrous ion binding site as visualized from the earlier crystallographic studies of Glusker (19). The apparent support by these spectroscopic studies provided a long life for the "ferrous wheel" mechanism.

Aconitase as an Fe-S Protein

Questions concerning the "ferrous-wheel" mechanism came soon after the Villafranca and Mildvan studies with the discovery in Gawron's laboratory that aconitase also contains labile sulfur (23), which, with iron, is indicative of the presence of Fe-S clusters.

At this point it may be valuable to digress a moment and discuss the state-of-the-knowledge in the field of Fe-S proteins by the mid-1970's. At this time there were three known structures found in nature, 1Fe as represented by rubredoxin, the [2Fe-2S] cluster as represented by plant ferredoxins, and the [4Fe-4S] cluster as found in many bacterial ferredoxins (24). The schematic structures and selected properties are listed in Table I.

Table I. Fe-S Structures and Some Electronic Properties

Fe-S Structure	Redox State	Spin	EPR g-average	Mössbauer isomer shift, δ, mm/s
 Fe	Fe^{3+}	5/2	4.3	0.3
	Fe^{2+}	2	--	0.7
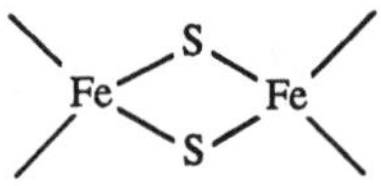 Fe–S₂–Fe	$[2Fe\text{-}2S]^{2+}$	0	--	0.3
	$[2Fe\text{-}2S]^{1+}$	1/2	1.95	0.3, 0.6
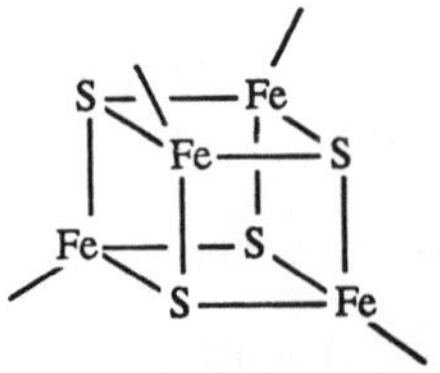 Fe₄S₄ cube	$[4Fe\text{-}4S]^{3+}$	1/2	2.05	0.3
	$[4Fe\text{-}4S]^{2+}$	0	--	0.45
	$[4Fe\text{-}4S]^{1+}$	1/2	1.95	0.6

In the three structures shown each iron coordinates four sulfur atoms with distorted tetrahedral geometry, and each structure is covalently attached to the polypeptide chain via four cysteines. For the Fe-S clusters all of the irons are high-spin and antiferromagnetically coupled resulting in S = 0 or S = 1/2 spin systems. For [2Fe-2S] clusters only two out of the possible three redox states are obtainable, while for [4Fe-4S] clusters three out of the possible five redox states are obtainable. In addition, individual Fe-S proteins only stabilize two redox states resulting in two classes of [4Fe-4S] proteins, namely, high potential Fe-S proteins (HiPIP) which operate between the 3+ and 2+ redox states, and the bacterial 4Fe (and 8Fe) ferredoxins which operate between the 2+ and 1+ redox states. Therefore, reduced HiPIP and oxidized 4Fe ferredoxin are electronically equivalent. The two spectroscopic techniques that were most useful at this time for identifying the types of clusters present were electron paramagnetic resonance (EPR) and Mössbauer spectroscopy. EPR spectroscopy is a sensitive technique for detection of unpaired electrons. The position of a resonance is characterized by its g-value, and the average of the 3 principal g-values (g_x, g_y, and g_z) is termed g-average. However, systems with zero or integer spin, including the antiferromagnetically coupled systems of Fe-S clusters, are generally EPR silent. Mössbauer spectroscopy, which involves the vibrationless absorption of gamma radiation by an atomic nucleus, has the advantage that it detects all iron regardless of its redox or spin state. However, only ^{57}Fe (2.2% natural abundance) is detected. In zero applied fields for integer spin systems or under conditions of fast relaxation for half-integer spin systems, each ^{57}Fe nucleus exhibits a doublet of peaks which are characterized by an isomer shift δ, and a quadrupole splitting ΔE_Q. While a wealth of additional information is available from Mössbauer spectroscopy, the isomer shift (a measure of the electron density at the Fe nucleus) exemplifies the discriminatory ability of this technique when applied to Fe-S proteins, as illustrated in Table I.

Aconitase was first determined to be an Fe-S protein in 1972 by Kennedy, Rauner and Gawron (23). Chemical analyses of inactive enzyme gave values of 2 Fe and 3 $S^{=}$/protein of 66,000 daltons. The observed molar relaxivity of water protons by this preparation of aconitase was 473 M^{-1} s^{-1} (25). This value was an order of magnitude lower than measured in the earlier preparation of Villafranca and Mildvan (21) and much closer to that of Fe-S proteins (26). One mole of Fe^{2+} per mole of protein was taken up by the enzyme upon activation in the presence of cysteine and ascorbate, or lost upon inactivation in the presence of the iron chelator ferrozine (27). Gawron's group also demonstrated a correlation between loss of one Fe and loss of enzyme activity, as well as the protection afforded by citrate against both losses. However, the presence of an Fe-S cluster in aconitase remained for the moment a curiosity, in particular because of the unusual Fe/$S^{=}$ stoichiometries. The essential Fe that is correlated with activity continued to be interpreted in terms of the "ferrous-wheel" model.

During this same period Suzuki *et al.* (28-31) published a series of papers on the properties of yeast aconitase purified from *Candida lipolytica* . This material remained active upon purification and was analyzed to contain 2 Fe and 1 $S^{=}$/protein molecule of 68,500 daltons (28). One Fe could be removed with chelators without loss of activity (30). Enzyme reconstituted with ^{57}Fe was studied with EPR and Mössbauer spectroscopy (31). Even though the measured Mössbauer parameters did not match those of other Fe-S proteins (ΔE_Q = 0.9 mm/s and δ = 0.36 mm/s for the dominant species), the spectra were interpreted as resulting from a [2Fe-2S] cluster. In addition chemical analyses on the reconstituted material now gave 2 Fe and 2 $S^=$/protein molecule. The EPR spectrum showed an isotropic signal near g = 2, which was interpreted by Suzuki to originate from a $[2Fe\text{-}2S]^{1+}$ cluster. (The same type of EPR signal had been associated with $[4Fe\text{-}4S]^{3+}$ clusters previously (32)). No signal quantitations or redox chemistry was reported. At this point there was no clear picture of what type of Fe-S cluster was present in aconitase.

Aconitase was first purified from beef heart mitochondria in 1974 by Ruzicka and Beinert (33) using its somewhat unique and nearly isotropic EPR spectrum (see Figure 1) as their only enzyme assay. Not realizing at the time that they had isolated aconitase, they named the protein mitochondrial HiPIP. Four years later, noticing the similarities between mitochondrial HiPIP and the published properties of aconitase, they were able to demonstrate that mitochondrial HiPIP is inactive aconitase and could be activated with Fe and reductant (34). Initially, the simple determination of the type of Fe-S cluster proved quite difficult. Chemical analyses gave values between 2 and 3 moles Fe or $S^=$/mol of protein. Because the EPR spectrum of inactive aconitase has a g-average of 2.01 and disappears upon reduction, it initially was identified as a [4Fe-4S] cluster (34). However, clusters removed from the enzyme in organic solvents were of the [2Fe-2S] type (35). This contradiction was resolved by the discovery of a new type of Fe-S cluster in *Azotobacter vinelandii* ferredoxin I containing three irons (36, 37), and the application of Mössbauer spectroscopy (38, 39). The spectrum of aconitase as isolated shows a single quadrupole doublet ($\Delta E_Q = 0.71$ mm/s, $\delta = 0.27$ mm/s) which is identifiable as an Fe-S cluster containing antiferromagnetically coupled ferric ions with a system spin of 1/2. Figure 2A-C show the Mössbauer spectra of three dithionite reduced proteins, ferredoxin I of *Azotobacter vinelandii* , ferredoxin II of *Desulfovibrio gigas*, and beef heart aconitase, respectively. The solid lines through the spectra of Figure 2A and 2C are the result of least-squares fitting two quadrupole doublets at a concentration ratio of 2:1. The smaller doublet retains an isomer shift of 0.30 mm/s indicating ferric ion while the larger doublet has an isomer shift of 0.46 mm/s suggesting a formal charge of +2.5. This indicates that upon reduction two irons share one electron. In addition, the application of a small external field (60 mT) results in a spectrum broadened by magnetic hyperfine interactions with the electronic spin, *i.e.*, $S \neq 0$. The Mössbauer parameters of these oxidized and reduced clusters clearly identify them as different from previously studied Fe-S clusters and require an odd number of antiferromagnetically coupled irons, specifically three. The detection of [2Fe-2S] clusters when removed from aconitase in organic solvents,as mentioned above, is explained by the fact that 3Fe clusters are unstable under these conditions and break down to form [2Fe-2S] clusters. At the time the only model for a 3Fe cluster was a [3Fe-3S] structure determined by X-ray crystallography on ferredoxin I by Stout *et al.* (37). Therefore, it was assumed that aconitase contained a similar [3Fe-3S] structure. A question that remained was: what relationship does the Fe-S cluster have to enzyme activity?

Fe-S Cluster Interconversions and the Role of Added Iron

One of the consistent results concerning aconitase had been its absolute requirements for ferrous ion and a reductant in the activation process. However, if the activation process is done under anaerobic conditions, surprising data are obtained (34, 40, 41). As displayed in Table II some reductants in the absence of added iron are quite good at activating aconitase.

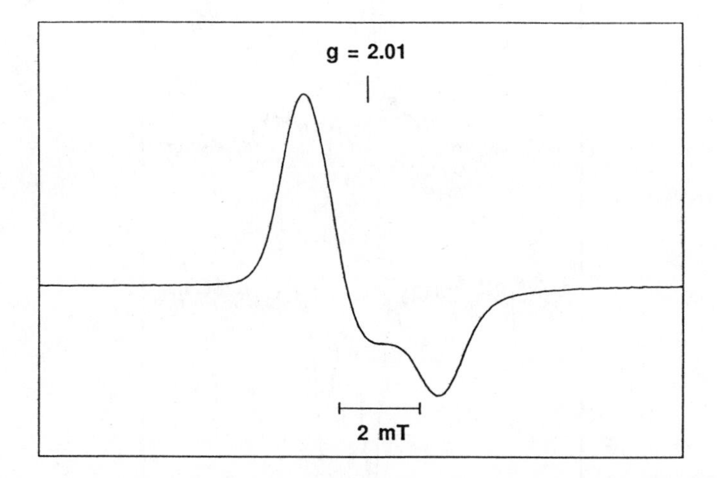

Figure 1. EPR absorption derivative spectrum of aconitase (as isolated from beef heart mitochondria) in 100 mM potassium phosphate, pH 7.0. Experimental conditions for obtaining the EPR spectrum were 10 K, 100 microwatts power, 0.8 milliTesla (mT) modulation amplitude, and 9.42 GHz microwave frequency.

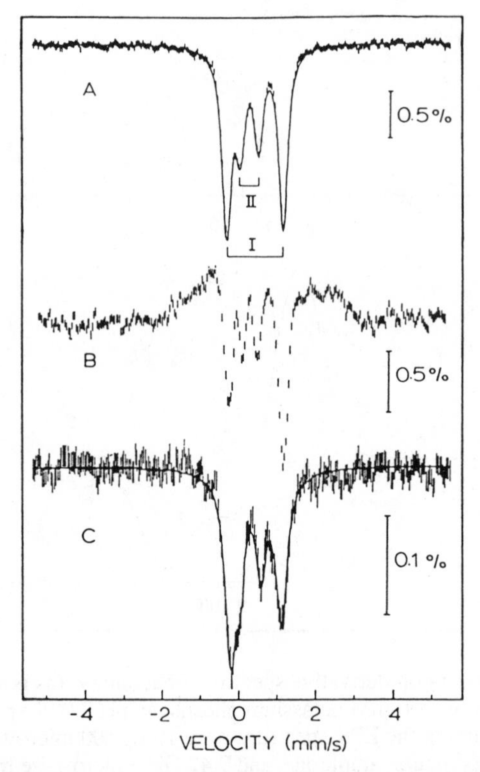

Figure 2. Zero-field Mössbauer spectra of dithionite reduced three-iron clusters of (A) *D. gigas* ferredoxin II, (B) *A. vinelandii* ferredoxin I, and (C) aconitase isolated from beef heart mitochondria. (Reproduced with permission from Ref. 38. Copyright 1982 Elsevier.)

Table II. Activation of Aconitase by Various Reagents

Reagent Added	% Maximum Activity Achieved
Fe^{2+}, dithiothreitol	100
Fe^{2+}	100
Dithiothreitol	75-80
$Na_2S_2O_4$	25-30
$Na_2S_2O_4$, dithiothreitol	60-65
Ascorbate	30-40
Ascorbate, dithiothreitol	60-70

At the time this suggested that the activation of aconitase required reduction of the Fe-S cluster, rather than the addition of iron. This would relegate iron to the role of an efficient reductant of the Fe-S cluster. Also, while reduction of the cluster by sodium dithionite was immediate, development of full enzyme activity required minutes (41, 42). This suggested that perhaps cluster reduction triggered a conformational change which led to activation. Fluorescence studies on the reduction and activation process of aconitase by Ramsay (43, 44) supported this idea. Again Mössbauer spectroscopy provided the key to understand the fate of the added iron (39).

The Mössbauer spectra shown in Figure 3A and 3B are from iron activated aconitase after reduction with sodium dithionite. Before the addition of iron the aconitase samples had the typical reduced 3Fe cluster Mössbauer spectrum as seen in Figure 2C. Figure 3A shows the Mössbauer spectrum of reduced aconitase after addition of ^{56}Fe (99.9% enriched). (The ^{56}Fe nucleus does not manifest the Mössbauer effect and therefore is not detectable by this technique). All of the intrinsic ^{57}Fe (2.2% natural abundance) of aconitase is now found in a new doublet with $\Delta E_Q = 1.30$ mm/s and $\delta = 0.45$ mm/s. These values are characteristic of $[4Fe\text{-}4S]^{2+}$ clusters. When ^{57}Fe (>90% enriched) is added to a second sample of reduced aconitase two new doublets are observed (Figure 3B). One doublet is characteristic of monomer ferrous ion, as seen in Figure 3C when ^{57}Fe is added to apoaconitase. The other doublet (labeled a) has parameters ($\Delta E_Q = 0.80$ mm/s, $\delta = 0.45$ mm/s) identifiable with $[4Fe\text{-}4S]^{2+}$ clusters. Thus, doublet *a* must belong to a fourth subsite (Fe_a) of a [4Fe-4S] cluster. The iron activated samples can be desalted anaerobically removing the monomer ferrous ion and leaving both the [4Fe-4S] cluster and activity intact. This was one of the first demonstrations of 3Fe to [4Fe-4S] cluster conversions in Fe-S proteins. (Previous to this Thomson *et al.* (45) using low-temperature magnetic circular dichroism (MCD) and EPR demonstrated that the 8Fe ferredoxin of *Clostridium pasteurianum* could undergo [4Fe-4S] to 3Fe cluster conversions upon oxidation). Later experiments using the radioisotope ^{59}Fe demonstrated the incorporation of 1 iron per Fe-S cluster (46), substantiating the earlier work from Gawron's laboratory (27) (but with a different interpretation). The facile cluster interconversion, as demonstrated for aconitase, clearly explained the role that added iron played in the activation process. However, there were still the problem of activation in the absence of added iron or added sulfide (for the presumed [3Fe-3S] $\rightarrow$ [4Fe-4S] cluster conversion), and the question of the relationship between cluster structure and enzyme activity.

One of the earlier observations concerning the activation process was that it was inhibited by iron chelators (9, 27). Kennedy *et al.* (46) demonstrated that the metal chelator ethylenediaminetetraacetic acid (EDTA) inhibits the activation process, even

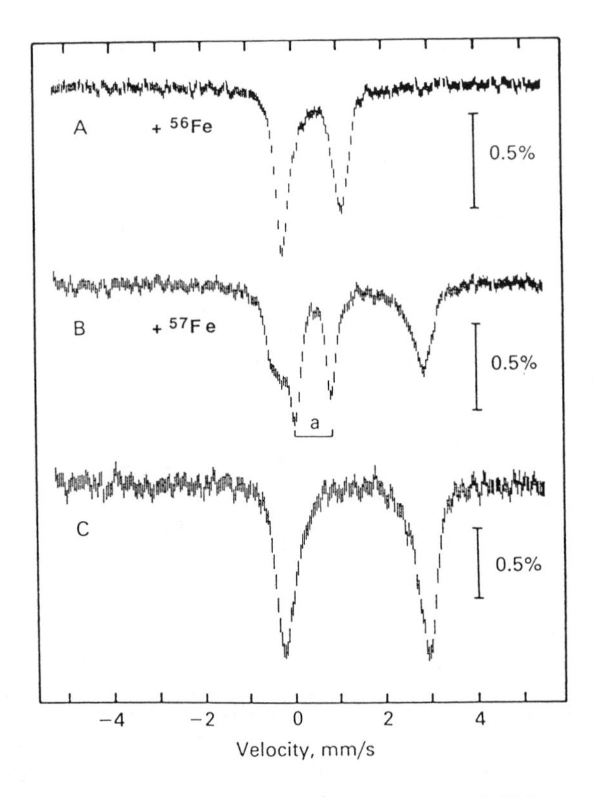

Figure 3. Mössbauer absorption spectra of aconitase at 4.2 K in zero field. (A) Dithionite reduced enzyme activated with 99.9% enriched ^{56}Fe. (B) Dithionite reduced enzyme activated with >90% enriched ^{57}Fe. (C) Apoaconitase incubated with dithionite and ^{57}Fe. (Reproduced with permission from ref. 39. Copyright 1982 the authors.)

though it is not an inhibitor of active aconitase. The inhibition of activation can be reversed by swamping the EDTA with excess iron or other metals such as zinc. A formation constant of 10^{10} for the addition of ferrous ion to the reduced 3Fe cluster was estimated from a competition reaction with EDTA . This compares to values of 10^{14} for EDTA and 10^{8} for the chelator nitrilotriacetic acid, which does not inhibit aconitase activation at similar concentrations. These results contradict the earlier idea that reduction of the 3Fe cluster alone was sufficient to activate aconitase and demonstrate the high affinity that the reduced 3Fe cluster has for iron. However, if added iron is essential for activity, how does one get the high activities seen in Table II upon activation without iron? The answer was obtained by Emptage *et al.* (47) by quantitating the clusters present after activation with either dithiothreitol plus iron or dithiothreitol alone. Initially, this was done indirectly by back-titrating the activated enzyme with the oxidant $K_3Fe(CN)_6$ (resulting in reversible inactivation) and determining the total number of 3Fe clusters by spin quantitations of the EPR signal at $g = 2.01$. The results of such a titration experiment are shown in Figure 4. The plots demonstrate that there is a concomitant loss of activity with the formation of 3Fe clusters. In addition, it requires 2.5 oxidizing equivalents per cluster to inactivate the enzyme. This is close to a value of 2, which would be expected for the conversion of a $[4Fe\text{-}4S]^{2+}$ cluster to a 3Fe cluster (all ferric) and 1 ferric ion. The experiment was then repeated using aconitase activated by reductant alone. The data comparing the two titration experiments on enzyme activated with or without iron are shown in Figure 5. In this plot the EPR spins/mg protein (*i.e.*, inactive aconitase) are compared directly with specific activity (units activity/mg protein). One sees that while the slopes of the two plots are the same, the enzyme activated with dithiothreitol alone has both x and y-axis intercepts that are about 25% smaller. Thus, the lower activity of the dithiothreitol activated enzyme correlates with fewer total clusters. Furthermore, the oxidizing equivalents/cluster for the inactivation process remained near 2.5 suggesting that the cluster in the activated aconitase is the same whether it is formed by activation with iron and reductant or reductant alone. The 25% lower activity and clusters fits a model in which, during cluster reorganization, 25% of the 3Fe clusters fall apart which, in turn, feeds the build-up of the other 75% into 4Fe clusters. Later Fe and $S^{=}$ analyses of activated aconitase anaerobically desalted to remove the activating reagents supported these conclusions.

From 1979 to 1982 the only model for 3Fe clusters was that derived from the X-ray diffraction data of *A.vinelandii* ferredoxin I by Stout's group (37, 48, 49), as shown schematically below.

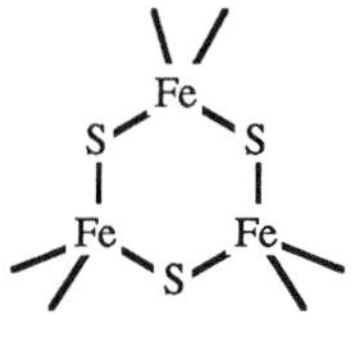

The cluster is a six-membered ring of alternating atoms of Fe and S. The unusual feature of this model is its openness, resulting in long Fe-Fe distances of 4.1 Å as compared to 2.7 Å for the [4Fe-4S] and [2Fe-2S] clusters. In the activation of aconitase, then, how does a [3Fe-3S] cluster convert to a [4Fe-4S] cluster without the addition of sulfide in the activation process? Beinert *et al.* (50) were able to

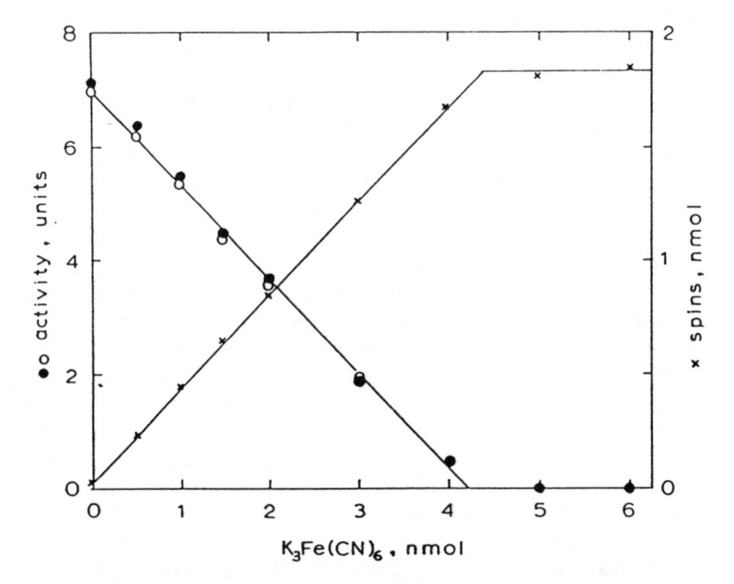

Figure 4. Oxidative titration of activated aconitase with potassium ferricyanide. Aliquots of activated aconitase and a measured amount of $K_3Fe(CN)_6$ were added to individual EPR tubes, assayed for enzymatic activity (●), and frozen. After determination of the number of spins by EPR at g = 2.01 (x), the samples were thawed and assayed again (o). (Reproduced with permission from Ref. 47. Copyright 1983 American Society of Biological Chemists.)

answer this by combining accurate microanalytic techniques for sulfide determinations (51), and extended X-ray absorption fine structure (EXAFS) spectroscopy of the cluster iron. As seen in Table III, the more accurate sulfide determinations placed the Fe/S= ratios very close to 0.75 for inactive aconitase and 1.0 for aconitase isolated in its active form, suggesting a $[3Fe\text{-}4S]^{1+} + e^- + Fe^{2+} \Rightarrow [4Fe\text{-}4S]^{2+}$ cluster conversion reaction. The stoichiometries of Fe and S= to protein also changed from near 2 and 3 moles/mole to near 3 and 4 moles/mole, respectively, for inactive aconitase when an accurate protein extinction coefficient was determined from total amino acid content (52).

Table III. Iron and Sulfide Content of Inactive and Active Aconitase

Aconitase Preparation number	Fe/mole	S=/mole	Fe/S=
Inactive			
3	2.49 ± 0.08 (4)	3.80 ± 0.06 (4)	0.66
10	2.64 ± 0.16 (4)	3.94 ± 0.28 (4)	0.67
11	2.90 ± 0.09 (3)	4.08 ± 0.12 (3)	0.71
12	2.80 ± 0.05 (3)	3.80 ± 0.11 (3)	0.74
13	2.76 ± 0.08 (3)	3.70 ± 0.10 (3)	0.74
14	2.57 ± 0.06 (4)	3.66 ± 0.09 (4)	0.70
Active[a]			
1	3.74 ± 0.14 (3)	4.18 ± 0.14 (6)	0.90
2	3.30 ± 0.07 (4)	3.48 ± 0.12 (4)	0.95
5	3.64 ± 0.22 (3)	3.54 ± 0.17 (4)	1.03
6	3.46 ± 0.11 (2)	3.49 ± 0.14 (5)	0.99
8	3.48 ± 0.25 (4)	3.46 ± 0.04 (4)	1.01
9	3.18 ± 0.12 (3)	3.30 ± 0.12 (3)	0.96
10	3.48 ± 0.08 (3)	3.59 ± 0.08 (4)	0.97
11	3.37 ± 0.13 (3)	3.40 ± 0.13 (4)	0.99

Values given as the mean ± SEM for (n) individual samples.
[a]The active aconitase preparations were ~90% active and ~80% pure. The Fe and S= contents were calculated on the assumption that the protein impurities do not contain Fe or S=.

The Fe EXAFS data from inactive aconitase and its Fourier transform are shown in Figure 6. EXAFS of metal complexes arises from the interference of outgoing photoelectron waves (generated from the absorption of X-rays by the 1s electrons of the metal) with waves back-scattered from neighboring atoms. Analyses of the phase, frequency and amplitude of EXAFS provide information about the number and type of atoms near the metal ion, and more importantly here, interatomic distances. The largest peak in Figure 6a results from Fe-S scattering and gives a distance of 2.24 Å, a typical value for Fe-S proteins. The next largest peak at higher R arises from Fe-Fe scattering and gives a distance of 2.71 Å. This distance is typical for a [4Fe-4S] cubane structure and much smaller than the 4.1 Å value for the [3Fe-3S] cluster model. These results require that the [3Fe-4S] cluster maintains a cubane structure as shown schematically below.

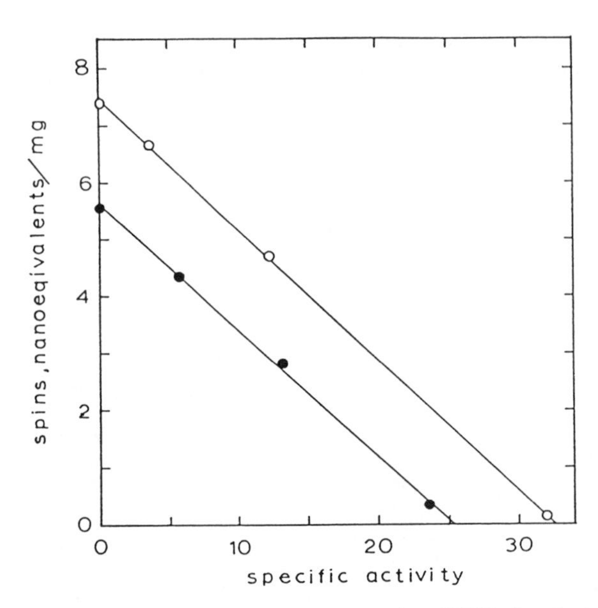

Figure 5. Relationship of 3Fe cluster concentration (EPR signal at g = 2.01) and enzymatic activity of aconitase. The plots result from the $K_3Fe(CN)_6$ titration of aconitase activated with 5mM dithiothreitol and 0.5 mM iron (o) or activated with 5 mM dithiothreitol, 0.05 mM safranin O and **no** added iron (●). (Reproduced with permission from ref. 46. Copyright 1983 American Society of Biological Chemists.)

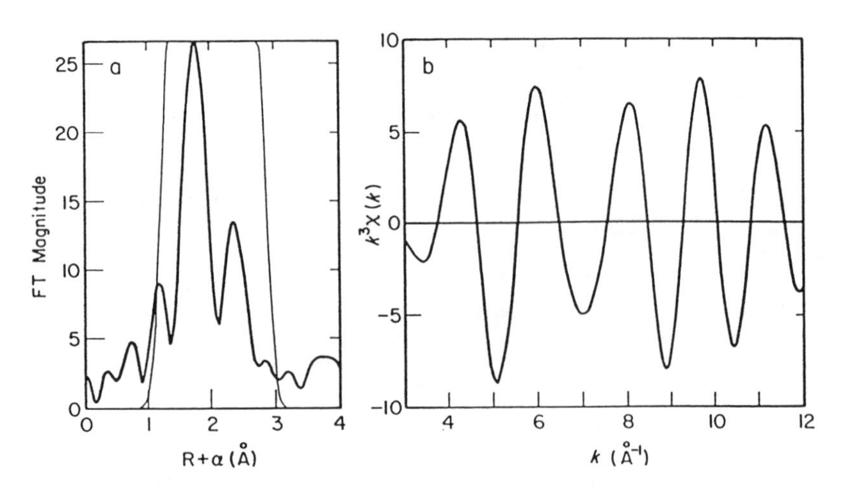

Figure 6. Fe EXAFS data for inactive aconitase. (a) FT of raw EXAFS data; (b) filtered EXAFS by using the filter window shown in a. (Reproduced with permission from ref. 50. Copyright 1983 the authors.)

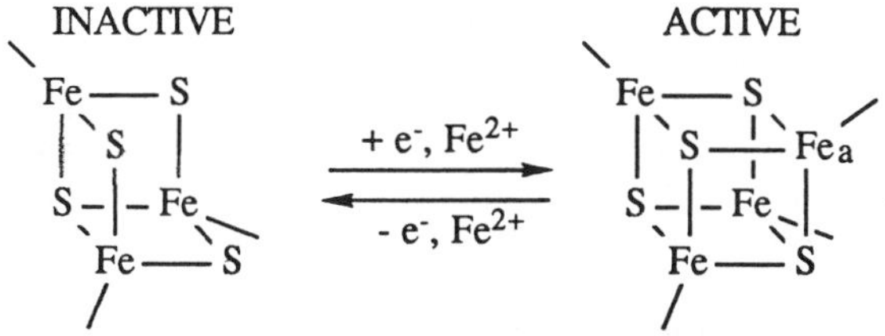

INACTIVE $\underset{-e^-, Fe^{2+}}{\overset{+e^-, Fe^{2+}}{\rightleftharpoons}}$ ACTIVE (4)

Furthermore, resonance Raman studies of aconitase by Johnson *et al.* (53) demonstrated homologous spectra for both inactive and active aconitase. This suggests similar vibrational modes and thus similar core structures for the two forms. Finally, a cubane structure for the [3Fe-4S] cluster is supported by recent protein crystallographic studies of inactive aconitase by Robbins and Stout (54). (Recent results from Jensen's group (55) on the redetermination of the crystal structure of the *Azotobacter* ferredoxin I clearly show that the 3Fe cluster does **not** have a [3Fe-3S] ring structure, as originally determined (37), but has a [3Fe-4S] cubane structure.)

By this time it was demonstrated that the $[3Fe\text{-}4S]^{0/+}$ form of aconitase is inactive, while the $[4Fe\text{-}4S]^{2+}$ form is active. How is the activity of the enzyme affected by the oxidation state of the [4Fe-4S] cluster? Because the active enzyme contains a $[4Fe\text{-}4S]^{2+}$ cluster, either the 3+ or 1+ oxidation states may also be stable. The 3+ state is unstable since oxidation of the $[4Fe\text{-}4S]^{2+}$ resulted in the immediate loss of a ferrous ion and conversion to a $[3Fe\text{-}4S]^{1+}$ cluster (46, 47). However, reduction of active aconitase by sodium dithionite or photoreduction in the presence of deazaflavin produced in high yields an EPR signal characteristic for $[4Fe\text{-}4S]^{1+}$ clusters (47). When active enzyme within an anaerobic assay cuvette was photoreduced, the activity of the enzyme dropped to 1/3 of its initial value. Further photoreduction resulted in cluster destruction. Then, if the enzyme is re-oxidized with air, the activity returned to its original value. This demonstrated that the redox state of the cluster can modulate the enzyme activity. A scheme summarizing the cluster interconversions and various redox states of the Fe-S cluster of aconitase is shown below.

Scheme I

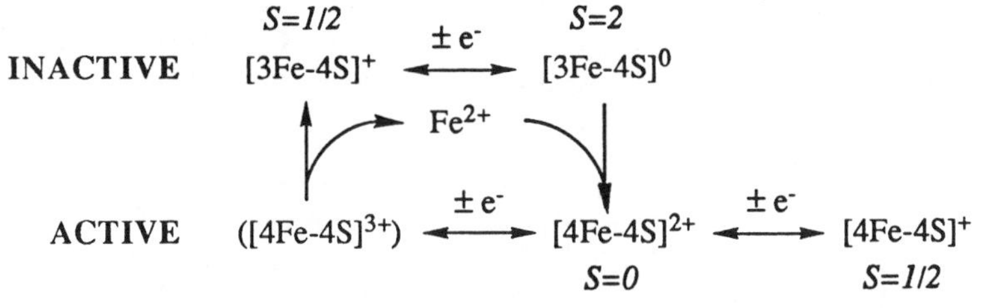

INACTIVE: $[3Fe\text{-}4S]^+$ ($S=1/2$) $\xleftrightarrow{\pm e^-}$ $[3Fe\text{-}4S]^0$ ($S=2$)

Fe^{2+}

ACTIVE: $([4Fe\text{-}4S]^{3+})$ $\xleftrightarrow{\pm e^-}$ $[4Fe\text{-}4S]^{2+}$ ($S=0$) $\xleftrightarrow{\pm e^-}$ $[4Fe\text{-}4S]^+$ ($S=1/2$)

A short comment should be made concerning some results by Johnson *et al.* (56) which seemed to show that the reduced [3Fe-4S] cluster in aconitase produces an active enzyme. Using low-temperature MCD they characterized the dithionite reduced inactive enzyme and established the presence of $[3Fe\text{-}4S]^0$ (S = 2) clusters only. However, when an aliquot of the sample was diluted 100-fold into anaerobic buffer for enzyme assays, it had 66% the activity of a control diluted into an activation mixture of ferrous ion and dithiothreitol. The explanation is as follows: The high protein concentrations required for MCD experiments probably help to stabilize the reduced [3Fe-4S] cluster and prevent cluster breakdown. At low protein concentrations, as required for enzyme assays, the stabilization is lost, which allows cluster breakdown and provides iron for formation of [4Fe-4S] clusters. We (Emptage, Kennedy and Beinert) have repeated such a procedure as outlined above and followed the change in activity over time after dilution of dithionite reduced inactive aconitase into anaerobic buffer. At the earliest time sampled (~1 min after dilution) the activity was 27% that of an iron activated control. The activity of the diluted sample continued to rise over a period of 60 min up to 70% of the control, which is quite close to the expected 75%. This result supports the model outlined in Scheme I and points out the difficulty of comparing experiments done at vastly different sample concentrations. In addition, the essential role played by the [4Fe-4S] cluster in catalysis, as outlined below, makes it highly unlikely that the [3Fe-4S] form of aconitase would be active.

Function of the Fe-S Cluster

One of the first clues to the function of the Fe-S cluster in aconitase came with the observation by Emptage *et al.* (47) that substrate has a dramatic effect on the EPR spectrum of reduced active aconitase. (Here, and that which follows, substrate will refer to the equilibrium mixture of substrates which contains 88% citrate, 8% isocitrate, and 4% *cis*-aconitate (20).) This can be seen by comparing the two spectra in Figures 7A (- substrate) and 7B (+ substrate). The top spectrum is that of reduced active aconitase plus tricarballylate (1,2,3-propanetricarboxylic acid) and is typical for proteins containing $[4Fe\text{-}4S]^+$ clusters. Tricarballylate is a weak competitive inhibitor of aconitase ($K_i = 2$ mM) and effects no change in the EPR spectrum upon binding in the active site. In the presence of substrate the spectrum (Figure 7B) shifts to g-values (2.04, 1.85, 1.78) lower than that seen previously for 4Fe clusters. Surprisingly, the competitive inhibitor *trans*-aconitate ($K_i = 0.1$ mM) also shifts the EPR spectrum (Figure 7C) to lower g-values. Other competitive inhibitors that have been tested generate one, or a mixture, of these three spectral types. The dramatic shift in g-average from 1.95 to 1.89 upon addition of substrate suggested a substantial change in the electronic properties of the cluster induced by the substrate.

Mössbauer studies of the substrate effects on the Fe-S cluster again provided indispensable information (57). As was done previously to obtain the experimental results shown in Figure 3, aconitase was activated with either ^{57}Fe to examine the activating iron site (Fe_a), or ^{56}Fe to examine the remaining 3 iron sites with intrinsic ^{57}Fe at natural abundance. Figure 8A shows the same results (the Fe_a site in activated aconitase) as seen in Figure 3B, except for the removal of excess iron by anaerobic desalting. The solid line traces the quadrupole doublet of Fe_a. Upon addition of substrate the spectrum changes to that seen in Figure 8B. The solid line was generated by superimposing three quadrupole doublets with relative intensities of 20% (for Fe_a), 40% (for S_1), and 40% (for S_2). The Mössbauer parameters of the two new species, S_1 and S_2, are $\Delta E_Q(S_1) = 1.26$ mm/s and $\delta(S_1) = 0.84$ mm/s, and $\Delta E_Q(S_2) = 1.83$ mm/s and $\delta(S_2) = 0.89$ mm/s. These values are much larger than that seen for typical Fe-S clusters (see Table I). The Mössbauer parameters most closely match that of high-spin 6-coordinate ferrous ion, but the Fe_a site is still

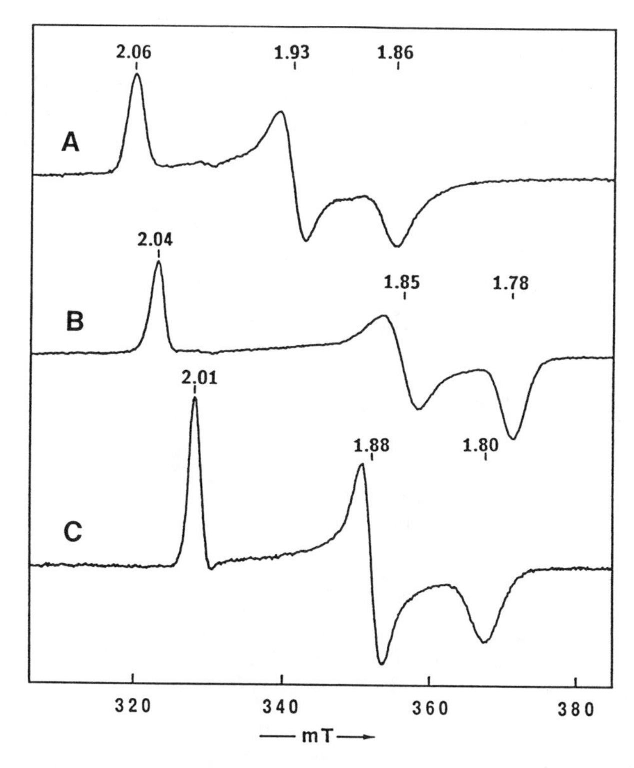

Figure 7. EPR absorption derivative spectra of photoreduced active aconitase. Enzyme (~5 mg/ml) in 100 mM Hepes, pH 7.5, plus 5 μM deazaflavin and 10 mM potassium oxalate was photoreduced in the presence of either A) 10 mM tricarballylate, B) 1 mM citrate, or C) 10 mM *trans*-aconitate. The numbers above each spectrum are the g-values of prominent features. Experimental conditions for obtaining EPR spectra were 13 K, 1 milliwatt microwave power, 0.8 mT modulation amplitude, and 9.24 GHz microwave frequency.

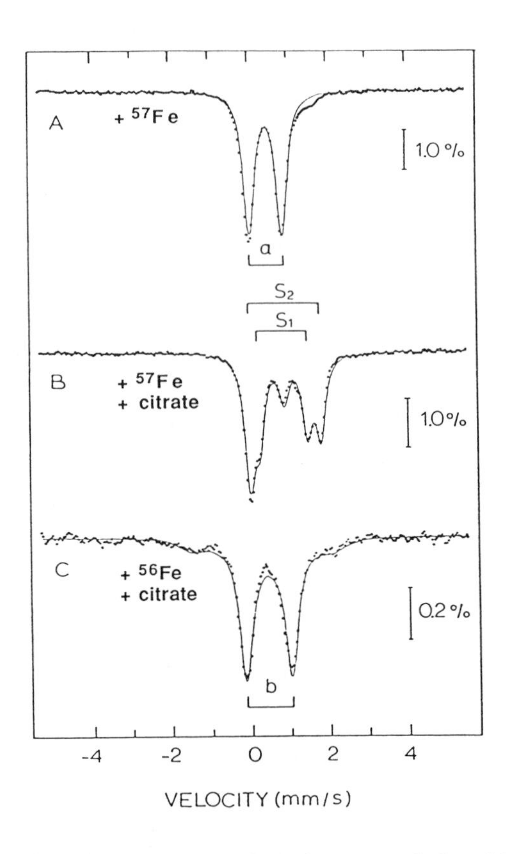

Figure 8. Mössbauer spectra of activated aconitase recorded at 4.2 K in zero field (A and B) or in a magnetic field of 60 mT applied parallel to the observed γ-radiation (C). (A) Enzyme activated with <90% enriched ^{57}Fe. (B) Enzyme activated with ^{57}Fe in the presence of 10 mM citrate. (C) Enzyme activated with 99.9% enriched ^{56}Fe in the presence of 10 mM citrate. (Reproduced with permission from ref. 57. Copyright 1983 the authors.)

part of a diamagnetic spin-coupled cluster. This suggested that binding of substrate generates a 5 or 6 coordinated iron at the Fe_a site. In the complementary experiment of activation by ^{56}Fe the remaining 3 irons of the Fe-S cluster show very little change (ΔE_Q shifts from 1.30 to 1.15 mm/s) upon substrate binding (compare Figures 3A and 8C). Thus, only the exchangeable iron (Fe_a), which is necessary for active aconitase, is affected by substrate binding.

To complement the results above, Emptage *et al.* (57) examined the substrate-cluster interaction from the perspective of the substrate. If the substrate interacts with the Fe-S cluster via oxygen coordination to the iron, then magnetic hyperfine interactions between the electronic spin of the reduced cluster (S = 1/2) and the nuclear spin of ^{17}O (I = 5/2) labeled substrate should manifest themselves as broadening of the EPR signal. Aconitase rapidly exchanges the citrate and isocitrate hydroxyl groups with solvent (18), so that [^{17}O-hydroxyl]citrate/isocitrate was formed by adding *cis*-aconitate to active aconitase in $H_2{}^{17}O$ (38% enriched). This was followed by photoreduction to generate the paramagnetic state of the cluster. The effects on the low-field lines of the EPR spectra by $H_2{}^{17}O$ with substrate or *trans*-aconitate are shown in Figures 9A and 9B, respectively. The 0.5 mT broadening of the EPR signal in Figure 9A was the first direct evidence for substrate and/or water coordination to an Fe-S cluster of an enzyme. Because *trans*-aconitate is not a substrate, the 0.3 mT broadening of its EPR signal must be caused by the coordination of $H_2{}^{17}O$ directly to the Fe-S cluster. No broadening of the EPR spectra was observed in the absence of substrate, and only a slight broadening in the presence of tricarballylate. Thus, there appeared to be a correlation between substrate and inhibitor induced high-field shift of the EPR spectrum and coordination of water and/or hydroxyl of substrate. When the carboxyl groups of citrate were enriched with ^{17}O, no measurable broadening of the EPR spectra were observed. This was initially interpreted to mean that only the hydroxyl groups of citrate and isocitrate or water can strongly coordinate to the Fe-S cluster, and that the carboxyl groups are not, or at best, only weakly coordinated. (More recent ENDOR results proved this latter interpretation to be in error, as shown below.) These results suggested that the Fe-S cluster in aconitase functions as a Lewis acid and facilitates the dehydration of citrate and isocitrate.

The exact nature of the two substrate-bound species seen in the Mössbauer spectrum of Figure 8B has not yet been determined. Later Mössbauer studies by Kent *et al.* (58) examined these species in greater detail using substrate analogues, rapid mix/rapid freeze techniques and cryoenzymology. The ratio of the two species remains the same whether Mössbauer samples were frozen in seconds with the enzyme-substrate mixture at chemical equilibrium or within 5 ms after mixing enzyme and any one of the three substrates. Since the turnover times for aconitase with the substrates is 25 to 75 ms at room temperature, internal equilibrium is reached much faster than product is released. Mixing substrates and enzyme at -60 ^{o}C produced no S_1 species but yielded a new species with $\delta \cong 0.7$ mm/s, intermediate in value between the substrate free and bound forms. This is suggestive of a different geometry at the Fe_a site. The addition of nitro analogues of citrate or isocitrate (see Equation 5 and accompanying text), which are thought to be carbanion intermediate structural analogues, produced the S_2, but not the S_1 species. This latter result suggested that the S_2 species may represent a carbanion intermediate bound to the Fe-S cluster. The Mössbauer spectrum of Fe_a in the reduced cluster is a single species (ΔE_Q = 2.24 mm/s and δ = 0.99 mm/s), in agreement with the observation of a single EPR species. As had been surmised from the Mössbauer parameters for the oxidized cluster, the change in parameters for the Fe_a site in the reduced cluster after substrate addition suggested an increased coordination number.

Substrate analogues have been very useful in enzymology for deducing enzyme mechanisms. The nitro analogues of citrate and isocitrate, 2-hydroxy-3-nitro-1,2-propanedicarboxylic acid and 1-hydroxy-2-nitro-1,3-propanedicarboxylic acid, respectively, have played that role for aconitase (59). The substitution of a nitro group for a carboxyl group lowers the pK of the adjacent carbon acid from ~25 to near 10. The fully deprotonated forms of the acids are strikingly potent inhibitors, probably because they structurally mimic a carbanion reaction intermediate as shown below for nitroisocitrate.

$$^{-}OOC\text{-}CH(H)(NO_2)\text{-}CH(OH)(H)\text{-}COO^{-} \underset{}{\overset{pK = 9.5}{\rightleftharpoons}} {}^{-}OOC\text{-}CH_2\text{-}C(=N^{+}(O^{-})O^{-})\text{-}CH(OH)(H)\text{-}COO^{-} + H^{+} \quad (5)$$

The final K_i for nitroisocitrate drops by 3 orders of magnitude to 0.7 nM when the inhibitor is fully deprotonated. (This compares to a K_m of 30 μM for isocitrate). Schloss *et al.* (59) argue that the much higher affinity for a carbanion structural analogue strongly implies a carbanion mechanism (*i.e.*, the proton is removed from citrate or isocitrate before the hydroxyl group) for aconitase. The advantage that the nitro analogues bring to the problem at hand is the ability to separate water and substrate hydroxyl coordination to the Fe-S cluster. As mentioned above aconitase catalyzes a rapid exchange between the hydroxyl group of substrate and solvent water preventing the labeling of each separately. The hydroxyl group of nitrocitrate and nitroisocitrate, on the other hand, do not exchange with solvent allowing the separate determination of water and substrate coordination to the Fe-S cluster.

The technique of ENDOR spectroscopy combines the inherent sensitivity of EPR with the capability of NMR to provide structural information at the molecular level. With this technique NMR transitions of nucleii, that are magnetically coupled to the electronic spin, are detected via the intensity change of a simultaneously irradiated EPR transition. The application of ^{17}O ENDOR corroborated the EPR broadening results and provided additional details on the the Fe-S cluster-substrate interactions (60). The ENDOR spectrum of reduced active aconitase with [^{17}O-hydroxyl]-citrate and $H_2{}^{17}O$ is shown in Figure 10. The ENDOR spectrum for ^{17}O ($I = 5/2$) consists of two 5 line patterns centered at $A^o/2$ and separated by $\nu_o/2$, where A^o is the hyperfine coupling constant in MHz and ν_o is the Larmor frequency for ^{17}O. The results of a series of labeling experiments including that of nitro-isocitrate are shown in Table IV. Two different ^{17}O ENDOR spectra were observed, one corresponding to $H_2{}^{17}O$ coordination as seen with the inhibitors *trans*-aconitate and nitroisocitrate, and the other corresponding to hydroxyl coordination as observed with [^{17}O-hydroxyl]-nitroisocitrate. When both [^{17}O-hydroxyl]-nitroisocitrate and $H_2{}^{17}O$ are present in the same sample, additional EPR line broadening is observed over the single labeled samples, as well as two over-lapping ^{17}O ENDOR signals. EPR spectral simulations using the measured

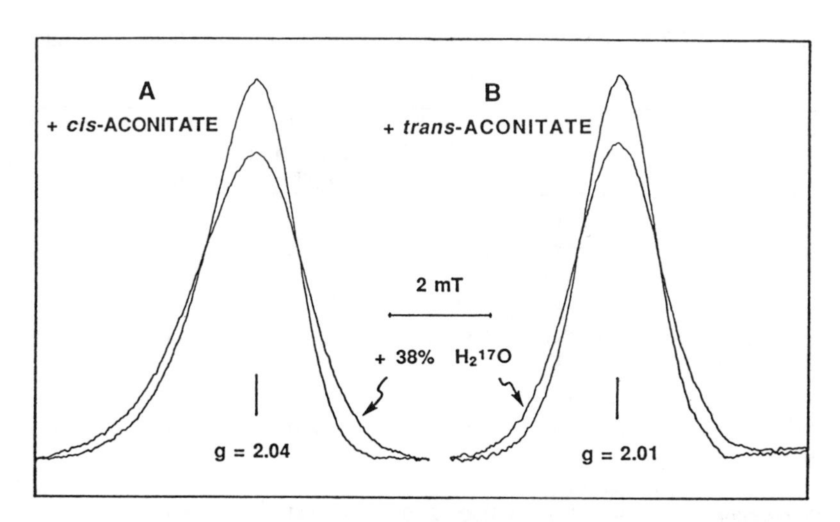

Figure 9. Superposition of the low-field resonances of the EPR spectra of photoreduced active aconitase incubated with 2 mM *cis*-aconitate (A) or 10 mM *trans*-aconitate (B). The narrow resonances were observed in $H_2{}^{16}O$ and the broader ones in 38% enriched $H_2{}^{17}O$. (Adapted from ref. 57.)

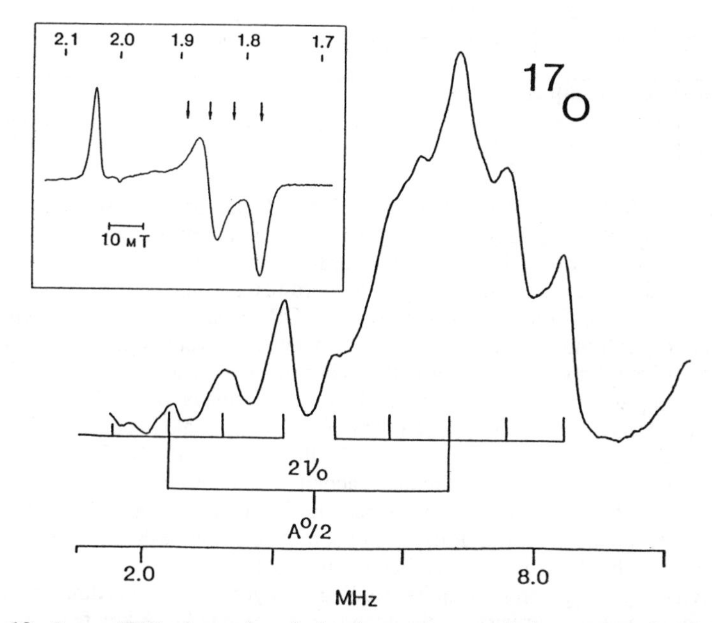

Figure 10. *Inset*, EPR absorption derivative spectrum of photoreduced active aconitase in the presence of citrate. Field positions at which ENDOR spectra were recorded are indicated by arrows. Numbers at the top are from the g-value scale. *Main figure*, ^{17}O ENDOR spectrum at g = 1.88 for photoreduced active aconitase in the presence of citrate and 38% enriched $H_2{}^{17}O$. (Reproduced with permission from Ref. 60. Copyright 1986 American Society of Biological Chemists.)

Table IV. EPR and ENDOR Results on Aconitase with ^{17}O Labeled Substrate, Analogues and Water

Compound Added	^{17}O Labeling (●)	EPR Linewidth (mT)	EPR Broadening[a] (mT)	Hyperfine Coupling Constant, A_3 (MHz) Water	Hydroxyl
Tricarballyate	COO^-, COO^-, COO^- + H_2●	2.35	0.16	0	0
cis-Aconitate[a] (88% citrate)	H●— COO^-, COO^-, COO^- + H_2●	2.01	0.50	8.4	0
trans-Aconitate	^-OOC— COO^-, COO^- + H_2●	1.65	0.31	9.2	0
Nitroisocitrate	HO— COO^-, NO_2, COO^- + H_2●	2.10	0.41	8.4	0
Nitroisocitrate	H●— COO^-, NO_2, COO^- + H_2O	2.10	0.27	0	9.2
Nitroisocitrate	H●— COO^-, NO_2, COO^- + H_2●	2.10	0.77	8.4	9.2

[a]After labeling with 40% ^{17}O.
[b]At equilibrium the substrate mixture is composed of 88% citrate, 8% isocitrate, and 4% *cis*-aconitate (20).

linewidths in ^{16}O samples and the hyperfine and quadrupole tensors determined by ENDOR spectroscopy agreed reasonably well with the measured spectra. This suggests that water and hydroxyl can coordinate the cluster iron site simultaneously, and thus, the Fe_a site is at least 5-coordinate, as had been deduced from the Mössbauer data. Surprisingly, the ^{17}O ENDOR spectrum of aconitase with substrate only exhibits signals originating from water coordination, rather than substrate hydroxyl coordination (or both). Thus the initial assessment of the EPR broadening was not entirely correct. However, it must be pointed out that hydroxyl coordination can take place as seen with the substrate analogue nitroisocitrate (see Table IV). The most likely explanation for the unexpected ENDOR results is that the dominant species bound to reduced aconitase at equilibrium is *cis*-aconitate. A second possible explanation is that, since nitroisocitrate is a carbanion intermediate analogue, substrate hydroxyl may only coordinate to the Fe-S cluster when the substrate is the short-lived carbanion intermediate.

Recently, Sargeson's group has studied the hydration of coordinated carboxyalkenes as model chemistry for the aconitase reaction (61). A key feature of their model is the activation of the double bond via carboxylate coordination to the metal and subsequent attack by a coordinated *cis*-hydroxo group. This was a feature of the "ferrous wheel" mechanism as well. The model chemistry suggested that a closer examination of carboxyl labeled substrate and substrate analogues was necessary. The initial observations of no measurable EPR broadening with [^{17}O-carboxyl]-citrate were interpreted to mean that carboxyl was not, or at best, only weakly coordinated to the cluster. However, ENDOR spectroscopy of this same sample resulted in a strong ^{17}O signal different from both the water and hydroxyl signals (61). Subsequently, each of the carboxyl groups of substrate was individually labeled with ^{17}O by enzymatic or chemical means. The resulting ENDOR spectra are shown in Figure 11. In addition, the 1-carboxyl group (adjacent to the hydroxyl group) of nitroisocitrate was labeled. The overall results are compiled in Table V.

Table V. ENDOR Results on Aconitase with ^{17}O Labeled Carboxylates of Substrates and Analogue

Compound Added	^{17}O Labeling (●)	Hyperfine Coupling Constant A_y (MHz)
Citrate	—C●●⁻ HO—C●●⁻ —C●●⁻	15
Citrate	—C●●⁻ HO—COO^- —COO^-	0
Citrate	—COO^- HO—C●●⁻ —COO^-	15
Isocitrate	HO—COO^- —COO^- —C●●⁻	0
Nitroisocitrate	H●—C●●⁻ —NO_2 —COO^-	9 , 13

The results were unexpected in that only the middle carboxyl group of substrate is coordinated to the reduced Fe-S cluster. In support of this result, a ^{13}C ENDOR spectrum is observable only when the middle carboxyl group is enriched with ^{13}C (~99%). This and the previous results suggest that at equilibrium on the reduced enzyme *cis*-aconitate is by far the major species, and is bound primarily in the citrate-forming orientation (see Scheme II). The labeled nitroisocitrate results demonstrate that substrate coordination to the reduced cluster in the isocitrate orientation can occur. These results also demonstrate that both hydroxyl and carboxyl can coordinate the cluster simultaneously. This, taken with the previous data showing that the nitroisocitrate samples also have simultaneous water coordination, suggest a six-coordinate Fe_a site having three S ligands from the cluster sulfide and three O ligands from the substrate analogue and water.

The details obtained from the ENDOR work, in addition to past results, allow for the following mechanism for aconitase. In Scheme II the R group on the substrate is CH_2COO^-, -B: represents the amino acid side-chain which stereospecifically transfers a proton between citrate and isocitrate, and X is either water or a protein ligand. In the presence of *cis*-aconitate, bound water can freely exchange

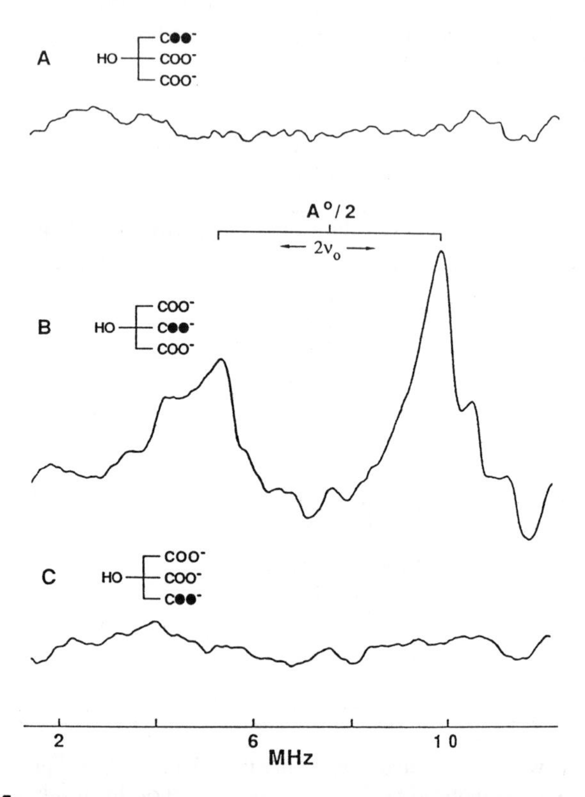

Figure 11. ^{17}O ENDOR spectra at g = 1.85 of photoreduced active aconitase in the presence of substrate whose carboxyl groups were individually labeled with ~50% enriched ^{17}O as shown by the solid oxygens (●) of the stick structures. The bar above spectrum B indicates $2\nu_o$ for ^{17}O centered at $A^o/2$. (Adapted from ref. 62.)

with solvent water. The Fe-S cluster acts both as a Lewis acid to facilitate the dehydration of citrate and isocitrate, and as an activator of the carbon adjacent to the bound carboxyl group of *cis*-aconitate for attack by a bound hydroxyl. Together, the stereochemistry of the reaction, the specific transfer of the proton from citrate to isocitrate, and the coordination of a single carboxyl and adjacent hydroxyl group of substrate require that *cis*-aconitate disengage from the active site and rotate 180° before completing the catalytic cycle. This model differs substantially from the previous "ferrous wheel" model (see Equation 2) in three ways: One, the substrate coordinates to a iron site in a [4Fe-4S] cluster rather than to monomer iron. Two, citrate and isocitrate coordinate as bidentate rather than tridentate ligands. Three, the two orientations of bound *cis*-aconitate differ by a 180° rotation (as originally proposed by Gawron *et al.* (14)), rather than a 90° rotation.

Scheme II

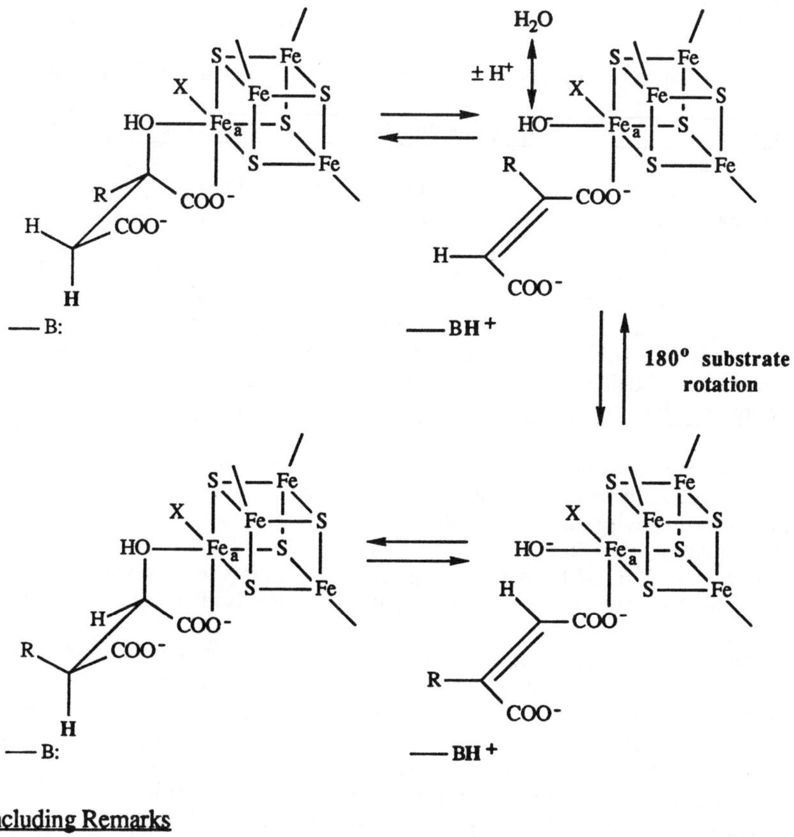

180° substrate rotation

Concluding Remarks

In the absence of a 3-dimensional crystal structure a rather detailed picture of the active site of aconitase has been generated by the application of a variety of spectroscopic techniques. Most important of these has been the melding of information from EPR, Mössbauer and ENDOR spectroscopy. The future determination of the crystal structure of the enzyme-substrate complex within the next few years by Robbins and Stout should fill in the details concerning the nature of the other catalytic groups and the the exact coordination geometry at the Fe-S cluster.

Why did nature use an Fe-S cluster to catalyze this reaction, when an enzyme such as fumarase can catalyze the same type of chemistry in the absence of any metals or other cofactors? One speculation would be that since aconitase must catalyze both hydrations and dehydrations, and bind substrate in two orientations, Fe in the corner of a cubane cluster may provide the proper coordination geometry and electronics to do all of these reactions. Another possibility is that the cluster interconversion is utilized *in vivo* to regulate enzyme activity, and thus, help control cellular levels of citrate. A third, but less likely, explanation is that during evolution an ancestral Fe-S protein, whose primary function was electron transfer, gained the ability to catalyze the aconitase reaction through random mutation.

Is aconitase unique or are there other hydro-lyases which utilize Fe-S clusters as catalytic groups? A list of enzymes of the hydro-lyase class that recently have been discovered to be Fe-S proteins is found in Table VI.

Table VI. Enzymes of the Hydro-Lyase Class that are Fe-S Proteins

Enzyme	Molecular Weight	Fe-S Clusters	Reference
Aconitase			
Mitochondrial	81,000	[4Fe-4S]	39
Cytoplasmic	86,000	[4Fe-4S]	63
6-Phosphogluconate dehydratase	126,000 (dimer)	? (6-8 Fe and $S^{=}$/mole)	64
Maleate dehydratase	68,000	[4Fe-4S]	65
Lactyl-CoA dehydratase	1,000,000 (multimer)	[4Fe-4S], [3Fe-4S]	66
Tartrate dehydratase	100,000 (tetramer)	? (3-5 Fe and $S^{=}$/mole)	67
Dihydroxyacid dehydratase	105,000 (dimer)	[2Fe-2S]	68, 69
Isopropylmalate isomerase	90,000	[4Fe-4S]	70

Yeast isopropylmalate isomerase of the leucine biosynthetic pathway, which catalyzes a totally analogous reaction to that of aconitase, converts 3-hydroxy-3-carboxy-4-methylpentanoate to 2-hydroxy-3-carboxy-4-methylpentanoate via an allylic intermediate. In its initial characterization by EPR spectroscopy, a high-field shift in its EPR signal from a g-average of 1.96 to 1.90 is seen upon addition of substrate (70). This result suggests that its mechanism is the same as that found for aconitase.

Dihydroxyacid dehydratase of the branched-chain amino acid biosynthetic pathway catalyzes the dehydration and tautomerization of 2,3-dihydroxy-3-methyl-(butyrate and pentanoate) to 2-keto-3-methyl(butyrate and pentanoate). The enzyme isolated from spinach recently has been shown to have **not** a [4Fe-4S] cluster, but rather a spectroscopically unusual [2Fe-2S] cluster in its active site (68,69). The EPR spectrum of the reduced enzyme is similar to that seen for Rieske Fe-S proteins (71) with a g-average of 1.91. Upon addition of substrate the g-average of the EPR spectrum shifts to 1.96 (opposite the effect of substrate on aconitase), and then reverts back to a g-average of 1.90 when only the product is present. The dramatic changes in the EPR spectra upon addition of substrate suggest, in analogy to aconitase, that the Fe-S cluster may be directly involved in catalysis.

Perhaps the most unusual of this group is lactyl-CoA dehydratase as isolated from *Clostridium propionicum*. Kuchta *et al.* (66) have found associated with this enzyme, in addition to [4Fe-4S] and [3Fe-4S] clusters, two flavins and the requirement for catalytic amounts of ATP. Only the spectrum identified as belonging to a [3Fe-4S] is affected by substrate binding, suggesting that it may be at the active site. Because the reaction involves the removal of a proton from a very non-acidic methyl group, it is likely that a free radical mechanism is involved.

The diversity in the type of Fe-S clusters associated with these enzymes, catalyzing apparently simple hydration and/or dehydration reactions, is striking. Taken together, these results suggest that some of the Fe-S clusters that have been assigned redox roles in various enzymes may actually be functioning as catalytic groups. Clearly the field of Fe-S proteins, which a decade ago seemed to be well understood, has developed into a dynamic and fertile area for future research endeavors.

Acknowledgments

I would like to thank Dr. H. Beinert for his support and tutelage, and without whom much of what has been presented here would not have been possible. I also thank Drs. M. C. Kennedy, E. Münck, T. A. Kent, B. M. Hoffman, J. V. Schloss and W. W. Cleland for many fruitful interactions and stimulating discussions.

Literature Cited

1. Martius, C.; Knoop, F. *Z. Physiol. Chem.* **1937**, *246*, I-II.
2. Martius, C. *Z. Physiol. Chem.* **1937**, *247*, 104-110.
3. Breusch, F. L. *Z. Physiol. Chem.* **1937**, *250*, 262-280.
4. Potter, V. R.; Heidelberger, C. *Nature* **1948**, *164*, 180-181.
5. Ogston, A. G. *Nature* **1948**, *162*, 4129.
6. Ochoa, S. *J. Biol. Chem.* **1948**, *174*, 133-157.
7. Buchanan, J. M.; Anfinsen, C. B. *J. Biol. Chem.* **1949**, *180*, 47-54.
8. Dickman, S. R.; Cloutier, A. A. *Arch. Biochem.* **1950**, *25*, 229-230.
9. Dickman, S. R.; Cloutier, A. A. *J. Biol. Chem.* **1951**, *188*, 379-388.
10. Morrison, J. F. *Biochem. J.* **1954**, *58*, 685-692.
11. Villafranca, J. J.; Mildvan, A. S. *J. Biol. Chem.* **1971**, *246*, 5791-5798.
12. Speyer, J. F.; Dickman, S. R. *J. Biol. Chem.* **1955**, *220*, 193-208.
13. Gawron, O.; Glaid, III, A.J. *J. Am. Chem. Soc.* **1955**, *77*, 6638-6640.
14. Gawron, O.; Glaid, III, A. J.; Fondy, T. P. *J. Am. Chem. Soc.* **1958**, *80*, 5856-5860.
15. Englard, S.; Colowick, S. P. *J. Biol. Chem.* **1957**, *226*, 1047-1058.
16. Englard, S. *J. Biol. Chem.* **1960**, *235*, 1510-1516.
17. Gawron, O.; Glaid, III, A. J.; LoMonte, A.; Gary, S. *J. Am. Chem. Soc.* **1960**, *83*, 3634-3640.
18. Rose, I. A. ;O' Connell, E. L. *J. Biol. Chem.* **1967**, *242*, 1870-1879.
19. Glusker, J. P. *J. Mol. Biol.* **1968**, *38*, 149-162.
20. Glusker, J. P. In *The Enzymes*; Boyer, P. D., Ed.; Academic Press: New York, **1971**; Vol. 5, pp 413-439.
21. Villafranca, J. J.; Mildvan, A. S. *J. Biol. Chem.* **1971**, *246*, 772-779.
22. Villafranca, J. J.; Mildvan, A. S. *J. Biol. Chem.* **1972**, *247*, 3454-3463.
23. Kennedy, C.; Rauner, R.; Gawron, O. *Biochem. Biophys. Res. Commun.* **1972**, *47*, 740-745.

24. Palmer, G. In *The Enzymes*; Boyer, P.D., Ed.; Academic Press: New York, **1975**, Vol. 12, pp 1-56.
25. Gawron, O.; Kennedy, M.C.; Rauner, R. A. *Biochem. J.* **1974**, *143*, 717-722.
26. Mildvan, A. S.; Estabrook, R. W.; Palmer, G. In *Magnetic Resonance in Biological Systems*; Ehrenberg, A., Malmström, B. G., Vänngård, T.; Pergamon Press: New York, **1967**, pp 175-179.
27. Gawron, O.; Waheed, A.; Glaid, III, A. J.; Jaklitsch, A. *Biochem. J.* **1974**, *139*, 709-714.
28. Suzuki, T.; Yamazaki, O.; Nara, K.; Akiyama, S.; Nakao, Y., Fukuda, H. *Agric. Biol. Chem.* **1973**, *37*, 2211-2212.
29. Suzuki, T.; Yamazaki, O.; Nara, K.; Akiyama, S.; Nakao, Y., Fukuda, H. *J. Biochem. (Tokyo)* **1975**, *77*, 367-372.
30. Suzuki, T.; Akiyama, S.; Fujimoto, S.; Ishikawa, M.; Nakao, Y., Fukuda, H. *J. Biochem. (Tokyo)* **1976**, *80*, 799-804.
31. Suzuki, S.; Maeda, Y.; Sakai, H.; Fujimoto, S.; Morita, Y. *J. Biochem (Tokyo)* **1975**, *78*, 555-560.
32. Sweeney, W. V.; Bearden, A. J.; Rabinowitz, J. C. *Biochem. Biophys. Res. Commun.* **1974**, *59*, 188-194.
33. Ruzicka, F. J.; Beinert, H. *Biochem. Biophys. Res. Commun.* **1974**, *58*, 556-563.
34. Ruzicka, F. J.; Beinert, H. *J. Biol. Chem.* **1978**, *253*, 2514-2517.
35. Kurtz, D. M.; Holm, R. H.; Ruzicka, F. J.; Beinert, H.; Coles, C. J.; Singer, T. P. *J. Biol. Chem.* **1979**, *254*, 4967-4969.
36. Emptage, M. H.; Kent, T. A.; Huynh, B. H.; Rawlings, J.; Orme-Johnson, W. H.; Münck, E. *J. Biol. Chem.* **1980**, *255*, 1793-1796.
37. Stout, C. D.; Ghosh, D.; Patthabi, B.; Robbins, A. H. *J. Biol. Chem.* **1980**, *255*, 1797-1800.
38. Kent, T. A.; Dreyer, J.-L.; Emptage, M. H.; Moura, I.; Moura, J. J. G.; Huynh, B. H.; Xavier, A. V.; LeGall, J.; Beinert, H.; Orme-Johnson, W. H.; Münck, E. In *Electron Transport and Oxygen Utilization*; Ho, C., Ed.; Elsevier: New York, **1982**; pp 371-374.
39. Kent, T. A.; Dreyer, J.-L.; Kennedy, M. C.; Huynh, B. H.; Emptage, M. H.; Beinert, H.; Münck, E. *Proc. Natl. Acad. Sci. USA* **1982**, *79*, 1096-1100.
40. Ruzicka, F. J.; Beinert, H. (1978) In *Frontiers of Biological Energetics*; Academic Press: New York, **1978**, Vol. 2, pp 985-995.
41. Beinert, H.; Ruzicka, F. J.; Dreyer, J.-L. In *Membrane Bioenergetics* , Lee, C. P.; Schatz, G.; Ernster, L., Eds.; Addison-Wesley: Reading, MA, **1979**, pp 45-60.
42. Ramsay, R. R.; Dreyer, J.-L.; Schloss, J. V.; Jackson, R. H.; Coles, C. J.; Beinert, H.; Cleland, W. W.; Singer, T. P. *Biochemistry* **1981**, *20*, 7476-7482.
43. Ramsay, R. R. *Biochem. J.* **1982**, *203*, 327-332.
44. Ramsay, R. R.; Singer, T. P. *Biochem. J.* **1984**, *221*, 489-497.
45. Thomson, A. J.; Robinson, A. E.; Johnson, M. K.; Cammack, R.; Rao, K. K.; Hall, D. O. *Biochim. Biophys. Acta* **1981**, *637*, 423-432.
46. Kennedy, M. C.; Emptage, M. H.; Dreyer, J.-L.; Beinert, H. *J. Biol. Chem.* **1983**, *258*, 11098-11105.
47. Emptage, M. H.; Dreyer, J.-L.; Kennedy, M. C.; Beinert, H. *J. Biol. Chem.* **1983**, *258*, 11106-11111.

48. Ghosh, D.; Furey, Jr., W.; O'Donnell, S.; Stout, C. D. *J. Biol. Chem.* **1981**, *256*, 4185-4192.
49. Ghosh, D.; O'Donnell, S.; Furey, Jr., W.; Robbins, A. H.; Stout, C. D. *J. Mol. Biol.* **1982**, *158*, 73-109.
50. Beinert, H.; Emptage, M. H.; Dreyer, J.-L.; Scott, R. A.; Hahn, J. E.; Hodgson, K. O.; Thomson, A. J.; *Proc. Natl. Acad. Sci. USA* **1983**, *80*, 393-396.
51. Beinert, H. *Anal. Biochem.* **1983**, *131*, 373-378.
52. Rydén, L.; Öfverstedt, L.-G.; Beinert, H.; Emptage, M. H.; Kennedy, M. C. *J. Biol. Chem.* **1984**, *259*, 3141-3144.
53. Johnson, M. K.; Czernuszewicz, R. S.; Spiro, T. G.; Ramsay, R. R.; Singer, T. P. *J. Biol. Chem.* **1983**, *258*, 12771-12774.
54. Robbins, A. H.; Stout, C. D. *J. Biol. Chem.* **1985**, *260*, 2328-2333.
55. Stout, G. H.; Turley, S.; Sieker, L. C.; Jensen, L. H. *Proc. Natl. Acad. Sci. USA* **1988**, *85*, In Press.
56. Johnson, M. K.; Thomson, A. J.; Richards, A. J. M.; Peterson, J.; Robinson, A. E.; Ramsay, R. R.; Singer, T. P.; *J. Biol. Chem.* **1984**, *259*, 2274-2282.
57. Emptage, M. H.; Kent, T. A.; Kennedy, M. C.; Beinert, H.; Münck, E. *Proc. Natl. Acad. Sci. USA* **1983**, *80*, 4674-4678.
58. Kent, T. A.; Emptage, M. H.; Merkle, H.; Kennedy, M. C.; Beinert, H.; Münck, E. *J. Biol. Chem.* **1985**, *260*, 6871-6881.
59. Schloss, J. V.; Porter, D. J. T.; Bright, H. J.; Cleland, W. W. *Biochemistry* **1980**, *19*, 2358-2362.
60. Telser, J.; Emptage, M. H.; Merkle, H.; Kennedy, M. C.; Beinert, H.; Hoffman, B. M. *J. Biol. Chem.* **1986**, *261*, 4840-4846.
61. Gahan, L. R.; Harrowfield, J. M.; Herlt, A. J.; Lindoy, L. E.; Whimp, P. O.; Sargeson, A. M. *J. Am. Chem. Soc.* **1985**, *107*, 6231-6242.
62. Kennedy, M. C.; Werst, M.; Telser, J.; Emptage, M. H.; Beinert, H.; Hoffman, B. M. *Proc. Natl. Acad. Sci. USA* **1988**, *85*, In Press.
63. Emptage, M. H.; Spoto, G.; Beinert, H. Unpublished Data.
64. Scopes, R. K.; Griffiths-Smith, K. *Anal. Biochem.* **1984**, *136*, 530-534.
65. Dreyer, J.-L. *Eur. J. Biochem.* **1985**, *150*, 145-154.
66. Kuchta, R. D.; Hanson, G. R.; Holmquist, B.; Abeles, R. H. *Biochemistry*, **1986**, *25*, 7301-7307.
67. Kelly, J. M.; Scopes, R. K. *FEBS Lett.* **1986**, *202*, 274-276.
68. Flint, D. H.; Emptage, M. H. *Fed. Proc.* **1987**, *46*, 2240.
69. Flint, D. H.; Emptage, M. H. *J. Biol. Chem.* **1988**, *263*, In Press.
70. Emptage, M. H. Unpublished Data.
71. Trumpower, B. L. *Biochim. Biophys. Acta* **1981**, *639*, 129-155.

RECEIVED January 19, 1988

Chapter 18

Nitrogenase

Overview and Recent Results

E. I. Stiefel, H. Thomann, H. Jin, R. E. Bare, T. V. Morgan, S. J. N. Burgmayer, and C. L. Coyle

Exxon Research and Engineering Company, Route 22 East, Annandale, NJ 08801

The biological nitrogen fixation process is introduced. Discussion focusses on the 'Dominant Hypothesis' of nitrogenase composition and functioning. The enzyme system catalyzes the six-electron reduction of N_2 to 2 NH_3 concomitant with the evolution of H_2. ATP hydrolysis drives the process. The two protein components of the enzyme, [Fe] and [FeMo], contain transition metal sulfide clusters. Recently, an alternative nitrogenase containing vanadium has also been reported. [FeMo] contains the unique FeMo-co site, which has been studied using microbiological, molecular genetic, biochemical, biophysical, chemical and synthetic modelling approaches. Alternative substrates for nitrogenase include a number of unsaturated small molecules. The inorganic chemical literature yields clues for the activation of relevant small molecules such as acetylenes. Substrate reactions of nitrogenase also implicate hydrogen activation as a key feature of nitrogenase turnover. The different ways in which hydrogen can interact with transition metal sulfide clusters are discussed. The need for application of sophisticated probes to distinguish structural and mechanistic possibilities is emphasized. Recent work is presented on the use of Electron Spin Echo Spectroscopy to probe the relation of the extracted cofactor to the center in the intact FeMo protein.

The process of nitrogen fixation is an essential part of the nitrogen cycle on the planet earth(1). It is estimated that greater than 60% of the N_2 that is ultimately converted to NH_4^+ is done so by the nitrogenase enzyme system. The availability of fixed nitrogen is often the limiting factor in plant growth(2). To

0097–6156/88/0372–0372$06.00/0

compensate for the inability of natural systems to provide enough nitrogen, man has developed processes to "fix nitrogen" chemically. By far the major process in use today is the Haber-Bosch process in which N_2 and H_2 are reacted at temperatures between 300-500° C and pressures over 300 atm using (usually) Fe-based catalysts(3). Literally hundreds of massive chemical plants are used, and these often produce 1,000 tons of NH_3/day. In contrast, in the biological process, N_2 is reduced locally as needed at room temperature and ~0.8 atm pressure.

Superficially, it might seem that the biological process is inherently simpler than the industrial one. That this is not the case is seen first by the genetic complexity of the biological process. Figure 1 shows that at least seventeen genes are required for effective nitrogen fixation in the bacterium Klebsiella pneumoniae(4,5). These *nif* genes specify proteins that are involved in regulation (*nif*A and L), pyruvate oxidation/flavin reduction (*nif*J), electron transfer (*nif*F for flavodoxin), the subunits of the structural proteins of the nitrogenase (*nif*H, D, K), Fe-S cluster assembly (*nif*M) and the biosynthesis of the iron molybdenum cofactor, FeMo-co (*nif*N, B, E, Q, V, H)(5a). It is the last two functions, involving the placement of unusual transition metal sulfide clusters into the nitrogenase proteins, that cause nitrogenase and its components to be appropriately included in this symposium.

In this chapter we first present what has been called(6) the 'Dominant Hypothesis' for the structure and function of molybdenum-based nitrogenases. We summarize the history of the newly discovered alternative vanadium-based nitrogenase. The properties of the vanadium-based enzyme point to the importance of the study of alternative substrate reactions in probing nitrogenase reactivity. These reactions are summarized and used as a starting point to discuss results from inorganic chemistry that are potentially relevant to understanding the enzyme. The possibilities for substrate and hydrogen activation revealed in inorganic studies point poignantly to the need for further structural, spectroscopic and mechanistic definition of the nitrogenase active sites themselves. Such definition is progressing on many fronts and we briefly describe some of our own efforts using Electron Spin Echo (ESE) Spectroscopy.

The Dominant Hypothesis for Molybdenum Nitrogenase(6,7,8,9)

The action of the nitrogenase enzyme requires the presence of two component proteins. The larger of the two proteins, sometimes incorrectly(10) designated(11) as dinitrogenase, is usually called the MoFe or FeMo protein ([MoFe] or [FeMo]). The smaller, called the Fe protein or [Fe] is sometimes incorrectly(10) referred to(11) as dinitrogenase reductase. Figure 2 shows a schematic diagram that summarizes the protein composition and functioning.

The Fe protein contains two identical subunits of M.W.≈ 30,000 which are products of the *nif*H gene(12). A single Fe_4S_4 center is present in the protein and appears to be bound between the subunits.(12a) The Fe protein is reducible by cellular reductants such as flavodoxin or ferredoxin or artificial

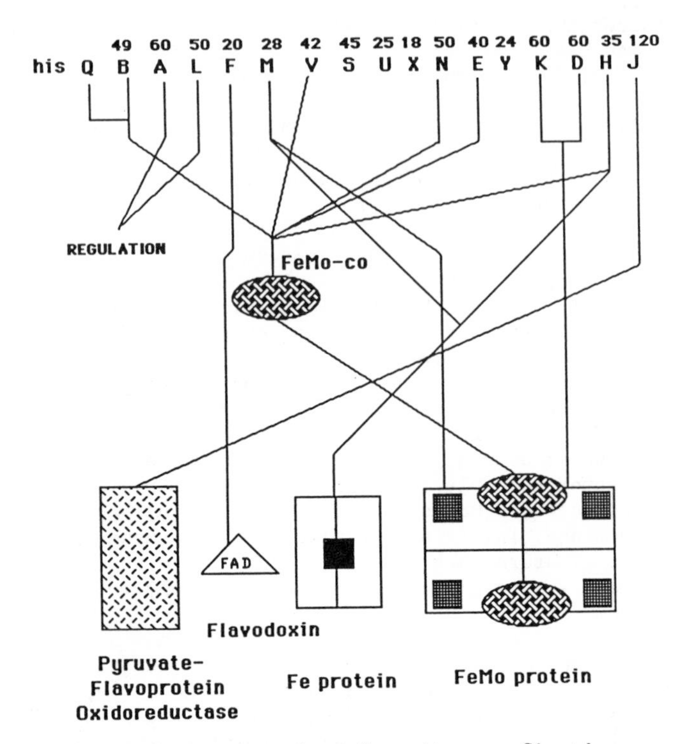

Figure 1. *Nif* genes required for nitrogen fixation as arranged in Klebsiella pneumoniae. The numbers in the top row are molecular weights in kilodaltons of the protein gene products of the respective *nif* genes whose letters are shown below them.

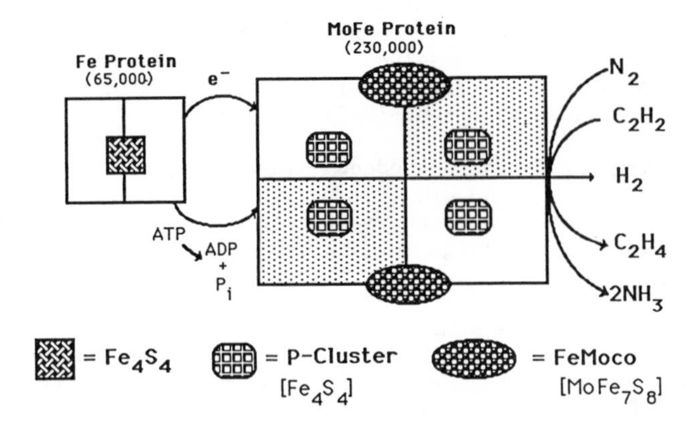

Figure 2. Schematic diagram of the nitrogen fixing enzyme.

reductants such as dithionite or viologens. Its Fe_4S_4 center undergoes a single one-electron redox process wherein the reduced form is EPR active and the oxidized form is diamagnetic. Until quite recently a major mystery involved the low EPR spin-quantitation of the Fe_4S_4 center compared to the analytically derived values for the number of such centers(6,13). The explanation for the discrepancy now seems to be in hand(14-17) as it has been clearly established that the Fe_4S_4 center exists in two spin states, $S = 1/2$ and $S = 3/2$. Only the former with its g values near 2 was considered in the earlier spin quantitations. When the $S = 3/2$ center with g between 4 and 6 is also taken into account, the spin-integration clearly shows one paramagnetic site for each Fe_4S_4 unit. Model systems(18) and theoretical studies(19) strongly support the notion that various spin states are possible for Fe_4S_4 cluster systems. During enzyme turnover this center transfers electrons to the FeMo protein in single electron steps.

The Fe protein binds two molecules of MgATP(20). In the active enzyme system, a minimum of two molecules of MgATP are hydrolyzed to MgADP and phosphate in conjunction with the transfer of a single electron to the FeMo protein(21). The ATP/$2e^-$ ratio is generally accepted to have a minimum value of 4 with higher numbers representing decreased efficiency, possibly due to "futile cycling" where back electron transfer from [FeMo] to [Fe] raises the effective ratio(13,21). Except for an unconfirmed report of reduction by thermallized electrons produced by pulse radiolysis(22), no evidence exists that the FeMo protein can be reduced to a catalytically capable form without the Fe protein present.

This situation calls to mind the nomenclatural proposal(11) of Hageman and Burris that [FeMo] be designated as dinitrogenase and [Fe] as dinitrogenase reductase. The incorrectness or, at best, prematurity(10) of this suggestion lies in the inability of either protein to function catalytically in the absence of the other. [FeMo] will not reduce N_2 or C_2H_2 in the absence of [Fe]. [Fe] will not hydrolyze ATP in the absence of [FeMo]. The nitrogen fixation process requires the presence of both proteins. Although mechanistic considerations(23) point to [FeMo] as the substrate binding and reducing protein and [Fe] as the ATP binding locus, catalytic reactions have never been consummated by one protein in the absence of the other (but vide infra for the uptake of H_2). Therefore, neither protein can be considered as an enzyme, which the proposed nomenclature implies. We therefore use the [FeMo] and [Fe] designations in accord with most workers in the area.

This caveat notwithstanding, the Dominant Hypothesis(6) designates [FeMo] as the protein responsible for substrate reduction. The FeMo protein contains an $\alpha_2\beta_2$ subunit structure due to expression of the *nif*D and *nif*K genes(24,25). Its overall M.W. of about 230,000 reflects the 50-60,000 M.W. of each of its four subunits. The nonprotein composition of 30 Fe, 2 Mo, and 30 S^{2-} betokens the presence of transition metal sulfide clusters, which are presumed to be the active centers of the protein. Without intending to make any spacial implications, figure 2 shows the distribution of major clusters according to the Dominant Hypothesis.

Four Fe_4S_4-like clusters appear to be present in [FeMo] and these have been designated as P-clusters(25). The P clusters are, however, by no means ordinary Fe_4S_4 clusters, if indeed they are Fe_4S_4 clusters at all. P clusters are conspicuous in UV-VIS and especially, MCD and Mössbauer spectra(6,13). The observed spectra are clearly not conventional. The clusters have a very high spin in their oxidized forms, probably $S = 7/2$ from recent(28a) EPR studies, and have decidedly inequivalent Fe populations(28). This implies that the putative Fe_4S_4 clusters are highly distorted, presumably by unsymmetrical coordination from the protein. Moreover, the P clusters do not appear to behave identically to each other under many circumstances. There is open disagreement as to the redox behavior of this set(6,26,27). Furthermore, an additional Mössbauer signal called S may in fact be part of the P-cluster signal(28).

Clearly, the spectroscopic properties of the P clusters in the proteins do not reveal their structural nature. However, extrusion of these clusters from the protein leads to the clear identification of 3-4 Fe_4S_4 clusters(13,29). Despite the uncertainties inherent in the extrusion procedure (due to possible cluster rearrangement) the extrusion result supports the Dominant Hypothesis, which designates the P centers as Fe_4S_4 units, albeit highly unusual ones. The P clusters are thought to be involved in electron transfer and storage presumably providing a reservoir of low potential electrons to be used by the M center (FeMo-co) in substrate reduction.

The FeMo-co or M center of the FeMo protein has been identified spectroscopically(6,13,30) within the protein and has been extracted from the protein into N-methyl formamide(31) and other organic solvents(32,33). Its biochemical authenticity can be assayed by its ability to activate FeMo protein from a mutant organism that produces protein that lacks the M center(31). The extracted cofactor resembles the M-center unit spectroscopically and structurally as shown in Table I. It seems reasonable to presume that the differences are due to variation in the ligation of the center between the protein and the organic solvent(34).

In the Dominant Hypothesis the FeMo-co center is presumed to be the substrate binding and reducing site. Strong evidence to support this idea comes from the study of *nif*V mutants(35-37). (Note that *nif*V has nothing to do with vanadium). *Nif*V mutants have altered substrate specificity insofar as they do not fix nitrogen *in vivo*. *In vitro* H_2 evolution by isolated *nif*V nitrogenase is inhibited by CO unlike the wild type where H_2 evolution is insensitive to CO. Most significantly, this behavior is transferred with the FeMo-co unit that is extracted from the *nif*V protein and used to reactivate the FeMo-co-deficient mutant. The reconstituted FeMo protein has CO sensitive H_2 evolution(37). The FeMo-co site is thus clearly implicated as a major participant in the substrate reactions of the nitrogenase enzyme complex.

The two-component enzyme catalyzes the reduction of N_2 to $2NH_4^+$ or of C_2H_2 to (exclusively) C_2H_4 and the evolution of H_2 with electrons supplied by the reductants named above. ATP

Table I

Comparison of the FeMo Protein and Isolated FeMo-co[a]

	FeMo Protein (M Center)	FeMo-co (in NMF)
EPR		
g values	4.3	4.8
	3.7	3.3
	2.01	2.0
EXAFS[a]		
Mo-S	2.37(4.5)[b,d]	2.37(3.1)[c,d]
Mo-Fe	2.68(3.5)[b,d]	2.70(2.6)[c,d]
Mo-O or N	2.12(1.7)[b,d]	2.10(3.1)[c,d]
Fe-S		2.25(3.4)[e]
Fe-Fe		2.66(2.3)[e]
Fe-Mo		2.76(0.4)[e]
Fe-O or N		1.81(1.2)[e]
XANES		
MoO_3S_3 fits best[f]		

[a]Distance in Å with number of atoms in parentheses.

[b]Earlier study by Cramer, S. P.; Hodgson, K. O.; Gillam, W. O.; and Mortenson, L. E.; *J. Am. Chem. Soc.*, **1978** *100*, 2748.

[c]Earlier study by Burgess, B. K.; Yang, S.-S.; You, C.-B.; Li, J.-G.; Friesen, G. Pan, W.-V.; Stiefel, E. I.; Newton, W. E.; Conradson S. D.; Hodgson, K. O. in ref. 8, p 71.

[d]Data from Conradson, S. D.; Burgess, B. K.; Newton, W. E.; Mortenson, L. E.; and Hodgson, K. O. *J. Am. Chem. Soc.* **1987** *109*, 7507.

[e]Antonio, M. R.; Teo, B.-K.; Orme-Johnson, W. H.; Nelson, M. J.; Groh, S. E.; Lundahl, P. A.; Kauzlarich S. M.; Averill, B. A. *J. Am. Chem. Soc.*, **1982** *104*, 4703.

[f]Conradson, S. D.; Burgess, B. K.; Newton, W. E.; Hodgson, K. O.; McDonald, J. W.; McDonald, J. W.; Rubinson, J. F.; Gheller, W. F.; Mortenson, L. E.; Adams, M. W. W.; Mascharak, P. K.; Armstrong W. A.; Holm, R. H. *J. Am. Chem. Soc.*, **1985**, *107*, 7935.

hydrolysis occurs concomitantly with electron transfer from [Fe] to [FeMo]. Dissociation of [Fe] from [FeMo] after this electron transfer is believed to be the rate-limiting step in the overall turnover of the enzyme(25).

Vanadium and Nitrogenase

The association of molybdenum with nitrogen fixation was first reported by Bortels in 1930(38). This seminal finding opened the door to the characterization of the molybdenum nitrogenases discussed in the preceding section. The Bortels work has been cited many times and is often referred to without citation. In addition to nitrogenase, many other Mo-containing enzymes were subsequently sought and found(39).

Subsequent to the classic 1930 paper, Bortels reported in 1936(40) that vanadium stimulated nitrogen fixation. The 1936 paper was hardly ever cited prior to 1986. Despite a few positive reinforcements, it was not until the 1970's that serious attempts were made to isolate a vanadium nitrogenase. In 1971, two groups reported the successful isolation of a vanadium-containing nitrogenase from Azotobacter vinelandii(41,42). The enzyme was similar to the Mo enzyme but was reported to have low activity and an altered substrate specificity. Driven in part by skepticism from the nitrogenase research community, one of the groups carefully reinvestigated their preparation and found small amounts of molybdenum presumed to be sufficient to account for the activity. (The selectivity difference was not addressed.)(43) The vanadium was thought to play a stabilizing role for the FeMo protein, allowing the small amount or active [FeMo] to be effectively isolated. Interestingly, the vanadium was said to "substitute" for Mo in the FeMo nitrogenase. Apparently, no thought was given to the possibility of a truly alternative nitrogenase system whose protein as well as metal content differed from that of the Mo nitrogenase.

The irreplaceable essentiality of molybdenum in nitrogenase went unchallenged until 1980 when Bishop and coworkers demonstrated(44) that an alternative nitrogen fixation pathway became active in Azotobacter vinelandii when this organism was starved for molybdenum(45). Despite continued skepticism, Bishop persevered and eventually showed that even in a mutant from which the *nif*H, D and K genes had been deleted, the alternative system could be elicited upon Mo starvation. The circle was finally closed in 1986 when two groups(46-51) isolated the component proteins of the alternative nitrogenase from different species of Azotobacter. They showed unequivocally that one component contained vanadium and that neither component contained molybdenum.

The purification and characterization of the V nitrogenase proteins has shown that one component is extremely similar to the Fe protein of nitrogenase. This evidence comes both from the isolation of the protein from Azotobacter vinelandii(46) and from the identification of genetic homology between *nif*H (the genetic determinant for the subunit of the Fe protein in the Mo nitrogenase system) and *nif*H* (the corresponding gene in the V based system). Both Fe proteins contain an α_2 subunit structure and a single Fe_4S_4 cluster, which is EPR active in its reduced state.

The FeV protein from Azotobacter vinelandii and Azotobacter chroococcum has an $\alpha_2\beta_2$ subunit structure with metal analysis and spectroscopic properties shown in Table II in comparison with properties of [FeMo]. Clearly, the major difference is the presence of V instead of Mo in the FeV protein. However, it is clear that the two nitrogenase systems are really quite similar. In each case, two highly O_2-sensitive proteins carry out an ATP-dependent reduction of N_2 and concomitant evolution of H_2. The Fe proteins have the same (α_2) subunit structure, cluster content and spectroscopic signature. The V versions of the larger protein, while somewhat lower in M.W., have the same subunit structure as their Mo analogs ($\alpha_2\beta_2$) and appear by MCD spectroscopy to contain P-like clusters(48). The FeV center differs from the FeMo system by the substitution of V for Mo. However, the FeV site still may be an S = 3/2 center (by EPR, although this differs from that of the FeMo center)(52) and seems to have V-S and V-Fe distances (by EXAFS)(53,54) similar to those in thiocubane VFe_3S_4 clusters. This observation recalls the finding that FeMo-co has Mo-S and Mo-Fe distances similar to $MoFe_3S_4$ thiocubanes. Likewise XANES(53) implicates VS_3O_3 type coordination in the V nitrogenase as it does MoS_3O_3 coordination in FeMo-co. Finally, the "FeV cofactor" is extractable into NMF and can reconstitute the *nif*B⁻ FeMo-co-less mutant of the Mo system(55). Clearly, despite the substitution of V for Mo, the properties of the proteins, including their respective M-Fe-S sites do not appear to differ drastically. However, the compositional changes do have a significant correlation in altered reactivity.

A major difference between the V and Mo enzymes lies in substrate specificity and product formation(51). As clearly shown in Table II, the FeV nitrogenase has a much lower reactivity toward acetylene than does the Mo system. Furthermore, while the FeMo system exclusively produces ethylene from acetylene the FeV system yields significant amounts of the four-electron reduction product, ethane(51). The detection of ethane in the acetylene assay may prove a powerful technique for detecting the presence of the V nitrogenase in natural systems. Moreover, this reactivity pattern is found in the *nif*B⁻ mutant reconstituted with FeV-co(55). The reactivity change upon going from Mo to V in otherwise similar protein systems clearly adds weight to the direct implication of the M-Fe-S center (M = V or Mo) in substrate reduction.

Substrate Reactions

In addition to the physiological reaction of N_2 reduction, nitrogenase catalyzes a wide variety of reactions involving small unsaturated molecules(56). Table III lists key reactants and products for FeMo nitrogenases. All substrate reductions involve minimally the transfer of two electrons. Multielectron substrate reductions may involve the accretion of such two-electron

Table II

Comparison of Nitrogenase Proteins[a]

Property	Av1[47]	Av1*[47]	Ac1*[50]
Molecular weight	240,000	200,000	210,000
Molybdenum[b]	2	<0.05	<0.06
Vanadium[b]	-	0.7	2
Iron[b]	30-32	9.3	23
Activity[c]			
H^+	2200	1400	1350
C_2H_2	2000	220	608
N_2	520	330	350
EPR g values	4.3	5.31	5.6
	3.7	4.34	4.35
	2.01	2.04	3.77
		1.93	1.93

[a]Av1 is the FeMo protein of Azotobacter vinelandii, Av1* is the FeV protein of A. vinelandii and Ac1* is the FeV protein of A. chroococcum.

[b]Atoms per molecule.

[c]nmole product/min/mg of protein.

Table III

Nitrogenase Substrate Reactions

Two-electron Reductions

$$2e^- + 2H^+ \longrightarrow H_2$$

$$C_2H_2 + 2e^- + 2H^+ \longrightarrow C_2H_4$$

$$N_3^- + 2e^- + 3H^+ \longrightarrow NH_3 + N_2$$

$$N_2O + 2e^- + 2H^+ \longrightarrow H_2O + N_2$$

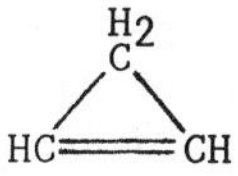

$$\text{cyclopropene } (HC{=}CH \text{ ring with } CH_2) + 2e^- + 2H^+ \longrightarrow CH_2 = CH - CH_3 + \text{cyclopropane } (H_2C - CH_2 \text{ ring with } CH_2)$$

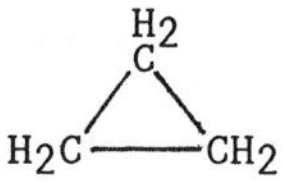

Four-Electron Reductions

$$HCN + 4e^- + 4H^+ \longrightarrow CH_3NH_2$$

$$RNC + 4e^- + 4H^+ \longrightarrow RNHCH_3$$

Six-Electron Reductions

$$N_2 + 6e^- + 6H^+ \longrightarrow 2NH_3$$

$$HCN + 6e^- + 6H^+ \longrightarrow CH_4 + NH_3$$

$$HN_3 + 6e^- + 6H^+ \longrightarrow NH_3 + N_2H_4$$

$$RNC + 6e^- + 6H^+ \longrightarrow RNH_2 + CH_4$$

$$RCN + 6e^- + 6H^+ \longrightarrow RCH_3 + NH_3$$

Multielectron Reductions

$$RNC \rightarrow (C_2H_6,\ C_3H_6,\ C_3H_8) + RNH_2$$

processes. Further, except for the reduction of N_3^- to NH_3 and N_2,(57) an equal number of protons and electrons is transferred to the substrate. In the absence of substrate, two electrons combine with two protons to form H_2. The active site of nitrogenase seems capable of delivering the elementary particles of H_2 to the substrate.

Of the alternative substrates of nitrogenase, many contain triple bonds in at least one of their resonance forms. As discussed in the next section, the reactivity of model molecules such as acetylene could give insights into the manner in which such unsaturated molecules bind to transition metal sulfur systems.

Much data point to an intimate connection between H_2 and the N_2 binding site in nitrogenase. Simpson and Burris(58) confirmed the important finding of Hadfield and Bulen(59) by showing that one H_2 is evolved for each N_2 "fixed" even at 50 ATM N_2, which is well above the pressure of N_2 at which saturation occurs. H_2 evolution is therefore a mandatory part of the N_2 fixation process(58-62) whose stoichiometry must be written as:

$$N_2 + 8\ H^+ + 8\ e^- \longrightarrow 2\ NH_3 + H_2$$

Additionally, Wang and Watt have shown that the FeMo protein alone can act as an uptake hydrogenase(63). Specifically, H_2 in the presence of [FeMo] causes the reduction of oxidizing dyes such as methylene blue or dichlorophenolindophenol <u>in the absence of Fe protein</u>. The hydrogen evolution and uptake behavior of nitrogenase proteins forces us to consider the ways in which hydrogen can interact with transition metal sulfur centers. This we discuss in the following section.

Insights from Inorganic Reactivity

Many recent studies on inorganic systems that do not directly model the nitrogen fixation process, nevertheless give potential insight into biological nitrogen fixation. We discuss two categories of relevant chemistry; first, acetylene binding and reactivity and second, dihydrogen binding and activation.

Classically, acetylene binds to metal centers by using its π and π^* orbitals to form, respectively, σ-donor and π-acceptor bonds to the metal. This situation can hold when the metal is in a sulfur coordination(64-65) environment such as in $MoO(S_2CNR_2)_2(RC{\equiv}CR)$(64), and $Mo(S_2CNR_2)_2(RC{\equiv}CR)_2$(65) Here, acetylene interacts directly with the metal center; a mode of binding that must be considered for nitrogenase substrates.

The work of M. Rakowski DuBois and coworkers(66,67) points to a totally different mode of acetylene binding. For example, $(Cp')_2Mo_2S_4$ reacts with acetylene to produce

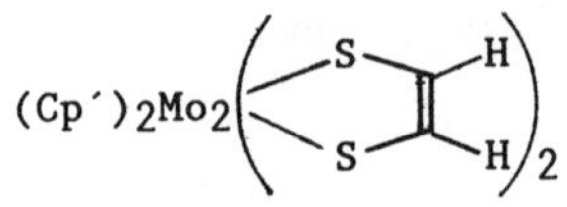

$$(Cp')_2Mo_2\left(\begin{array}{l} S - C - H \\ \quad\ \ \| \\ S - C - H \end{array}\right)_2$$

containing a bridging ethylene-1,2-dithiolate. In this case, the acetylene has bound directly to the sulfur atoms in the molybdenum sulfur complex. In related studies, acetylenes or substituted (activated) acetylenes are found to displace ethylene from bridging or terminal 1,2-dithiolate ligands(67,68). These reactions present examples of where the sulfur sites rather than the metal sites are reactive towards small unsaturated molecules of the cluster.

A final example illustrates the versatility that transition metal sulfur systems may provide. An activated acetylene has been shown(68) to insert into a metal-sulfur bond in $Mo_2O_2S_2(S_2)_2^{2-}$ forming a vinyl disulfide chelating ligand,

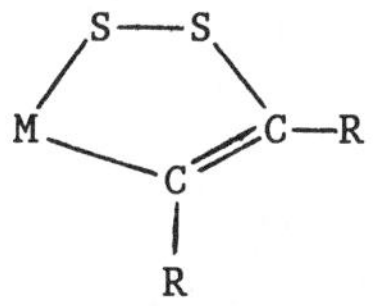

S—S
M C—R
C
R

on an $Mo_2O_2S_2$ core(68). The rich chemistry of acetylene reacting with transition metal sulfur systems is summarized in Figure 3 which clearly suggests a number of possibilities for nitrogenase reactivity.

Similarly, recent years have brought new insights into the way dihydrogen can be bound at a transition metal site. Kubas and others(69,70,71) have shown that H_2 itself can form simple complexes with a variety of transition metal sites in which the H-H bond is largely maintained. This finding contrasts with the classical situation in which H_2 interacts with a transition metal site by oxidative addition to form a dihydride complex(72). In certain cases the dihydrogen and dihydride complexes exist in simple equilibrium:(71)

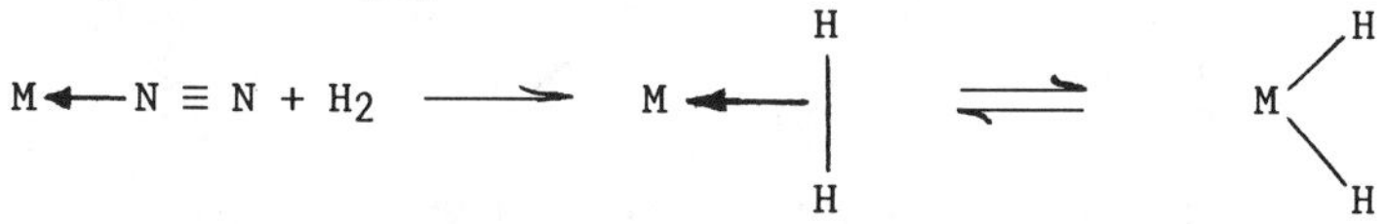

$$M \leftarrow N \equiv N + H_2 \longrightarrow M \leftarrow \begin{matrix} H \\ | \\ H \end{matrix} \rightleftharpoons M \begin{matrix} H \\ \\ H \end{matrix}$$

Crabtree(73) has suggested that the presence of a dihydrogen complex is required for H_2 to be displaced by N_2 to form a dinitrogen complex. This reaction would explain the required stoichiometry of N_2 reduction and H_2 evolution.

In other studies, Rakowski DuBois et al.(67) have shown reactivity of $(Cp')_2Mo_2S_4$ with H_2 to form $(Cp')_2Mo_2S_2(SH)_2$ in which the dihydrogen is cleaved but shows no direct interaction with the metal center. The resulting complex contains bridging SH groups but no direct metal-H bonding.(67) Finally, in recent work, Bianchini et al.(73) reported that the dinuclear rhodium-sulfur complex $\{RhS[P(C_6H_5)_2CH_2CH_2]_3CH\}_2$ reacts with two equivalents of H_2 to yield a complex $\{Rh(H)(SH)[P(C_6H_5)_2CH_2CH_2]_3CH\}_2$ in which two SH groups bridge the two Rh centers each of which contains a single hydrido ligand. Figure 4 illustrates some of these possibilities for hydrogen activation. Clearly, as in the case for acetylene reactivity, metal-based, S-ligand based and M-S based reactivity

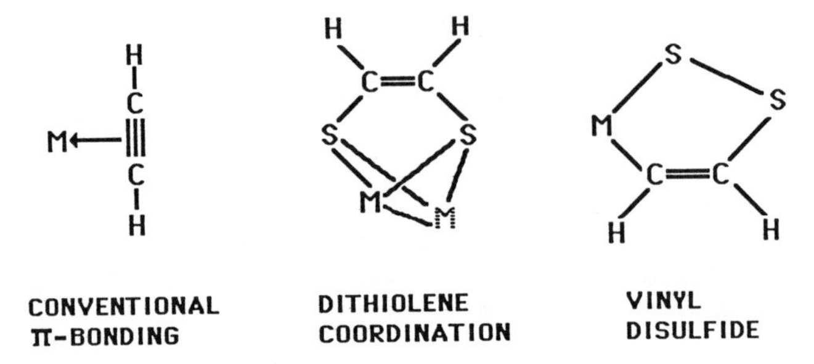

Figure 3. Possible modes of acetylene binding to transition metal sulfur sites.

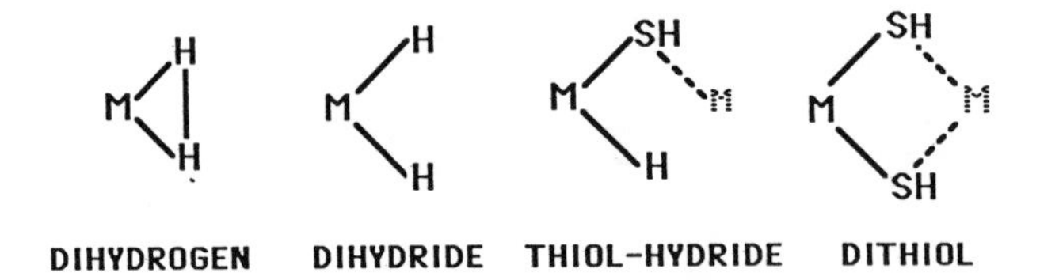

Figure 4. Possible modes of hydrogen activation on transition metal sulfur sites.

must each be considered as possibilities for the hydrogen (or substrate) activation process of nitrogenase.

The type of activation process that is at work in the enzyme is currently unknown. Clearly we need greater structural definition of the active site and this will be forthcoming through the continued application of sophisticated diffraction and spectroscopic probes. Diffraction alone, however, will be incapable of locating protons and possibly other low M.W. ligands. Therefore, spectroscopic probes such as ENDOR(32), and ESEEM(75-78), which are based on EPR spectroscopy, may become crucial in elucidating mechanistically significant structural details. In the remaining section, we briefly discuss the results of our recent studies using electron spin echo spectroscopy.

Electron Spin Echo Envelope Modulation (ESEEM) Spectroscopy

ESEEM is a pulsed EPR technique which is complementary to both conventional EPR and ENDOR spectroscopy(74,75). In the ESEEM experiment, one selects a field (effective g value) in the EPR spectrum and through a sequence of microwave pulses generates a spin echo whose intensity is monitored as a function of the delay time between the pulses. This resulting echo envelope decay pattern is amplitude modulated due to the magnetic interaction of nuclear spins that are coupled to the electron spin. Cosine Fourier transformation of this envelope yields an ENDOR-like spectrum from which nuclear hyperfine and quadrupole splittings can be determined.

In nitrogenase, the $S = 3/2$ signal associated with FeMo-co provides the opportunity for detailed ESEEM investigation(76,77). Such investigations will be presented in detail elsewhere(78), but here we summarize some of the qualitative results. In particular, we find that ^{14}N hyperfine splitting is observed in the ESEEM of the M center (FeMo-co in [FeMo]). This ^{14}N splitting is not from the substrate, or from an intermediate or product of nitrogen fixation since allowing the enzyme to turn over under $^{15}N_2$ does not remove the splitting. Nor does the elimination of N-containing buffers remove the splitting. We conclude that the observed ^{14}N-splitting is due to a nitrogen atom associated with FeMo-co in the protein. This nitrogen is either part of the cofactor itself or is provided to the cofactor from a protein side chain. To distinguish these possibilities ESEEM was carried out on FeMo-co removed from the protein in NMF solution.

ESEEM of FeMo-co does not show the nitrogen quadrupole frequencies observed for the M center in the protein. This shows that the observed splitting is likely due to an ^{14}N atom on the protein. Further, this nitrogen is probably not of the deprotonated amide type, which is the mode in which NMF is thought to bind to FeMo-co.

Interestingly, the splitting parameters for the nitrogen are quite similar to those involving imidazole nitrogen directly bound to low-spin heme(79). We conclude that a protein N binds directly to FeMo-co in the FeMo protein. This conclusion is reinforced by the determination of a non-zero isotropic hyperfine term for the splitting, which unequivocally shows direct (Fermi contact) interaction of the ^{14}N nuclear and electron spins. Time should reveal the particular significance that this nitrogen has for the structure or functioning of the M center.

Conclusion

Although we know a great deal about nitrogenase, we still have much to learn in this pre-structural phase of nitrogenase enzymology. We do not yet know the structure of any of the metal clusters in the FeMo or FeV proteins. Nor do we know the arrangement of these clusters in the proteins. However, even with structural definition, which is in progress(80-82), we will have to continue to apply the most powerful tools of physical bioinorganic chemistry to determine how hydrogen and substrates are handled by this remarkable enzyme.

Acknowledgments

We thank Drs. Graham George and Roger Prince and the reviewers for useful comments and are grateful to Drs. Brian Hales and Graham George for informing us of results prior to publications.

Literature Cited

1. Blackburn, T. H. in *Microbial Geochemistry*; W. E. Krumbein, Editor, Blackwell Scientific, Oxford, **1983**; p 63.
2. Postgate, J. R. *Fundamentals of Nitrogen Fixation*; Cambridge University Press, Cambridge, New York, **1982**.
3. Hardy, R. W. F. *Treatise on Dinitrogen Fixation*; Wiley and Sons, New York, *1979*; Section I.
4. Brill, W. J. *NATO Adv. Sci. Inst., Ser. A*, **1983**, *63*, 231.
5. Haselkorn, R. *Ann. Rev. Microbiol.* **1986**, *40*, 525.
5a. Robinson, A. C.; Dean, D. R.; Burgess, B. K. *J. Biol. Chem.* **1987**, *262*, 14,327.
6. Stephens, P. J. in *Molybdenum Enzymes*; T. G. Spiro, Editor; J. Wiley and Sons, **1985**; p 117.
7. Gibson, A. H.; Newton, W. E., Editors, *Current Perspectives in Nitrogen Fixation*; Australian Academy of Science, Canberra, **1981**.
8. Veeger, C.; Newton, W. E., Editors, *Advances in Nitrogen Fixation Research*; Nijhoff/Junk, The Hague, **1984**.
9. Evans, H. J.; Bottomley, P. J.; Newton, W. E., Editors, *Nitrogen Fixation Research Progress*; Martinus Nijhoff, Boston, **1985**.
10. Stiefel, E. I. in ref. 7, p 55.
11. Hageman, R. V; Burris, R. H. *Proc. Natl. Acad. Sci. USA* **1978**, *75*, 2699.
12. Sundaresan, V.; Ausubel, F. M. *J. Biol. Chem.* **1981**, *256*, 2808.
12a. Hausinger, R. P.; Howard, J. B. *J. Biol Chem.* **1983**, *258*, 13,486.
13. Orme-Johnson, W. H.; Davis, L. C.; Henzl, M. T.; Averill, B. A.; Orme-Johnson, N. R.; Münck, E.; Zimmerman, R. *Recent Developments in Nitrogen Fixation*; W. E. Newton, J. R.

Postgate and C. Rodriguez Barrueco, Editors; Academic Press, **1977**, p 131.
14. Lindahl, P. A.; Day, E. P.; Kent, T. A.; Orme-Johnson, W. H.; Münck, E. *J. Biol. Chem.* **1985**, *260*, 11160.
15. Watt, G. D.; McDonald, J. W. *Biochemistry* **1985**, *24*, 7226.
16. Hagen, W. R.; Eady, R. R.; Dunham, W. R.; Haaker, H. *FEBS Letters* **1986**, *189*, 250.
17. Morgan, T. V.; Prince, R. C.; Mortenson, L. E. *FEBS Letters* **1986**, *206*, 4.
18. Carney, M. J.; Holm, R. H.; Papaefthymiou, G. C.; Frankel, R. B. J. Am. Chem. Soc. 1986, 108, 3519.
19. Noodleman, L.; Norman, J. G., Jr.; Osborne, J. H.; Aizman, A.; Case, D. A. *J. Am. Chem. Soc.* **1985**, *107*, 3418.
20. Watt, G. D.; Wang, Z.-C.; Knotts, R. R. *Biochemistry* **1986**, *25*, 8156; Cordewener, J.; Haaker, H.; Van Ewijk, P.; Veeger, C. *Eur. J. Biochem.* **1985**, *148*, 499.
21. Mortenson, L. E.; Thorneley, R. N. F. *Am. Rev. Biochem.* **1979**, *48*, 387.
22. Kulikov, A. V.; Syrtsova, L. A.; Likhtenshtein, G. I.; Popko, E. V.; Druzhinin, S. Y.; Uznskaya, A. M. *Dokl. Akad. Nauk SSR* **1981**, *262*, 1177.
23. Thorneley, R. N. F.; Lowe, D. J. in *Molybdenum Enzymes* T. G. Spiro, Editor; J. Wiley and Sons, New York, **1985**; p 221.
24. Brigle, K. E.; Newton, W. E.; Dean, D. R. *Gene* **1985**, *37*, 37.
25. Dixon, R. A. *J. Gen. Micro.* **1984**, *130*, 2745.
26. Watt, G. D.; Burns, A.; Tennent, D. L. *Biochemistry* **1981**, *20*, 7272; Watt, G. D.; Wang, Z. C.; *Biochemistry* **1986** *25*, 5196.
27. Zimmermann, R.; Münck, E.; Brill, W. J.; Shah, V. K.; Henzl, M. T.; Rawlings, J.; Orme-Johnson, W. H. *Biochim. Biophys. Acta.* **1978**, *537*, 185.
28. McLean, P. A.; Papaefthymiou, V.; O'Hagen, F.; Orme-Johnson; Münck, E. *J. Biol. Chem.* **1987** *262*, 12,900.
29. Kurtz, D. M.; McMillan, R. S.; Burgess, B. K.; Mortenson, L. E.; Holm, R. H.; *Proc. Nat. Acad. Sci. USA*, **1979**, *76*, 4986.
30. Venters, R. A.; Nelson, M.; McLean, P. A.; True, A. E.; Levy, M. A.; Hoffman, B. M.; Orme-Johnson, W. H.; *J. Am. Chem. Soc.* **1986**, *108*, 3487.
31. Shah, V.; Brill, W. J. *Proc. Nat. Acad. Sci. USA* **1977**, *74*, 3249.
32. Lough, S. M.; Jacobs, D. L.; Lyons, D. M.; Watt, G. D.; McDonald, J. W. *Biochem. Biophys. Res. Comm.* **1986**, *139*, 740.
33. Stiefel, E. I.; S. P. Cramer in *Molybdenum Enzymes*; T. G. Spiro, Editor; J. Wiley and Sons, **1985**, p 88.
34. Walters, M. A.; Chapman, S. K.; Orme-Johnson, W. H. *Polyhedron* **1986**, *5*, 561.
35. McLean, P. A.; Dixon, R. A.; *Nature*, **1981**, *292*, 655.
36. McLean, P. A.; Smith, B. E.; *Biochem. J.* **1983**, *211*, 589.
37. Hawkes, T. R.; McLean, P. A.; Smith, B. E. *Biochem. J.* **1984**, *217*, 317.
38. Bortels, H. *Arch. Mikrobiol.*, **1930**, *1*, 333.
39. Burgmayer, S. J. N.; Stiefel, E. I. *J. Chem. Educ.*; **1985**, *62*, 943.
40. Bortels, H. *Zentbl. Bakt. Parasiten Abt. II* **1935**, *95*, 193.

41. McKenna, C. E.; Benemann, J. R.; Traylor, T. G. *Biochem. Biophys. Res. Comm.* **1970**, *41*, 1501.
42. Burns, R. C.; Fuchsman, W. H.; Hardy, R. W. F. *Biochem. Biophys. Res. Comm.* **1971**, *42*, 353.
43. Benemann, J. R., McKenna, C. E.; Lie, R. F.; Traylor, T. G.; Kamen, M. D. *Biochim. Biophys. Acta*, **1972**, *264*, 25.
44. Bishop, P. E.; Jarlenski, D. M. L.; Hetherington, D. R. *Proc. Nat. Acad. Sci. USA*, 1980, 77, 7342.
45. Bishop, P. E., Premakumar, R.; Dean, D. R.; Jacobson, M. R.; Chisnell, J. R.; Rizzo, T. M.; Kopczynski, J. *Science* **1986**, *232*, 92.
46. Hales, B. J.; Langosch, D. J.; Case, E. E. *J. Biol. Chem.* **1986**, *261*, 15301.
47. Hales, B. J.; Case, E. E.; Morningstar, J. E.; Dzeda, M. F.; Mauterer, L. A. *Biochemistry* **1987**, *26*, 1795.
48. Morningstar, J.; Johnson, M. K.; Case, E. E.; Hales, B. J. *Biochemistry* **1987**, *26*, 1795.
49. Robson, R. L.; Eady, R. R.; Richardson, T. H.; Miller, R. W.; Hawkins, M. H.; J. R. Postgate *Nature* **1986**, *322*, 388.
50. Eady, R. R.; Robson, R. L.; Richardson, T. H.; Miller, R. W.; Hawkins, M. *Biochem. J.* **1987**, *244*, 197.
51. Dilworth, M. J.; Eady, R. R.; Robson, R. L.; Miller, R. W. *Nature* **1987**, *327*, 167.
52. Morningstar, J. E.; Hales, B. J. *J. Am. Chem. Soc.* **1987**, *109*, 6854.
53. Arber, J. M.; Dobson, B. R.; Eady, R. R.; Stevens, P.; Hasnain, S. S.; Garner, C. D.; Smith, B. E. *Nature* **1987**, *372*, 325.
54. George, G. N.; Cramer, S. P.; Hales, B. J.; personal communications, submitted for publication.
55. Eady, R. R.; Robson, R. L.; Smith, B. E.; Lowe, D. J.; Miller, R. W.; Dilworth, M. J. *Recueil des Travaus Chim. des Pays-Bas* **1987**, *106*, 175.
56. Burgess, B. K. in *Molybdenum Enzymes*; T. G. Spiro, Editor; J. Wiley and Sons, New York, **1985**; p 161.
57. Rubinson, J. F.; Burgess, B. K.; Corbin, J. L.; Dilworth, M. J. *Biochemistry* **1985**, *24*, 273.
58. Simpson, F. B.; Burris, R. *Science* **1984**, *224*, 1095.
59. Hadfield, K. L.; Bulen, W. A. *Biochemistry* **1969**, *8*, 5103.
60. Stiefel, E. I.; Newton, W. E.; Watt, G. D.; Hadfield, K. L.; Bulen, W. A. *Advances in Chemistry Series, No. 162, Bioinorganic Chemistry II*; **1977**, p. 353.
61. Burgess, B. K.; Wherland, S.; Newton, W. E.; Stiefel, E. I. *Biochemistry* **1981**, *20*, 5140.
62. Wherland, S.; Burgess, B. K.; Stiefel, E. I.; Newton, W. E. *Biochemistry* **1981**, *20*, 5132.
63. Wang, Z.-C.; Watt, G. D. *Proc. Natl. Acad. Sci. USA*, **1984**, *81*, 376.
64. Newton, W. E.; McDonald, J. W.; Corbin, J. L.; Ricard, L.; Weiss, R. *Inorg. Chem.* **1980**, *19*, 1997.
65. Herrick, R. S.; Templeton, J. L. *Organometallics*, **1982**, *1*, 842.
66. Rakowski DuBois, M.; R. C.; Haltiwanger, R. C.; Miller, D. J.; Glazmaier, G. *J. Am. Chem. Soc*; **1979**, *101*, 5245.

67. Rakowski DuBois, M.; VanDerveer, M. C.; DuBois, D. L.; Haltiwanger, R. C.; Miller, W. K. *J. Am. Chem. Soc.* **1980**, *102*, 7456.
68. Halbert, T. R.; Pan, W.-H.; Stiefel, E. I. *J. Amer. Chem. Soc.* **1983**, *105*, 5476.
69. Kubas, G. J.; Ryan, R. R.; Swanson, B. I.; Vergamini, P. J.; Wasserman, H. J. *J. Am. Chem. Soc.* **1984**, *106*, 451.
70. Kubas, G. J.; Ryan, R. R. *Polyhedron* 1986, **5**, 473.
71. Kubas, G. J.; Unkefer, C. J.; Swanson, B. I.; Fukushima, E. *J. Am. Chem. Soc.* **1986**, *108*, 7000.
72. Cotton, F. A.; Wilkinson, G. *Advanced Inorganic Chemistry*; J. Wiley and Sons, **1980**.
73. Crabtree, R. H. *Inorg. Chim. Acta* **1986**, *125*, L7.
74. Bianchini, C.; Mealli, C.; Meli, A.; Sabat, M. *Inorg. Chem.* **1986**, *25*, 4617.
75. Mims, W. B.; Peisach, *J. Biological Magnetic Resonance*; Vol. 3; J. Berliner and J. Reuben, Eds.; Plenum, New York, **1981**; p 213.
76. Mims, W. B.; Peisach, J. *Biological Applications of Magnetic Resonance*; R. G. Shulman, Editor; Academic Press, **1980**; p 221.
77. Orme-Johnson, W. H.; Lindahl, P.; Meade, J.; Warren, W.; Nelson, M.; Groh, S.; Orme-Johnson, N. R.; Münck; Huynh, B. H.; Emptage, M.; Rawlings, J.; Smith, J.; Roberts, J.; Hoffman, B.; Mims, W. B. in ref. 8, p 79.
78. Thomann, H.; Morgan, T. V.; Burgmayer, S. J. N.; Bare, R. E.; Jin, H.; Stiefel, E. I. *J. Am. Chem. Soc.*, **1987**, *109*, 7913.
79. Peisach, J.; Mims, W. B.; Davis, J. L. *J. Biol. Chem.* **1979**, *254*, 12,379.
80. Yamane, T.; Weininger, M. S.; Mortenson, L. S.; Rossmann, M. G. *J. Biol. Chem.* **1982**, *257*, 1221.
81. Rees, D. C.; Howard, J. B. *J. Biol. Chem.* **1983**, *258*, 12,733.
82. Sosfenov, N.; Adrianov, V. I.; Vagin, A. A.; Strokopytov, B. V.; Vainstein, B. K.; Shilov, A. E.; Gvozdev, R. I.; Likhtenshtein, G. I.; Mitsova, I. Z.; Blazhchuk, I. S. Dokl. Akad. Nauk. SSR 1986, 291, 1123.

RECEIVED February 24, 1988

Chapter 19

Synthetic Analogs for the Fe–Mo–S Site in Nitrogenase

Chemistry of the $[Fe_6S_6(L)_6(M(CO)_3)_2]^{n-}$ Clusters

D. Coucouvanis

Department of Chemistry, University of Michigan, Ann Arbor, MI 48106

A unique Fe/Mo/S cluster in the nitrogenase enzymes has been identified as a site intimately involved in the biological fixation of N_2. The synthesis and characterization of analogs for this site has been the object of ongoing research in bioinorganic chemistry. A limited survey of progress in this research is presented. The synthesis, structural characterization, and electronic structures of a new class of heteropolynuclear clusters of the type $[Fe_6S_6(L)_6(M(CO)_3)_2]^{n-}$ (L=halides, p-substituted aryloxides) are described in detail. These clusters are obtained from the preformed $[Fe_6S_6(L)_6]^{n-}$ clusters and are considered precursors in the synthesis of new Fe/Mo/S clusters, with stoichiometries and structures resembling those of the Fe/Mo/S site in nitrogenase.

The nitrogenases are a class of enzymes that catalyze the six-electron reduction of N_2 to ammonia (eq.1).

$$N_2 + 6H^+ + 6e^- \dashrightarrow 2NH_3 \qquad (1)$$

Extensive studies on nitrogenases, and in particular on the Fe-Mo protein (1) component of these complex enzymic systems, have revealed the presence of a unique Fe/Mo/S aggregate intimately involved in biological N_2 fixation.

The Fe/Mo/S aggregate in nitrogenase

Nitrogenases from various nitrogen fixing organisms seem to contain the same Fe/Mo/S structural unit that occurs as an extractable cofactor (FeMoco) (2). Extracts of the Fe-Mo component protein from inactive mutant strains of different microorganisms that do not contain the Fe/Mo/S center are activated upon addition of the FeMoco.

0097–6156/88/0372–0390$06.00/0

Intense interest in the basic coordination chemistry of Fe/M/S clusters (M=Mo,W) has emerged in parallel with the advances in understanding the nature of the Fe/Mo/S center in the nitrogenases. Studies directed toward the synthesis of at least a structural analog for the Fe/Mo/S center have been guided by the broadly defined analytical and spectroscopic data on the Fe-Mo protein and the FeMoco.

Originally the FeMoco was reported to contain iron, molybdenum and sulfide in a 8:1:6 atomic ratio (2). Subsequent analytical determinations gave Fe/Mo ratios of 7:1 (3) and 8.2 ± 0.4:1 (4). The most recent analytical data on functional FeMoco, isolated by a mild procedure, indicates (5) a Fe/Mo ratio of 5 ± 0.5:1. A similar fluctuation in analytical results is also apparent in several determinations of sulfur content, and ratios of S/Mo as low as 4:1 (3) and as high as 9 or 8:1 (4) have been reported.

Spectroscopic studies on the Fe-Mo protein by EPR and Mössbauer spectroscopy have shown six iron atoms each in a distinctive magnetic environment coupled to an overall S=3/2 spin system (6,7,8) and electron nuclear double resonance (ENDOR) studies suggest one molybdenum per spin system (8). The ^{57}Fe signals (five or six doublets) observed in the ENDOR spectra (8) indicate a rather asymmetric structure for the Fe/Mo/S aggregate in which the iron atoms roughly can be grouped into two sets of trios, each set having very similar hyperfine parameters.

Extended X-ray absorption fine structure (EXAFS) studies on the Fe/Mo/S aggregate in nitrogenase have made available structural data that are essential in the design of synthetic analog clusters. Analyses of the Mo K-edge EXAFS of both the Fe-Mo protein and the FeMoco (9) have shown as major features 3-4 sulfur atoms in the first coordination sphere at 2.35 Å and 2-3 iron atoms further out from the Mo atom at ~ 2.7 Å. The Fe EXAFS of the FeMoco (10,11) shows the average iron environment to consist of 3.4 ±1.6 S(Cl) atoms at 2.25(2) Å , 2.3 ±0.9 Fe atoms at 2.66(3) Å, 0.4± 0.1 Mo atoms at 2.76(3) Å and 1.2 ±1.0 O(N) atoms at 1.81(7) Å . In the most recent Fe EXAFS study of the FeMoco (11) a second shell of Fe atoms was observed at a distance of 3.75 Å.

Collectively the available structural and spectroscopic data on the Fe/Mo/S aggregate indicate an asymmetric arrangement of the Fe atoms relative to the Mo, in a cluster that may contain as structural features the FeS_2Mo rhombic units. The Fe EXAFS data further suggest that about half of the iron atoms are at close proximity to the Mo atom (at 2.76 Å) and terminal ligation to the Fe atoms may involve O or N donor ligands.

Syntheses and Structures of Fe-M-S Clusters (M=Mo,W)

Prior to 1978, only one Fe-Mo-S compound was known to contain the FeS_2Mo structural unit. In this compound, $(Cp)_2Mo(S\text{-}n\text{-}Bu)_2FeCl_2$, the Mo and Fe atoms are bridged by two n-BuS^- ligands (12). Since then, numerous Fe-Mo-S complexes have been synthesized and without exception they all contain FeS_2Mo units. These complexes have been obtained by a) the use of the MoS_4^{2-} anion as a ligand in MoS_4^{2-}-$Fe(L)_n$ ligand exchange reactions (13), b) by spontaneous self assembly reactions that employ MoS_4^{2-} as one of the reagents (14,15), and c) the use of preformed Fe/S clusters ($[Fe_6S_6(L)_6]^{3-}$ (16), $[Fe_2S_2(CO)_6]^{2-}$ (17)) and Mo reagents other than MoS_4^{2-}.

Oligonuclear Fe-M-S Complexes. The syntheses, structures, and electronic properties of the oligonuclear Fe-M-S (M=Mo,W) complexes have been described and reviewed in detail (13,14). These complexes originally were conceived as potential "building-blocks" for the synthesis of higher nuclearity aggregates. The simple, mixed ligand complexes of the type $[L_2FeS_2MS_2]^{2-}$ (L=PhS^- (18); L=Cl^- (19); PhO^- (20), NO^+ (21); L_2= S_5^{2-} (22)) have been obtained and characterized in detail. Other oligonuclear complexes, derivatives of the MS_4^{2-} thioanions, include $[Cl_2FeS_2MS_2FeCl_2]^{2-}$ (19,23), $[Fe(MoS_4)_2]^{3-}$ (24), $[Fe(WS_4)_2(DMF)_2]^{2-}$ (25), $[(S_2)OMoS_2FeCl_2]^{2-}$ (26), $[S_2MoS_2FeS_2Fe(S\text{-}p\text{-}C_6H_4Me)_2]^{3-}$ (20,27), $[Mo_2FeS_6(SCH_2CH_2S)_2]^{3-}$ (28), and $[(S_2WS_2)_3Fe_3S_2]^{4-}$ (29). The intrinsic structural and chemical value of these compounds notwithstanding, there has been little success in using them to obtain higher nuclearity Fe/Mo/S clusters that could qualify as synthetic analogs on the basis of stoichiometry.

Polynuclear Fe-M-S Complexes from "spontaneous self assembly" reactions. Synthetic analog clusters for the Fe_2S_2 and Fe_4S_4 centers in the Fe/S proteins (ferredoxins) have been obtained by procedures that are based on the concept of "spontaneous self assembly". The latter (30) assumes that the cores of the Fe/S centers are thermodynamically stable units that should be accessible from appropriate reagents even in the absence of a protein environment.

A great number of polynuclear Fe-M-S clusters have been obtained (14) by "spontaneous self-assembly" procedures that employ simple reactants and the $[MS_4]^{2-}$ anions as a soluble source of Mo or W sulfides. These clusters contain as a common structural feature the MFe_3S_4 cubane type units (M=Mo,W), and can be classified into two general structural types: the double-cubanes that contain two bridged MFe_3S_4 cubane units (31-36), and the single cubanes that contain the MFe_3S_4 cores. The former include $[M_2Fe_6S_9(SR)_8]^{3-}$ (31,33), $[M_2Fe_6S_8(SR)_9]^{3-,4-}$ (35), $[M_2Fe_6S_8(OMe)_3(SR)_6]^{3-}$ (33), $[M_2Fe_7S_8(SR)_{12}]^{3,4-}$ (35), and $[M_2Fe_6S_8(SR)_6(3,6\text{-}R'_2cat)_2]^{4-}$ (36). The reactivity of some of the double-cubanes, in bridge cleavage reactions, has led to the synthesis of single cubane clusters that include $[MFe_3S_4(SR)_3(3,6\text{-}R_2'cat)L]^{2-,3-}$ (37) (L=$p\text{-}ClC_6H_4S^-$, THF, N_3^-, and CN^-) and $[MoFe_3S_4(SEt)_3Fe(cat)_3]^{3-}$ (38).

In the spontaneous self assembly approach to synthesis, the inherent kinetic lability of the Fe/Mo/S system allows for the formation of various thermodynamically stable, heteronuclear products that, as of now, do not include one with the desirable Fe/Mo/S atom ratio of 6-7:1:6-8.

Polynuclear Fe-M-S Complexes from preassembled Fe/S units. Outstanding among the plethora of proposed structural models (14) for the nitrogenase active site are certain clusters that can be envisioned as derivatives of already known Fe/S compounds. One of these models (Fig. 1) contains the cubic $MoFe_7S_6$ core structurally analogous to that in the mineral pentlandites (39) and the known $[Co_8S_6(SPh)_8]^{4-}$ (40) and $[Fe_8S_6(I)_8]^{3-}$ (41) clusters.

An entry to the synthesis of heteronuclear clusters related in core structure to the M_8S_6 pentlandite unit (Fig.1) was suggested by the recently discovered $[Fe_6S_6L_6]^{2-,3-}$ clusters (42) and derivatives (42). The Fe_6S_6 cores in the $[Fe_6S_6L_6]^{2-,3-}$ prismanes are related topologically, and may be converted to the cubic M_8S_6 unit. We have demonstrated that the hexagonal Fe_3S_3 cross sections of the Fe_6S_6 prismatic cage can serve as ligands for coordinatively unsaturated metal

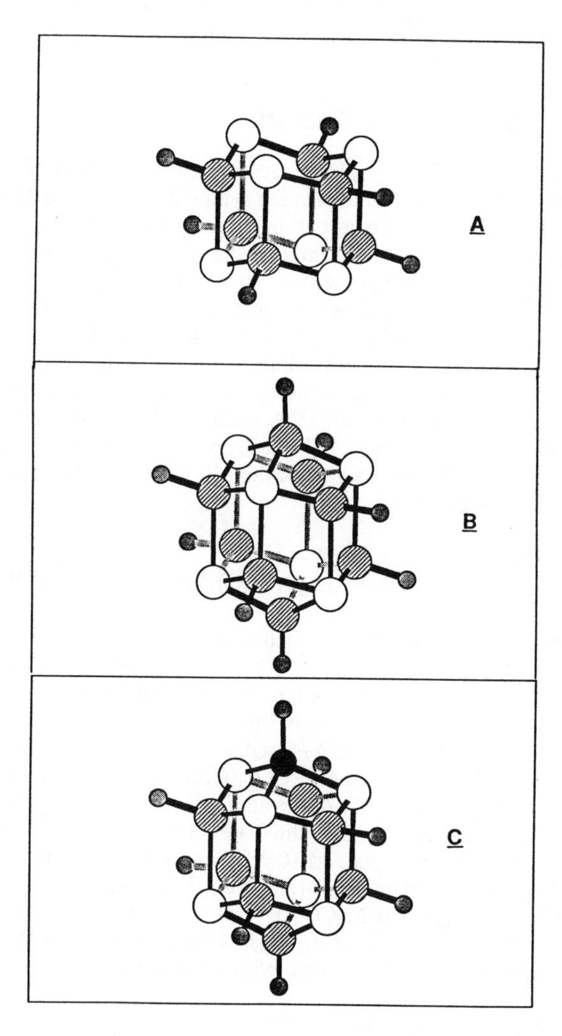

Figure 1. Schematic drawings of: **A**, the $[Fe_6S_6L_6]^{n-}$ clusters (42). **B**, the $[M_8S_6L_6]^{n-}$ pentlandite type of clusters (M = Fe, n = 3, L = I, ref. 41; M = Co, n = 4, L = PhS, ref. 40). **C**, a variant of a proposed model (40) for the nitrogenase Fe/Mo/S center that contains the Fe_7MoS_6 core.

complexes (16). In this paper we report in detail on the properties of the $[Fe_6S_6(L)_6(Mo(CO)_3)_2]^{n-}$ clusters ($L=Cl^-$, Br^-, RO^-; n=3,4). The latter conceivably can serve as precursors for the synthesis of new clusters with a 6:1 Fe:Mo ratio and the simultaneous presence of structural features relevant to those in the Fe-Mo-S site of nitrogenase.

$Mo(CO)_3$ adducts of the $[Fe_6S_6L_6]^{n-}$ Prismanes

Synthesis. The metastable $[Fe_6S_6L_6]^{n-}$ prismanes represent a class of preformed Fe/S clusters that may be ideally suited for the synthesis of Fe/Mo/S aggregates, of at least an acceptable stoichiometry, as possible analogues for the site in nitrogenase. Geometric considerations indicate that the triangular array of sulfur donors in the Fe_3S_3 "face" of the $[Fe_6S_6L_6]^{n-}$ prismanes, (Fig. 1), with S-S distances of ~ 3.8 Å, can accommodate well both tetrahedral or octahedral geometries for the M in a Fe_3S_3-M(L') type of coordination. With M-S bond lengths between 2.3-2.4 Å, a tetrahedral coordination geometry would be less strained by comparison to octahedral geometry. The latter would be more appropriate for Fe_3S_3-M(L') interactions that result in M-S bond lengths >2.4 Å.

The synthesis of the $[Fe_6S_6(L)_6(M(CO)_3)_2]^{n-}$ adducts (n=3,4) is accomplished (16) readily by the reaction of the $M(CO)_3(CH_3CN)_3$ complexes with the $[Fe_6S_6L_6]^{n-}$ prismanes (eq.2).

$$[Fe_6S_6L_6]^{3-} + 2M(CO)_3(CH_3CN)_3 \dashrightarrow [Fe_6S_6(L)_6(M(CO)_3)_2]^{3-} + 6CH_3CN \quad (2)$$

Pure products are obtained only when an approximate twofold molar excess of the $M(CO)_3(CH_3CN)_3$ reagent is used. When stoichiometric amounts of $M(CO)_3(CH_3CN)_3$ are used, mixtures of the 1:1 and 2:1 adducts are obtained for L= halide or RO^-. The final oxidation level (n=3 or 4) of the isolated adducts depends greatly on the potential of the 3-/4- redox couple for the $[Fe_6S_6(L)_6(M(CO)_3)_2]^{n-}$ clusters. The $M(CO)_3(CH_3CN)_3$ complexes that are employed as reagents in the synthesis of the adducts show reversible redox waves in cyclic voltammetry that correspond to the 0/+1 couples. The $E_{1/2}$ values for M=Mo and W respectively, are found at +0.30 V and +0.24 V. The $M(CO)_3(CH_3CN)_3$ "reagent" complexes (that are used in excess) can serve as reducing agents for the $Fe_6S_6(L)_6(M(CO)_3)_2]^{3-}$ clusters when the redox couples of the latter are more positive than -0.30 V. Such is the case when L=Cl, Br or I, and as a result only the tetraanions are obtained with halides as terminal ligands (eq.3).

$$[Fe_6S_6(X)_6(M(CO)_3)_2]^{3-} + M(CO)_3(CH_3CN)_3 \dashrightarrow [Fe_6S_6(X)_6(M(CO)_3)_2]^{4-} + [M(CO)_3(CH_3CN)_3]^{+} \quad (3)$$

The $[Fe_6S_6(L)_6(M(CO)_3)_2]^{4-}$ adducts (L=Cl, Br, I) also can be obtained in very good yields in reactions between $M(CO)_3(CH_3CN)_3$ and either the $[Fe_2S_2(X)_4]^{2-}$ or the $[Fe_4S_4(X)_4]^{2-}$ clusters. In the latter reactions the $M(CO)_3$ units apparently serve as templates for the assembly of $[Fe_2S_2(X)_2(Sol)_2]^-$ fragments ($Sol=CH_3CN$) that very likely are present under the prevailing reducing synthetic conditions.

Characterization. Collective properties for selected $[(Fe_6S_6X_6)(M(CO)_3)_2]^{n-}$ clusters are shown in Table I. The $[Fe_6S_6(L)_6(M(CO)_3)_2]^{3-}$ clusters show two one-electron reversible waves in cyclic voltammetry that correspond to the 3-/4- and 4-/5- redox couples (Table I). The redox potentials for the $Fe_6S_6(L)_6(M(CO)_3)_2]^{n-}$ clusters, (L= p-R-$C_6H_4O^-$), are greatly affected by the nature of the para substituents on the aryloxide ligands. With electron releasing R groups such as -CH_3 or -OCH_3, the redox couples are found (vide infra) at potentials more negative than -0.30 V and consequently the trianionic adducts are obtained readily. With the electron withdrawing acetyl group as a para substituent, the 3-/4- couple is found at -0.20 V, and as a result only the tetraanionic adduct can be obtained. Chemical oxidation by $Fe(Cp)_2^+$ or reduction by BH_4^- can be used for the synthesis of the trianionic and tetraanionic clusters from their one electron redox counterparts, respectively.

The infrared spectra of the pure 2:1 prismane adducts resemble those of the 1,4,7-trithiacyclononane $Mo(CO)_3$ complex (43) (C=O, 1915 cm^{-1}, 1783 cm^{-1}) and show two intense C=O bands in the infrared spectra. These can be assigned to the asymmetric E and the symmetric A vibrational modes for the $M(CO)_3$ units in C_{3v} local symmetry. In the $[Fe_6S_6(X)_6(M(CO)_3)_2]^{n-}$ adducts (X=Cl,Br,I; n=3) these vibrations are found between 1948 cm^{-1} and 1905 cm^{-1}. As expected a bathochromic shift is observed when these clusters are reduced, and, for n=4, the C=O frequencies are found between 1912 cm^{-1} and 1847 cm^{-1}. In general, slightly lower C=O stretching frequencies are observed for the $[Fe_6S_6(OR)_6(Mo(CO)_3)_2]^{n-}$ adducts. In the trianions (n=3) the C=O frequencies are observed between 1912 cm^{-1} and 1864 cm^{-1} while in the tetraanions (n=4) they are found between 1895 cm^{-1} and 1830 cm^{-1}. In the absense of an excess of $Mo(CO)_3(CH_3CN)_3$ the product obtained in the synthesis of $[Fe_6S_6(p\text{-}CH_3C_6H_4O)_6(Mo(CO)_3)_2]^{3-}$ shows an additional doublet of C=O stretching vibrations at 1880 cm^{-1} and 1800 cm^{-1}. An examination of the latter product by proton NMR spectroscopy shows three sets of isotropically shifted resonances for the o-, m- and p-CH_3 protons. One set is due to the $[Fe_6S_6(p\text{-}CH_3C_6H_4O)_6(Mo(CO)_3)_2]^{3-}$. The other two are tentatively assigned to the 1:1 adduct, $[Fe_6S_6(p\text{-}CH_3C_6H_4O)_6(Mo(CO)_3)]^{3-}$ (16a).

The electronic spectra of the $[Fe_6S_6(ArO)_6(Mo(CO)_3)_2]^{n-}$ adducts are dominated by an intense ($\varepsilon \sim 1x10^4$) absorption between 400 nm and 500 nm that corresponds to a L-Fe charge transfer process. This absorption is bathochromically shifted (by c.a. 30-50 nm) when compared to the same absorption in the "parent" prismanes. The shift to lower energies is expected for the adducts where the $M(CO)_3$ units deplete electron density from the core sulfido ligands and indirectly from the iron atoms. The charge transfer absorption is affected in a predictable manner by the oxidation level of the adducts and by the electronic characteristics of the terminal ligands L. Within the $[Fe_6S_6(ArO)_6(M(CO)_3)_2]^{n-}$ clusters the L-Fe transition occurs at higher energies (by c.a. 20-40 nm) for n=4 than for n=3. Similarly, in the $[Fe_6S_6(p\text{-}R\text{-}C_6H_4O)_6(Mo(CO)_3)]^{n-}$ adducts, the charge transfer absorptions shift to lower energies with increasing electron releasing strength of the para substituents (Table I).

The $[(Fe_6S_6X_6)(M(CO)_3)_2]^{3-}$ trianions are characterized by S=1/2 ground states and their EPR spectra, that can be obtained only at temperatures <15 K resemble the spectra of the $[(Fe_6S_6X_6)]^{3-}$ "parent" prismanes (42). The corresponding tetraanions are EPR silent and possess diamagnetic ground states.

TABLE I Collective properties of selected $[(Fe_6S_6L_6)(M(CO)_3)_2]^{n-}$ adducts (A, M=Mo, n=4, L=Cl; B, M=Mo, n=3, L=Cl; C, M=Mo, n=4, L=Br; D M=Mo, n=3, L=Br; E, M=Mo, n=3, L=p-CH_3-PhO; F M=W, n=3, X=p-CH_3-PhO; G, M=Mo, n=4, X=p-CH_3-PhO; H, M=Mo, n=3, L=p-OCH_3-PhO; I, M=Mo, n=3, L=p-$COCH_3$-PhO; J, M=Mo, n=4, L=p-$COCH_3$-PhO

	L-Fe_{CT} (nm)	$\nu_{C\text{-}O}$ (cm^{-1})	EPR $(g_x)^a$	NMR (ppm)[b] -o	-p	-m	$E_{1/2}(V)^c$ 3-/4-	4-/5-
A		1896,1834	---				+0.05	-0.55
B		1921,1865	2.064				+0.05	-0.55
C		1906,1852					+0.08	-0.50
D		1937,1902					+0.08	-0.50
E	468	1915,1860	2.08	-3.00	14.47	11.88	-0.39	-0.93
F	470	1909,1851	2.07	-3.42	14.79	13.31	-0.38	-0.87
G	410						-0.39	-0.93
H[d]	480	1912,1858					-0.84	-1.38
I[d]	455	1925,1874					-0.10	-0.62
J[d]	442						-0.10	-0.62

[a]Broad resonance observed at T<15 K.
[b]Isotropically shifted ^1H resonances associated with the aryloxide ligands and obtained in CH_3CN-d_3 solution, they represent resonances associated with the 2:1 adducts that are predominant in solution. Doublets that correspond to the o- p- and m- protons of the 1:1 adduct, that exists in equilibrium with the 2:1 adduct, are not listed.
[c]The electrochemical measurements were carried out on a Pt electrode vs SCE with Bu_4NClO_4 as the supporting electrolyte. The scan rate was 200 mV/s and the reported waves are all reversible.
[d]Ref. 44.

The six high spin Fe sites in the $[(Fe_6S_6X_6)(M(CO)_3)_2]^{n-}$ anions are antiferromagnetically coupled as evidenced by the reduced ambient temperature magnetic moments, and the temperature dependence of the isotropically shifted resonances in the aryloxide derivatives.

Structures of the $[Fe_6S_6(L)_6(M(CO)_3)_2]^{n-}$ clusters. The crystal structures of the centrosymmetric $[Fe_6S_6(L)_6(M(CO)_3)_2]^{n-}$ anions (L=Cl^-, M=Mo (16b), n=4 ; L= p-$CH_3(CO)$-$C_6H_4O^-$, M=Mo (44), n=4; L=Cl^-, Br^-, M=Mo (16d) n=3; L=PhO^-, M=W (16c), n=3) have been determined. All of these clusters show very similar $[Fe_6M_2S_6]^{2+,3+}$ cores (Fig. 2). These cores are best described as M_2Fe_6 rhombohedral units of nearly exact D_3d symmetry, with the two M atoms located in two opposite corners on the idealized 3 axis of the rhombohedron. A quadruply bridging S^{2-} ligand is located on each of the six faces of the rhombohedron (Fig. 2). By comparison to the structures of the $[Fe_6S_6Cl_6]^{n-}$ clusters the $[Fe_6S_6]$ subunits in the $[Fe_6M_2S_6]^{2+,3+}$ cores are slightly elongated along the idealized 3 axis and show slightly shorter Fe-Fe distances and Fe-S bond lengths. The Fe-Mo and Mo-S distances determined by EXAFS analyses for the Fe/Mo/S center in nitrogenase are significantly shorter than the corresponding distances in the $[Fe_6S_6(L)_6(M(CO)_3)_2]^{n-}$ clusters (Table II).

TABLE II Selected Distances (Å) and Angles (deg.) in the $(Et_4N)^+$ Salts of the $[(Fe_6S_6L_6)(M(CO)_3)_2]^{n-}$ Ions[a]

(A , M=Mo, n=4,L=Cl; B, M=Mo, n=3, L=Cl ; C, M=Mo, n=3, L=Br; D, M=Mo, n=4, L=p-(CO)CH_3-PhO-; E, M=W, n=3, X=p-CH_3-PhO)

	A[b]	B[c]	C[c]	D[d]	E[e]
Distances					
Fe-Mo,W	3.005 (3,11)	2.929 (2,2)	2.946 (3,17)[f]	3.000(3,17)	2.958(3,11)
Fe-Fe	3.785 (3,10)	3.761 (2,3)	3.755 (3,6)	3.784(3,17)	3.807(3,7)
Fe-Fe	2.761 (3,10)	2.744 (2,3)	2.733 (3,10)	2.79(3,3)	2.791(3,3)
M-S	2.619 (3,3)	2.579 (2,3)	2.572 (3,7)	2.649(3,6)	2.589(3,3)
Fe-S[h]	2.333 (3,3)	2.314 (2,4)	2.292 (3,13)	2.351(3,13)	2.338(3,5)
Fe-Sg	2.286 (6,5)	2.282 (4,4)	2.290 (6,10)	2.291(6,3)	2.285(6,4)
Fe-X,O	2.245 (3,3)	2.225 (2,3)	2.355 (3,4)	1.918(3,8)	1.843(3,10)
M-C	1.958 (3,11)	2.01 (2,2)	1.98 (3,2)	1.93(3,2)	1.954(3,16)
Angles					
S-M-S	93.4 (3,2)	95.8 (2,1)	95.2 (3,7)	93.4(3,5)	94.9(3,5)
Fe-M-F[g]	78.07 (3,2)	79.9 (2,1)	79.2 (3,2)	78.2(3,2)	80.1(3,1)
Fe-S-Fe[g]	111.8 (3,5)	111.0 (2,3)	110. (3,1)	109.9(3,9)	112.9(3,7)
Fe-S-Fe[h]	73.4 (6,2)	73.3 (4,3)	73.3 (6,2)	74.2(6,4)	74.3(6,1)
S-Fe-S[g]	113.0 (3,4)	114.1 (2,3)	112.2 (3,9)	113.0(3,1)	113.2(3,4)

[a]The mean values of chemically equivalent bonds are given. In parentheses the first entry represents the number of independent distances or angles averaged out , the second entry represents the larger of the standard deviations for an individual value estimated from the inverse matrix or of the standard deviation, $\sigma = [\Sigma(x_i - x)^2 / N(N-1)]^{1/2}$. [b]From ref.16b. [c]From ref. 16d. [d]From ref. 44. [e]From ref.16c [f]Range: 2.913 (2) Å- 2.954(2) Å. [g]Distances or Angles within the Fe_3S_3 structural units. [h]Distances or Angles within the Fe_2S_2 rhombic units.

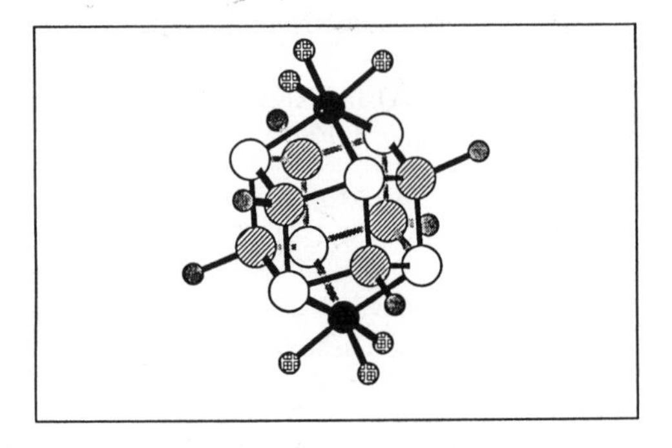

Figure 2. Schematic structures of the $[Fe_6S_6L_6(M(CO)_3)_2]^{n-}$ clusters (16).

These differences are not unexpected in view of the low formal oxidation state of the Mo atoms in the $[Fe_6S_6(L)_6(M(CO)_3)_2]^{n-}$ clusters, furthermore they argue in favor of a higher oxidation state for the Mo atom in nitrogenase. A structural comparison of the $[Fe_6M_2S_6]$ cores in the trianions and the tetraanions (Table II) shows no significant differences within the Fe_6S_6 subunits. However, a significant shortening of the Mo-Fe and the Mo-S distances is found in the trianions. The crystallographic and Mössbauer data (vide infra) suggest that the highest occupied M.O. in the tetraanions consists mainly of Mo and S atomic functions and is antibonding in character.

Mössbauer Spectra. The ^{57}Fe Mössbauer spectra of the $[Fe_6S_6(L)_6(Mo(CO)_3)_2]^{n-}$ clusters (Table III) show similar isomer shift (I.S.) and quadrupole splitting (ΔE_Q) parameters. Thus the I.S. and ΔE_Qvalues for the trianions (0.56 mm/sec., and 1.00 mm/sec. for L=Cl^- and 0.59 mm/sec., and 1.06 mm/sec. for L=Br^-; I.S. vs. Fe, T=92 K) are quite similar to those of the corresponding tetraanions at 0.63 mm/sec., 1.00 mm/sec., and 0.62 mm/sec., 1.02 mm/sec. respectively for L= Cl^- and Br^-. The lack of significant differences in these values, for the two different oxidation levels of the clusters, is consistent with the structural data (Table I) that also indicate that the Fe_6S_6 cores are not significantly perturbed as a result of a change in the oxidation level. These results further reinforce the conclusion that the oxidation of the tetraanionic $Mo(CO)_3$ adducts is centered primarily on the Mo atoms.

TABLE III Mössbauer spectra[a] of the $[(Fe_6S_6L_6)(M(CO)_3)_2]^{n-}$ Complexes

Complex	I.S. (mm/sec)	ΔE_Q (mm/sec)	T(K)
A	0.63(1)	1.00(1)	92
B	0.56(1)	1.00(1)	120
C	0.62(1)	1.02(1)	92
D	0.59(1)	1.06(1)	92
E	0.57(1)	0.88(1)	120
F	0.58(1)	0.74(1)	120
G	0.49(1)	1.08(1)	125
H	0.49(1)	1.13(1)	125
I	0.48(1)	1.02(1)	125
J	0.44(1)	1.04(1)	125
K	0.42(1)	0.62(1)	125

[a]Spectra were obtained on solids and the I.S. values are reported vs. Fe. A, M=Mo, n=4, L=Cl; B, M=Mo, n=3, L=Cl; C, M=Mo, n=4, L=Br; D M=Mo, n=3, L=Br; E, M=Mo, n=3, L=p-CH_3-PhO; F M=W, n=3, X=p-CH_3-PhO; G, $[Fe_6S_6Cl_6]^{3-}$; H, $[Fe_6S_6Br_6]^{3-}$; I, $[(Fe_6S_6(p\text{-}CH_3\text{-}PhO)_6]^{3-}$; J, $[(Fe_6S_6(p\text{-}CH_3\text{-}PhS)_6]^{3-}$; K, $[Fe_6S_6Cl_6]^{2-}$.

The isomer shifts in the $[Fe_6S_6(L)_6(Mo(CO)_3)_2]^{n-}$ clusters are significantly larger by comparison to those in the "parent" $[Fe_6S_6(L)_6]^{3-}$ clusters, (I.S. ~0.49 mm/sec, ΔE_Q ~1.10 vs. Fe, T=125 K). The data suggest that, in the $[Fe_6S_6(L)_6(Mo(CO)_3)_2]^{n-}$ adducts the Fe_6S_6 cores assume an oxidation level lower than the one in the $[Fe_6S_6(L)_6]^{3-}$ clusters. In a formal sense, both oxidation levels of the adducts can be described as containing the $[Fe_6S_6]^{2+}$ core. In the 3- adducts this core is bound to Mo in a formal +0.5 oxidation state, while in the 4- adducts it is bound to Mo(0). It appears that a reduction of the Fe_6S_6 core is important for the stability of either the 3- or the 4- levels of the adducts.

An evaluation of the $[Fe_6S_6(L)_6(M(CO)_3)_2]^{n-}$ clusters with respect to the nitrogenase problem.

On the basis of stoichiometric, spectroscopic, and magnetic ground state criteria the $[(Fe_6S_6X_6)(M(CO)_3)_2]^{n-}$ clusters cannot be considered as satisfactory analogues for the Fe/Mo/S center in nitrogenase. The $[(Fe_6S_6(OAr)_6)(M(CO)_3)]^{3-}$ monocapped prismanes, however, have a stoichiometry and metal coordination environments (at least for the Fe atoms) more closely resembling that of the Fe/Mo/S aggregate in nitrogenase. The monocapped $[(Fe_6S_6(OAr)_6)(M(CO)_3)]^{3-}$ clusters exist in equilibrium with the $[(Fe_6S_6(OAr)_6)(M(CO)_3)_2]^{3-}$ clusters in solution (16a), and thus far have been difficult to isolate in pure crystalline form. Recently we have discovered (44) that, with electron releasing substituents at the para position of the aryloxide ligands, the $[(Fe_6S_6(OAr)_6)(Mo(CO)_3)]^{3-}$ clusters are as stable as the corresponding 2:1 adducts and could be isolated in crystalline form.

With an acceptable Fe/Mo ratio of 6:1 the $[(Fe_6S_6(OAr)_6)(Mo(CO)_3)]^{3-}$ clusters are still inappropriate as models for the nitrogenase Fe/Mo/S site. The large size of the higly reduced Mo atom in the synthetic clusters results in Fe-Mo and Mo-S distances (Table II) that are considerably longer than those found by EXAFS in nitrogenase and the nitrogenase cofactor. Clearly, the $Mo(CO)_3$ appendix must be oxidatively decarbonylated with the Mo atom raised to higher oxidation levels before any acceptable comparisons can be made. Synthetic and reactivity studies addressing the problems described above currently are in progress in our laboratory.

Acknowledgments

The author wishes to acknowledge the invaluable contributions of his associates: M. G. Kanatzidis, A. Salifoglou, S. Al Ahmad, R. W. Dunham and W. R. Hagen who have been the prime contributors in this research. Most of the Mössbauer spectroscopy results were obtained by A. Simopoulos, A. Kostikas and V. Papaefthymiou of the solid state physics group in the nuclear research center Demokritos, Aghia Paraskevi, Attiki, Greece. Generous support of this work by a grant from the National Institutes of Health (GM-26671) also is gratefully acknowledged.

Literature Cited

1. a) Orme-Johnson, W. H.; *Ann. Rev. Biophys. Biophys. Chem.*, **1985**, *14*, 419-459 and references therein. b) Nelson, M. A.; Lindahl, P. A.; Orme-Johnson, W. H. *Adv. Inorg. Biochem.* G. L. Eichhorn and L.B. Marzilli, eds., Elsevier Biomedical, **1982**, *4*, 1.

2. Shah, V. K.; and Brill, W. J., *Proc. Natl. Acad. Sci. USA* , **1977**, *74*, 3249.
3. a)Burgess, B. K.; Jacogs, D. B.; and Stiefel, E. I., *Biochim. Biophys. Acta*, **1980**, *614*, 196. b) Newton, W. E.; Burgess, B. K.; and Stiefel, E. I., in, *Molybdenum Chemistry of Biological Significance*, Newton W. E., and Otsuka, S. eds. p. 191, Plenum Press, **1980**, New York. c) Burgess, B. K.; and Newton, W. E., in *Nitrogen Fixation,* Mueller; A. and Newton W.E., eds, p. 83, Plenum Press, **1983**, New York.
4. Nelson, M. J.; Levy, M. A.; and Orme-Johnson, W. H., *Proc. Natl. Acad. Sci. USA*, **1983**, *80*, 147.
5. Orme-Johnson, W. H.; Wink, D. A.; Mclean, P. A.; Harris, G. S.; True, A. E.; Hoffman, B.; Münck, E.; Papaefthymiou, V., *Rec. Trav. Chim. Pays-Bas.*, **1987**, *106*, 299.
6. Rawlings, J.; Shah, V. K.; Chisnell, J. R.; Brill, W. J.; Zimmerman, R.; Münck, E. and Orme-Johnson, W. H., *J. Biol. Chem.*, **1978**, *253*, 1001.
7. Huynh, B. H.; Münck, E.; Orme-Johnson, W. H., *Biochim. Biophys. Acta*, **1979**, *527*, 192.
8. a) Hoffman, B. M.; Roberts, J. E.; Orme-Johnson, W. H., *J. Am. Chem. Soc.*, **1982**, *104*, 860. b) Hoffman, B. M.; Venters, R. A.; Roberts, J. E.; Nelson, M.; Orme-Johnson, W. H., *J. Am. Chem. Soc.*, **1982,** *104*, 4711.
9. a) Cramer, S. P.; Gillum, W. D.; Hodgson, K. O.; Mortenson, L. E.; Stiefel, E. I.; Chisnell, J. R.; Brill, W. J.; and Shah, V. K., *J. Am. Chem. Soc.* , **1978**, *100*, 4630. b) Cramer, S. P.; Hodgson, K. O.; Gillum, W. O.; Mortenson, L. E., *J. Am. Chem. Soc.* **1978**, *100*, 3398.
10. Antonio, M. R.; Teo, B. K.; Orme-Johnson, W. H.; Nelson, M. J.; Groh, S. E.; Lindahl, P. A.; Kauzlarich, S. M.; and Averill, B. A., *J. Am. Chem. Soc.,* **1982**, *104,* 4703.
11. Arber, J. M.; Flood, A. C.; Garner, C. D.; Hasnain S. S.; Smith B. E., *Journal de Physique*, **1986**, *47*, C8-1159.
12. Cameron, T. S.; Prout, C. K., *Acta Cryst.*, **1972**, *B28*, 453.
13. Coucouvanis, D., *Acc. Chem. Res.*, **1981**, *14*, 201-209.
14. Holm, R. H.; Simhon, E. D., In *Molybdenum Enzymes,* 104, Spiro, T. G.; Ed., Wiley-Interscience: New York, **1985;** Chapter 1 and references therein.
15. Christou, G.; Garner, C. D.; Mabbs, F. E.; and King, T. J., *J. Chem. Soc. Chem. Commun.*, **1978**, 740.
16. a) Coucouvanis, D.; Kanatzidis, M. G., *J. Am. Chem. Soc.*, **1985**, *107*, 5005. b) Kanatzidis, M. G.; Coucouvanis, D., *J. Am. Chem. Soc.*, **1986**, *108*, 337. c) Salifoglou, A.; Kanatzidis, M. G.; Coucouvanis, D.; *J. Chem. Soc., Chem. Commun.*, **1986**, *559*. d) Coucouvanis, D.; Salifoglou, A.; Kanatzidis, M. G.; Simopoulos, A.; Kostikas, A., *J. Am. Chem. Soc.*, **1987**, *109*, 3807.
17. Bose, K. S.; Lamberty, P. E.; Kovacs, J. E.; Sinn, E.; Averill, B. A., *Polyhedron*, **1986**, *5*, 393-398.
18. a) Coucouvanis, D.; Simhon, E. D.; Swenson, D.; Baenziger, N. C., *J. Chem. Soc., Chem. Commun.*, **1979**, *361*. b) Tieckelmann, R. H.; Silvis, H.C.; Kent, T. A.; Huynh, B. H.; Waszczak, J. V.; Teo, B. -K.; Averill, B.A., *J. Am. Chem. Soc.*, **1980**, *102*, 5550. (c) Coucouvanis, D.; Stremple, P.; Simhon, E. D.; Swenson, D.; Baenziger, N. C.; Draganjac, M.; Chan, L. T.; Simopoulos, A.; Papaefthymiou, V.; Kostikas, A.; Petrouleas, V., *Inorg. Chem.*, **1983**, *22*, 293.

19. a) Coucouvanis, D.; Baenziger, N. C.; Simhon, E. D.; Stremple, P.; Swenson, D.; Simopoulos, A.; Kostikas, A.; Petrouleas, V.; Papaefthymiou, V., *J. Am. Chem. Soc.*, **1980**, *102*, 1732. b) Muller, A.; Tolle, H.-G.; Bogge, H., *Z. Anorg. Allgem. Chem.*, **1980**, *471*, 115.
20. a) Silvis, H. C.; Averill, B. A., *Inorg. Chim. Acta*, **1981**, *54*, L57, b) Antonio, M. R.; Tieckelmann, R. H.; Silvis, H. C.; Averill, B. A., *J. Am. Chem. Soc.*, **1982**, *104*, 6126.
21. Coucouvanis, D.; Simhon, E. D.; Stremple, P.; Baenziger, N. C., *Inorg. Chim. Acta*, **1981**, *53*, L135.
22. Coucouvanis, D.; Baenziger, N. C.; Simhon, E. D.; Stremple, P.; Swenson, D.; Kostikas, A.; Simopoulos, A.; Petrouleas, V.; Papaefthymiou, V., *J. Am. Chem. Soc.*, **1980**, *102*, 1730.
23. Muller, A.; Sarkar, S.; Dommrose, A. -M.; Filgueira, R., *Z. Naturforsch*, **1980**, *35b*, 1592.
24. a) Coucouvanis, D.; Simhon, E. D.; Baenziger, N. C., *J. Am. Chem. Soc.*, **1980**, *102*, 6644. b) McDonald, J. W.; Friesen, G. D.; Newton, W. E., *Inorg. Chim. Acta*, **1980**, *46*, L79. c) Muller, A.; Hellmann, W.; Schneider, J.; Schimanski, U.; Demmer, U.; Trautwein, A.; Bender, U., *Inorg. Chim. Acta*, **1982**, *65*, L41. d) Friesen, G. D.; McDonald, J. W.; Newton, W. E.; Euler, W. B.; Hoffman, B. M., *Inorg. Chem.*, **1983**, *22*, 2202.
25. Stremple, P.; Baenziger, N. C.; Coucouvanis, D., *J. Am. Chem. Soc.*, **1980**, *103*, 4601.
26. Muller, A.; Sarkar, S.; Bogge, H.; Jostes, R.; Trautwein, A.; Lauer, U., *Angew. Chem. Int. Ed. Engl.*, **1983**, *22*, 561; *Angew. Chem. Suppl.*, **1983**, 747.
27. Tieckelmann, R. H.; Averill, B. A., *Inorg. Chim. Acta*, **1980**, *46*, L35.
28. Dahlstrom, P. L.; Kumar, S.; Zubieta, J., *J. Chem. Soc., Chem. Commun.*, **1981**, *411*.
29. Muller, A.; Hellman, W.; Bogge, H.; Jostes, R.; Romer, M.; Schimanski, U., *Angew. Chem. Int. Ed. Engl.*, **1982**, *21*, 860; *Angew Chem. Suppl.*, **1982**, *1757*.
30. a) Ibers, J. A.; Holm, R. H., *Science*, **1980**, *209*, 223. b) Berg, J. M.; Holm, R. H., *Iron Sulfur Proteins*, Spiro, T. G., Ed.; Wiley-Interscience: New York, **1982**, Chapter 1.
31. Wolff, T. E.; Berg, J. M.; Warrick, C.; Hodgson, K. O.; Holm, R. H.; Frankel, R. B., *J. Am. Chem. Soc.*, **1978**, *100*, 4630.
32. a) Wolff, T. E.; Berg, J. M.; Hodgson, K. O.; Frankel, R. B.; Holm, R. H., *J. Am. Chem. Soc.*, **1979**, *101*, 4140. b) Wolff, T. E.; Berg, J. M.; Power, P. P.; Hodgson, K. O.; Holm, R. H.; Frankel, R. B., *J. Am. Chem. Soc.*, **1979**, *101*, 5454.
33. a) Acott, S. R.; Christou, G.; Garner, C. D.; King, T. J.; Mabbs, F. E.; Miller, R. E., *Inorg. Chim. Acta*, **1979**, *35*, L337. b) Christou, G.; Garner, C. D.; Mabbs, F. E., *Inorg. Chim. Acta*, **1978**, *28*, L189. c) Christou, G.; Garner, C. D.; Mabbs, F. E.; King, T. J., *J. Chem. Soc., Chem. Commun.*, **1978**, *740*.
34. Christou, G.; Mascharak, P. K.; Armstrong, W. H.; Papaefthymiou, G. C.; Frankel, R. B.; Holm, R. H., *J. Am. Chem. Soc.*, **1982**, *104*, 2820.
35. Wolff, T. E.; Power, P. P.; Frankel, R. B.; Holm, R. H., *J. Am. Chem. Soc., 1980*, *102*, 4694.

36. a) Armstrong, W. H.; Holm, R. H., *J. Am. Chem. Soc.*, **1981**, *103*, 6246. b) Armstrong, W. H.; Mascharak, P. K.; Holm, R. H., *J. Am. Chem. Soc.*, **1982**, *104*, 4373.
37. Palermo, R. E.; Singh, R.; Bashkin, J. K.; Holm, R. H., *J. Am. Chem. Soc.*, **1984**, *106*, 2600.
38. Wolff, T. E.; Berg, J. M.; Holm, R. H., *Inorg. Chem.*, **1981**, *103*, 6246.
39. Rajamani, V.; Prewitt, C. T., *Can. Mineral*, **1973**, *12*, 178.
40. Christou, G.; Hagen, K. S.; Holm, R. H., *J. Am. Chem. Soc.*, **1982**, *104*, 1744.
41. Pohl, S.; Saak, W., *Angew. Chem. Int. Ed. Engl.*, **1984**, *23*, 907.
42. a) Kanatzidis, M. G.; Dunham, W. R.; Hagen, W. R.; Coucouvanis, D., *J. Chem. Soc., Chem. Commun.*, **1984**, *356*. b) Coucouvanis, D.; Kanatzidis, M. G.; Dunham, W. R.; Hagen, W. R., *J. Am. Chem. Soc.*, **1984**, *106*, 2978. c) Kanatzidis, M. G.; Hagen, W. R.; Dunham, W. R.; Lester, R. K.; Coucouvanis, D., *J. Am. Chem. Soc.*, **1985**, *107*, 953. d) Kanatzidis, M. G.; Salifoglou, A.; Coucouvanis, D., *J. Am. Chem. Soc.*, **1985**, *107*, 3358. e) Kanatzidis, M. G.; Salifoglou, A.; Coucouvanis, D., *Inorg. Chem.*, **1986**, *25*, 2460.
43. Ashby, M. T.; Lichtenberger, D. L. *Inorg. Chem.*, **1985**, *24*, 636.
44. Coucouvanis, D.; Al-Ahmad, S.; Salifoglou, A., manuscript in preparation.

RECEIVED February 22, 1988

Author Index

Affiliation Index

Subject Index

L

Production by Cara Aldridge Young
Indexing by Deborah H. Steiner
Jacket design by Carla L. Clemens

Elements typeset by Hot Type Ltd., Washington, DC
Printed and bound by Maple Press, York, PA